Lectures in
Microbiology

As per the CBME curriculum

Lectures in Microbiology

Nandita Shinkre Banaji
Professor and Head, Department of Microbiology
Indira Gandhi Medical College & Research Institute
Puducherry

Universities Press

LECTURES IN MICROBIOLOGY

UNIVERSITIES PRESS (INDIA) PRIVATE LIMITED

Registered Office
3-6-747/1/A & 3-6-754/1, Himayatnagar, Hyderabad 500 029, Telangana, India
info@universitiespress.com, www.universitiespress.com

Distributed by
Orient Blackswan Private Limited

Registered Office
3-6-752, Himayatnagar, Hyderabad 500 029, Telangana, India

Other Offices
Bengaluru, Chennai, Guwahati, Hyderabad, Kolkata
Mumbai, New Delhi, Noida, Patna, Visakhapatnam

ISBN 978 81 954009 1 1

Typeset in Life 10/12.6 *by*
ELITE Graphics, Hyderabad 500 039

501764

Printed in India by
Rasi Graphics Private Limited
Chennai 600 014

Published by
Universities Press (India) Private Limited
3-6-747/1/A & 3-6-754/1, Himayatnagar, Hyderabad 500 029, Telangana, India

Contents

SECTION IX: ZOONOSES AND OTHER INFECTIONS (MI 8.1–8.4)

SECTION X: HOSPITAL-ACQUIRED INFECTIONS (MI 1.5, 8.5–8.8)

Preface

Dear Friends,

It is with great enthusiasm that I welcome the introduction of 'Competency-Based Medical Education' for microbiology. This is so because I have been naturally inclined, from the very start, to teach microbiology in the manner that is now proposed by the National Medical Commission, and have done so throughout my teaching career. In 2019, when the CBME guidelines were introduced, I decided to put together my notes in a structured manner to prepare a textbook that is tailored to the CBME curriculum. It is because of my understanding of the subject and my belief in this teaching approach that I am now keen and motivated to formally implement it by writing this book.

The new curriculum follows an outcome-based approach that lays emphasis on acquiring specific competencies. The intended outcome is for the student to acquire the relevant knowledge, skills, attitudes, values and responsiveness to be able to function as a **physician of first contact**.

The primary aim of education in microbiology (in the MBBS course) is to help the learner become competent in the microbiologic aspects involved in medical practice. This 'practice' could range from being a doctor in a primary health centre or a family physician to being a doctor in a specialised tertiary care centre. The level of competence in microbiology that is expected of a medical graduate includes the ability to make independent decisions when managing common infections and infectious diseases.

From this perspective, it is important that the microbiology curriculum addresses the following objectives:

1. To understand the interactions between microbes and humans, including the immune response to microbial infection
2. To apply the understanding towards prevention of infection by general measures, immunoprophylaxis or chemoprophylaxis
3. To prevent hospital-acquired infections by safe practices in healthcare settings
4. To develop the skill required to collect an optimal sample from a patient for microbiological investigation
5. To be able to interpret the result of a microbiology investigation to diagnose an infectious disease in accordance with national programmes, if any
6. To be able to choose an appropriate antimicrobial agent to treat an infection and prevent the emergence of antimicrobial resistance in accordance with the National Treatment Guidelines

With the explosion of research and information available in microbiology, it has become imperative that we determine the extent of theoretical details a medical graduate needs to possess to meet the above objectives. This book has originated from the notes I prepared for my interactive classes, both theory and practical, with medical students over the last 30 years.

In the new curriculum prescribed by the National Medical Commission, a total of 190 hours has been allotted to teaching microbiology, which includes 70 hours of didactic lectures, 110 hours of small group discussions and 10 hours of self-directed learning. Care has been taken to address this reduction in teaching time without compromising on the competency or content. This book has 46 chapters across 10 sections that include the required practical competencies and AETCOM content relevant to microbiology.

Though I wrote the text single-handedly, there are colleagues who have contributed towards it. My colleague Dr Jayavarthinni M., Assistant Professor, Department of Microbiology, Indira Gandhi Medical College & Research Institute (IGMC & RI), Puducherry, helped me source images online and provided me with much needed technical help.

Some parts of this textbook have been reviewed by two of my colleagues: Section A by Dr G Jyothi Lakshmi, Professor and Head, Department of Microbiology, Osmania Medical College, Hyderabad, and Chapter 9 on 'Bacterial Genetics and Antimicrobial Resistance' by Dr V. Balaji, Professor, Department of Microbiology, Christian Medical College (CMC), Vellore. Their suggestions and recommendations have been valuable. I am also indebted to them for honouring my view that the content of the reviewed portion be restricted to the needs of a medical graduate and not to those of a student pursuing a postgraduate degree in microbiology. Given their vast experience and depth of knowledge in this subject it must not have been easy for them to restrict the scope of the subject only to medical graduates.

Since this text is meant for graduate students, I was also keen to have the perspective/feedback of a graduate student. Dr Lokesh Siva, an MBBS graduate from IGMC & RI, eagerly fulfilled this wish. His reading and suggestions have helped in moderating the real case scenarios and ensuring that the text is clinically relevant. I shared the section on immunology with another student from IGMC & RI, Dr Gaurang Narayan, who has undergone training in advanced immunology. He read this section and assured me that it is easy to understand. I am grateful to these young minds for stepping up to the task.

I have been presenting the subject of microbiology to medical students keeping the patient and not the microbe as central in my approach much before the introduction of the CBME. In fact, I had often discussed this paradigm shift in the teaching of the subject with my husband, Rukhshad Banaji, in the hope of someday publishing my lecture notes. It is his constant encouragement and experience in publishing that have enabled me to develop my lecture notes into a book. He has also been my advisor at various stages—in my interactions with the publisher, in the preparation of the manuscript, and in reviewing the edited text. His support has been invaluable.

I am grateful to Madhu Reddy, Director, Universities Press, and President, Orient BlackSwan, for graciously accepting my manuscript for publication. It is indeed an honour to have my book be published by the very publisher whose book on microbiology was one of my favourites. I thank the publishing team at Universities Press—especially Dr Sudha Ganesan and Anusha Vincent—as well as illustrator Chand Singh Naagar for their contribution in transitioning this manuscript to a printed book.

It is said that a book is to be read, first and foremost, because one enjoys reading it, and when that happens, learning naturally follows. I sincerely hope that students and faculty enjoy reading this book just as much as I enjoyed writing it.

Finally, I would welcome any comments or suggestions the reader may have that would help improve the readability and usefulness of this book.

Nandita Shinkre Banaji, MD, PhD
Professor and Head, Department of Microbiology
Indira Gandhi Medical College & Research Institute
(a Govt. of Puducherry Institution)
nandita.banaji@hotmail.com

This book is dedicated to my students—past and present, for providing me the opportunity to learn. After all, the best way to learn is to teach!

And to my parents, Suman Joshi and Balkrishna Shinkre, who, by their example, taught me the value of integrity. Without integrity, a teacher has no value.

Introduction to Microbiology

A cell is the basic structural component that defines life and living beings. Cells can be **prokaryotic** or **eukaryotic**. Prokaryotic cells are not well-organised and lack a membrane around their nucleus and cell organelles. Eukaryotic cells are well-organised and possess a membrane-bound nucleus and cell organelles. **Bacteria are prokaryotic**, with free DNA in the cytoplasm. The free DNA is not bound by a nucleus. All plants, animals, fungi and protists (includes protozoans) are eukaryotes.

Microbes pathogenic to man include prokaryotic bacteria, eukaryotic fungi and protozoans. Viruses occupy the twilight zone between the living and the non-living as they lack cellular organisation and use host cells to grow and multiply. They possess either DNA or RNA, never both. Prions are proteins that lack both DNA and RNA and are infectious to man.

Bacteria, protozoans and yeasts (fungi) are unicellular, while mycelia (fungi) and helminths are multicellular organisms pathogenic to man. Amongst these, protozoans and helminths are parasites on human hosts.

A virus particle is known as a **'virion'**. Viruses are obligate, intracellular organisms that cannot be grown on inanimate media, unlike bacteria and fungi. Antimicrobial agents are very specifically antibacterial, antiviral, antifungal or antiparasitic.

Growth and reproduction define the evolution of life. However, sexual reproduction is not seen in bacteria, viruses and most protozoans. Amongst protozoans, sexual reproduction appears in sporozoans and some fungi. Helminths can reproduce sexually but the sexes may or may not be separate.

The average size of bacteria varies from 0.2 to 10 μm, viruses from 18 to 200 nm, yeasts from 7 to 20 μm and protozoans from 3 to 100 μm. With the exception of adult forms of helminths, these microbes cannot be visualised by the naked eye and, therefore, necessitate the use of a compound light microscope. The unit of measurement in microscopy is **'microns'** or micrometre (μm) in light microscopy, and **nanometre** (nm) and **Angstrom units** (Å) in electron microscopy.

1 μm = 1 mm/1000
1 nm = 1 μm/1000
1 Å = 1 nm/10

The magnification power of a microscope is its ability to enlarge the image of an object.

Total magnification power = Magnification power of eye piece × Magnification power of objective

The total magnification power of a low power (LP of 10) lens is 100X, high power (HP of 40) lens is 400X and oil immersion (OI of 100) lens is 1000X when using an eye piece magnification of 10X.

The **resolution power** of a microscope is its ability to distinguish two closely situated points as distinct and separate. The resolution power of a compound light microscope is 0.2 μm, while that of the naked eye is 0.2 mm or 200 μm. The resolution power of the electron microscope is 0.1 nm.

With the exception of viruses, other microbes can be visualised by light microscopy. For an object to be visualised by light, the light has to be reflected from the surface of that object. The smallest virus (parvovirus) measures 18 nm and the largest (pox virus) 200 nm, with the wavelength of light being 400 nm. As the wavelength of light is larger than the size of these viruses, light does not reflect; therefore, viruses cannot be visualised by light microscopy. To overcome this, an electron beam with a wavelength of 1.23 nm is used in electron microscopy to visualise a virus; however, this is not used for diagnostic purposes as there are better serologic and molecular diagnostic methods for viral infections.

Parts, use and care of a light microscope

Certifiable Practical Competency — Use and care of a light microscope

The light microscope has a base to stand on, a light source at the bottom, a stage to place the slide on, a set of lenses, a condenser and an iris diaphragm to control the light flow into the lenses (Fig. 1.1).

The **eyepiece** is the replaceable lens that we see through. It has a magnification of 10X and can be replaced with a micrometry lens, if required, for measuring the size of microscopically observed microbes. The **objective lenses** are of four magnifications, each with a varied use— 4X (scanner), 10X (low power, LP), 40X (high power, HP) and 100X (oil immersion, OI). The **scanner** is used to identify the specific area to be focused upon for examination before choosing to examine the specimen. It has limited use in microbiology, as it is predominantly useful for the histopathologic examination of human tissue. In microbiology, the objective lenses 10X LP and 40X HP are used to examine a wet mount while the 100X OI is used to examine fixed stained smears. The magnification power of each is stated on each of the

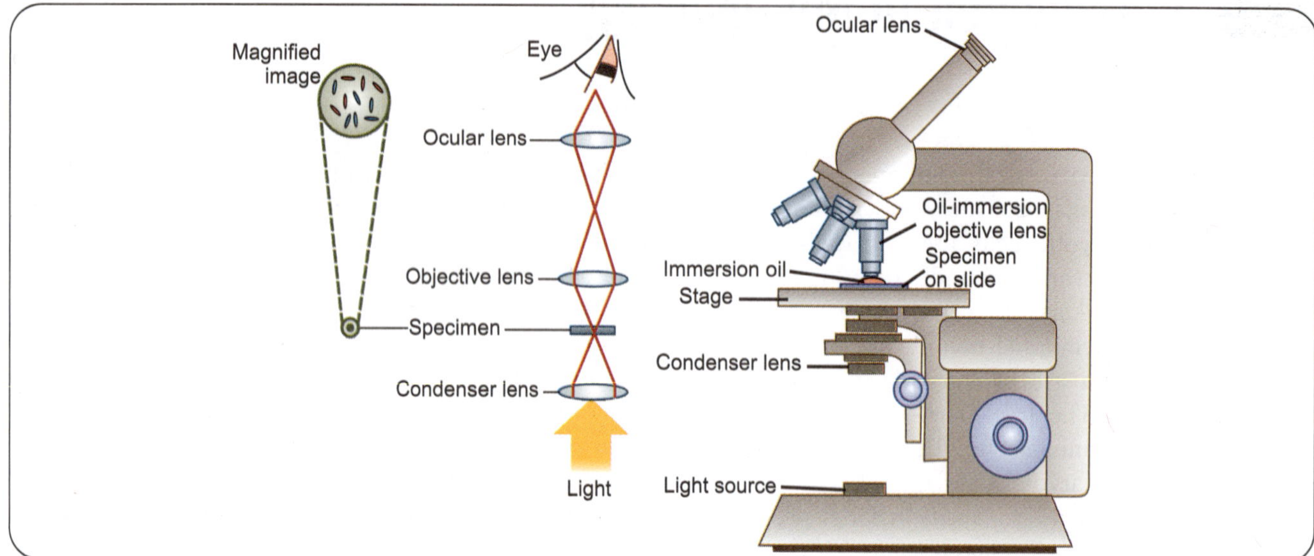

Fig. 1.1 Light microscope

respective lenses. Depending upon the size (microns) of the microbe to be observed, the correct magnification lens is chosen. **In general, bacteria and protozoans are visible on 100X and fungi and helminth forms on 40X.** Viruses cannot be visualised by light microscopy, other than viral inclusion bodies that can be observed on histopathologic examination.

The **condenser** can be maintained at two extreme positions—fully raised to touch the bottom of the glass slide placed on the stage and fully lowered, as far away from the stage. The **iris diaphragm** can be maintained at two extreme positions—fully open, allowing maximum light to pass through and fully closed with no light passing through. Both of these are utilised to control the amount of light required for optimal visualisation of the microscopic object.

When examining a fixed, stained smear for bacteria, fungi or parasites, a drop of oil (cedar wood) is placed on the smear, and the smear is observed with the 100X objective, with the condenser fully raised and the iris diaphragm allowing optimal light to visualise the structures. It is important that a continuous column of light be maintained from the light source to the eyepiece lens without refraction or the loss of any light.

As opposed to a stained smear, when examining a wet mount, optimal visuality is obtained by increasing the contrast between the object and its background. Too much light, as in the case above, is likely to render the structures invisible. Therefore, the condenser needs to be lowered and the iris diaphragm partially open to optimally view the structures. When adjusting a wet mount, it is ideal to visualise the structures first under the LP 10X, and once in focus, to move the lens to HP 40X. **Following microscopic examination, the slide should be disposed in a disinfectant solution of 7% phenol or any other suitable disinfectant.**

> A microscope should be taken care of in the following ways:
> 1. The lens should be cleaned with a lens cloth or tissue paper
> 2. The microscope should be covered when not in use.
> 3. It should always be kept in an upright position.
> 4. When carrying the device, one hand should hold the stem and the other should hold the base of the microscope.

Types of microscopy and applications

Besides compound light microscopy, a variety of other microscopy techniques are used for specific purposes. Dark field and phase contrast microscopy enable the cork screw motility of spirochetes to be visualised but are not employed in laboratory diagnosis. Nor is electron microscopy used for the laboratory diagnosis of viral infections. Fluorescent microscopy using fluorescent dyes to visualise *Mycobacterium tuberculosis* in sputum and malarial parasites in blood is widely accepted. Fluorescent microscopy can also be used for immunofluorescence, wherein the corresponding antigen or antibody molecule is tagged with a fluorescent conjugate for the detection of its antibody or antigen, respectively.

HISTORY OF MICROBIOLOGY

The Nobel awards were instituted in 1901 and many of them have been won for discoveries in infectious diseases (Table 1.1). However, the first observations of microbes were made as early as in 1683, when **Anton van Leeuwenhoek** described 'animalcules'. Subsequently, the theory of spontaneous generation believed that the living arose from the non-living, until the late 1800s, when **Louis Pasteur** proposed the germ theory of disease, stating that all life forms arose from their like and that microbes are the cause of infectious disease (Fig. 1.2). Later, he also introduced the technique of steam sterilisation to prevent microbial growth and a vaccine for anthrax.

In 1846, **Ignaz Semmelweiss** discovered that puerperal sepsis could be prevented in postpartum women by introducing simple handwashing techniques for the attending staff. This was followed by **Joseph Lister** using carbolic acid as an antiseptic in surgery in 1867.

In 1798, **Edward Jenner** introduced the first vaccine for smallpox. He observed that milkmaids who suffered from cowpox were immune to smallpox and introduced cowpox lesional fluid by variolation on the forearm as a vaccine. The word 'vaccine' is derived from the Latin word '*vacca*', meaning cow. Years later, in 1980, vaccination with the vaccinia virus led to the eradication of smallpox.

Pasteur's germ theory was furthered in 1882 by **Robert Koch** who discovered microorganisms causing a variety of infectious diseases, the foremost being tuberculosis (Fig. 1.3). In addition, he made significant contributions to the understanding of plague, cholera, sleeping sickness, anthrax, malaria, amoebic dysentery and bacterial conjunctivitis. He was the first to culture *M. tuberculosis* and devised the tuberculin test using old tuberculin, which is an extract of the tubercle bacillus in culture, for which he was awarded the Nobel Prize in 1905.

Koch's concepts that microbes occur as a mixed population and that they need to be isolated in pure culture using solid and not liquid media, that one single bacterial cell gives rise to one colony on a solid medium and the use of aniline dyes to stain bacteria established him as the **father of microbiology**.

Koch also proposed the criteria for establishing an organism as the cause of a specific infectious disease; these are known as the Koch's postulates:

Fig. 1.2 Louis Pasteur

Fig. 1.3 Robert Koch

Koch's postulates

1. The organism must be found first in the diseased but not in the normal host
2. The organism must first be isolated from the diseased animal and grown in pure culture
3. The pure culture must produce the disease when injected into a normal animal
4. The organism must be re-isolated from the experimentally infected animal

Tuberculosis was the first disease that he applied the Koch's postulates to in order to establish *M. tuberculosis* as the causative agent. In 1922, the **Bacillus Calmette Guerin (BCG)** vaccine was produced using a live attenuated strain of *M. tuberculosis*. A fifth postulate stating that antibody to the causative agent should be demonstrable in the patient's serum was later added.

In 1900, **Paul Ehrlich** discovered Salvarsan as the first effective drug in syphilis (Fig. 1.4). In 1932, prontosil was first used as an agent to treat streptococcal infection. By the mid-1950s, sulphonamides, penicillin, streptomycin and tetracycline had been introduced for the treatment of infectious diseases. Despite the discovery of these antimicrobial agents, even today we find that infectious diseases continue to remain a major cause of death, worldwide. Ehrlich also laid down the foundations for the **concept of the complement system**.

The discovery of the first virus was by Dmitry Ivanovsky in 1890; it was the virus causing mosaic disease in tobacco plants. However, it was only in 1946 that **Wendell Stanley** crystallised the tobacco mosaic virus. That a living organism could be crystallised blurred the distinction between the living and non-living, furthered in complexity by the discovery of prions by **Stanley Prusiner** in 1997. The prion is filterable and self-reproducing but unlike the virus lacks nucleic acid and hence does not code for its structure.

Since the inception of the award in 1901, a total of **33** Nobel awards for Physiology or Medicine have been awarded for contributions in the fields of immunology, bacteriology, virology, parasitology and antimicrobial therapy. With the latest one, in 2020, awarded to H. J. Alter, M. Houghton and C. M. Rice for their discovery of the hepatitis C virus.

Emil Von Behring received the first Nobel award for Physiology or Medicine in 1901 for his discovery of antitoxic immunity in the treatment of diphtheria and tetanus. In 1919, it was Jules Bordet for his

Fig. 1.4 Paul Ehrlich

discovery of 'factors' in blood that protected against infectious diseases. The 'factors' were later proven to be antibody and complement. Although it was in 1960 that Nobel Prize recipient Peter Medawar demonstrated that tissue rejection was an immunological process, it was only in 1996 that the concept of major histocompatibility complex restriction brought the Nobel to **Peter Doherty** and **Rolf Zinkernel**.

The 1905 Nobel Prize was awarded to Koch for his discovery of *M. tuberculosis* as the cause of tuberculosis and in 1928 to Charles Nicolle for his discovery that epidemic typhus is transmitted by the louse.

The 1951 award went to **Max Theiler** for the discovery of a vaccine for yellow fever and, in 1954, to John Franklin Enders, Thomas Huckle Weller and Frederick Chapman Robbins for cultivating poliovirus in vitro in cell culture. In 1966, Peyton Rous received the Nobel Prize for his discovery that sarcomas in chicken are caused by 'filterable' agents, followed by Renato Dulbecco for the mechanism of tumour production by DNA viruses. Howard Temin and David Baltimore discovered the reverse transcriptase enzyme, and **Baruch Blumberg** discovered the hepatitis B virus, for which the scientists won the Nobel Prize in 1975 and 1976, respectively. In 1997, Stanley Prusiner received the Nobel for his studies on prions and, in 2008, **Luc Montaigner** for his discovery and investigation of the human immunodeficiency virus (HIV).

Two Nobel awards were given for research on malaria—to Ronald Ross in 1902 for understanding the life cycle of the parasite in mosquitoes and Charles Laveran in 1907 for his discovery of the malarial parasite in man.

The landmark discoveries in antimicrobial agents that were recognised with a Nobel Prize were those of prontosil in 1939, penicillin in 1945 (Alexander Fleming) and streptomycin in 1952.

Table 1.1 Some notable Nobel laureates and their contributions to the science of microbiology

Year	Nobel laureate	Work
1902	Ronald Ross	Malaria
1905	Robert Koch	Tuberculosis
1908	P. Ehrlich and E. Metchnikof	Immunity
1928	Charles Nicolle	Typhus
1930	Karl Landsteiner	Discovery of human blood groups
1945	A. Fleming, E. Boris Chain and Howard Walter Florey	Penicillin
1951	Max Theiler	Yellow fever
1952	Selman A. Waksman	Streptomycin
1954	Franklin Enders, T. H. Weller and F. C. Robbins	Poliomyelitis growth in tissue
1960	F. M. Burnet and P. B. Medawar	Acquired immunological tolerance
1966	Peyton Rous	Tumour-inducing viruses
1969	M. Delbrück, A. D. Hershey and S. E. Luria	Replication mechanism and the genetic structure of viruses
1975	D. Baltimore, R. Dulbecco and H. Martin Temin	Interaction between tumour viruses and the genetic material of the cell
1976	Baruch S. Blumberg and D. Carleton Gajdusek	New mechanisms of infectious disease dissemination and Australia antigen
1980	Baruj Benacerraf, Jean Dausset and George D. Snell	Immunological regulation by cell surface
1984	Niels K. Jerne, Georges J. F. Köhler and César Milstein	Control of immune system and monoclonal antibodies
1987	Susumu Tonegawa	Generation of antibody diversity
1989	J. Michael Bishop and Harold E. Varmus	Origin of retroviral oncogenes
1996	Peter C. Doherty and Rolf M. Zinkernagel	Specificity in cell-mediated immune defence
1997	Stanley B. Prusiner	Prions
2005	Barry J. Marshall and J. Robin Warren	*Helicobacter pylori*
2008	Harald Hausen and Françoise Barré-S. and L. Montagnier	Human papillomaviruses and human immuno-deficiency virus
2011	Bruce A. Beutler, Jules A. Hoffmann and Ralph M. Steinman	Activation of innate immunity, and the dendritic cell and its role in adaptive immunity
2015	William C. Campbell and Satoshi Ōmura	Discovery of a novel therapy against roundworm infection
2015	Youyou Tu	Discovery of a novel therapy against malaria
2018	James P. Allison and Tasuko Honjo	Though not directly related to microbiology, their work explained the process of negative inhibition of immunity in cancer therapy

Questions

1. Classify microbes based on their cellular and nuclear structure.
2. Elaborate on the use and limitations of the light microscope in visualising the microbial world.
3. Highlight the contributions of Edward Jenner, Louis Pasteur and Robert Koch to the field of microbiology.

Classification and Characteristics of Bacteria

BACTERIAL MORPHOLOGY

Bacteria are prokaryotic cells as they do not possess a well-defined nucleus or cell organelles. They possess both a double-stranded DNA (dsDNA) and RNA. Fig. 2.1 shows the structure of a bacterial cell.

A bacterial cell can vary from 0.2 to 10 μm in size and is covered by an outer cell wall. The cell wall confers the bacterium its shape, protects the cell and permits transport of molecules across the cell. The structure of the cell wall is complex and variess with the Gram reaction of the bacterium. The shape of a bacterium is a spherical 'coccus' or rod 'bacillus', with some being spiral.

The cytoplasm within the cell wall does not show mobility or streaming as in eukaryotes and is covered by a permeable plasma membrane. A single double-stranded chromosomal DNA molecule lies free in the cytoplasm, with some bacteria possessing an extrachromosomal piece of DNA called 'plasmid'. Unlike the cellular DNA, plasmids are not essential for the survival of the bacterial cell but may code for toxin production or antimicrobial resistance that provide a survival advantage. Besides the free DNA, the cytoplasm also has cell organelles like ribosomes with RNA for protein synthesis and cellular inclusions for nutrients.

Additional structures that may or may not be present in all bacteria are flagella, pili or fimbriae, and capsules.

Bacterial cell wall

The cell wall is the outermost rigid covering of the bacterial cell that protects the cell and gives the bacterium its shape (Fig. 2.2). It is antigenically distinct in specific strains of bacteria and referred to as **O antigen**. A cell wall-deficient form of a bacterium is called a **protoplast;** it consists of a plasma membrane and cytoplasm with its constituents.

Christian Gram developed a staining technique for bacteria, following which the bacteria appear either violet (Fig. 2.3) or pink (Fig. 2.4). The ability of bacteria to stain either gram-positive or gram-negative is, amongst other factors, largely dependent on the bacterial cell wall that differs in both these Gram reactions. The technique and principle of Gram staining have been elaborated on in Chapter 8.

The gram-positive cell wall consists of a thick outer layer of peptidoglycan, teichoic acid and lipoteichoic acid. The gram-negative cell wall is more complex and consists of a thin inner layer of peptidoglycan, without teichoic acid and lipoteichoic acid. The outer membrane of the gram-negative cell wall consists of outer membrane proteins (OMPs) or porin proteins and lipopolysaccharide (LPS). The LPS is virulent and contributes to pathogenicity; it is also antigenic, giving rise to anti-LPS antibodies in the immune response. Bacterial LPS is released following death of the bacterium in gram-negative bacteremia, leading to endotoxic shock. It has the ability to activate the alternate complement pathway.

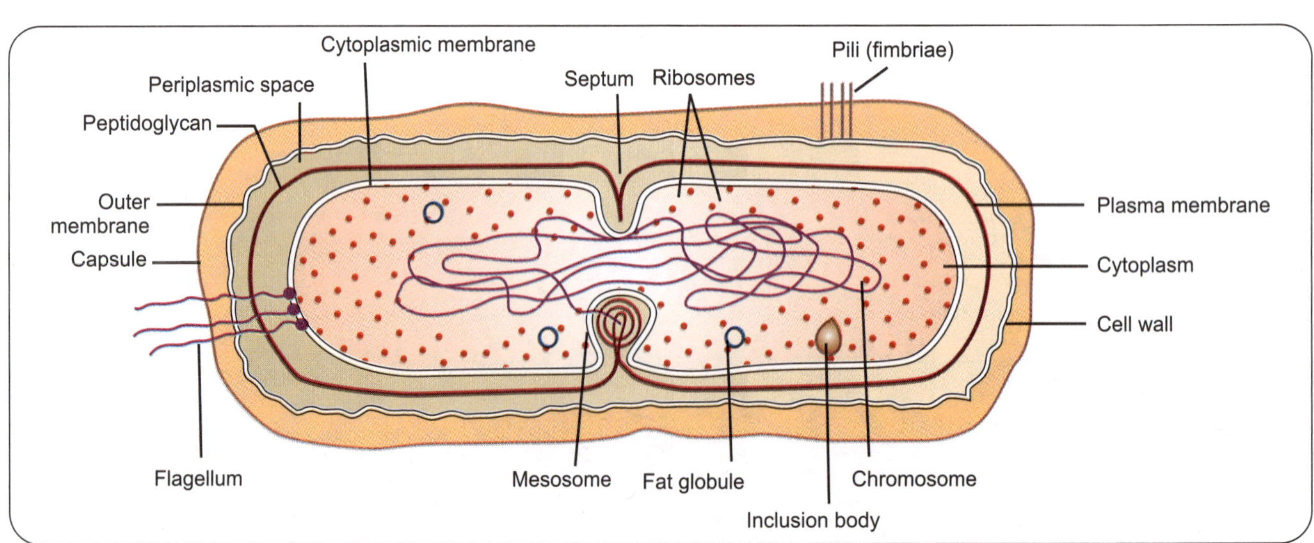

Fig. 2.1 Structure of a bacterial cell

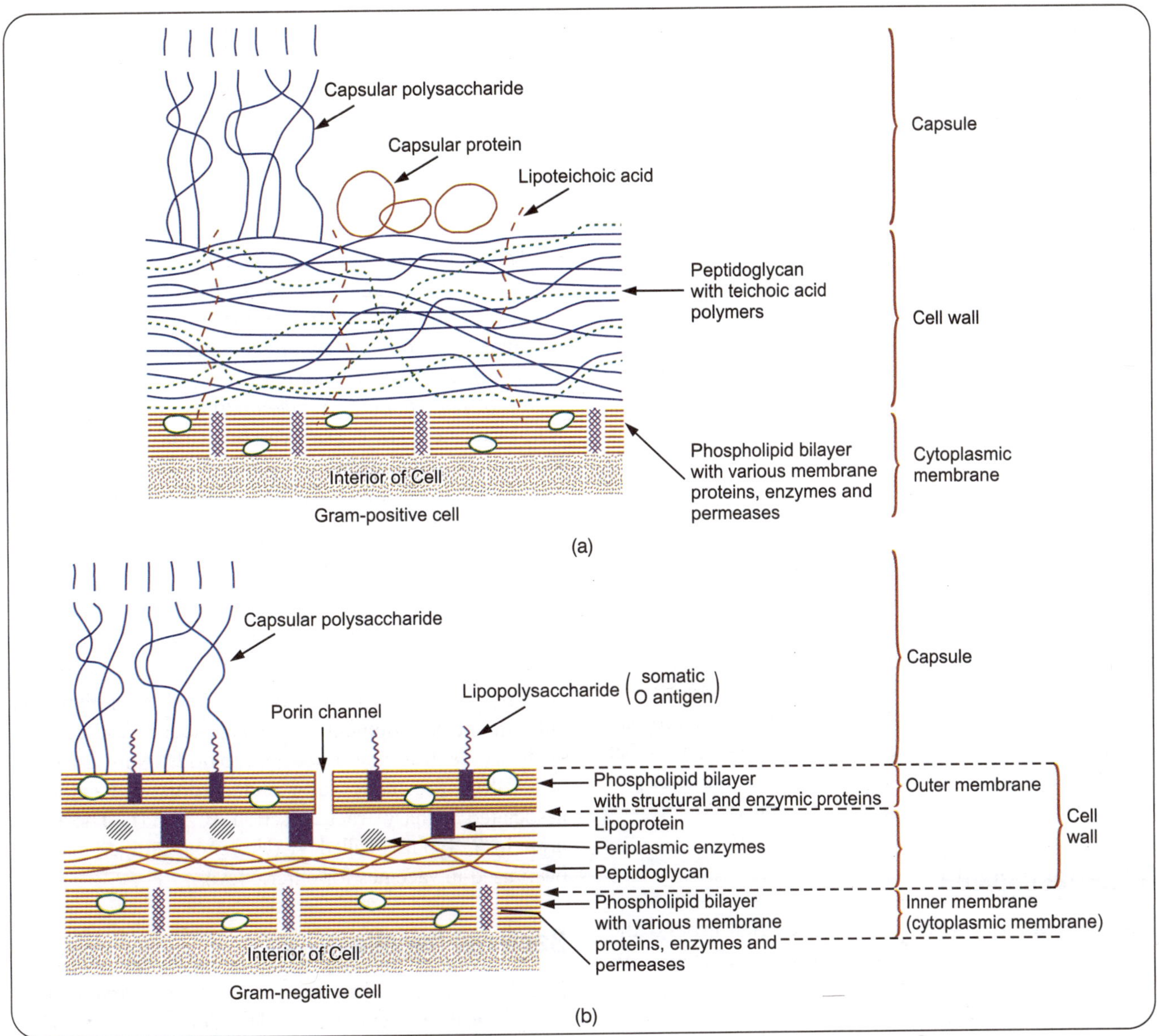

Fig. 2.2 Structure of gram-positive and gram-negative bacterial cell wall

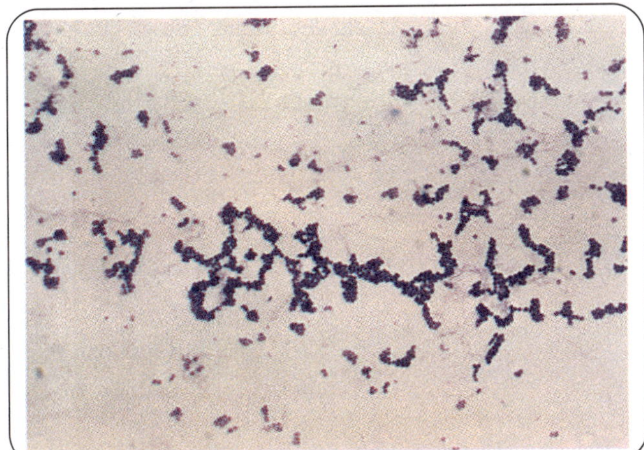

Fig. 2.3 Gram-positive cocci (*Source*: Public Health Image Library, ID 2296 / Centers for Disease Control and Prevention [CDC] / Dr. Richard Facklam)

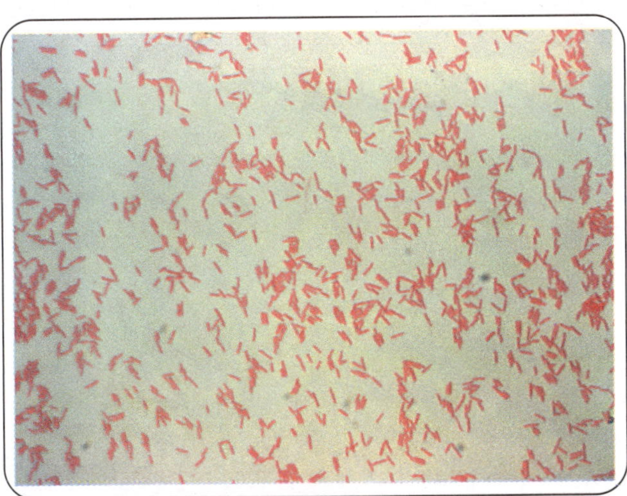

Fig. 2.4 Gram-negative bacilli (*Source*: Public Health Image Library, ID 17370 / CDC / Dr. W.A. Clark)

Bacterial plasma membrane, cytoplasm and its constituents

The space between the cell or plasma membrane and the cell wall is called the periplasmic space. The cell or plasma membrane is a semi-permeable membrane of phospholipids over the cytoplasm that allows the selective transport of ions and molecules across it. The cytoplasm is 70% water in which exist ribosomes, storage granules called intracytoplasmic inclusions and mesosomes. Ribosomes composed of rRNA and ribosomal proteins are the site for protein synthesis. Mesosomes are invaginations of the plasma membrane into the cytoplasm, analogous to mitochondria in eukaryotes, and are involved in bacterial respiration.

Bacterial DNA

Bacteria do not possess a true nucleus, nucleolus or nuclear membrane. They possess a single, circular dsDNA coil. Some bacteria may possess extrachromosomal DNA called plasmids. Bacteria reproduce by binary fission, wherein DNA division is followed by cytoplasmic and cell wall division.

Besides the cell wall, plasma membrane, cytoplasm and its constituents, some bacteria possess additional structures: a capsule, slime layer, flagella, fimbriae or pili, while others exhibit the ability to form spores.

Bacterial capsule

The well-organised viscid material secreted around the cell surface of bacteria is known as 'capsule' (Fig. 2.5). When it is diffuse and unorganised, it is called 'slime layer'. It is antigenically distinct in specific strains of bacteria and referred to as **K antigen**. Examples of capsulated bacteria are *Streptococcus pneumoniae, Neisseria meningitidis, Haemophilus influenzae, Bacillus anthracis* and *Klebsiella pneumoniae.* The chemical nature of the capsule is polysaccharide in most, except for *B. anthracis,* in which it is polypeptide. It protects bacteria by preventing the action of lytic enzymes as well as preventing phagocytosis and complement-mediated cell lysis. The capsule of *Streptococcus pyogenes* (hyaluronic acid) and *Streptococcus mutans* plays a role in their virulence. The capsule is also antigenic, a property that is utilised in the rapid diagnosis of pyogenic meningitis using the latex agglutination test that detects the specific capsular antigen of *S. pneumoniae, N. meningitidis* and *H. influenzae.* The capsular antigen is the vaccine content of the *S. pneumoniae, N. meningitidis* and *H. influenza (Hib)* vaccines. The capsule can be demonstrated by light microscopy with negative staining by India ink, where the capsule is seen as a clearing or halo around the cell. This technique is used in the laboratory diagnosis of cryptococcal meningitis and not bacterial infections.

Examples of bacteria that produce a polysaccharide slime layer are coagulase-negative staphylococci (CoNS) and *Salmonella* Typhi. In CoNS, this layer leads to the adhesion of bacteria to medical devices and prosthetic surfaces, resulting in the formation of a biofilm and subsequent development of hospital-acquired infections, such as central line-associated bloodstream infection (CLABSI), prosthetic valve endocarditis and catheter-associated urinary tract infection (CAUTI). The slime layer in *S.* Typhi is called Vi (virulence) antigen and has been used in vaccines for enteric fever.

Bacterial flagella

Flagella are thread-like organs of locomotion arising from the cell wall that render bacteria motile (Fig. 2.6). They are antigenically distinct in specific strains of bacteria and referred to as **H antigen**. Depending upon the number

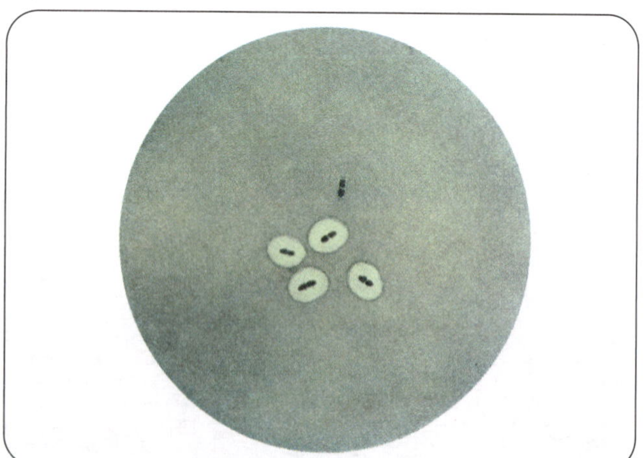

Fig. 2.5 *Streptococcus pneumoniae* capsule, India ink, 100X (*Source*: Public Health Image Library, ID 2113 / CDC)

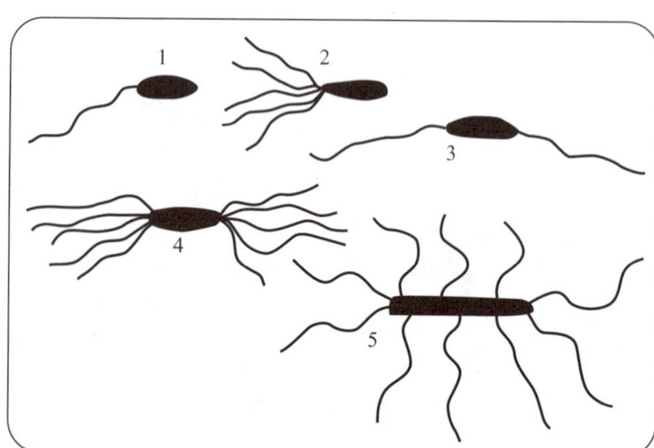

Fig. 2.6 Types of flagellar arrangement: 1. monotrichous; 2. lophotrichous; 3. amphitrichous; 4. amphilophotrichous; 5. peritrichous flagella

and arrangement of flagella, a bacterium is monotrichous (single, polar as in *Vibrio cholerae*), amphitrichous (single, bipolar), lophotrichous (multiple, polar) or peritrichous (all over the cell as in *S.* Typhi and *Escherichia coli*). Structurally, the flagellum consists of a basal body that is embedded in the plasma membrane and cell wall, and a hook that connects the basal body to the filament, which is the thread-like extension from the cell surface. The filament is made of a protein called **flagellin**. Flagella are difficult to stain but their presence can be indirectly inferred by demonstrating the motility of bacteria. Depending upon the character of the flagella, bacteria produce distinct types of motility. *Klebsiella* and *Shigella* do not possess flagella and are known to be non-motile. The most distinct amongst these is the darting (gnat-like) motility of *V. cholerae* and corkscrew flexion and extension movement of spirochetes due to the presence of endoflagellum. The darting motility of *V. cholerae* can be demonstrated by the hanging drop method while that of spirochetes by dark-field microscopy.

Bacterial fimbriae or pili

These are short, fine organs of adhesion made of protein that are antigenic. They help bacteria to adhere to epithelial surfaces, leading to colonisation, or to red blood cells, causing hemagglutination. In gram-negative bacteria, specialised pili called 'sex pili' are also involved in the transfer of antimicrobial resistance through the process of conjugation. The adhesion of *Neisseria gonorrhoeae* pili to urethral epithelium is the virulence mechanism causing gonococcal urethritis.

Bacterial spores

Sporulation in bacteria, unlike that in fungi and protozoans, is not a method of reproduction, as bacteria do not divide during sporulation. On depletion of nutrients, in unfavourable conditions, certain gram-positive bacteria form spores. Spores are the dormant state of these bacteria that can germinate into active vegetative bacteria under favourable conditions.

Bacterial spores are the most resistant to chemical disinfection and can only be removed by physical methods of sterilisation. Their heat-resistant property is utilised in the quality control of heat sterilisation using biological spore strip indicators. Spores can be demonstrated by Gram staining as unstained bodies and are weakly acid-fast.

The position, shape and size of the spore provide clues to the identity of gram-positive bacilli. The spores of *Bacillus* are smaller than the bacillary body and therefore do not bulge, while those of *Clostridia* are larger than the bacillary body and cause a bulge in the bacterium. The position of the spore can be central, subterminal or terminal, the shape being either spherical or oval (Figs. 2.7a and b).

BACTERIAL PHYSIOLOGY

Bacterial growth and metabolism

The minimum bacterial growth requirement is water (80% of the cell) with a source of carbon and nitrogen for non-fastidious bacteria. Nutritional requirements of bacteria vary and can get more complex, with fastidious bacteria requiring inorganic salts and vitamins. Other environmental factors that affect bacterial growth are gas, temperature, pH, light and moisture.

Depending upon the requirement of oxygen, bacteria are obligate aerobes that grow only in the presence of oxygen, e.g., *M. tuberculosis.* Obligate anaerobes are those that grow only in the absence of oxygen,

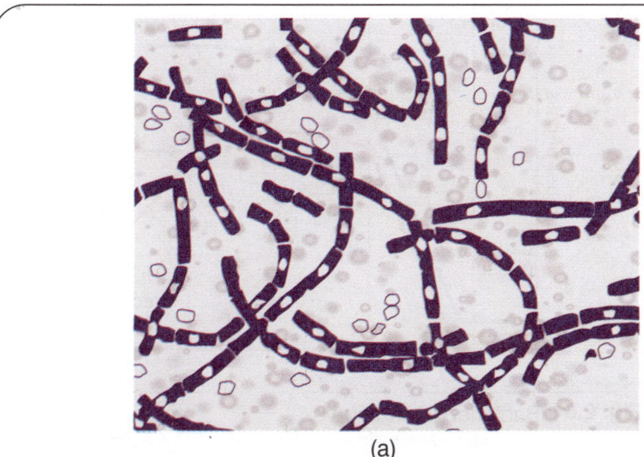

(a)

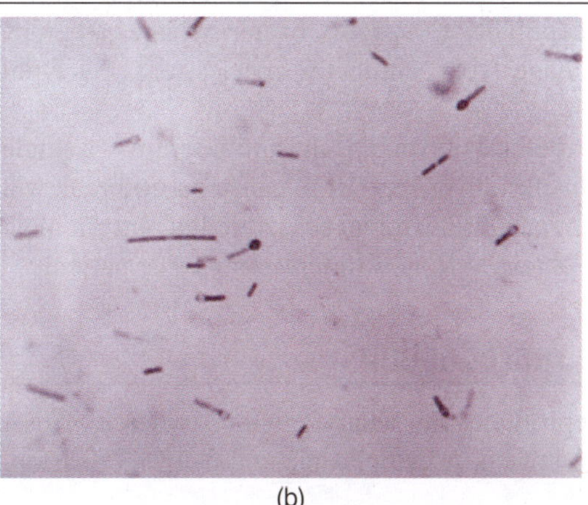

(b)

Fig. 2.7 (a) Gram-positive bacilli with central spores: *Bacillus anthracis*, Gram stain, 100X; (b) Gram-positive bacilli with spherical terminal spores: *Clostridium tetani*, Gram stain, 100X (*Source*: Public Health Image Library, ID 12056 / CDC / Dr. Holdeman)

e.g., *Clostridium* and facultative anaerobes are those that can grow in both the presence and absence of oxygen, e.g., *E. coli*. Most bacteria pathogenic to man are facultative anaerobes. Microaerophilic bacteria grow in the presence of low oxygen and capnophilic bacteria require CO_2. The bacterial metabolism depends on whether the bacterium is aerobic or anaerobic. Aerobic bacteria utilise glucose by oxidation via Kreb's cycle while anaerobes by fermentation via the Embden-Meyerhof-Parnas pathway.

Most pathogenic bacteria grow at 37°C (human body temperature) and a pH of 7.4 (human pH), exceptions being *Lactobacillus* that prefers an acidic pH and *Vibrio* that prefers an alkaline pH.

Bacterial culture, identification and typing

Unlike viruses, bacteria can be cultivated in vitro in the laboratory on inanimate culture media by providing nutritional requirements depending upon the nature of the bacterium, aerobically and anaerobically. The laboratory diagnosis of bacterial infections rests upon demonstrating the causative microbe by microscopic examination of the smear, culture on media and identifying the bacterial growth either by its metabolic process (biochemical reactions) or serologic (agglutination reactions) characteristics.

For epidemiologic purposes and outbreak investigations, bacteria can also be 'typed' to discern identical or variable strains of a particular species. Bacterial typing methods include biotyping (based on biochemical tests), phage typing (based on susceptibility to bacteriophages) and serotyping (based on antigenic composition).

Biotyping is applied in *V. cholerae* to differentiate the classical and El Tor biotypes. Serotyping is also useful in cholera to distinguish the O1 and O139 serogroups and to distinguish the serotypes Inaba, Ogawa and Hikojima within the O1 serogroup. Bacteriophages are viruses that infect bacterial cells. Certain strains of bacterial species exhibit susceptibility to specific bacteriophages, thereby rendering phage types within the species, e.g., *S.* Typhi phage types.

A simple method of typing is the antibiogram—a single strain of a bacterium suspected to cause an outbreak will have an identical antimicrobial susceptibility pattern in all the isolates obtained from different samples or patients.

Bacterial reproduction

Bacteria reproduce by binary fission, wherein DNA division is followed by cytoplasmic and cell wall division. The two strands of the parent dsDNA separate and two new complementary strands are formed. A transverse septum arises from the mesosome across the cell, separating it into two daughter cells each with a dsDNA.

The time taken for one bacterial cell to undergo binary fission and become two cells is called **generation time**. The generation time varies from 15 minutes in *E. coli* and most other pathogenic bacteria to 15 hours in *M. tuberculosis* and 15 days in *M. leprae*. The generation time reflects the time taken by the bacterium to grow visibly on in vitro culture as well as on the incubation period of the disease. For example, urinary tract infection caused by *E. coli* has an incubation period of 3 days and the bacterium grows in 12–24 hours, pulmonary tuberculosis has an incubation period of 2–12 weeks and *M. tuberculosis* takes 3–4 weeks to grow on solid media, while leprosy has an incubation period of 2–10 years!

Bacterial growth curve

If a bacterium is inoculated in a liquid culture medium and the bacterial count measured, the bacterial growth curve shows four phases (Fig. 2.8).

Phases of bacterial growth curve

1. Lag phase, in which the bacteria increase in size to reach their maximum size but do not multiply.
2. Log phase, in which the bacteria begin to multiply exponentially but are smaller in size and stain uniformly.
3. Stationary phase, in which the bacterial growth and multiplication stop due to the depletion of nutrients and accumulation of toxic products. The viable count (live bacteria) stagnates but the total count rises (live + dead bacteria). Sporulation occurs in this phase.
4. Decline phase, in which multiplication stops completely and cells die due to a lack of nutrients and accumulation of toxic products. The total count stagnates and viable count decreases.

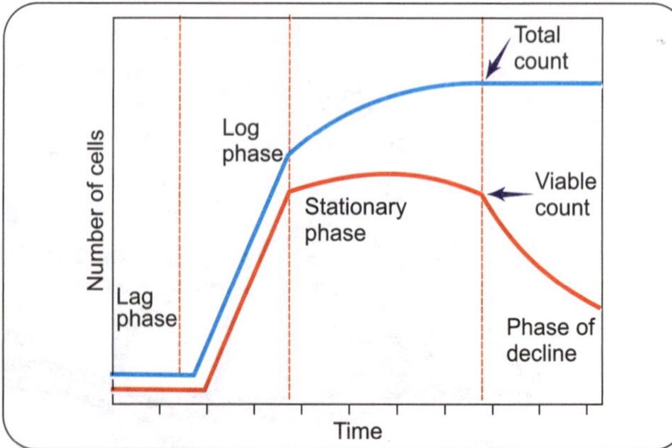

Fig. 2.8 Bacterial growth curve

QUESTIONS

1. Describe in brief the structure and function of the bacterial cell wall and its role in disease.
2. What are the applications of bacterial capsule in laboratory diagnosis and immunoprophylaxis?
3. How do flagella indicate the identity of a bacterium?
4. What is the role of pili in infections?
5. Discuss bacterial sporulation and its role in the morphologic identification of bacteria.
6. Define generation time. What is its importance in bacterial infection?

Classification and Characteristics of Viruses

A virus possesses only one nucleic acid–either DNA or RNA (never both)—surrounded by one or more proteins with or without an outer membrane envelope. Amongst microorganisms, they are neither prokaryotes nor eukaryotes. As they do not possess a cellular organisation, the nucleic acid encodes only those proteins necessary for replication and assembly within host cells and does not encode the enzymes of metabolism. Hence, viruses are obligate intracellular parasites within specific host cells. Once infected, the host cell machinery is utilised by the virus for metabolism and multiplication.

Viruses exhibit host cell specificity via host cell receptors and, therefore, all viral infections are tissue-specific, e.g., hepatitis B virus (HBV) infects hepatocytes and human immunodeficiency virus (HIV) infects CD4 T cells.

Virusoids are nucleic acids that depend on helper viruses for replication, e.g., delta virus and HBV. **Viroids** are naked double-stranded RNAs restricted to plants. Prions are abnormal protein molecules without nucleic acids that can spread and have been implicated in neurodegenerative diseases.

MORPHOLOGY OF VIRUS

Viruses are the smallest known infective agents capable of infecting humans, animals, plants and bacteria. The virus size varies from 10 nm to 300 nm, whereas bacteria have an average size of 1,000 nm. Therefore, viruses can pass through bacterial filters.

The smallest virus causing human infection—parvovirus—is 20 nm in size and the largest—vaccinia (causing smallpox)—is 300 nm in size. Due to their size being in 'nm', viruses cannot be visualised by light microscopy, as the wavelength of light is 400–700 nm.

As the wavelength of light, which is 400–700 nm, is larger than the size of a virus, light cannot be reflected from its surface; therefore, light microscopy is not used to visualise a virus. To overcome this obstacle, an illumination source with a shorter wavelength than light is used as an electron beam. Viruses, thus, can be visualised by electron microscopy alone.

However, when a virus infects and multiplies within a specific cell, the aggregation of virions or their effect on the cell can be visualised as a distinct morphologic change, termed as an **'inclusion body'**, in the histopathological examination of virus-infected tissue.

Structure of a virus

A virus consists of one nucleic acid that is either DNA or RNA, never both. The DNA can be single- or double-stranded, while the RNA can be single or segmented. A positive-sense RNA can be directly translated into proteins, i.e., the virus RNA itself functions as messenger RNA (mRNA). A negative-sense RNA needs to be copied into translatable RNA, i.e., the virus RNA transcribes into mRNA.

Therefore, a viral genome consists of any one of the following:

1. Single-stranded DNA (ssDNA), e.g., parvovirus B19
2. Double-stranded DNA (dsDNA), e.g., HBV, herpes group of viruses
3. Single-stranded, positive-sense RNA, e.g., HIV, poliovirus, hepatitis A virus (HAV), hepatitis C virus (HCV), hepatitis E virus (HEV), coronaviruses
4. Single-stranded, negative-sense RNA, e.g., rabies virus
5. Segmented, negative-sense RNA, e.g., influenza virus, rotavirus

Viral nucleic acid is usually associated with nucleoproteins, both of which are enveloped in a protein shell called **capsid,** composed of identical capsomeres (Fig. 3.1). **Capsomeres** have a helical or an icosahedral symmetry. The structural core of nuclei acid, nucleoprotein and capsid is called a **nucleocapsid**.

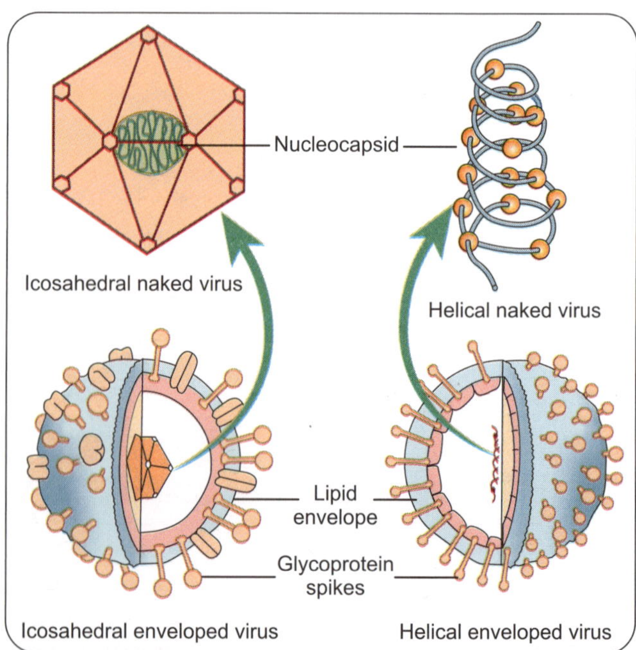

Fig. 3.1 Structure of a virus

Most human viruses are nucleocapsids, wherein the capsid interacts with a specific host cell receptor. Other viruses have another outer lipid envelope that is derived from the virus-modified host cell membrane, that establishes contact with infective host cells. The space between the envelope and capsid is filled with proteins.

In general, enveloped viruses are more sensitive to lipid solvents and disinfectants than non-enveloped viruses. Non-enveloped viruses are more resistant to chemical disinfection than enveloped viruses. Some active antiviral disinfectants are hydrogen peroxide, sodium hypochlorite, halogens (iodine, chlorine) and aldehydes. Most viruses are enveloped, except parvovirus, adenovirus, papovavirus, picornavirus, reovirus, HAV and HEV.

Viruses are heat-labile, being denatured at 56°C in seconds, 37°C in minutes and 4°C in days. To store virus samples over an indefinite period of time, they would have to be maintained at –70°C.

Antibiotics and antimicrobial agents effective against bacteria are ineffective against viruses. Antiviral agents are recommended in specific viral infections that are not self-limiting.

CLASSIFICATION OF VIRUSES

Viruses are either classified as RNA or DNA (Table 3.1).

Depending on the pathogenicity, viruses can be classified as:

1. Viruses causing respiratory infections: influenza virus, parainfluenza virus, respiratory syncytial virus (RSV), coronaviruses including severe acute respiratory syndrome (SARS) virus, Middle East respiratory syndrome (MERS), and severe acute respiratory syndrome coronavirus 2 (COVID-19)
2. Viruses causing diarrhea: rotavirus, astroviruses, adenoviruses, Norwalk virus

Table 3.1 Classification of viruses

Genome	Family	Representative	Genome	Envelope
RNA viruses	Picornaviridae	Poliovirus Hepatitis A virus Rhinovirus	(+) RNA	No
	Caliciviridae	Norwalk virus Hepatitis E virus	(+) RNA	No
	Togaviridae	Rubella virus	(+) RNA	Yes
	Flaviviridae	Dengue virus Hepatitis C virus	(+) RNA	Yes
	Coronaviridae	Coronavirus	(+) RNA	Yes
	Rhabdoviridae	Rabies virus	(-) RNA	Yes
	Orthomyxoviridae	Influenza virus	(-) RNA,8 segs	Yes
	Paramyxoviridae	Parainfluenza virus Respiratory syncytial virus Mumps virus Measles virus	(-) RNA	Yes
	Reoviridae	Rotavirus Reovirus	dsRNA, 12 segs	No
	Retroviridae	Human T-cell lymphotropic virus 1 and 2 Human immunodeficiency virus 1 and 2	(+) RNA	Yes
DNA viruses	Hepadnaviridae	Hepatitis B virus	dsDNA	Yes
	Herpesviridae	Herpes simplex virus 1 and 2 Varicella zoster Epstein–Barr virus Human herpesvirus 6 and 7	dsDNA	Yes
	Papovaviridae	Human papillomavirus	dsDNA	No
	Adenoviridae	Adenovirus	dsDNA	No
	Parvoviridae	Parvovirus B19	ssDNA	No
	Poxviridae	Smallpox virus Molluscum contagiosum virus	dsDNA	Yes

(+) positive sense, (-) negative sense

3. Viruses causing exanthema: chickenpox virus, measles virus, rubella virus, herpes simplex virus, parvovirus B19
4. Viruses causing hepatitis: hepatitis A, B, C, D and E viruses (HAV, HBV, HCV, HDV and HEV)
5. Viruses causing arboviral fevers: dengue virus, chikungunya virus
6. Zoonotic viruses: rabies virus
7. Oncogenic viruses: HIV, human T-cell lymphotropic virus (HTLV)
8. Teratogenic viruses: rubella virus, cytomegalovirus
9. Viruses transmitted perinatally: HBV, HCV, HIV

VIRAL REPLICATION

A virus infects a specific susceptible host cell, followed by the reproduction of nucleic acid and proteins using host cell machinery. It then assembles and releases infectious progeny from the host cell (Fig. 3.2). The stages of viral replication are as follows:

1. **Attachment and penetration:** Viral surface proteins bind with specific host cell receptors. For enveloped viruses, these proteins are spikes, e.g., hemagglutinin in influenza. In non-enveloped viruses, surface peptides function as the viral attachment protein. Following adsorption, the virion is translocated across the plasma membrane into the host cell.
2. **Uncoating:** Once inside the host cell, the virion is uncoated by release of the capsid to make the viral nucleic acid accessible for protein synthesis. Viruses with dsDNA synthesise mRNA in the same way as the host cell, using DNA-dependent RNA polymerase. RNA viruses make mRNA and translate it into protein through a variety of mechanisms, depending upon the nature of the RNA (ss or ds, + or - sense or antisense).
3. **Assembly and release:** Once the replication of the viral genome and synthesis of viral protein are complete, the virions are assembled and released from the host cells. Non-enveloped viruses are released when the cell lyses. Enveloped viruses are released by budding out of the host cell. During budding, the virus may carry virus-modified host cell membrane to form its envelope.

There are three outcomes of infection of a host cell by a virus: lytic infection, latent infection or chronic infection. In lytic infection, the virus replicates multiple times and causes death of the host cell, e.g., polio and influenza. In latent infection, the virus integrates into the host cell

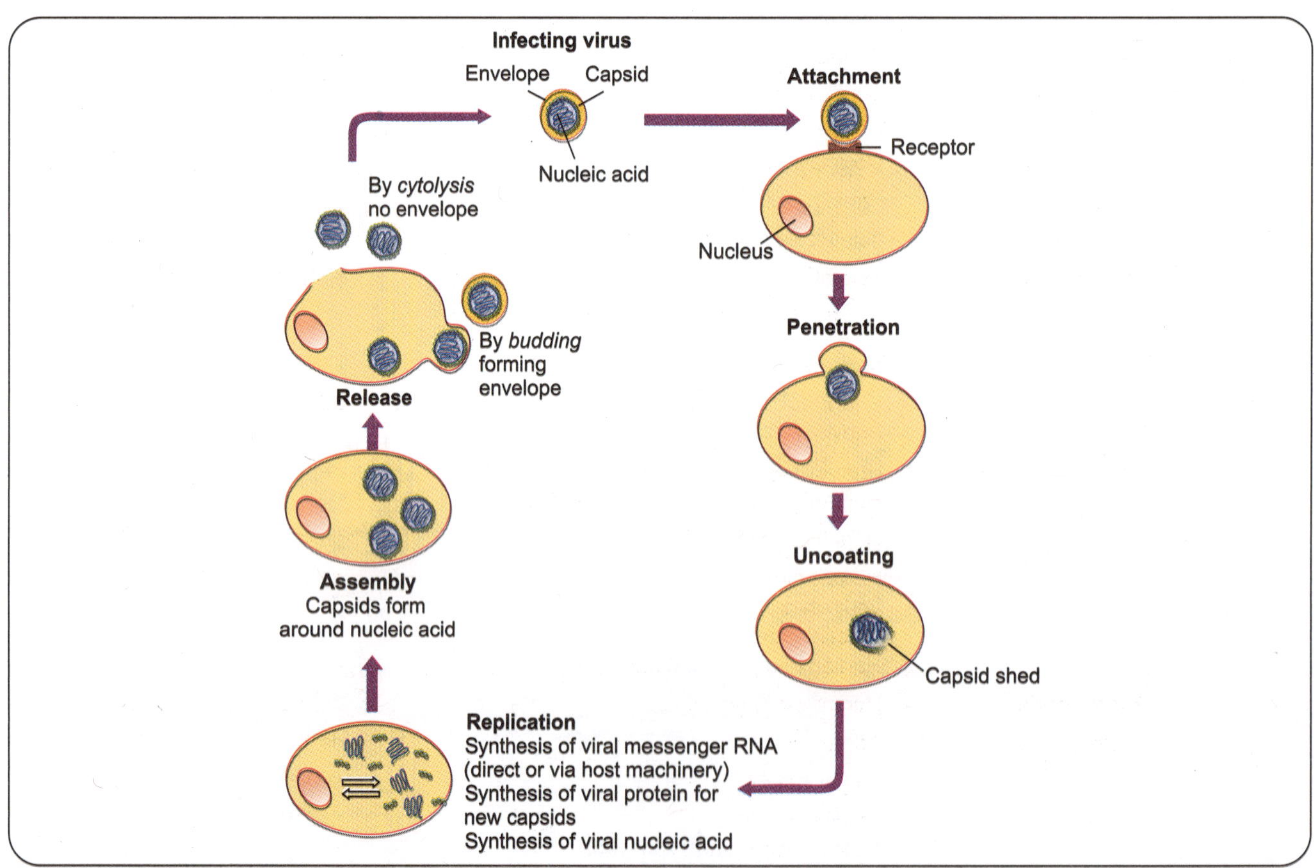

Fig. 3.2 Replication of virus in a host cell

but does not replicate immediately, e.g., HIV and some *Herpesviridae*. Later, the virus may be reactivated or result in transformation of the cell, causing malignancy. In chronic infection, the virus continues to replicate after a period of acute illness. There occurs a low release of the virus with host cell death, e.g., HBV.

VIRAL PATHOGENESIS AND IMMUNE RESPONSE

A virus can enter through any route of infection:

1. Respiratory – Influenza virus, measles virus, varicella zoster virus
2. Gastrointestinal tract – Epstein–Barr virus (EBV) (oral), poliovirus (intestinal)
3. Skin – HPV
4. Conjunctiva – Adenovirus
5. Sexual contact – HBV, HCV, HIV
6. Congenital – Cytomegalovirus (CMV), rubella virus
7. Blood transfusion – HBV, HCV,HIV
8. Arboviral – Dengue virus, chikungunya virus
9. Zoonotic – Rabies virus

Following local multiplication within the specific host cell, the virus enters the lymphatics and the bloodstream (primary viremia) to reach the liver and spleen. After multiplying in the liver and spleen, the virus re-enters the blood (secondary viraemia) with the onset of clinical prodromal symptoms. Subsequently, the virus reaches the specific target organ. The incubation period of the disease is from the entry of the virus till the appearance of symptoms in the target organ. For some viruses, tissue tropism—entry, primary replication and target replication—occurs at the same anatomic site, e.g., HPV warts.

Other viruses enter at one site but target a distant site. Poliovirus enters through the intestine but targets the central nervous system. Other than viruses that spread through the bloodstream, some viruses can spread via the nerves, e.g., rabies virus, poliovirus, herpes simplex virus (HSV) and varicella zoster (VZ) virus. The rabies virus enters at the motor end plate of the neuromuscular junction using the acetylcholine (Ach) receptor and HSV via receptors on the synaptic ends. These viruses then use intraneuronal mechanisms to travel along the axon fibres.

The olfactory pathway represents a special category of neural spread in the case of aerosol exposure to rabies and HSV in HSV encephalitis—the localisation of HSV in the orbitofrontal and medial temporal complex. Some viruses may use multiple pathways, e.g., VZ virus disseminates to skin by the hematogenous route and then spreads along the nerves from the skin to the dorsal root ganglion where it remains latent. Viruses like HBV and HIV also have oncogenic potential, while rubella virus and CMV display teratogenicity.

The immune response to viruses is humoral and cellular as viruses are obligate intracellular pathogens. All classes of antiviral immunoglobulins (Igs) are secreted, but the effect of the antibodies varies with the specificity of the virus. Neutralising antibodies to surface antigens are seen in hepatitis B and influenza. Anti-hepatitis B surface antibody is able to neutralise hepatitis B surface antigen (HBsAg), thereby preventing hepatitis B and anti-hemagglutinin neutralises influenza virus. However, not all viral antibodies, especially those to internal antigens, neutralise virus infectivity. Some antibodies contribute to pathologic consequences as in viral hemorrhagic fevers, especially dengue.

Cell-mediated immunity with T_H and T_C cells is very important in dealing with virus-infected cells. Viruses that infect the T_H cells themselves, such as HIV, cause severe immunosuppression. In general, viral infections confer lifelong immunity to re-infection by the same strain of the virus but not an antigenically different strain. Interferons (IFNs) are cytokines produced by virus-infected cells that can act on other uninfected cells and prevent them from being infected by the virus. They are not virus-specific, but are species-specific, i.e., any virus-infected human cell secretes IFNα, IFNβ and IFNγ, that prevent viral infections in humans.

VIRAL CULTURE

A virus needs host cells and cannot be grown on inanimate media; therefore, viral culture is not employed for the laboratory diagnosis of viral infections. **Viruses are cultivated for producing vaccines** and for **research purposes**. They can be cultivated in chick embryo, in vivo animal models or in vitro cell line cultures.

Viruses cultivable in chick embryo for producing vaccines include influenza, yellow fever and Japanese B encephalitis. In vitro cell lines or tissue culture can be of human, animal or cell origin using either normal or malignant cells. Human fibroblasts and monkey kidney cells are examples of normal cell lines that have a finite number of cell divisions for a limited period, e.g., Vero (vervet monkey kidney for rabies vaccine). Malignant cell lines are capable of infinite cell divisions, easy to maintain and called “continuous cell lines”. They are more widely used, e.g., human cervical carcinoma (HeLa), human laryngeal epithelioma (HEp-2).

The growth of a virus in a cell culture can be detected by the specific morphologic, cytopathic effect of the virus or by detecting the viral antigen or genome.

LABORATORY DIAGNOSIS OF VIRAL INFECTIONS

1. Microscopy

As viruses can only be visualised by electron microscopy and not light microscopy, examining samples under a light microscope is not advocated. The light microscope can, however, visualise inclusion bodies in the histopathological examination of tissues affected by viruses.

Inclusion bodies are virus-specific intracellular aggregates of virions of a definite shape, size and location present in virus-infected cells. They can be demonstrated by histopathological examination of virus-infected tissue by light microscopy for diagnosis. Inclusion bodies can be intracytoplasmic or intranuclear or both. Intranuclear inclusion bodies are seen in CMV (Fig. 3.3), HSV, yellow fever virus, adenovirus and poliovirus. Intracytoplasmic inclusion bodies are seen in rabies virus (Negri bodies) and in molluscum contagiosum virus.

Other microscopic diagnostic methods include direct immunofluorescent microscopy to detect rabies in corneal smears and the Tzanck smear examination of skin lesions in varicella, herpes zoster and herpes simplex.

2. Serological diagnosis

The detection of an antigen or antibody remains the mainstay for the laboratory diagnosis of most viral infections. The technique employed is either rapid immunochromatography (ICT), that can be performed at the bedside, or the enzyme-linked immunosorbent assay (ELISA) or chemiluminescent immunoassay (CLIA), that is done in the laboratory.

In general, viral antigen can be detected during the acute phase of the illness with viremia and its level declines with the appearance of antibodies. IgM antibodies precede the appearance of IgG and an increase in the specific IgG titre denotes recent infection over past infections or vaccination.

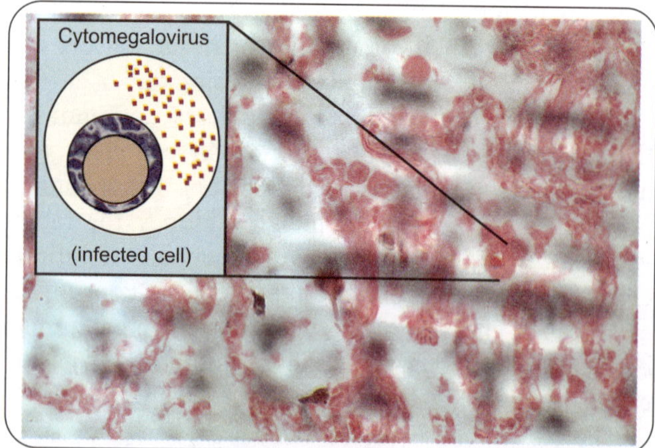

Fig. 3.3 Cytomegalovirus showing intranuclear inclusion bodies-owl's eye appearance, H&E, 40X

The natural history of a specific viral infection determines the appearance and disappearance of specific serological viral markers across the timeline of the infection. Some infections like HIV exhibit a window period wherein anti-HIV antibodies may be absent even though the patient is infected.

The structure of a virus also determines the antigens and antibodies associated with the disease, e.g., active HBV infection shows the presence of HBsAg and HBeAg in blood but not HBcAg. These are often referred to as serologic "viral markers" as their presence denotes viral infection (Table 3.2).

Table 3.2 Serologic viral markers tested for the diagnosis of viral infections

Virus	Marker
HBV	HBsAg, HBeAg, anti-HBs, anti-HBe, anti-HBc
HIV	p24, gp120, anti-HIV1, anti-HIV2
Dengue virus	NS1, anti-DENG IgM and IgG
Chikungunya virus	IgM
Japanese B encephalitis virus	IgM
Rotavirus	Antigen in stool

3. Molecular diagnosis

Detection of the presence of viral genome RNA or DNA is considered as the confirmatory test of viral infection. For this, sample from the site of infection is collected in viral transport medium (VTM) at the bedside or OPD and sent to the laboratory. The molecular method used to detect viruses is the **polymerase chain reaction** (PCR). It involves lysis of the virus, extraction of the viral DNA, amplification of a specific region of the viral DNA using specific 'primer' sequences and detection of the presence of the specific amplified region. Reverse transcriptase PCR (RT-PCR) is used for detecting RNA viruses, wherein the viral RNA is reverse-transcribed to DNA before amplification and detection of the amplified product. Real-time PCR permits quantification of the 'viral load' to determine the prognosis and response to antiviral therapy (in the case of HBV and HIV) with the amplification being visualised on a real-time basis during PCR and is, therefore, less time-consuming. Viral infections diagnosed by PCR are hepatitis B, HIV, H1N1, COVID-19, HSV, CMV, EBV and many others.

Irrespective of whether viral serologic markers are detected by ELISA or CLIA or molecular markers by PCR, the result of the test is made available on the day of testing within 4–5 hours.

VIRAL IMMUNOPROPHYLAXIS

Immunoprophylaxis in viral infections includes both active and passive strategies. Active immunisation with live, killed and subunit vaccines is available against many viral infections (Table 3.3), while passive post-exposure prophylaxis with pre-formed antibodies is seen in rabies and perinatal hepatitis B.

Table 3.3 Viral vaccines

Vaccine	Viral disease
Live	Chickenpox Mumps Measles Rubella Polio (Sabin) Influenza
Killed	Rabies Polio (Salk) Japanese encephalitis
Subunit	Hepatitis B (HBsAg) Influenza

Advantages of live vaccines

Live vaccines **"simulate"** an infection and therefore more potently stimulate cell-mediated immunity as well as humoral immune response. Moreover, in the case of viral infections, the vaccine is able to provide antigenic stimulation for long periods and therefore reduces the need for boosters or secondary doses of the vaccine. For example, viral vaccines like the measles, mumps and rubella (MMR) vaccine and chickenpox vaccine are single-dose vaccines providing lifelong protection.

Disadvantage of live vaccines

These cannot be administered to immunocompromised individuals. As opposed to live vaccines, killed or subunit vaccines can be administered to immunocompromised individuals, but necessitate the need for booster doses for long-term protection.

ANTIVIRAL AGENTS

A large number of viral infections are self-limiting and antiviral therapy is indicated only in some. Amantadine and rimantadine block viral penetration and the uncoating of the virus in a cell, uridine and thymidine analogues interfere with viral DNA synthesis while acyclovir and gancyclovir are nucleoside analogues. Viral infections associated with an indication for antiviral therapy are HIV, hepatitis B, herpes zoster, HSV, CMV and influenza. The mother-to-child transmission of neonatal HIV infection is prevented by prophylactic antiviral nevirapine.

QUESTIONS

1. How do viruses differ from bacteria?
2. Describe the morphological structure of a virus and its role in virus replication in a human cell.
3. Describe the mechanisms of pathogenesis in viral infections.
4. Why is viral culture not routinely done for the diagnosis of viral infections? Discuss the alternatives available to diagnose viral infections, with examples of each.
5. Discuss strategies for the active and passive immunoprophylaxis of viral infections, giving examples of each.

Classification and Characteristics of Fungi

Fungi are eukaryotes with a cell wall, a cell membrane, cell organelles and a well-defined nucleus. The cell wall is composed of chitin and polysaccharides. Fungi may be unicellular (yeast) or multicellular (hyphae) and can reproduce both asexually and sexually. They can be classified according to their morphological characteristics or on the basis of the site and spectrum of disease.

Factors predisposing fungal infections

Fungi, endogenous or saprophytic in the environment, cause human disease under certain predisposing conditions. These predisposing factors are characteristic of the disease and the causative agent:

1. Superficial skin infections are seen to occur in moist areas with humidity, e.g., tinea cruris in the groins
2. Oral mucosal candidiasis in the elderly following use of steroid inhalers for bronchial asthma.
3. Subcutaneous infections following traumatic implantation of saprophytic fungi from the soil in farmers.
4. Aspergilloma caused by *Aspergillus* following cavitary tuberculosis.
5. *Candida* infection following long-standing indwelling central line and urinary catheters.
6. *Cryptococcus neoformans* meningitis in the immunocompromised, especially in the HIV-infected.
7. Mucormycosis is mostly seen in patients with diabetic ketoacidosis.
8. Usage of broad-spectrum antibiotics and total parenteral nutrition in intensive care units.
9. Immunosuppression following severe neutropenia, blood malignancy, chemotherapy and radiotherapy predispose one to candidemia.

CLASSIFICATION OF FUNGI

Classification of fungi based on morphology

1. *True yeast:* Spherical or oval unicellular fungi that reproduce by the process of budding, e.g., *Cryptococcus neoformans.*
2. *Yeast-like*: Spherical or oval unicellular fungi that reproduce by the process of budding. The budding can form elongated pseudohyphae with constrictions at the septa, e.g., *Candida* spp.
3. *Molds:* Multicellular hyphae or mycelia that are either septate or non-septate and show branching. The mycelia that grow on the surface of the culture media are vegetative and those that grow above are aerial and generally reproduce with asexual or sexual spores that are infectious, e.g., dermatophytes, *Aspergillus, Rhizopus.*
4. *Dimorphic fungi*: These exist as hyphae in the environment or culture and as yeasts in human tissue, e.g., *Coccidioides* spp., *Paracoccidioides* spp., *Histoplasma capsulatum, Sporothrix schenckii.*

Figure 4.1 shows the morphologic forms of fungi.

Classification of fungal infections based on the site and spectrum of disease

1. *Superficial mycoses*: They involve the skin, hair, nails and mucosa, e.g., *Malassezia furfur, Pityriasis versicolor,* dermatophytes, *Candida*.
2. *Subcutaneous mycoses*: They are infections of the skin and subcutaneous tissues caused by saprophytic soil fungi, e.g., mycotic mycetoma, chromoblastomycosis, sporotrichosis, rhinosporidiosis.
3. *Systemic mycoses*: They are infections of deeper systemic tissues or organs, e.g., histoplasmosis, blastomyces, cryptococcosis, candidiasis, aspergillosis, mucorales.
4. *Opportunistic mycoses*: Here, the fungi are pathogenic only under certain opportunities, like low immunity, e.g., cryptococcosis, candidiasis, aspergillosis, mucorales.
5. *Mycotoxicoses*: They are diseases caused by consuming fungal toxins, e.g., the aflatoxin of *Aspergillus* that can contaminate grains and spices.

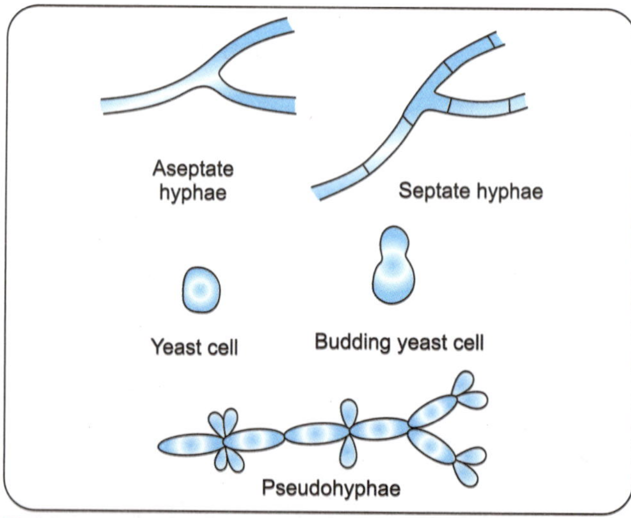

Fig. 4.1 Morphologic forms of fungi

Laboratory diagnosis of fungal infections

1. Sample collection

A sample can be collected from any site suspected to be infected, such as skin or corneal scraping, hair, nail, sputum, urine, blood or other body fluids. As fungi are known to be commensals (*Candida*) and saprophytes (*Aspergillus*), proper sampling and repeated isolation are important in certain clinical situations to establish a fungal etiology, especially in pulmonary and bloodstream infections.

2. Direct microscopy

Potassium hydroxide (KOH) is known to digest keratin and is used to treat clinical samples of skin, hair and other biopsy samples before they are examined under the light microscope as a wet mount (with 10% KOH). For nail specimens, 40% KOH is required to dissolve the keratin. Gram stain is used in yeast infections and India ink negative staining to demonstrate the capsule of *Cryptococcus neoformans*.

Histopathological examination of biopsy and tissue specimens is also used for subcutaneous, organ and deep fungal infections. The staining techniques used are periodic acid-Schiff, Gomori methenamine silver and hematoxylin and eosin (H&E), e.g., mycetomas, aspergillosis in lung (Fig. 4.2) and mucormycosis in brain. Mucicarmine stain is used to stain the cell wall and capsule of *Cryptococcus neoformans*.

3. Fungal culture

Fungi can be cultured in vitro in a variety of culture media, but the time taken for visible growth depends on the identity of the fungus. Rapid growth (1–3 days) is seen in *Mucorales, Candida* and *Cryptococcus*, while dermatophytes and subcutaneous fungi take 1–4 weeks. Blood culture for yeasts is done by conventional blood culture and automated methods.

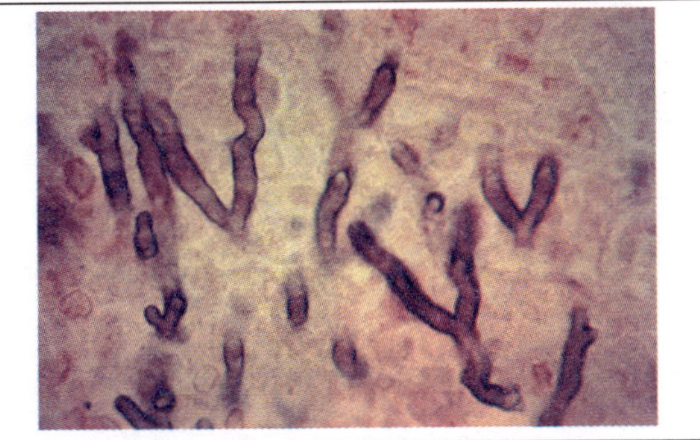

Fig. 4.2 *Aspergillus* spp. showing dichotomous branched hyphae, H&E stain, 40X

The commonest culture medium used for fungi is Sabouraud's dextrose agar. The examples of other media used are chlamydospore agar and Niger seed agar.

The growth is identified by its macroscopic and microscopic morphologic features. The macroscopic features include speed of growth, pigmentation and texture. The microscopic features are examined with a lactophenol cotton blue (LPCB) wet mount to identify the fungus.

The microscopic examination of fungal growth demonstrates the yeast cells, mycelia and the reproductive structures. Fungi reproduce asexually by fragmentation of hyphae, budding of yeast cells or by asexual spores, e.g., conidiospores, sporangiospores. Sexual reproduction results in the formation of sexual spores, e.g., ascospores, basidiospores.

Further identification can be done by biochemical tests. In vitro antifungal susceptibility testing is also feasible in the laboratory. The antifungal agents used are azole compounds (fluconazole, voriconazole), griseofulvin, amphotericin B, nystatin, caspofungin, flucytosine and terbinafine.

4. Serologic diagnosis

The role of serologic tests in fungal infections is limited. A latex agglutination test (LAT) is available to detect cryptococcal antigens in the cerebrospinal fluid in the case of cryptococcal meningitis. Blood IgE levels are measured for the diagnosis of allergic bronchopulmonary aspergillosis. The beta D glucan antigen is detected in a wide range of fungal infections including aspergillosis. The serum galactomannan assay is available for invasive pulmonary aspergillosis.

5. Wood's lamp examination

When a skin lesion is examined under ultraviolet light, a characteristic fluorescence is produced by fungi, that provides clues to the identity of the organism causing the infection, e.g., *Microsporum* produces a bright green fluorescence and *Malassezia furfur* a golden one.

Questions

1. Enumerate the predisposing factors for fungal infections with an example of each.
2. Classify fungi on the basis of their morphology with an example of each.
3. Classify fungi on the basis of the spectrum of disease with an example of each.
4. Discuss the laboratory diagnosis of fungal infections with an example of each method.

Classification and Characteristics of Parasites

A parasite is a living organism that lives in and feeds on a host, causing harm to the host. Parasites that thrive on human hosts do so in order to complete a part of their life cycle, which is essential for their survival. Man serves as a biological vector for the transmission of infection and survival of the parasite in nature.

Hosts are of two types—**definitive** and **intermediate.** Definitive hosts are those in whom the development of the sexual stage of the parasite occurs as opposed to intermediate hosts in whom the asexual stage develops. Man is the definitive host for most parasites wherein sexual reproduction is seen to occur, except *Plasmodium* spp. and *Echinococcus granulosus*, for which man is the intermediate host.

The existence of parasites in man can either be an infection or an infestation. **Infection** implies the invasion of tissue by the parasite, e.g., *Entamoeba histolytica*. **Infestation** implies its existence either on the surface (ectoparasite) or in a mucosal lumen (endoparasite). *Sarcoptes scabei* is an ectoparasite that exists on the skin, while *Trichuris trichiura* is an endoparasite in the intestine.

Taxonomically, invertebrates from the animal kingdom are classified into nine phyla. Of these nine phyla, three are known to parasitise humans—*Protozoa*, *Platyhelminthes* and *Nemathelminthes*. Protozoans are unicellular organisms while metazoans or helminths are multicellular, bilaterally symmetrical parasites.

Protozoans pathogenic to man are further classified in Table 5.1 and helminths pathogenic to man are further classified in Table 5.2.

Morphologic characteristics

The morphologic forms of unicellular protozoans are either cysts or trophozoites. A cyst is the vegetative but infective state, whereas a trophozoite is the actively multiplying, reproductive state. The organ of locomotion in protozoans is pseudopodium in *Entamoeba histolytica*, flagellum in *Giardia lamblia* and *Trichomonas vaginalis*, and cilium in *Balantidium coli*.

The morphologic forms of helminths are ova, larvae and adult (generally macroscopic). The ova and larvae are reproductive forms and the adults may not be of separate sexes. The general features of helminths are stated in Table 5.3.

Methods of reproduction

1. Asexual reproduction by binary fission, e.g., *Entamoeba histolytica* trophozoite divides into two cells with nuclear and cytoplasmic division.
2. Asexual reproduction by schizogony, e.g., *Plasmodium* schizont undergoes multiple divisions giving rise to many merozoites.
3. Sexual reproduction by syngamy, e.g., *Plasmodium* male and female gametocytes form an oocyst that gives rise to sporozoites.
4. Sexual reproduction in adult helminths, e.g., in nematodes, the sexes are separate (male and female) for fertilisation. In hermaphrodite cestodes and trematodes, fertilisation occurs within the same adult.

Table 5.1 Protozoan parasites pathogenic to man and prevalent in India

Subphylum	Genus species	Habitat	Disease
Sarcomastigophora	*Entamoeba histolytica*	Large intestine	Dysentery, liver abscess
	Giardia lamblia	Small intestine	Diarrhea
	Trichomonas vaginalis	Urethra, vagina	STD
	Leishmania donovani	MPS	Kala-azar
	Leishmania tropica	Skin	Oriental sore
Sporozoa	*Plasmodium* spp.⁺	RBC	Malaria
	Cryptosporidium parvum	Intestinal epithelium	Diarrhea
	Isospora belli	Intestinal epithelium	Diarrhea
	Toxoplasma gondii	MPS	Brain cysts, chorioretinitis
Ciliophora	*Balantidium coli*	Large intestine	Dysentery

Abbreviations: MPS, mononuclear phagocytic system; STD, sexually transmitted disease
⁺*Plasmodium vivax, falciparum* and *malariae. P. ovale* infection has not been reported in India.

Table 5.2 Helminth parasites pathogenic to man and prevalent in India

Class	Genus species	Habitat	Disease
Cestodes	*Taenia saginata*	Jejunum	Adult in intestine
	Taenia solium	Jejunum	Adult in intestine, cysticercosis
	Diphyllobothrium latum	Ileum	Adult in intestine, anemia
	Echinococcus granulosus	Liver	Hydatid disease
Trematodes (Flukes)	*Fasciola hepatica*	Liver	Adult in liver
	Fasciola buski	Small intestine	Adult in intestine
	Paragonimus westermanii	Lung	Adult in lung
	Schistosome haematobium	Urinary bladder	Hematuria
Nematodes (Intestinal)	*Ascaris lumbricoides*	Jejunum	Adult in intestine
	Stongyloides stercoralis	Jejunum	Adult in intestine
	Ancylostoma duodenale	Jejunum	Adult in intestine
	Necator americanus	Jejunum	Adult in intestine
	Trichuris trichiura	Cecum, appendix	Adult in intestine
	Enterobius vermicularis	Cecum, appendix	Adult in intestine
	Trichinella spiralis	Skeletal muscle	Cyst in muscle
Nematodes (Tissue)	*Wuchereria bancrofti*	Lymphatics	Filariasis
	Brugia malayi	Lymphatics	Filariasis
	Dracunculus medinensis	Lower extremity	Subcutaneous infection

Table 5.3 General morphologic features of cestodes, trematodes and nematodes

Feature	Cestode	Trematode	Nematode
Shape	Flat tape, segmented	Flat leaf, unsegmented	Cylindrical, unsegmented
Sexes	Not separate	Not separate, except schistosomes	Separate
Head	Suckers and hooks	Suckers, no hooks	No suckers, no hooks
Body cavity	Absent	Absent	Present

Modes of infection

1. Ingestion of infective cysts and ova, e.g., *Entamoeba histolytica* and *Ascaris lumbricoides,* respectively
2. Penetration of skin by larvae, e.g., *Ancylostoma duodenale*
3. Bite of a biological vector, e.g., Plasmodia and *Anopheles* mosquito
4. Blood transfusion, e.g., Plasmodia
5. Sexually transmitted, e.g., *Trichomonas vaginalis*

Mechanisms of pathogenesis

1. Trauma or mechanical damage, e.g., rupture of amoebic liver abscess, hematuria caused by *Schistosome haematobium* in the urinary bladder, destruction of villi by hookworm attachment, mechanical obstruction by adult parasites, e.g., *A. lumbricoides.*
2. Lytic necrosis by intracellular parasites, e.g., Plasmodia causing red blood cell lysis
3. Destruction of tissue by secretion of toxin, e.g., histolysin secreted by *Entamoeba histolytica*
4. Provocation of hyperplasia where they reside, leading to carcinoma, e.g., trematodes.
5. Allergic skin manifestations, e.g., *A. lumbricoides*
6. Anaphylaxis due to contact with hydatid fluid on rupture of the hydatid cyst.

Immune response

As parasites go through different stages in their life cycle, there exist antigenic variations in each stage, leading to the production of stage-specific immune response.

1. Role of antibodies: The immune response to protozoans produces IgM and IgG antibodies. In addition, intestinal protozoans produce IgA and helminths IgE. Non-specific hypergammaglobulinemia is seen in leishmaniasis.
2. Specific T cell response directed against stage-specific antigens, e.g., sporozoite, merozoite antigens of *Plasmodium.*
3. Eosinophilic response is associated with parasitic infection and seen as Charcot–Leyden crystals in stool

in certain intestinal infections, especially amoebiasis and hookworm.

Amongst parasitic infections, solid immunity against re-infection is seen only in *Leishmania tropica* causing oriental sore. In malaria, premunition to *Plasmodium* spp. develops on repeated infection, that protects against a large number of parasites and severe illness without eliminating the disease

LABORATORY DIAGNOSIS OF PARASITIC INFECTIONS

1. Examination of stool for parasites

Certifiable Practical Competency

i. Macroscopic examination

The stool sample is examined for its consistency and the presence of adult parasites, blood or mucus.

> Adult helminths commonly demonstrable in stool in India are:
> Cestodes — *Taenia solium, Taenia saginata, Diphyllobothrium latum*
> Nematodes — *Ascaris lumbricoides, Ancylostoma duodenale, Necator americanus, Enterobius vermicularis, Trichuris trichiura*

Morphologic characteristics of adult nematodes for gross identification

Ascaris lumbricoides: It is the largest intestinal nematode parasitising man. It is round, white in colour and tapers at both ends. The male worm is about 15–25 cm in length and female worm about 25–40 cm. The posterior end of the male worm is curved ventral like a hook and in the female, it is conical and straight (Fig. 5.1a).

Ancylostoma duodenale: Small greyish-white cylindrical worm measuring about 8–13 mm in length (Fig. 5.1b).

Enterobius vermicularis: Spindle-shaped worm, 2–4 mm in length (Fig. 5.1c).

Trichuris trichiura: The adult worm resembles a whip, measuring 2–4 cm in length. The anterior three-fifth of the worm is very thin and hair-like and the posterior two-fifth is thick and stout (Fig. 5.1d).

ii. Microscopic examination of stool

Prior to examining the stool wet mount, the stool sample can be concentrated to increase the sensitivity of recovering the parasites. The concentration methods for stool are:

A. Floatation techniques: Zinc sulphate floatation, saturated salt solution floatation
B. Sedimentation techniques: Simple sedimentation, formol ether concentration

Procedure for saline floatation

Two mL of saturated salt solution is taken in a test tube and 1 g of feces is emulsified in it. The test tube is filled completely to the brim with the saturated salt solution. The ova are collected by placing a coverslip on the surface of this solution. The ova stick to the coverslip. The coverslip is then placed on a slide to examine under the microscope. A solution of high specific gravity is used so that the ova of the parasites float on the surface. This method is useful for concentrating hookworm and roundworm ova. Unfertilised ova of *A. lumbricoides* and cysts do not float in saline. Hence, floatation techniques are not useful for detecting them.

Stool specimen is examined for parasites by preparing two wet mounts using saline and iodine each. A drop of saline is placed on the left side of the slide and a drop of Lugol's iodine on the right. Using a stick or wire loop, a small amount of feces is mixed and a smooth suspension is made. A coverslip is placed on each of these suspensions. The preparations are examined with a low-power objective lens (10X) first followed by a high-power objective lens (40X) (Figs. 5.2, 5.3 and 5.4).

Charcot–Leyden crystals are seen in stool in certain intestinal infections, especially amoebiasis and hookworm. They are microscopic crystals that are 50 μm in length and are composed of eosinophil protein galectin-10 (Fig. 5.5).

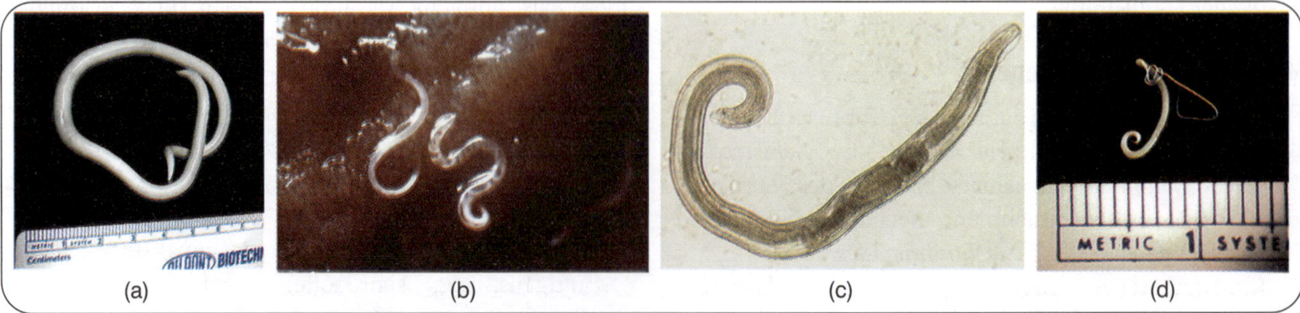

(a) (b) (c) (d)

Fig. 5.1 (a) Adult of *Ascaris lumbricoides* (*Source:* Centers for Disease Control and Prevention [CDC], DPDx), (b) Adult of *Ancylostoma duodenale* (*Source:* Public Health Image Library, ID 5205 / CDC), (c) Adult of *Enterobius vermicularis* (*Source:* CDC, DPDx) and (d) Adult of *Trichuris trichiura* (*Source:* CDC, DPDx)

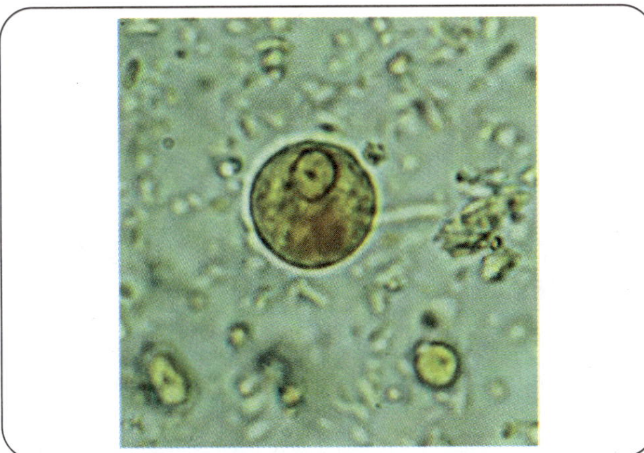

Fig. 5.2 Cyst of *Entamoeba histolytica*, Iodine mount, 40X (*Source:* CDC, DPDx)

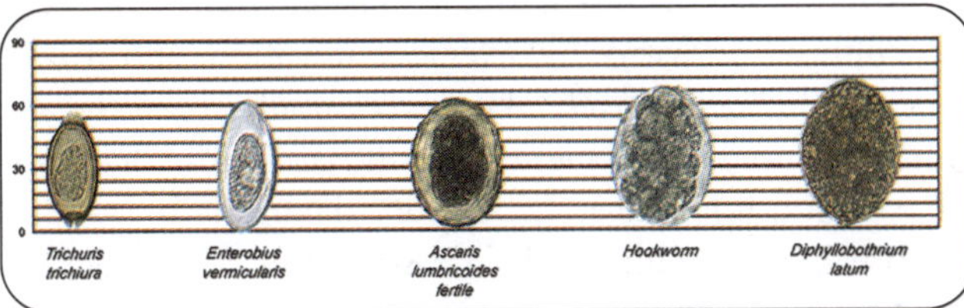

Fig. 5.3 Ovum of *Trichuris trichiura, Enterobius vermicularis, Ascaris lumbricoides*, hookworm and *Diphyllobothrium latum* (*Source:* CDC, DPDx)

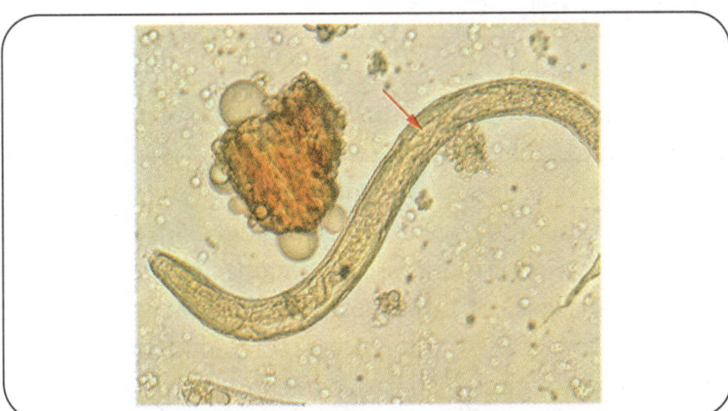

Fig. 5.4 Larvae of *Strongyloides stercoralis*, Iodine mount, 40X (*Source:* CDC, DPDx)

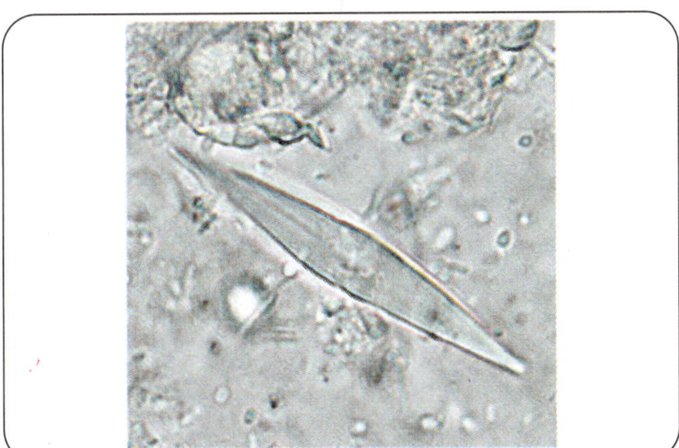

Fig. 5.5 Charcot–Leyden crystals in stool (*Source:* CDC, DPDx)

2. Microscopic examination of blood for parasites

In India, the commonly encountered diseases in which parasites are seen in blood are malaria, filaria and kala-azar. *Plasmodium* spp. and *Leishmania donovani* are intracellular protozoans that are small in size and can be examined only under an oil immersion microscope, while the microfilariae of filarial worms are much larger and can be examined under 40X magnification (Figs. 5.6, 5.7 and 5.8).

Whole peripheral blood collected with anticoagulant EDTA or a drop of blood by a finger prick and touched

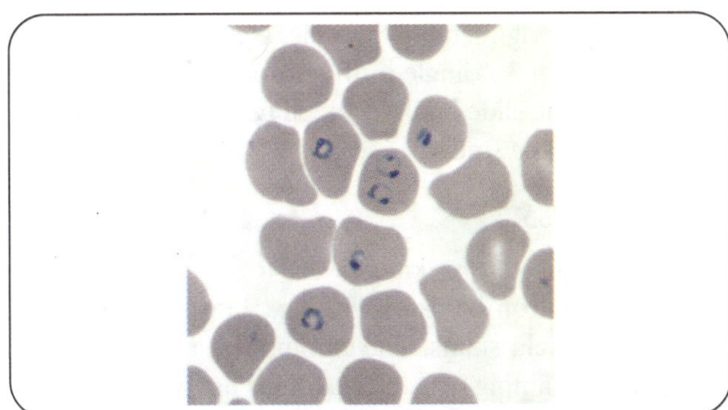

Fig. 5.6 Rings of *Plasmodium falciparum* in a thin blood smear, 100X (*Source:* CDC, DPDx)

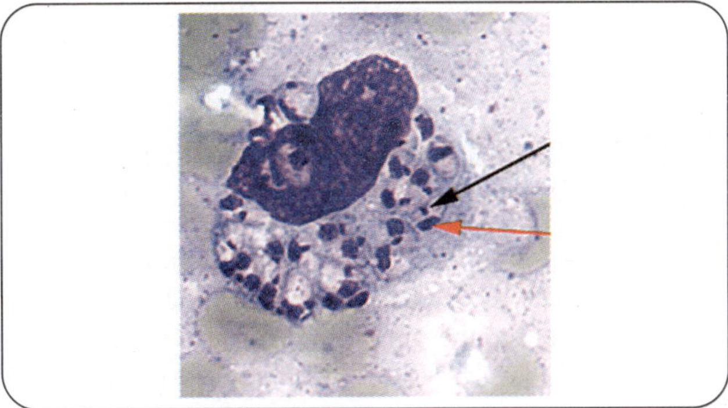

Fig. 5.7 Bone marrow macrophages showing amastigotes of *Leishmania donovani*, 100X (*Source:* CDC, DPDx)

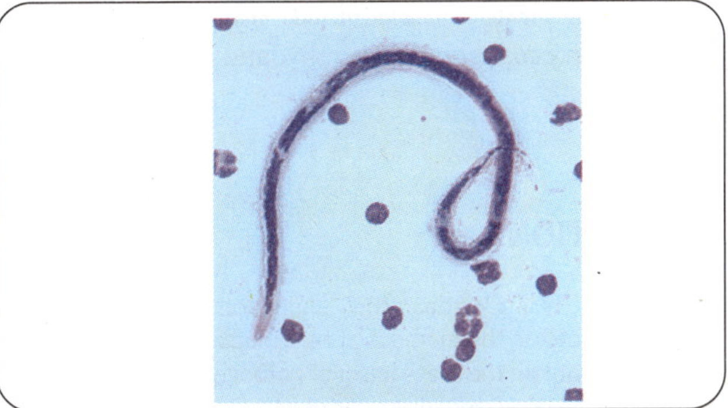

Fig. 5.8 Peripheral blood smear showing microfilaria of *Wuchereria bancrofti*, 40X (*Source:* CDC, DPDx)

onto a glass slide is used to make a peripheral blood smear. Peripheral blood smears are of two types—thick and thin. The thick smear is dense and used for screening suspects for the presence of parasitic infection, while the thin smear permits species identification of the parasite.

Method of preparing a thick smear

A big drop of blood is taken on a slide and spread with a needle to cover a half inch square area and air dried. The thickness should be such as to enable a print to be read through it.

Method of preparing a thin smear

Place a small (pin head size) drop of blood half an inch away from the right end of the slide. Next, hold a spreader (coverslip) at a 30° angle in the drop and gently push to the left over the slide until the blood tails off and exhausts. Then, air dry the smear.

Staining of PBS

The stains used are known as Romanowsky stains. They contain iron hematoxylin, e.g., Leishman, Giemsa, Wright and Field stains. The commonest method used is Leishman's staining. The procedure is as follows.

Fix the air-dried blood film in absolute methanol for 30 sec. Allow the slide to air dry and pour Leishman's staining solution to immerse the smear for 30 mins. Wash off the solution, air dry and observe under a microscope.

3. Microscopic examination of biopsy material for parasites

i. A biopsy specimen of spleen, lymph node or bone marrow can be examined by Leishman's stain for demonstrating the amastigote forms of *Leishmania donovani* in kala-azar or *Leishmania tropica* in skin biopsy.
ii. Liver aspirate is examined for *Entamoeba histolytica* trophozoites in amoebic liver abscess.
iii. Liver aspirate is examined for the scolices of *Echinococcus granulosus* in hydatid disease.
iv. Lymph node biopsy specimen is examined for the presence of adult filarial parasites.
v. Fine needle aspiration cytology or biopsy is done for histopathologic examination of cysticercus cellulosae in brain and other tissues.

4. Microscopic examination of urine for parasites

Urine wet mount can show the trophozoites of *Trichomonas vaginalis* and microfilariae of filarial parasite in chyluria.

5. In vitro culture for parasites

Protozoans *E. histolytica* and *Plasmodium* spp. can be cultured in vitro by axenic cultivation and RPMI inoculation, respectively. However, unlike bacteria and fungi, culture is not a method for the laboratory diagnosis of parasitic infections. Laboratory diagnosis by in vitro culture is done only for kala-azar on Novy Nicolle and McNeal medium.

6. Serologic and molecular diagnosis of parasitic infections

Specific antibody detection is feasible in amoebiasis, leishmaniasis, toxoplasmosis, cysticercosis, hydatid disease and filaria. Antigen detection is feasible in amoebiasis, giardiasis, leishmaniasis, malaria and filaria. Molecular methods have been employed in amoebiasis, giardiasis, trichomoniasis, leishmaniasis, malaria and hydatid disease. Commercial molecular panels are available for the simultaneous screening of multiple intestinal pathogens or protozoans and for vaginitis.

7. Hypersensitivity skin test

Intradermal skin tests for immediate and delayed hypersensitivity are demonstrable in a number of parasitic infections but are rarely used. Casoni's test for the diagnosis of hydatid disease is an example.

Questions

1. Define the terms 'ectoparasite', 'endoparasite', 'infestation', 'intermediate host' and 'definitive host'.
2. Write a note on the methods of reproduction in parasites, citing examples.
3. Write a note on the mechanism of pathogenesis in parasitic infections, citing examples.
4. Discuss the laboratory diagnosis of parasites found in stool.
5. Discuss the laboratory diagnosis of parasites found in blood.

Chapter 6

Infection and Epidemiologic Principles

Microbes are everywhere in the animate and inanimate world and are not always harmful. Human disease caused by microbes is a consequence of poor sanitation, lack of vector control, unhygienic food or practices and inadequate preventive measures such as vaccination. Unlike other human diseases, infectious diseases are largely preventable.

The microbes that can cause human disease are pathogens, commensals, saprophytes and parasites. Pathogens are microbes that can cause human disease in the immunocompetent, while opportunistic pathogens can cause disease only in the immunocompromised, e.g., the protozoans *Entamoeba histolytica* and *Isospora belli*, respectively. Commensals are microbes that normally exist on human skin and mucous membranes without causing harm, e.g., *Staphylococcus epidermidis* on the skin. Saprophytes are free-living microbes that are found in dead matter like soil, e.g., *Clostridium* spp. and fungi causing subcutaneous infections. Parasites are microbes that can exist only on a living host; these include all human pathogens—bacteria, fungi, viruses, protozoans and helminths.

Parasitic microbes can either colonise, infect or infest humans. Colonisation is microbial existence without causing harm or deleterious immune response, e.g., *Escherichia coli* in the intestines. Infection is the multiplication of a pathogen causing an immune response, e.g., *Mycobacterium tuberculosis*. An infection may or may not result in disease, depending on the efficacy of the immune response. Both innate and acquired immune responses are defense arms against microbial invasion and infection. In the event of their failure to curtail the microbe, infectious disease sets in. Infestation implies the disease-causing existence of an ectoparasite or endoparasite on the surface of the skin or in the intestinal lumen, respectively.

Normal microbial flora

The total population of microbial cells normally found on humans far exceeds the number of human cells in the body! This indigenous or commensal microflora colonises the skin and mucosal surfaces. Deeper tissues, organs, blood and body fluids are essentially sterile. The normal microbial flora predominantly comprises bacteria, with a larger proportion of anaerobes than aerobes, and some fungi. Colonisation begins at birth and persists lifelong.

Commensal bacteria are beneficial to humans as they prevent the entry of and colonisation by pathogens. Enteric commensal bacteria produce vitamins K and B12 that are absorbed. The sites that are colonised are skin, the gastrointestinal tract (oral cavity, stomach, intestine), the respiratory tract (nasopharynx) and the female genital tract (vagina). Table 6.1 shows the distribution of the normal flora in humans.

A commensal is defined as a microbe that colonises or whose presence is considered normal as opposed to a pathogen whose presence is harmful. However, a commensal may play the role of a pathogen or be the source of infection, e.g., coagulase-negative Staphylococci (CoNS) causing catheter-associated bloodstream infections, endogenous *E. coli* causing urinary tract infections, *Streptococcus viridans* causing endocarditis, nasopharyngeal bacteria causing lower respiratory tract infections and *Streptococcus pneumoniae* or *Neisseria meningitidis* causing pyogenic meningitis.

All antibiotics, except parenteral aminoglycosides, sulfonamides and vancomycin, alter the intestinal microflora and can induce pseudomembranous colitis. The most common are clindamycin, ampicillin, amoxicillin and the cephalosporins. Antibiotics suppress other intestinal bacteria and promote the selective overgrowth

Table 6.1 Distribution of normal flora in humans

Site	Total bacteria/mL	Anaerobes: Aerobes ratio	Microorganisms
Skin	10^3	3:2	*Staphylococcus epidermidis*, diphtheroids, streptococci, *Staphylococcus aureus*, *Propionibacterium*
Oral cavity	10^{10}	1:1	*Streptococcus viridans*, non-sporing anaerobes
Stomach	10^3	1:1	Lactobacilli, *Helicobacter pylori*
Intestines	10^{12}	10^3:1	*Bacteroides, Fusobacterium, Prevotella, Clostridium difficile* Anaerobic cocci, *Escherichia coli*, enterococci, streptococci, *Candida*
Nasopharynx	–		Streptococci, *Neisseria, Fusobacterium, Prevotella*
Vagina	10^9	10:1	Lactobacilli, streptococci, *Neisseria, Candida*

of *Clostridium difficile.* The cytotoxin of *C. difficile* has been identified as the cause of pseudomembranous colitis.

Host-agent relationship

In spite of the availability of vaccines and antimicrobial agents, infectious diseases continue to be the leading cause of death in the world, with epidemics and pandemics arising from time to time. The most recent pandemic being the COVID-19 respiratory illness.

The epidemiologic factors that favour the emergence of infectious agents

1. Microbial adaptation and change, e.g., the extensive and irrational use of antimicrobials has resulted in the emergence of multi-drug resistant bacteria.
2. Human susceptibility to infection, e.g., the advent of HIV resulted in the rise of infections with uncommon pathogens like *Cryptococcus* and *Pneumocystis*.
3. Climate, weather, changing ecosystems, e.g., rainfall and water stagnation breeds mosquitos that are vectors for dengue.
4. Human demographics and behaviour, e.g., the existence of commercial sex workers, multiple sexual partners and unprotected sex.
5. International travel and commerce, e.g., spread of respiratory viruses that resulted in pandemics—H1N1 influenza and COVID-19.
6. Potential for bioterrorism, e.g., anthrax spores, plague
7. Poor public health and sanitation, lack of safe drinking water, poverty, war and famine.

Infection results from an encounter with a potentially pathogenic agent gaining entry into a susceptible human host. As the source usually lies outside the human host, exposure to the agent is crucial. Disease is one possible outcome of infection, depending upon the host and the agent. An infection does not necessarily result in disease and instead could be asymptomatic or result in immunity. If the agent overpowers the host's response to the infection, it will result in disease. This relationship between the agent, route of transmission and host is known as the ***chain of infection***.

An **agent** is any microbe that can infect a human or animal and produce disease. For some agents, humans are the only reservoir (source) of infection while others have the capacity to infect non-human hosts, such as animals, birds and insects, that can serve as reservoirs. For a disease to be eradicated, the agent should not have a non-human reservoir and have an effective vaccine.

The agent can be exogenous (external source) or endogenous (with the host being the source). An agent that exists as normal flora and colonises the host does not produce disease under normal circumstances but may be pathogenic under altered circumstances, e.g., staphylococci on skin cause bloodstream infections following intravenous catheterisation.

Infectiousness is a characteristic of the agent that describes the ease of its transmission or spread to non-immune hosts. For example, a respiratory infection is more infectious than a sexually transmitted one.

Infectivity describes the ability of a microbe to enter, survive and multiply in the host.

Secondary attack rate is the probability that infection can occur in a susceptible host after exposure to an infected individual.

Pathogenicity determines the extent to which overt disease is produced in an infected population and is measured by the number of persons with clinical disease to the number infected. The gradient of infection or biologic gradient is the range of manifestations of illness in the host, from death at one extreme to inapparent or sub-clinical infection at the other.

Virulence is the quantitative measure of the disease-producing potential of a pathogen or the ratio between the number of serious disability-producing infections to the total number of infections. If death is the only criterion for severity, the virulence is referred to as *case fatality rate*. The virulence of an agent may also be dependent on the host, e.g., the course of hepatitis B ranges from sub-clinical infection to fulminant hepatitis depending on the host response and genetics.

The **host factors** that influence exposure, infection and disease are varied, with the two most important being the immune status of the host and the age at the time of infection. Both extremes of age—early life and old age—have a greater disposition for disease following infection. Other factors that influence exposure are pets, drug abuse, blood transfusion, living conditions and practices, occupation, gender, sexual activity, travel and vector endemicity. Factors that influence infection and disease are age, alcoholism, anatomic defect, antibiotic use, vaccine status, genetic (HLA) make-up and nutrition.

The microbe-related epidemiologic factors that favour the potential eradication of an infectious disease are the **absence** of animal reservoirs, sub-clinical cases, carrier state, reinfection and the **presence** of an effective vaccine. These factors have enabled the eradication of small pox and the elimination of poliomyelitis.

Route of transmission is the mechanism by which an agent can spread from the environment/an infected host to a non-immune host. The route of transmission can be direct or indirect. Direct transmission occurs by touching, kissing, sexual intercourse, inhaling large respiratory droplets (limited to a 1-metre distance), bites (as in rabies) or contact with soil (inoculation), or occurs transplacentally.

The indirect mechanisms of infection are vehicle-borne, vector-borne and airborne. Vehicle-borne infections are those transmitted by fomites, water, food, blood, organs or surgical instruments. Vector-borne infections can spread through mechanical (house flies) or biologic (mosquito bite) means. Airborne infections are spread through small aerosols of size 1–5 µm that remain suspended in air and can travel distances, e.g., measles.

Prevention strategies for infectious diseases are by immunoprophylaxis, chemoprophylaxis, or long-term measures in improving water, sanitation and hygiene.

Source and route of transmission of infection

1. *Sexual transmission (infected persons): Treponema pallidum, Neisseria gonorrhoea,* HBV, HIV, HCV
2. *Indirect contact (fomites): Shigella, Salmonella*
3. *Inhalation (aerosols):* Respiratory droplets (size > 10 µm) travel short distances and infect those who come in close contact with a patient, e.g., diphtheria, pertussis, plague, meningococci, streptococci in pharynx, COVID-19, influenza, mumps, rubella. Respiratory aerosol droplets (size < 10 µm) travel longer distances and can infect any person they come in contact with, e.g., tuberculosis (TB), measles, chickenpox.
4. *Ingestion (food and water):* The infective dose varies with the pathogen. *Shigella* can cause infection with as low a dose as 10 bacilli, while *Salmonella* and *Vibrio* need larger doses of 10^6. *Shigella* survives gastric acidity while *Vibrio* is acid-labile.
5. *Inoculation (drug abuse, tattooing):* HBV, HIV
6. *Iatrogenic (injections, dialysis):* HBV, HIV
7. *Vectors (mosquito): Plasmodium* spp., dengue virus, chikungunya virus
8. *Mother-to-child transmission* can be transplacental – *Toxoplasma gondii* infection, rubella; occur during birth – Group B *Streptococcus* infection, chlamydia, hepatitis B, HIV infection; or occur postpartum by breastmilk.
9. *Blood transfusion*: HBV, HIV, HCV
10. *Organ transplantation:* CMV, HBV, HIV, HCV

Mechanisms of pathogenicity and virulence

The ability of a microbe to cause disease is defined as its 'pathogenicity', whereas the degree of its pathogenicity defines its 'virulence', e.g., pathogenic *E. coli* can cause diarrhea, amongst which the strain *E. coli* O157:H7 is more virulent than the other strains and can cause hemolytic uremic syndrome.

Pathogenic microbes possess virulence factors—molecules or mechanisms—that enable and enhance their ability to cause disease. The virulence of a microbe can be exalted (increased) or attenuated (decreased). Live vaccines are formulated on the concept of attenuation—retaining the antigenicity of the microbe and eliminating its virulence, e.g., the enteric fever vaccine Ty21a is an avirulent strain of *Salmonella* Typhi. "Species" refers to a genetically related group of bacteria and "strain" refers to isolates within a given species with specific characteristics, e.g., *E. coli* is a species, *E. coli* O157: H7 is a strain of *E. coli*, *Salmonella* Typhi is a species, Ty21a is a strain of *S.* Typhi.

Serotyping, biotyping and antibiograms of bacterial strains play a role as epidemiologic markers in outbreaks of infections.

The mechanisms of virulence in microbial pathogenicity are

1. *Adhesion*: Attachment of bacterium to cell surfaces mediated through specific receptors, fimbriae/pili (*N. gonorrhea*) or adhesins like lipoteichoic acid (*Staphylococcus aureus*) and M protein (*Streptococcus pyogenes*).
2. *Invasion*: Entry into host cells, mucosa and tissue, e.g., *Shigella* and *E. histolytica*. This may also occur by the production of enzymes like hyaluronidase and fibrinolysin that encourage the spread of infection, as in *S. pyogenes*.
3. *Toxigenicity*: Exotoxins are extracellular proteins secreted mostly by gram-positive and some gram-negative bacteria. They are potent in minute amounts and very specific in their action. Those secreted by gram-positive bacilli are antigenic and are used in vaccines as a toxoid, e.g., diphtheria and tetanus. Antibodies to exotoxins are called antitoxins as they have the ability to neutralise the toxin, e.g., diphtheria and tetanus. Gram-negative bacilli that secrete exotoxins are *Vibrio cholerae, Shigella dysenteriae* type-1 and diarrheagenic *E. coli*.

   Endotoxins are the lipid A of the gram-negative cell wall lipopolysaccharide (LPS) that is released only on lysis of the cell. Endotoxins can activate the alternative complement pathway, macrophages, platelets, endothelium, mast cells and the coagulation pathway to cause disseminated intravascular coagulation and death following gram-negative sepsis. LPS is antigenic and antibodies to it are used in gram-negative sepsis.
4. *Ability to escape the host immune system:*
   a. Antiphagocytic mechanisms: The presence of an extracellular slime layer or capsule prevents phagocytosis. The M protein of *S. pyogenes* binds to serum factor H, thus preventing opsonisation by C3b.
   b. Intracellular survival: E.g., *M. tuberculosis* prevents fusion of the phagosome-lysosome in macrophages.

S. Typhi can invade epithelial cells, M cells, macrophages and dendritic cells and survive within them.

c. Antigenic variations: E.g., *Borrelia recurrentis*, influenza virus. The influenza virus genome has eight RNAs, causing antigenic variations due to the re-assortment of the RNA segments from time to time. RNA re-assortment occurs either when two different influenza viruses (possibly those of human and non-human origin) infect and replicate in the same host cell or due to spontaneous mutation in the virus.

d. Viral latency: E.g., Varicella zoster (VZ) virus causing herpes zoster. Following chickenpox, the latency of the VZ virus in the affected sensory dorsal root ganglion may reactivate to cause zoster in the affected dermatome.

5. *Genetic factors*: Plasmids and bacteriophages carrying genes for toxin production and drug resistance, e.g., *Corynebacterium diphtheriae* and *E. coli,* respectively.
6. *Immune-mediated damage,* e.g., delayed hypersensitivity in TB and chronic hepatitis B as a result of the immune response against HBV in the hepatocytes. The host immune response responds to the intracellular virus multiplying in the hepatocytes with a cell-mediated attack on the hepatocytes to contain the virus. Therefore, the quality of the immune response determines the course of the infection and disease outcome.

Questions

1. Explain the terms "commensal", "pathogen", "saprophyte" and "parasite" with reference to microbes.
2. What is the difference between colonisation, infection and infestation?
3. Name sterile and colonised sites in the human body. What is the distribution of human normal microbial flora?
4. What is the role of normal microbial flora in health and disease?
5. Enumerate the epidemiologic factors that favour the emergence of infectious agents, giving examples.
6. What are the sources and routes of transmission?
7. Elaborate on the mechanisms of pathogenicity and virulence of microbes, giving examples for each.

Collection of Samples in Infections

The quality of a sample from a patient is the basis for an optimal test result. Therefore, utmost care needs to be taken to obtain a proper sample. Microbiological sampling depends on which system is affected by disease and the probable causative agent. In general, the investigation is for microscopic examination of smear, culture, serology or molecular testing.

> The general considerations for collection of a good sample are:
> 1. The sample should be obtained prior to antimicrobial therapy. In the event of a patient being on antimicrobials, the sample can be collected prior to the dose.
> 2. Contamination of a specimen with commensal flora should be avoided. The use of antiseptics on skin prior to collecting blood or body fluids under aseptic conditions is vital. Cleaning mouth or genitalia prior to collecting sputum or urine, respectively, must be ensured.
> 3. The sample should be transported to the laboratory without any delay. In the event of an expected delay, the sample should not be refrigerated.
> 4. A complete requisition form should be sent, with each sample stating the site of the sample and provisional diagnosis of the infection.

Steps for collection of blood for culture (manual method)

Under aseptic precautions, blood is collected by venepuncture and transferred into a blood culture bottle that contains the culture broth (Fig. 7.1). The cap of the bottle needs to be cleaned with spirit. Blood should be transferred through a hole in the cap made by inserting the needle of the syringe and not by opening the bottle.

The blood-to-broth ratio must be maintained as 1:5, i.e., in adult sample, 10 mL of blood in 50 mL of broth, in pediatric sample, 5 mL of blood in 25 mL of broth and in neonate sample, 1 mL of blood in 5 mL of broth.

Fig. 7.1 Blood culture bottle

In the case of automated blood culture systems, user recommendations need to be complied with as per their manual.

Steps for collection of blood for serology/parasitic infections

Five mL of blood is to be collected aseptically into a sterile test tube without anticoagulant. Serological tests are performed with serum and not plasma. Blood for parasitic infections is collected with anticoagulants for peripheral blood smears.

Steps for collection of urine for culture

1. An early-morning, first-pass urine sample of 2–5 mL is preferred as the bladder acts as a natural incubator for bacteria.
2. Before collecting the sample, the external genitalia should be cleaned with soap and water; men are to be instructed to retract the foreskin of the glans of the penis whereas women are to keep the labia apart.
3. The initial urine flow is allowed to pass. Urine is collected mid-stream without interrupting the flow of urine flow at any point.
4. Ensure that urine is collected in a sterile, wide-mouth container.

Steps for collection of stool for culture/parasites

The sample must be collected in a wide-mouth, clean/sterile, screw-capped, disposable container.

Steps for collection of pus for culture

1. Collect pus, surgically, in a sterile container. Use a syringe to aspirate.
2. Collect pus swab in a sterile test tube.

Steps for collection of pus for anaerobic culture

1. Aspirated pus/tissue biopsy is the preferred sample.
2. This is to be transported in anaerobic transport media like Robertson's cooked meat (RCM)/thioglycollate culture medium.
3. Sputum samples are unsuitable for anaerobic culture.

Steps for collection of sputum for culture

1. Early-morning sample, preferably.
2. Rinse the mouth with water before collection.
3. Collect sample in a nearby open space.

4. Inhale deeply 2–3 times and cough out deeply from the chest.
5. Open and bring the sterile screw-capped, wide-mouth, disposable container close to the mouth and bring out the sputum into it.
6. In the case of suspected tuberculosis, two sputum samples are to be collected for microscopy—one on the spot and the other early morning.
7. Sputum sample is to be collected in a Falcon tube for CBNAAT (Fig. 7.2).

Collection and transport of a stool sample

1. The stool sample should be collected in a clean, screw-capped, disposable container.
2. The stool sample must be immediately transported to the laboratory for processing.
3. If there is a delay in the transportation of samples, transport media should be used.

Samples that need to be reported urgently are:
- Stool in cholera and shigellosis
- Cerebrospinal fluid (CSF) in meningitis
- Throat swab for diphtheria
- Tissue for gas gangrene

Transport media

Transport media preserve the viability of microbes without allowing them to multiply and maintain the sample as close to the original state, i.e., at the time of collection, as possible. The specific content of a transport medium depends on the microbe it serves to maintain viable.

Options for specimen transport in various mediums and their purpose:

1. **Cary and Blair medium:** This semi-solid, white-coloured transport medium is used for feces that may contain *Salmonella, Shigella, Vibrio* or *Campylobacter*.
2. **Amies medium with charcoal:** Charcoal helps eliminate the metabolic products of bacterial growth, which may be especially useful in the isolation of fastidious organisms like *Neisseria gonorrheae*. However, it is suggested that some other pathogens like *Campylobacter* can also survive in such media.
3. **Amies medium without charcoal:** This is ideal for the isolation of *Mycoplasma* and *Ureaplasma*.
4. **Stuart's medium:** Commonly used for transporting specimens suspected of having gonococci. It is also used for transporting throat, wound and skin swabs that may contain fastidious organisms.
5. **Venkatraman Ramakrishnan (VR) medium:** This medium is used to transport feces from suspected cholera patients.
6. **Sach's buffered glycerol saline:** This one is used to transport feces from patients suspected to be suffering from bacillary dysentery.
7. **Viral transport medium (VTM)** (Fig. 7.3): This is ideal for samples to be used in the diagnosis of viral infection. Ocular, respiratory and tissue swabs can be submitted in this medium. Fluid samples, such as tracheal wash specimens or peritoneal fluid, should be submitted in "as is" condition, i.e., in sterile vials that prevent desiccation. In the absence of VTM, swabs must be submitted in sterile, sealed vials with several drops of saline added to prevent desiccation. Cotton, plastic, wood-handled, and Dacron and other synthetic swabs are all acceptable. Calcium alginate swabs should be avoided. Bacterial transport media are not appropriate for virology.

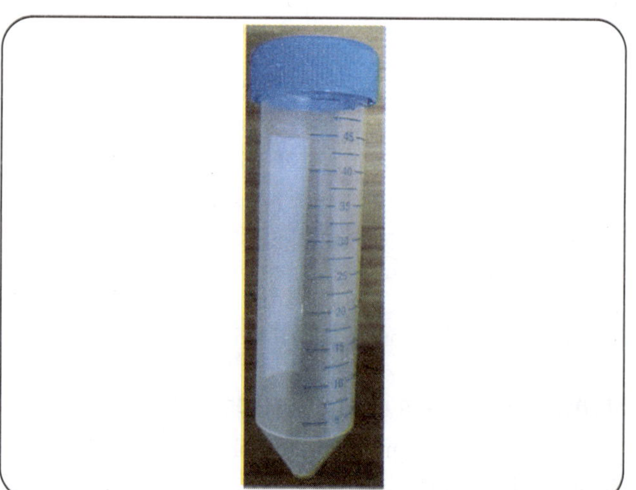

Fig. 7.2 Falcon tube for CBNAAT

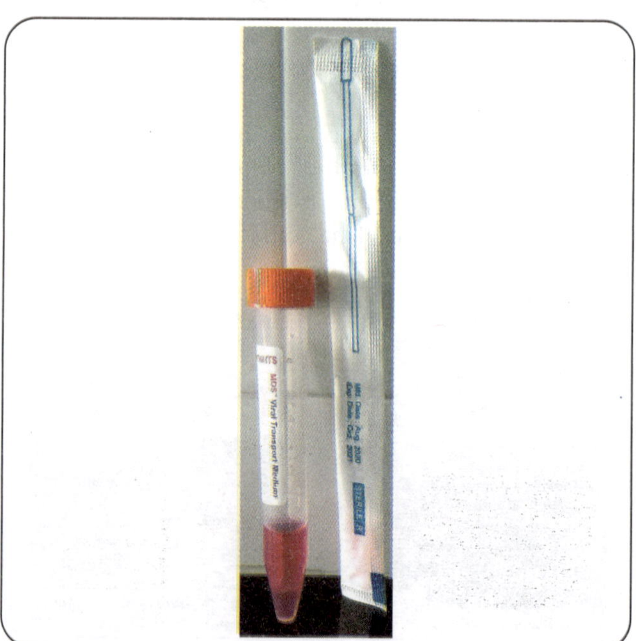

Fig. 7.3 Viral transport medium

8. **Anaerobic transport medium:** This mineral salt-based semi-solid medium with reducing agents is designed as a holding medium for maintaining the viability of anaerobic bacteria. It contains buffered mineral salts in a semi-solid medium with sodium thioglycollate and cysteine added to provide a reduced environment. Resazurin may also be added as a redox indicator to reveal exposure to oxygen by turning pink. It provides an environment which maintains the viability of most microorganisms without significant multiplication and allows for dilution of the inhibitors present in clinical material. Examples include thioglycollate broth and RCM.
9. **Dried blood spot:** This is used for the transport of whole blood, e.g., HIV in neonates. A drop of blood from a heel prick is absorbed on a filter paper and the paper is transported to the reference laboratory. Bedside testing using rapid immunochromatography (ICT), tests can be done with a drop of whole blood from a finger prick, e.g., ICT for malaria.

Sample storage

Patient samples are required to be stored either pre-test, in case of any delay in processing after sample collection or post-test, for retrieval of the sample if required. Test sera or bacterial isolates may also be required to be stored for epidemiological reasons or for the purpose of teaching and research. Pre-test samples can be stored at 2–8°C in the refrigerator with the exception of CSF (to be maintained at 20–25°C) and blood culture bottles can be stored at room temperature.

Post-test samples of body fluids and exudates as well as sera can be stored in a refrigerator for up to 1 week. Longer storage of sera and isolates need freezing, avoiding repeated thawing and freezing. Generally, the lower the storage temperature, the longer the culture will maintain viable cells.

Freezers are of three types: laboratory, ultra-low and cryogenic. The laboratory and ultra-low freezers maintain temperatures at –20 to –40°C and –70 to –80°C, respectively. Cryogenic freezers rely on liquid nitrogen and operate at –130°C. However, the best method of storage of cultures is by freeze drying or lyophilisation. This method removes water from the cells. It has the advantage of not relying on electric power or liquid nitrogen, and has ease of transport following lyophilisation.

Packaging of samples for transport

Infectious biomaterial (sera, cultures or VTMs) that is transported by post or courier or that is hand-delivered to one institute from another requires specific precautions and packaging (Fig. 7.4).

The specimen tube should be screw-capped, made of plastic and contain no cracks/leaks. It needs to be wrapped in cotton (to absorb in the event of an accidental leak) and placed in a sealed plastic bag. This, in turn, is to be placed in a box along with the request slip details in another sealed plastic bag. For mailing, this triple-packed box should be placed in another package with the mailing address and a biohazard sign (Fig. 7.5).

Notifiable infections

A notifiable disease is any disease that is required by law to be reported to government authorities. The collation of data allows public authorities to monitor the disease and warn of an impending outbreak in the community.

In India, the following infections are notifiable:

- Bacterial: Cholera, diphtheria, pertussis, plague, leprosy, tuberculosis, shigella dysentery, enteric fever and tetanus.
- Viral: HIV, hepatitis A, hepatitis B, hepatitis C, dengue, chikungunya, Japanese B encephalitis, influenza and rabies.
- Parasitic: Malaria, filaria and kala-azar.

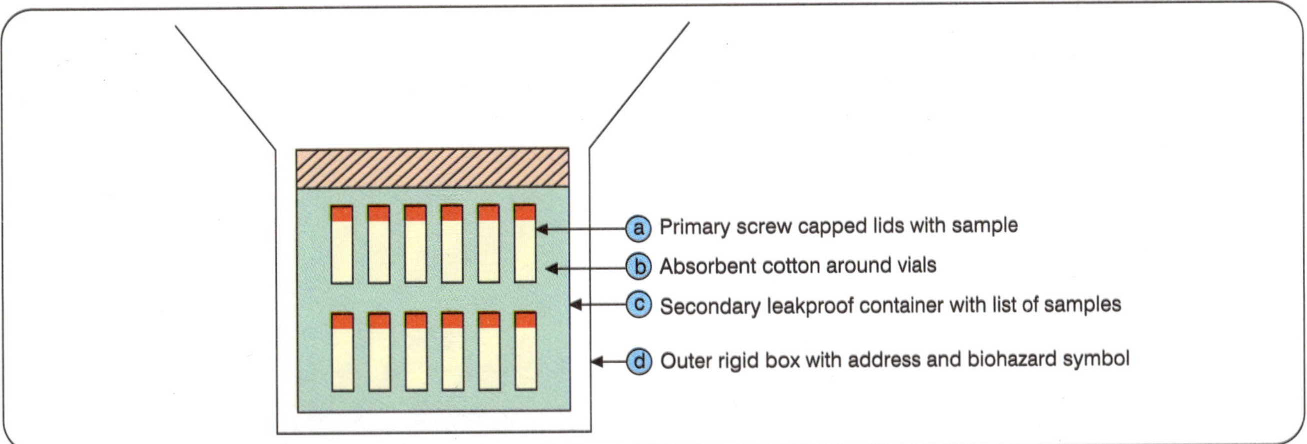

Fig. 7.4 Triple packaging system

Fig. 7.5 Biohazard sign

Respect for the sample and confidentiality of reports–ISO 15189

The National Accreditation Board for Laboratories (NABL) follows strict guidelines as per the ISO 15189 with respect to pre-examination procedures and the reporting of test results. The requisition form shall contain information sufficient to identify the patient and the authorised requester (medical practitioner) as well as pertinent clinical data relevant for the interpretation of results. The clinical data must include the type of primary sample and the anatomic site of origin where appropriate, the test required, date and time of collection, and time of receipt in the laboratory.

The requisition form and reporting format can be either electronic or in paper form ideally and documented if by telecommunication. The laboratory and the requester share the responsibility for confidentiality of the report and authorisation of the release of the report directly to a patient or only to authorised receivers. Immediate notification to a treating physician, in the event of an alert or a critical value or result, within the stipulated turn-around time is important.

The National Medical Commission's Code of Ethics Regulations states that "the physician shall not disclose the secrets of a patient that have been learnt in the exercise of his/her profession except in a court of law under orders of the Presiding Judge; in circumstances where there is a serious and identified risk to a specific person and/or community; or in the case of notifiable diseases."

The definition of patient confidentiality is: The law whereby a doctor or medical practitioner cannot reveal patient information except in a court of law.

The measures to ensure that confidentiality of patient information is maintained by all laboratory personnel are to:

1. Maintain a signed 'Confidentiality Agreement Form' stating the obligation of each personnel towards confidentiality of patient information.
2. Prohibit unauthorised access to test results.

Patient confidentiality is of utmost importance in sexually transmitted infections like HIV. Unlike all other laboratory tests that have an 'implied consent', the HIV test mandates an 'informed consent' wherein pre-test and post-test counselling is provided, with the test results being strictly confidential.

Questions

1. What are the general considerations you will ensure while collecting a sample for the diagnosis of an infection?
2. Outline the steps to perform blood culture.
3. Outline the instructions given to a patient to collect a sample of urine for culture.
4. Outline the instructions given to a patient to collect a sample of sputum for culture for tuberculosis.
5. State the practical options for specimen transport in medium with the purpose of each.
6. How will you store a serum sample in the laboratory? What precautions would you take to mail the sample to a reference centre?
7. What is patient confidentiality? How will you ensure that confidentiality of a patient's test result is not breached?

Laboratory Diagnosis of Infections

Once an optimal sample/specimen is received in the microbiology laboratory, the choice of test depends on the nature of the microorganism, the sample and the clinical suspicion of the likely causative agent.

> The test methodology applicable in the microbiology laboratory can be broadly classified as:
> 1. Demonstration of the microorganism by microscopy.
> 2. In vitro culture of bacteria and fungi (not applicable to viruses and parasites).
> 3. Detection of antigen, antibodies or inflammation markers by serological tests.
> 4. Detection of nucleic acids by molecular methods.

1. DEMONSTRATION OF MICROORGANISMS BY MICROSCOPY

Microorganisms can be seen under a light microscope either in a wet mount or direct stained smear, or on histopathologic examination.

Wet mount

Wet mount preparation implies that a drop of liquid or liquified sample is placed on a glass slide and a coverslip placed over it. This is followed by its examination under a low- or high-power objective lens.

The applications of a wet mount preparation include the microscopic examination of:

1. Urine for the presence of pus cells, bacteria, clue cells or trophozoites of *Trichomonas vaginalis*.
2. Stool for protozoan cysts, trophozoites, and helminth ova or larvae.
3. Liver aspirate for trophozoites of *Entamoeba histolytica*.
4. Potassium hydroxide (KOH) mount for fungal elements in a sample, e.g., skin or corneal scraping
5. India ink staining of cerebrospinal fluid (CSF) in cryptococcal meningitis.
6. Lactophenol cotton blue mount for identification of fungi in culture.
7. Motile microfilariae of filariasis in lymph node aspirate or blood.
8. Motile promastigotes of *Leishmania donovani* from Novy Nicolle and McNeal (NNN) medium.

Direct or primary smear

A direct or primary smear is one that is prepared from the patient's sample. The microscopic examination of a direct/primary smear is important as it enables us to:

1. Visualise the pattern, arrangement and morphology of the microbes present in it.
2. Detect the presence of inflammatory cells, e.g., pus cells in urine, sputum or body fluids. A properly collected sputum sample contains a larger number of pus cells and few epithelial cells. The presence of pus cells in urine and body fluids that are otherwise sterile provides clues to infection.
3. Choose an appropriate medium for the culture of the sample, which is based on the presumptive identity of the bacterium.
4. Rationalise empiric antibiotic therapy based on the Gram reaction of the infecting bacterium.

> ***Preparation of a heat fixed smear from a patient's sample for staining***: Pencil mark an area on a clean glass slide. On its reverse, using a sterile bacteriological loop, place a drop of the patient's sample (pus, urine, sputum, body fluids) and spread to cover a half-to-one-inch area by spreading the material in a circular motion. If the material is too thick, normal saline can be used to liquefy the sample. In case of a biopsy, the sample needs to be crushed in a ground glass tissue grinder prior to making the smear. Adequate material should be visible to the naked eye. Once air dried, the smear should be heat-fixed by passing thrice through a Bunsen burner flame with the smear side facing upwards.

A heat fixed smear needs to be stained and examined under the oil immersion objective, with the staining technique corresponding to the microorganism to be observed.

The applications of a microscopic examination of a stained preparation are:

1. Gram stain for bacteria in urine, pus, sputum, bronchoalveolar lavage, CSF and other body fluids and visceral aspirates.
2. Ziehl–Neelsen (ZN) or auramine stain for *Mycobacterium tuberculosis* and ZN for *Mycobacterium leprae*.
3. Specialised staining techniques like Albert stain for *Corynebacterium diphtheriae*.
4. Fontana stain for *Treponema pallidum*.
5. Wayson stain for *Yersinia pestis*.
6. Gram stain for yeast cells, especially *Candida* spp.
7. Leishman's stain for peripheral blood smear examination for malaria, filaria and kala-azar.

Gram staining

Certifiable Practical Competency—Staining and identification of common pathogens in clinical samples

Principle

Gram-positive cells have a more acidic protoplasm, which accounts for retaining the basic dye more strongly than gram-negative bacteria. Iodine makes the protoplasm more acidic and serves as a mordant, i.e., iodine combines with dye to form a dye-iodine complex and fixes the dye in the bacterial cell. The gram-positive cell wall or cytoplasmic membrane being less permeable, the dye-iodine complex gets trapped within the cell. The gram-negative cell wall has increased permeability to acetone or alcohol, permitting the outflow of the dye-iodine complex during decolourisation. This is not an all-or-none phenomenon. The integrity of the cell wall is essential for a positive stain.

Procedure

1. Cover the fixed smear with primary stain (crystal violet) solution for 1 minute.
2. Pour Gram's iodine over the smear for 1 minute.
3. Wash the smear with water.
4. Decolourise the smear with absolute alcohol for 10 seconds, taking care not to over-decolourise.
5. Immediately wash with water to remove the decolouriser.
6. Cover the smear with a counterstain (dilute carbol fuchsin/safranin/neutral red).
7. Wash the smear with water.
8. Blot-dry the smear.

On staining of a primary smear from a clinical sample, the morphologic appearance of the bacteria clues to the identity of the causative agent (Figs. 8.1 to 8.11).

Ziehl–Neelsen staining

Certifiable Practical Competency

Principle

Acid fastness is due to the high content of lipids (mycolic acid), fatty acids and alcohols found in the cell wall of *Mycobacterium* resisting decolourisation with acids and alcohol.

Procedure

1. Pour strong carbol fuchsin stain on the slide so as to cover the fixed smear and gently heat the underside of the slide. Leave the carbol fuchsin on the slide for 10 minutes with intermittent heating during that period. Care must be taken to ensure that the stain does not dry and remains steaming hot.
2. Wash with tap water.
3. Decolourise with 25% sulphuric acid until the smear becomes colourless. Wash with water.
4. Counterstain with methylene blue until the smear becomes blue.
5. Wash with water and dry.
6. Examine under oil immersion (100X) objective.

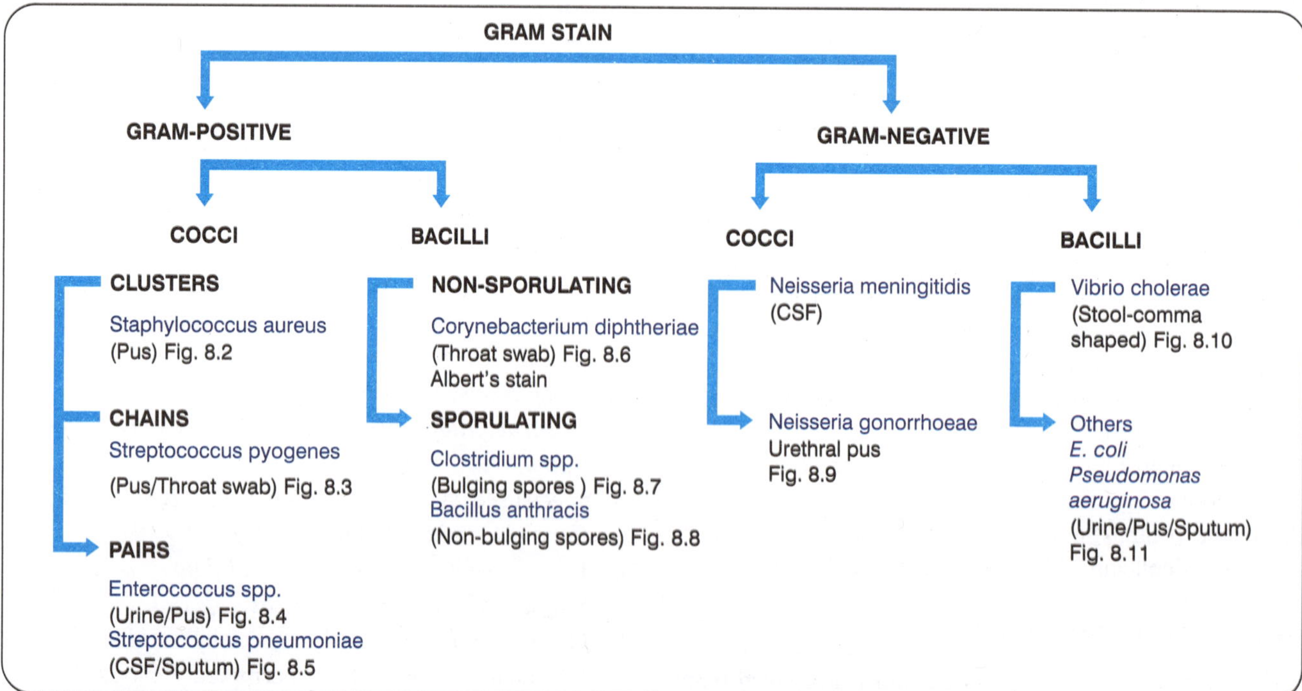

Fig. 8.1 Algorithm for the morphologic identification of bacteria in direct smears from clinical samples

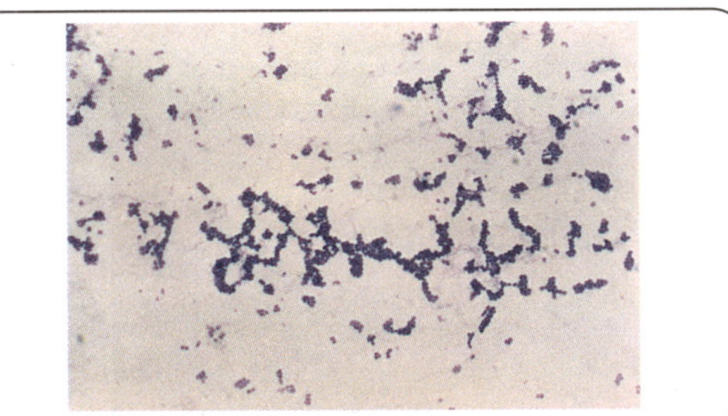

Fig. 8.2 Pus sample showing gram-positive cocci in clusters, Gram stain, 100X, e.g., *Staphylococcus aureus* (*Source*: Public Health Image Library, ID 2296 / Centers for Disease Control and Prevention [CDC] / Dr. Richard Facklam)

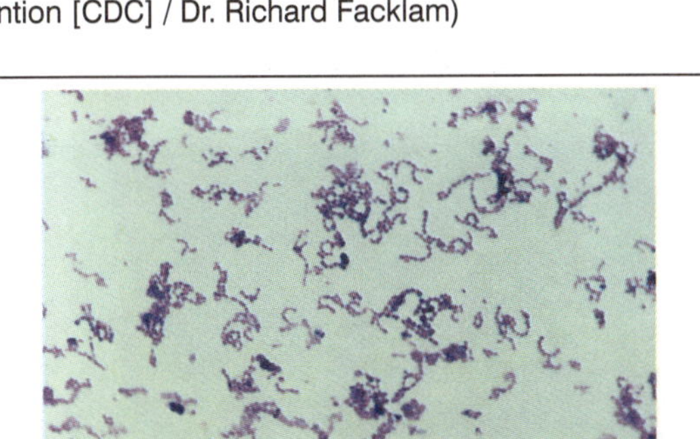

Fig. 8.3 Throat swab showing gram-positive cocci in chains, Gram stain, 100X, e.g., *Streptococcus pyogenes*

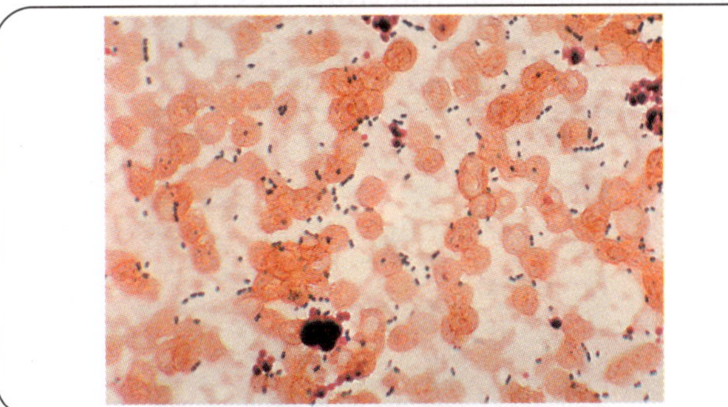

Fig. 8.4 Urine sample showing gram-positive cocci in pairs, Gram stain, 100X, e.g., *Enterococcus* spp. (*Source*: Public Health Image Library, ID 2899 / CDC / Dr. Mike Miller)

Examples

Some acid-fast organisms are *Mycobacterium tuberculosis* (Fig. 8.12), *Nocardia* spp., *Cryptosporidium parvum* and *Isospora belli*.

Special staining techniques are available for *Corynebacterium diphtheriae* (Albert stain) (Fig. 8.6b) and *Treponema pallidum* (Steiner stain) (Fig. 8.13).

Histopathologic staining

Microbes can also be demonstrated by the histopathologic examination of tissues.

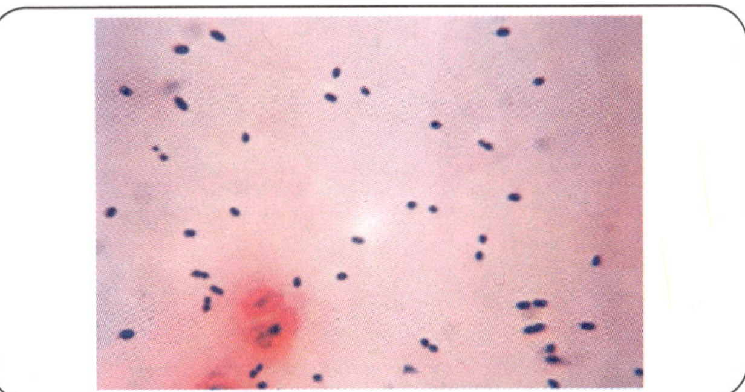

Fig. 8.5 Cerebrospinal fluid sample showing gram-positive cocci in pairs, Gram stain, 100X, e.g., *Streptococcus pneumoniae* (*Source*: Public Health Image Library, ID 18081 / CDC / Dr. Thomas Sellers, Emory University, Atlanta, Georgia)

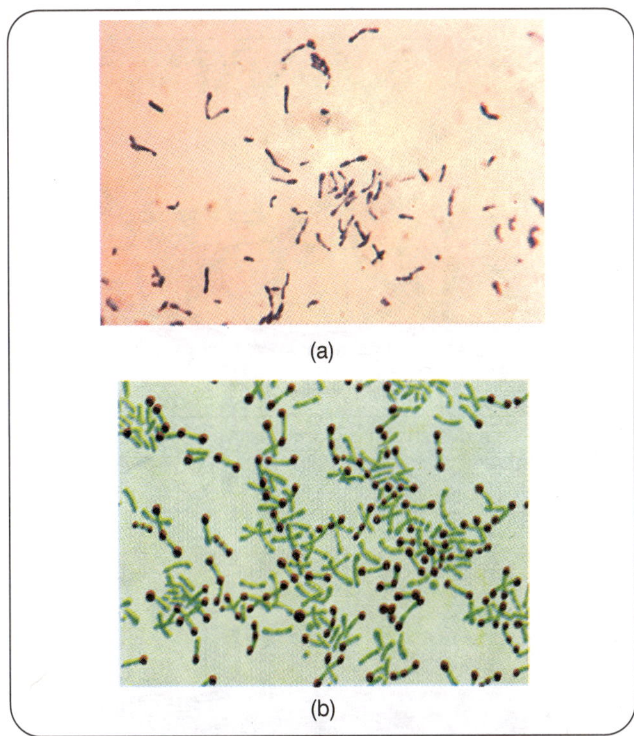

(a)

(b)

Fig. 8.6 (a) Throat swab showing gram-positive bacilli, Gram stain, 100X, e.g., *Corynebacterium diphtheriae* (*Source*: Public Health Image Library, ID 7323 / CDC / Dr. P.B. Smith),
(b) Throat swab showing gram-positive bacilli with metachromatic granules of *Corynebacterium diphtheriae*, Albert stain, 100X

1. Bacteria: *Helicobacter* pylori in gastric mucosa by Warthin—Starry stain, *M. tuberculosis* and *M. leprae* by Fite-Faraco stain and *Spirochetes* by Levaditi stain.
2. Inclusion bodies of viruses: Rabies virus in brain, cytomegalovirus in kidney and molluscum contagiosum virus in skin.
3. Fungal elements: *Cryptococcus* in brain, *Aspergillus* in lung and mucormycetes in brain.
4. Parasitic lesions: Hydatid cysts in liver and larval cysts of *Taenia solium* in brain.

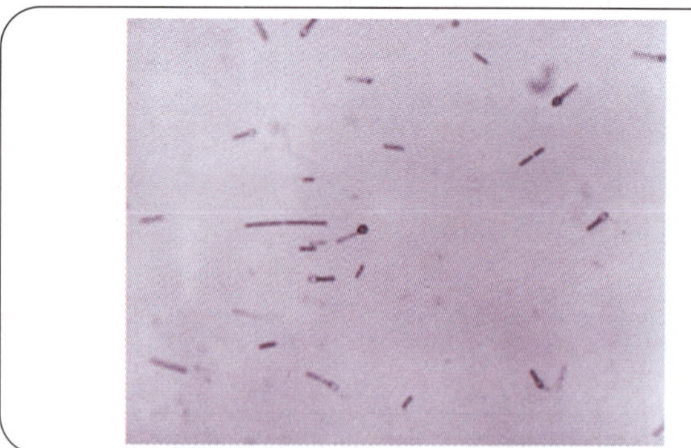

Fig. 8.7 Pus swab showing gram-positive bacilli with bulging terminal spores, Gram stain, 100X, e.g., *Clostridium* spp. (*Source*: Public Health Image Library, ID 12056 / CDC / Dr. Holdeman)

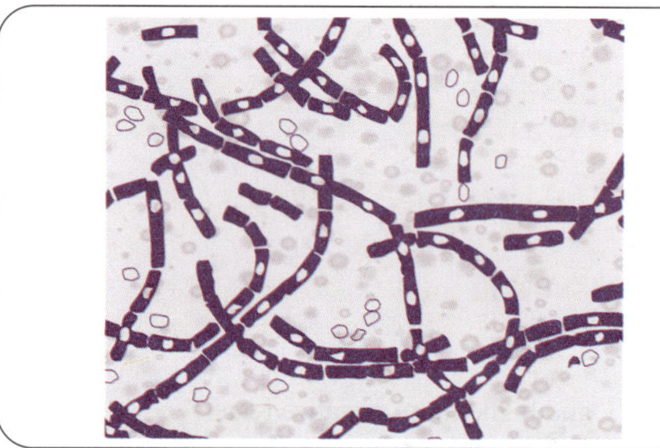

Fig. 8.8 Pus swab showing gram-positive bacilli with non-bulging central spores, Gram stain, 100X, e.g., *Bacillus anthracis*

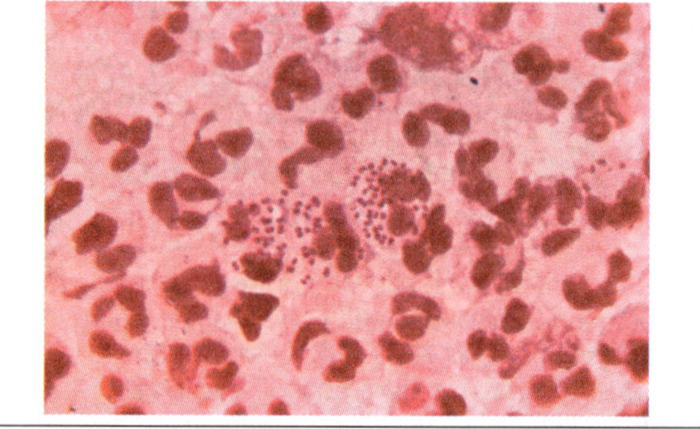

Fig. 8.9 Vaginal swab showing intracellular gram-negative cocci in pairs, Gram stain, 100X, e.g., *Neisseria gonorrhea* (*Source*: Public Health Image Library, ID 15018 / CDC / Dr. Bill Schwartz)

2. IN VITRO CULTURE OF MICROBES

Bacteria and fungi can be cultured in a diagnostic microbiology laboratory using inanimate media but viruses and parasites cannot (except *Leishmania donovani* in Novy Nicolle and McNeal [NNN] medium). The culture of viruses requires animate cell lines and is not routinely done for laboratory diagnosis.

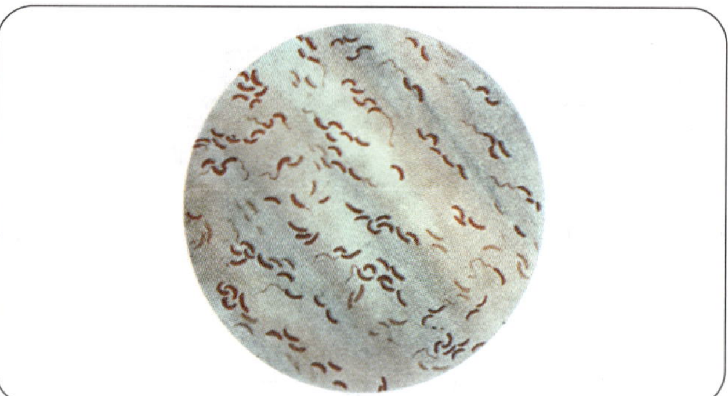

Fig. 8.10 Curved, comma-shaped gram-negative bacilli, *Vibrio cholerae*, Gram stain, 100X (*Source*: Public Health Image Library, ID 5324 / CDC)

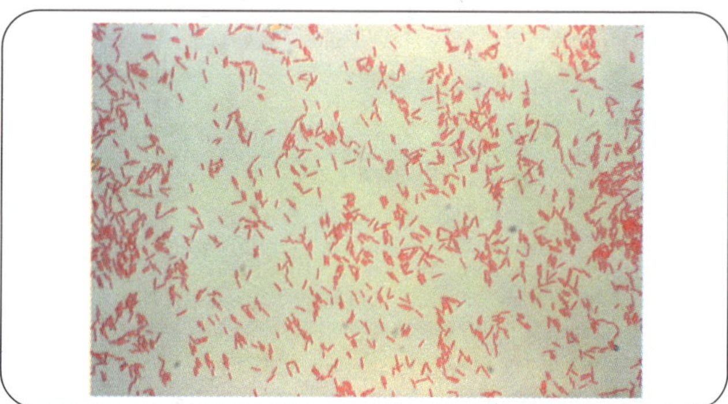

Fig. 8.11 Sputum sample showing gram-negative bacilli, Gram stain, 100X, e.g., *Pseudomonas aeruginosa* (*Source*: Public Health Image Library, ID 17370 / CDC / Dr. W.A. Clark)

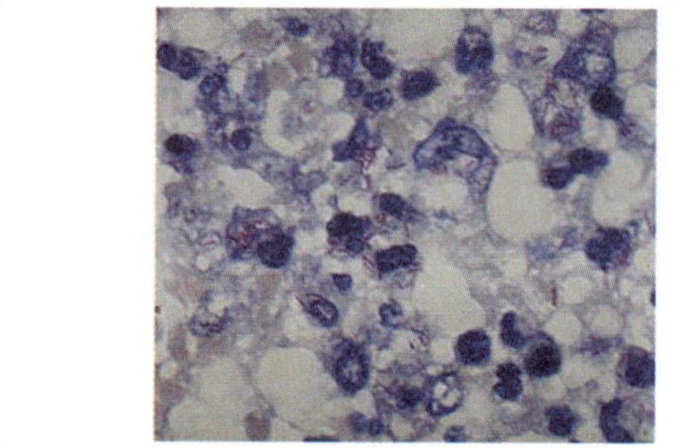

Fig. 8.12 Sputum sample showing acid-fast bacilli, Ziehl–Neelsen stain, 100X, e.g., *Mycobacterium tuberculosis*

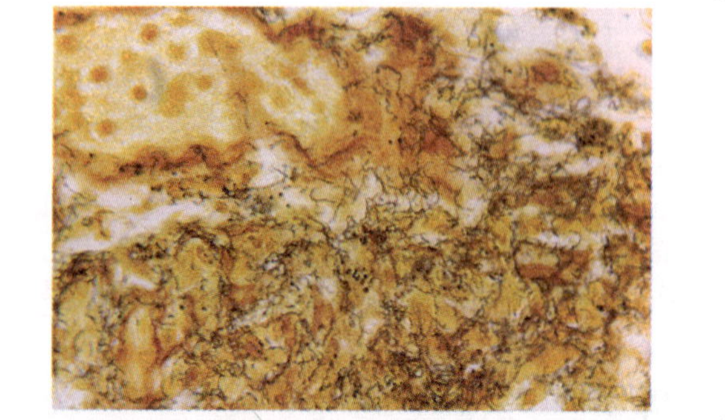

Fig. 8.13 Genital ulcer swab showing spiral bacilli of *Treponema pallidum*, Steiner silver stain, 100X (*Source*: Public Health Image Library, ID 21621 / CDC / Skip Van Orden)

1. Non-fastidious bacteria like *Staphylococcus, Streptococcus, Pneumococcus, Salmonella, Shigella, Vibrio, Escherichia coli, Klebsiella, Proteus, Pseudomonas* and *Acinetobacter* can be routinely cultured. In general, blood agar and chocolate agar are used to culture gram-positive cocci while MacConkey's agar is used for gram-negative bacilli. Special media are also available for specific non-fastidious gram-negative enteric bacilli, e.g., Wilson and Blair for *Salmonella* and thiosulfate-citrate-bile salts-sucrose (TCBS) for *Vibrio cholerae*.
2. Fastidious bacteria like *Corynebacterium diphtheriae, Neisseria* spp., *Haemophilus* spp. and *Bordetella* can be cultured with specialised media and techniques with specific, exacting growth requirements. *Spirochaetes, Mycoplasma* and *Ureaplasma* have very complex in vitro growth requirements, while *Rickettsiae* and *Chlamydiae* cannot be cultivated on inanimate media.
3. Culture for anaerobic bacteria is achieved by using gas in place of oxygen in a closed jar or in anaerobic liquid media like Robertson's cooked meat or thioglycollate. Samples not to be processed for anaerobic infections are sputum, stool, urine, throat swab and samples obtained from superficial wound infections.
4. Culture methods have also evolved from the manual to the automated in the critical areas of blood and body fluid culture, e.g., BACTEC, BacT/Alert for non-fastidious bacteria and mycobacteria growth indicator tube (MGIT) for tuberculosis. In general, they work on the principle of the presence of bacteria producing metabolites that can be detected by colorimetry. They are expensive but provide faster results.
5. Culture of *Mycobacterium tuberculosis* on solid (Lowenstein-Jensen [LJ]) medium or liquid (MGIT) medium.
6. Culture of fungi on Sabouraud's dextrose agar and other specific media.
7. Culture on NNN medium for *Leishmania donovani*.

The culture of bacteria is followed by testing for antimicrobial susceptibility of the bacterium isolated, the methods for which are discussed in Chapter 9.

3. DETECTION OF ANTIGENS OR ANTIBODIES—INDIRECT EVIDENCE

Both antigens and antibodies can be detected in vitro in a variety of samples or tissues (blood, CSF, body fluids, biopsies) for either qualitative or quantitative estimation. In clinical practice, such detection has two uses:

1. To detect or quantitate specific antigens/antibodies in the diagnosis of infections and tumours.
2. To detect immune responses and their pathology: immunodeficiency, autoimmunity and hypersensitivity.

Two important parameters applied in the tests to measure the levels of antigens/antibodies are **sensitivity** and **specificity**. Sensitivity refers to the ability of the test to detect very minute quantities of the antigen or antibody. When a test is highly sensitive, the possibility of false-negative results is minimised. Specificity refers to the ability of the test to detect reactions between homologous antigens and antibodies only, without cross-reactions. In a highly specific test, the possibility of false-positive reactions is minimised.

Positive predictive value (PPV) is the probability that patients with a positive screening test truly have the disease. Negative predictive value (NPV) is the probability that subjects with a negative screening test truly don't have the disease. PPV and NPV are the proportions of positive and negative results in diagnostic tests that are true positive and true negative results, respectively. They describe the performance of a diagnostic test.

The relationship between test performance, predictive values and disease prevalence is important. Test specificity is an issue at low prevalence; a low prevalence means lower PPV and higher false positive results. Conversely, sensitivity is a concern at high prevalence; a high prevalence means lower NPV and higher false negative results.

Thus, an ideal test should be specific enough to minimise the proportion of cases erroneously diagnosed as positive in low prevalence settings, and sensitive enough to avoid missing a diagnosis as the prevalence increases.

The term **"titre"** refers to the quantity or level of an antibody detected. In vitro, it is the maximum dilution at which the test reaction is positive. To demonstrate a rise or fall in the antibody titre, two samples of the patient are tested at an interval of 7 to 10 days (paired sera). A four-fold rise of the antibody titre, especially IgG, implies the presence of an active, acute infection. During convalescence, the antibody titres either remain stable or fall.

The formation of the antigen-antibody complex is determined by the relative concentrations of the antigen and antibody. Maximal precipitation forms in the zone of equivalence, with decreasing amounts in the zones of antigen excess or antibody excess.

The prozone phenomenon refers to the sub-optimal precipitation that occurs in the region of antibody excess. Prozone can result in false-negative serological tests when the antibody titre in a test sample is very high. Thus, dilutions of the test sample must be reacted with a fixed amount of antigen. Therefore, the purpose of diluting a test sample is two-fold: one, to estimate the antibody titre and two, to avoid false-negatives due to the prozone phenomenon.

In the laboratory, antigens/antibodies can be detected in vitro by various methods:

1. *Precipitation assays:* Flocculation test, e.g., venereal disease research laboratory (VDRL) test for syphilis.
2. *Agglutination assays:* Tube agglutination (e.g., Widal test for enteric fever), slide agglutination (e.g., antistreptolysin O [ASO] test for rheumatic fever).
3. *Enzyme-linked immunosorbent assay (ELISA),* e.g., *Leptospira*, HIV, HBV, HCV.
4. *Rapid immunochromatography (ICT) tests,* e.g., HIV, malaria.

Precipitation assays

When a soluble antigen combines with its antibody, an insoluble antigen-antibody complex precipitates. This precipitation can take place in liquid media or in gels such as agar, agarose or polyacrylamide.

In liquid, when the precipitate remains suspended as floccules, the reaction is called flocculation, e.g., VDRL or rapid plasma reagin test for syphilis.

When an antigen and antibody are allowed to react in a gel passively along a concentration gradient, the process is called **immunodiffusion**. Immunodiffusion is cumbersome and time-consuming and therefore, has limited use in diagnostic laboratories. Additionally, with the exception of the VDRL test, it has largely been replaced by more convenient methods.

The applications of precipitation assays are:

1. They are useful in demonstrating the identity of antigens from various infectious agents, e.g., Ascoli's thermoprecipitin test for anthrax, grouping of streptococci by the Lancefield technique, test for C-reactive protein (CRP) and Elek's test for diphtheria toxin.
2. Measurement of serum protein levels, e.g., Ig concentrations. Anti-IgG, anti-IgA or anti-IgM (directed against the F_c portion) is incorporated into the agar. Owing to the relatively low concentrations of IgD and IgE in human serum, this technique is not used.

   Serum immunoglobulin levels are important in the diagnosis of immunodeficiency disorders (hypogammaglobulinemia), certain infections (hypergammaglobulinemias) HIV, EBV infection, subacute bacterial endocarditis (SABE), TB, leishmaniasis and collagen disorders, such as systemic lupus erythematosus (SLE) and rheumatoid arthritis (RA).
3. Antibody detection in viral infections like rubella and influenza (single radial immune hemolysis [SRIH]).
4. Estimation of Ig levels in other body fluids. e.g., VDRL on CSF sample in syphilis.

Agglutination assays

When a particulate antigen combines with its antibody, the reaction is agglutination (visible clumping of the particles). Agglutination techniques are semi-quantitative. High sensitivity and the ability to assess end-points visually are important advantages of this technique.

The test may be performed on a slide, in a test tube or in a microtitre well.

The applications of agglutination reactions are:

1. Whole-cell agglutination on a slide, e.g., Serotyping of bacterial isolates with specific antisera for the identification of *Salmonella, Vibrio* and *Shigella* in culture.
2. For the detection of antibodies that are not cell-associated, purified antigen can be coated onto a particulate matter (latex or bentonite or tanned RBCs) and then used as a reagent for the detection of the corresponding antibody, e.g., detection of ASO and rheumatoid factor (RF).
3. Reverse agglutination: Agglutination is said to be reversed when particulate matter coated with an antibody is used for the detection of antigen. For example, the detection of CRP, and the detection of the antigens of *Streptococcus pneumoniae, Haemophilus influenzae, Neisseria meningitidis* and *Cryptococcus neoformans* in the CSF in meningitis.
4. Hemagglutination and hemagglutination inhibition for the detection of specific viral antigens and antibodies, respectively, e.g., influenza virus, *Treponema pallidum* hemagglutination for syphilis.
5. Tube agglutination tests, e.g., Widal test for the detection of *Salmonella* antibodies, Weil-Felix reaction for the diagnosis of typhus fevers and Paul-Bunnell test for infectious mononucleosis (EBV).

Enzyme-linked immunosorbent assay

In this assay, an antigen/antibody is detected in the test serum using an enzyme-tagged anti-human γ globulin; the enzymes catalyse a reaction on the addition of substrate, yielding a coloured end-product, suggesting the presence of the antigen or antibody (Fig. 8.14).

Standard ELISA involves an indirect, sandwich assay for antibody detection. This can be reversed for antigen detection. The solid phase component is an antigen coated onto a microtitre well. The antibody to be detected (patient's sample) binds to this component and then a second enzyme-labelled antibody directed against the antibody to be detected is added. The substrate of the enzyme is added, producing a coloured reaction product that is measured spectrophotometrically. An advantage of ELISA is the semi-automated or fully automated

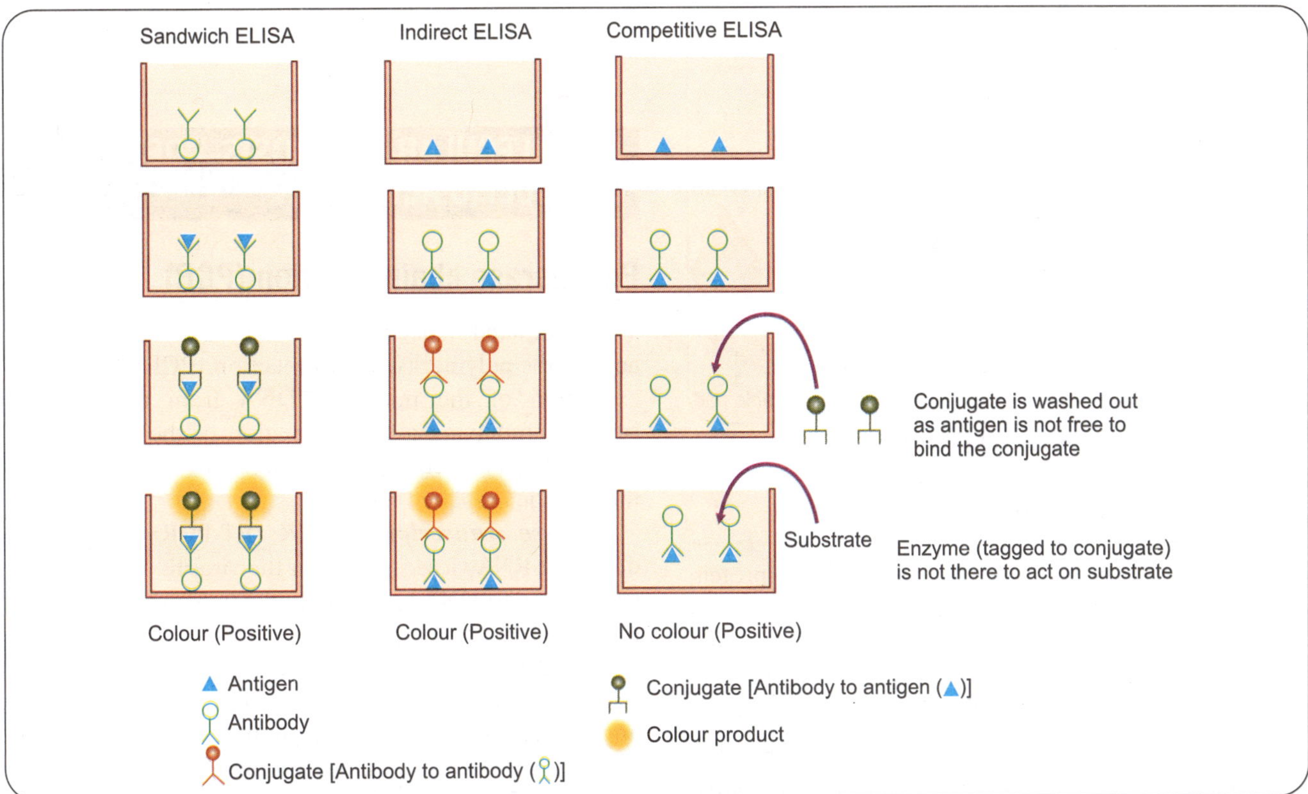

Fig. 8.14 Types of ELISA

methodology and the ability to run a large number of test samples at the same time.

The high degree of sensitivity has made ELISA popular in the qualitative and quantitative detection of antigens and antibodies. Numerous modifications are now available to increase the sensitivity and specificity. An example is μ-capture ELISA for the detection of IgM-specific antibodies to infectious agents.

The most important application of ELISA in microbiology is in the detection of the antibodies or antigens of viruses: HIV, HBV, HCV, herpes group of viruses, dengue virus and rotavirus, bacteria like *Chlamydia*, *Mycoplasma* and *Leptospira*, as well as protozoans like *Toxoplasma*.

> The applications of ELISA are:
> 1. **Detection of antigens:** Rotavirus in stool, p24 in HIV, hepatitis B surface antigen (HBsAg) in HBV, NS1 in dengue, *Chlamydia* and *Toxoplasma*.
> 2. **Detection of antibody:** HIV, HBV, HCV, herpes group of viruses, *Toxoplasma*, rubella virus, cytomegalovirus, herpes simplex virus (TORCH), dengue virus, chikungunya virus, Japanese B encephalitis virus and Leptospira.
> 3. Detection of inflammatory markers like procalcitonin.

Immunochromatography

The principle of ICT is similar to that of ELISA, but the solid phase on which the known antigen or antibody is coated is a nitrocellulose paper (Fig. 8.15). The advantage is that it is a rapid test that can be performed manually and read visually at the patient's bedside or a primary healthcare centre or for the community surveillance of an infection, e.g., HIV (antibody) and *Plasmodium* (antigen).

The sensitivity of serological assays, in detecting an antigen or antibody in a sample, is precipitation 20 μg/mL, agglutination 0.01 μg/mL and ELISA 0.0005 μg/mL.

Limitations of serological tests:

1. False-negative or false-positive result depending upon the sensitivity and specificity of the test used.
2. Variability in result depending upon the course of infection and appearance of the serological marker.
3. Biological false-positives due to antigenic cross-reaction with unrelated antigens.
4. Interference due to the presence of non-specific proteins and auto-antibodies in plasma, e.g., rheumatoid factor (RF).

Auto-antibodies in low titres are frequently encountered in normal individuals. RF is predominantly an IgM antibody directed against the F_c fragment of IgG molecules. RF is seen in autoimmune disorders like rheumatoid arthritis, other rheumatoid diseases, non-autoimmune diseases like leprosy and TB as well as in normal individuals. The presence of RF in the blood of a patient may change a negative test to give a false-positive result.

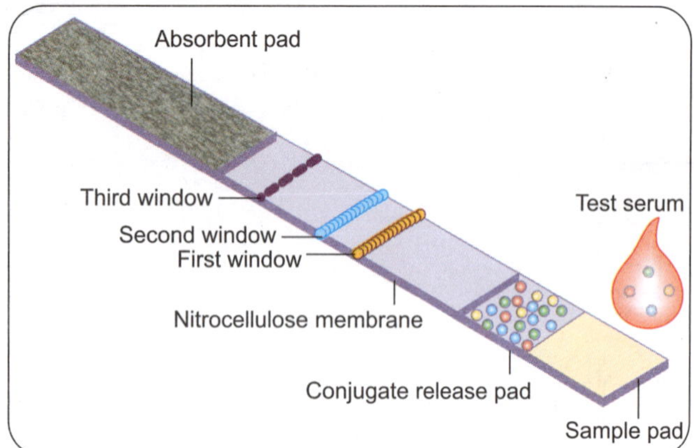

Fig. 8.15 Immunochromatography card test

Western blotting (WB)

It is a technique wherein the sample undergoes protein separation followed by gel electrophoresis to detect specific antibodies. It is more specific than ELISA and ICT and is used to detect HIV antibodies in cases that are indeterminate on ELISA and ICT. Another application is to detect *Borrelia burgdorferi* antibodies in Lyme's disease.

Inflammatory markers or acute phase reactants: Inflammatory markers include CRP, procalcitonin, ESR, plasma fibrinogen, ferritin, D-dimer and other acute phase reactants. Besides microbe-specific antigens or antibodies, serologic methods are also used to detect the presence of inflammatory markers—C-reactive protein (CRP) and procalcitonin (PCT).

CRP is called so because it complexes with the capsular polysaccharide (C) antigen of *S. pneumoniae* but is not an antibody to it. It is secreted by the liver, endothelium and other cells in response to acute inflammation and injury. The normal CRP level is < 0.2 mg/dL. The CRP level is markedly raised up to > 10 mg/dL in the event of acute bacterial infections. It can be detected by latex agglutination assay, ELISA or nephelometry

PCT is a precursor of the hormone calcitonin and is produced by the neuroendocrine cells of the lungs and intestine in response to bacterial endotoxins and cytokines. It is used to diagnose bacterial meningitis and sepsis and monitor the response to antimicrobial therapy in severe bacterial infections. It is 80% specific in distinguishing bacterial infections from non-bacterial infections and conditions. A PCT level of 0.10–0.49 ng/mL is considered insignificant or indicative of local infection, 0.50–1.99 ng/mL denotes a moderate risk of progression to sepsis, 2.00–9.99 ng/mL signifies a high risk of progression to sepsis and > 10 ng/mL implies severe sepsis or septic shock. D-dimers are protein fragments produced when plasmin cleaves fibrin in blood clots. Usually employed in thromboemboli disease, they are useful in COVID-19 along with CRP and ferritin levels.

4. MOLECULAR DIAGNOSIS—DETECTION OF NUCLEIC ACIDS

Polymerase chain reaction (PCR)

The automated molecular method used to detect nucleic acids is the **polymerase chain reaction** (PCR). It involves extraction of the microbial DNA from the infected sample, amplifying a specific region of the DNA using specific 'primer' sequences and detecting the presence of the specific amplified region.

Reverse transcriptase PCR (RT-PCR) is used for detecting RNA viruses, wherein the viral RNA is reverse-transcribed to DNA before amplification and detection of the amplified product.

Real-time PCR allows the amplification to be visualised on a real-time basis while performing PCR, and is therefore less time-consuming. In HBV and HIV disease, it is used to quantify the 'viral load' in order to determine the prognoses and response to antiviral therapy.

Nested PCR has two reactions of amplification using two primers that target two different DNA sequences of the same microorganism. Multiplex PCR uses multiple primers to detect specific DNA sequences of many microorganisms in a single reaction, e.g., Biofire FilmArray is an automated multiplex nested PCR system that yields fast results but is expensive. A single reaction is designed to detect a specific pathogen from a panel of respiratory infections, gastrointestinal infections, meningitis, encephalitis or bloodstream infections. Each panel includes common pathogens expected to cause the particular infective syndrome.

Cartridge-based nucleic acid amplification test (CBNAAT)

The CBNAAT detects the specific genes that code for a pathogen or drug resistance. It is a simple, fully automated, closed system with no contamination risk and a turnaround time (TAT) of 2 hours. All samples other than blood, stool and urine can be processed by CBNAAT.

Real-time PCR allows the amplification to be visualised on a real-time basis while performing PCR, and is therefore less time-consuming.

Currently, in tuberculosis, the Revised National TB Control Programme (RNTCP) recommends the use of CBNAAT for the detection of rifampicin resistance in smear-negative presumptive tuberculosis, extrapulmonary tuberculosis, pediatric tuberculosis and tuberculosis with concomitant HIV infection.

The CBNAAT is also available to detect:

1. *S. aureus* and MRSA from surveillance samples (nasal swabs), skin and soft tissue samples and positive blood culture bottles
2. Multi-drug resistant organisms and vancomycin-resistant enterococci using carbapenemase and van A and van B genes, respectively.
3. *Clostridium difficile* toxin B in antibiotic-associated diarrhea and colitis.
4. Influenza A and B as well as RSV in respiratory samples

Line probe assay

Line probe assay (LPA) is a rpoB/inhA gene amplification and hybridisation technique used to detect *M. tuberculosis* and drug resistance simultaneously with a TAT of 5 hours. All sputum smear-positive pulmonary cases are tested for first-line drugs. If rifampicin resistance is detected by LPA or CBNAAT (sputum smear-negatives), second-line drugs are tested by LPA viz; all fluoroquinolones, kanamycin, capreomycin and amikacin. When rifampicin resistance cannot be detected by CBNAAT, LPA is used for first-line drugs

The laboratory diagnosis of a number of infectious diseases is required to be as per the national guidelines and is to be reported regularly to the respective national programme—tuberculosis (RNTCP), HIV (National AIDS Control Organisation) and HBV, HCV, typhoid, shigellosis, cholera, dengue, influenza and COVID-19 (Integrated Disease Surveillance Programme [IDSP]).

Questions

1. Enumerate the applications of a microscopic examination of a wet mount preparation.
2. What is a direct or primary smear? What is the importance of the microscopic examination of a direct/primary smear?
3. Enumerate the applications of a microscopic examination of a stained preparation.
4. Define the specificity of a serologic test and its effect on test results.
5. What is the relationship between test performance, predictive values and disease prevalence?
6. Define the sensitivity of a serologic test and its effect on test results.
7. Define "titre". How can it affect the performance of a serologic test?
8. Define precipitation assay and list its applications.
9. Define agglutination assay and list its applications.
10. Explain ELISA and list its applications.
11. Explain immunochromatography assay and list its applications.
12. What is the role of inflammation markers in the diagnosis of infections?
13. What is PCR? Describe the types of PCRs and their applications.
14. Elaborate on the molecular methods applicable in the diagnosis of infections, giving examples.

Bacterial Genetics and Antimicrobial Resistance

BACTERIAL GENETICS

Bacteria, being prokaryotes, lack a true nucleus, nucleolus or nuclear membrane and possess a coiled circular double-stranded DNA. Some bacteria possess an additional extrachromosomal double-stranded circular DNA free in the cytoplasm called plasmid or integrated with the chromosomal DNA, called **episome**. Mobile genetic elements are DNA sequences that can move around the genome, changing their location or number of copies; they include transposable elements, plasmids and bacteriophage elements. **Integrons** are vital mobile genetic elements that acquire new genes and can express the function of that foreign gene. They act as a reservoir of antimicrobial genes, causing dissemination of resistance amongst bacteria.

Each DNA strand is composed of deoxyribose sugar, a nitrogenous base and phosphate, which makes a nucleotide. The nitrogenous bases are the purines—adenine (A) and guanine (G) and the pyrimidines—thymine (T) and cytosine (C). The two DNA strands are bridged by hydrogen bonds pairing A with T and G with C.

The DNA replication during bacterial reproduction by cell division or binary fission takes place as in eukaryotes—by DNA strand separation, synthesis of the complementary strand of nucleotides by DNA polymerase and joining by DNA ligase.

Bacterial RNA is similar to the DNA except that the sugar is ribose and thymine is replaced by uracil. Each cell possesses messenger RNA (mRNA), ribosomal RNA (rRNA) and transfer RNA (tRNA) for protein synthesis.

Plasmids have the ability to be transferred from one bacterial cell to another by the process of conjugation. They code for this property of conjugation (fertility or F plasmids), antibiotic resistance (R plasmids), bacteriocins (Col plasmids) or other virulence factors like adherence or toxins. Plasmids are also being used as vectors to insert requisite genes in recombinant DNA technology and gene therapy.

Genetic variations

Genetic variations in a bacterium can occur either by a mutation within the chromosomal DNA of the bacterium or by transfer of genetic material from another bacterium.

Mutation

Mutation is a random, heritable change in the nucleotide sequence of the chromosomal DNA, rarely plasmid. It is a natural event that occurs as cells divide and multiply, the frequency of mutants being 10^{-2} to 10^{-10} per bacterium per division. Bacterial mutations play a role in antimicrobial resistance in tuberculosis and in constructing artificial mutant strains for use as live vaccines, e.g., the enteric fever vaccine Ty21a is a mutant strain of *Salmonella Typhi*.

The rationale behind the use of a multi-drug regimen in tuberculosis is due to the occurrence of drug-resistant mutants in *Mycobacterium tuberculosis* when on drug therapy.

Methods of gene transfer in bacteria

1. *Transduction* (Fig. 9.1)

Transfer of a portion of chromosomal or plasmid DNA from one bacterium to another by a bacteriophage (virus infecting the bacterium) during multiplication of the virus. Transduction may be generalised when it involves any segment of donor DNA at random, or restricted when a specific bacteriophage transduces (carries) only a particular genetic trait, e.g., the lambda phage of *Escherichia coli* determines galactose utilisation by *E. coli*.

2. *Lysogenic conversion* (Fig. 9.2)

A bacteriophage is a virus that can infect a bacterium. Bacteriophages exhibit two types of cycles in bacteria—

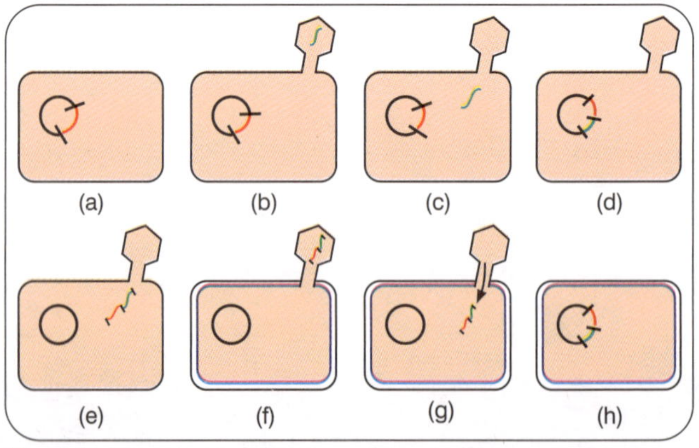

Fig. 9.1 Transduction: (a) Bacterial cell with resistance gene, (b) Bacteriophage infects the bacterium, (c) Phage DNA enters cell, (d) Phage DNA integrates with bacterial DNA during phage multiplication, (e) Phage DNA carries resistance gene, (f) Phage infects susceptible bacterium, (g) Phage DNA and resistance gene enters bacterium, (h) Resistance gene incorporated in bacterium

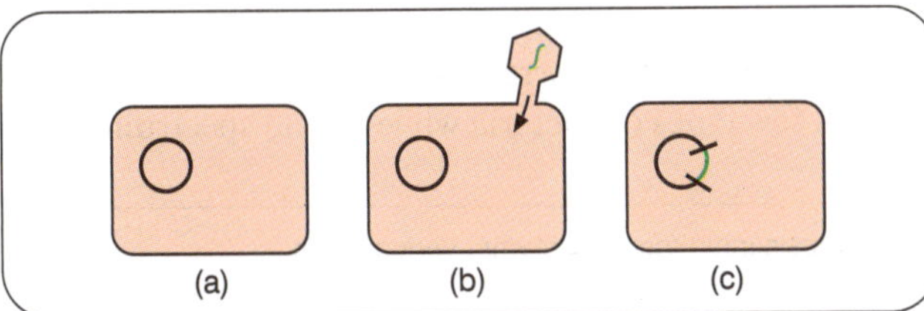

Fig. 9.2 Lysogenic conversion: (a) Susceptible bacterial cell, (b) Bacteriophage infects cell, (c) Phage DNA integrates into bacterium and itself codes for exotoxin

the lytic cycle and the non-lytic cycle (lysogenic). In the lytic cycle, the viruses multiply in the bacterium and cause its rupture/lysis. In the non-lytic cycle, the phage DNA itself acts as the prophage and confers its property to the bacterium, e.g., *Corynebacterium diphtheriae,* when infected by a beta phage, exhibits virulence by producing exotoxin.

3. Conjugation (Fig. 9.3)

This refers to the transfer of plasmid DNA (sometimes with chromosomal DNA) from a donor (male) bacterium to a recipient (female) bacterium through a conjugation tube (sex pilus) formed by physical contact between the two bacteria. The F factor is the fertility factor that confers the ability to form the conjugation tube. The resistance transfer factor (RTF) plasmid along with r determinants carry antimicrobial resistance across gram-negative bacteria.

4. Transposition (Fig. 9.4)

A transposon (jumping gene) is a segment of DNA with one or more genes in the centre and the two ends carrying 'inverted repeat sequences' complementary to each other but in the reverse order. This permits the formation of a single stranded gene as a loop carried by a double-stranded stem, e.g., genes for antimicrobial resistance as transposons.

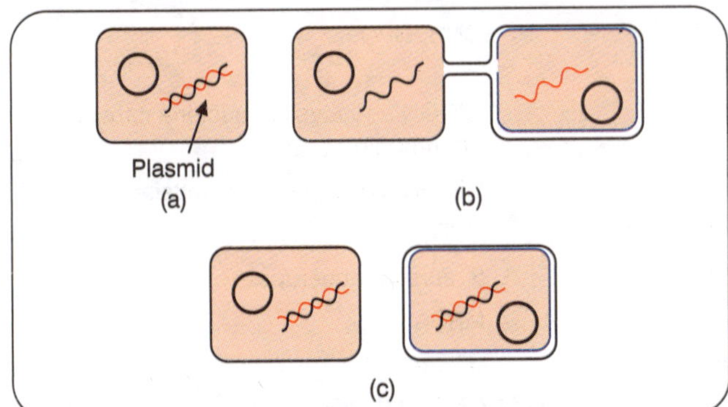

Fig. 9.3 Conjugation: (a) Donor bacterium plasmid codes for F factor, RTF and multiple r determinants, (b) Forms conjugation tube between resistant donor and susceptible recipient, (c) Plasmid coding for F factor, RTF and multiple r determinants make the recipient

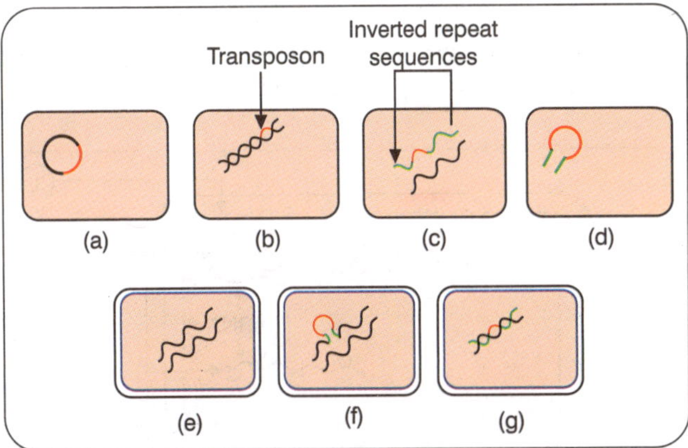

Fig. 9.4 Transposition: (a) and (b) Resistant gene or transposon in resistant bacterium, (c) Separation of strands with complementary nucleotides on either side of the resistant gene, (d) Loop of resistant gene with stem of inverted sequences, (e) Susceptible bacterium, (f) Transposon jumps into susceptible DNA strand, (g) Resistant gene integrated into susceptible bacterium

5. Transformation (Fig. 9.5)

If the DNA incorporated by any of the above mechanisms contains genes that encode for resistance to an antibiotic, a previously susceptible bacterium can be "transformed" to now be resistant. Transformation can also occur by transfer of free DNA.

ANTIMICROBIAL AGENTS IN BACTERIAL INFECTIONS

Definitions

'**Antibiotics**' are naturally secreted molecules that are produced by microorganisms that are known to selectively inhibit the growth of (static) or kill (cidal) other microorganisms, especially bacteria, e.g., penicillin and streptomycin are antibiotics produced by the fungi *Penicillium* and *Streptomyces*. The term '**antimicrobials**' covers natural, semi-synthetic and synthetic agents that act against microorganisms. Antimicrobial agents are selectively antibacterial, antiviral, antifungal, antiprotozoal or anti-helminthic. Antibacterial agents and their mechanism of resistance are included herein for further study; other antimicrobial agents have not

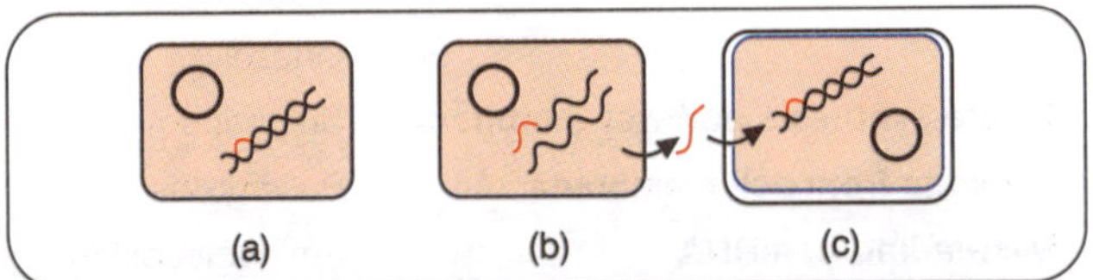

Fig. 9.5 Transformation: (a) Capsulated dead bacterial cell, (b) Free DNA coding for capsule released and (c) Free DNA integrates into live non-capsulated cell, rendering it capsulated

been elaborated on. The classification of antibacterial agents based on their mechanism of action (Fig. 9.6) and spectrum of activity is given in Table 9.1.

The spectrum of activity of an antibacterial can be narrow or broad. 'Narrow spectrum' refers to activity against few types of bacteria while 'broad spectrum' refers

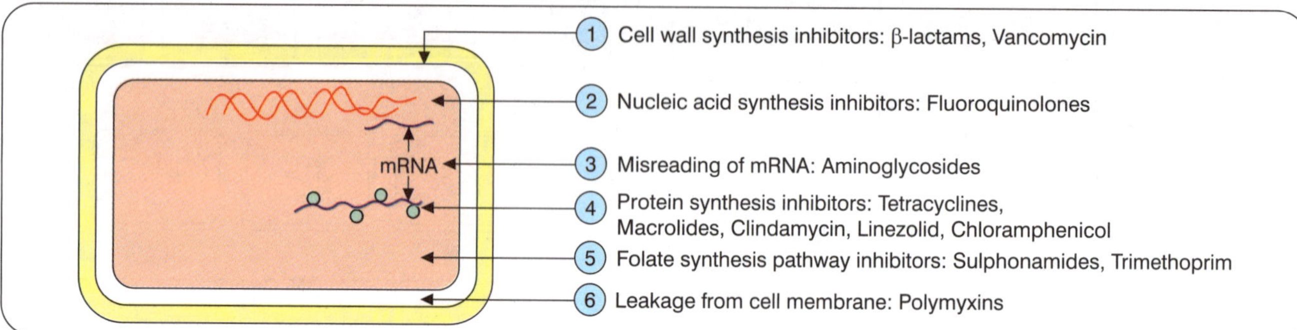

Fig. 9.6 Mechanism of action of antibacterial agents

Table 9.1 Mechanism of action and spectrum of activity of antibacterial agents

Mechanism of action	Antibacterial agent	Spectrum of activity
Cell wall synthesis inhibitors		
Penicillins	Penicillin	GPC, GNC, anaerobes
	Cloxacillin	Anti-staphylococcal
	Amoxicillin	GPC, GNC, GNB
	Piperacillin, ticarcillin	Anti-pseudomonal
Cephalosporins	1st generation: cefazolin, cephalexin	GPC
	2nd generation: cefuroxime	GPC, GNB
	3rd generation: cefotaxime, ceftazidime	Extended GNB
	4th generation: cefepime	GP and GN bacteria
	5th generation: ceftaroline	GP including MRSA, VISA, VRSA
Carbapenems	Imipenem, meropenem, faropenem	Broad-spectrum
	Monobactams—aztreonam	Aerobic GNB only
β-lactamase inhibitors	Clavulanic acid, sulbactum, tazobactum, avibactam, relebactam, vaborbactam	GPC and GNB
Glycopeptides	Vancomycin	MRSA
	Teicoplanin	MRSA
Nucleic acid synthesis inhibitors	Nalidixic acid	GNB
	Fluoroquinolones—norfloxacin, ciprofloxacin	GNB, GPC
Protein synthesis inhibitors	Tetracyclines	GP, GN and atypical bacteria causing pneumonia
	Chloramphenicol	GP, GN, spirochetes, rickettsiae
	Macrolides—erythromycin, azithromycin	GPC, GPB
	Clindamycin	*S. aureus*, anaerobes
	Linezolid	MRSA
Folate synthesis pathway inhibitors	Sulphonamide—trimethoprim	GP and GN
Leakage from cell membrane	Polymyxins	GNB—MDROs
Misreading of mRNA	Aminoglycoside—gentamicin, amikacin	GNB

Abbreviations: GPC, gram-positive cocci; GNC, gram-negative cocci; GPB, gram-positive bacilli; GNB, gram-negative bacilli; MRSA, methicillin-resistant *Staphylococcus aureus*; VISA, vancomycin-intermediate *Staphylococcus aureus*; VRSA, vancomycin-resistant *Staphylococcus aureus*; MDRO, multi-drug resistant organism

to activity against many types of bacteria, e.g., cefazolin can act against gram-positive cocci and cefoperazone can act against gram-positive cocci and gram-negative bacilli.

The selection of an antimicrobial agent depends on the identity of the bacterium, susceptibility of that bacterium to the agent, the site of infection and other pharmacologic- or patient-related factors. Immediate administration of an antimicrobial prior to bacterial identification or susceptibility testing is called 'empiric' therapy. The in vitro antimicrobial susceptibility of a bacterial pathogen can be determined by either phenotypic or genotypic methods.

Methods of measuring in vitro susceptibility

1. **Phenotypic methods:** Kirby–Bauer disc diffusion, minimum inhibitory concentration (MIC) measurement, minimum bactericidal concentration (MBC) measurement, epsilometer or E test or automated technologies (Vitek2, Phoenix, MALDI TOF, MicroScan) that combine bacterial identification with antimicrobial susceptibility testing.
2. **Genotypic methods:** PCR, CBNAAT or LPA, e.g., mecA gene in methicillin-resistant *Staphylococcus aureus* (MRSA), van A or B gene in vancomycin-resistant enterococci (VRE), rpo gene in TB (discussed in Chapter 8).

The MIC is the lowest concentration of an antibiotic that inhibits bacterial growth while the **MBC** is the lowest concentration that kills 99.9% of bacteria. As both these methods are tedious to perform on a routine diagnostic basis, the Kirby–Bauer disc diffusion method is the most commonly used. It is an internationally standardised method performed either as per the Clinical Laboratory Standards Institute (CLSI), USA or as per the European Committee for Antimicrobial Susceptibility Testing (EUCAST) guidelines.

CLSI guidelines are more commonly used in India, wherein susceptibility is reported as S, I or R. The E test by strip diffusion provides the exact MIC value of the antibiotic against the pathogen.

Antimicrobial Susceptibility Testing Report	Clinical relevance
Susceptible (S) — a given concentration of the antibiotic inhibited the growth of the organisms in vitro	Use with recommended dose
Intermediate (I) — lower response rate than for susceptible isolates, but clinical efficacy if the drug accumulates at the site of infection.	Use with dose modification
Resistant (R) — no reliable clinical efficacy	Do not use

The automated bacterial identification and antimicrobial susceptibility technologies are faster, with a wide range of bacteria and antimicrobial agents, but are expensive. Vitek 2, Phoenix, MicroScan and MALDI-TOF are the commercially available automated systems. No matter which phenotypic method is employed, there can be a gap between the in vitro and in vivo susceptibility of a bacterium to an antibiotic.

Antimicrobial susceptibility testing is done to:

1. Administer, check and sometimes modify the empiric therapy in individual patients. The rational choice of antibiotic depends on the identity of the bacterium, the site of infection and the resistance reported.
2. Epidemiological follow-up for the evolution of bacterial resistance in hospitals, regions or countries as part of an antimicrobial stewardship program (AMSP).

ANTIMICROBIAL RESISTANCE

Spectrum of action—The natural 'spectrum' of antibacterial activity is the list of bacterial species which, in their native state, are susceptible to an antibiotic. Bacterial species not listed in this spectrum are naturally resistant.

The phenotypic mechanisms of antimicrobial resistance fall into four main categories (Fig. 9.7):

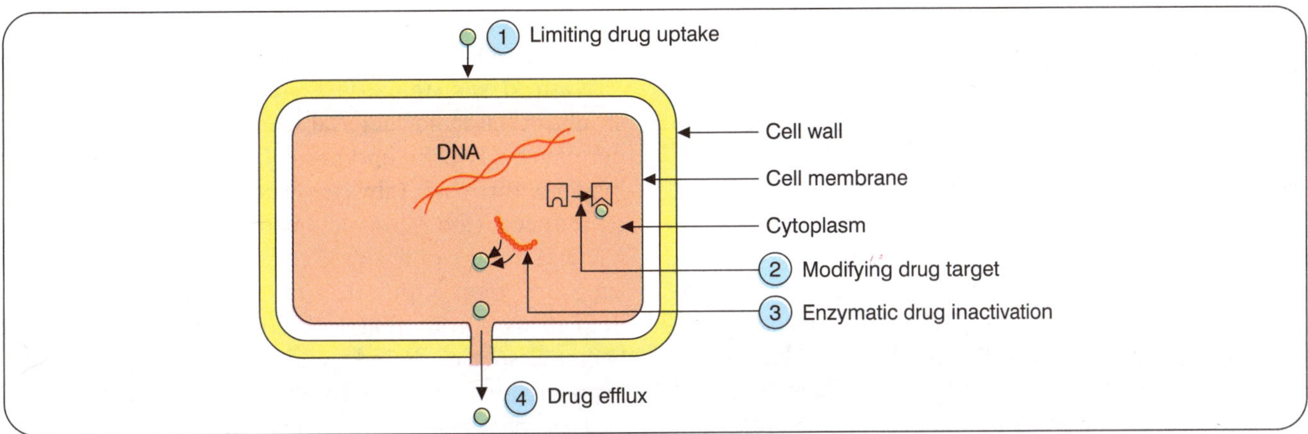

Fig. 9.7 Phenotypic mechanisms of antimicrobial resistance

(1) limiting uptake of a drug; (2) modifying a drug target; (3) inactivating a drug; and (4) active drug efflux.

1. Limiting drug uptake

The gram-negative cell wall LPS and lack of a cell wall in *Mycoplasma* provide innate resistance to certain drugs—vancomycin in gram-negative bacilli and β-lactams and glycopeptides in *Mycoplasma*. The high lipid content in the outer membrane in mycobacteria makes it accessible to hydrophobic drugs like rifampicin and fluroquinolones and inaccessible to hydrophilic drugs. Aminoglycosides find it difficult to penetrate enterococci.

Porins are proteins that cross a cell membrane, through which selective drugs can passively diffuse. A decrease in the number of porins present, or mutations that change the selectivity of the porin channel causes resistance, e.g., *Klebsiella* spp. reduce the porin numbers to become resistant to carbapenems, while *Enterobacter* spp. mutate the porin channel to become resistant to carbapenems.

Production of a biofilm by bacteria, e.g., *Pseudomonas aeruginosa,* makes it difficult for drugs to reach the bacterium.

2. Modification of drug targets

Any target of an antimicrobial agent can be modified by the bacterium to promote resistance. Gram-positive bacteria alter the number or structure of penicillin-binding proteins (PBPs) and become resistant to β-lactams. A change in the structure of PBP2a in *S. aureus* by acquisition of the *mecA* gene decreases or inhibits binding of the drug. In MRSA and VRE, the van genes decrease the ability to bind vancomycin.

Drugs that target ribosomal subunits (aminoglycosides), nucleic acid synthesis (fluoroquinolones) or metabolic pathways (sulphonamide-trimethoprim) can similarly be rendered resistant by mutations affecting these targets.

3. Drug inactivation

Bacteria can inactivate drugs by actual degradation of the drug or by transfer of a chemical group to the drug. Beta-lactamases are a very large group of drug-hydrolysing enzymes that inactivate β-lactams. Resistance to β-lactams is due to PBPs, β-lactamases and the presence of efflux pumps. There are three functional β-lactamase groups, based on their substrate specificity: the cephalosporinases, the serine β-lactamases and the metallo (zinc-dependent) β-lactamases. These enzymes may be innately found on the bacterial chromosome or may be acquired via a plasmid.

The first β-lactamase to be characterised was from *E. coli* and is chromosomally encoded by the *ampC* gene (so named for ampicillin resistance).The AmpC β-lactamases are most effective against the penicillins and some first-generation cephalosporins. If these β-lactamases confer resistance to later-generation cephalosporins, they are designated as extended spectrum β-lactamases (ESBLs). ESBL producers may also be resistant to multiple drug classes but are generally sensitive to β-lactamase inhibitors. Commonly used β-lactamase inhibitor/drug pairings include amoxicillin/clavulanic acid (amoxy-clav), ampicillin/sulbactam and piperacillin/tazobactam.

The emergence of β-lactamases that are active against carbapenems (carbapenemases) is found primarily in the *Enterobacteriaceae*. There are two types of carbapenemases—the *Klebsiella pneumoniae* carbapenemases (KPCs) and those designated as carbapenem-resistant *Enterobacteriaceae* (CRE) enzymes. KPCs are resistant to all β-lactam drugs but may still be affected by β-lactamase inhibitors. CRE strains are capable of hydrolysing all β-lactam drugs but are not inactivated by β-lactamase inhibitors. A new metallo-β-lactamase has recently been identified, mainly in strains of *E. coli*. It has been designated as New Delhi MBL (NDM-1).

Novel β-lactamase inhibitor drug combinations have been developed in an effort to combat CRE strains, e.g., ceftolozane/tazobactam against *P. aeruginosa* and ceftazidime/avibactam against gram-negative bacteria. In addition, avibactam is being tested for use with aztreonam and vaborbactam against gram-negative bacteria to combat the CREs.

4. Drug efflux

Most bacteria possess many different types of efflux pumps functioning primarily to rid the bacterial cell of toxic substances. Many of these efflux pumps can also transport a large variety of drugs out of the bacterial cell, e.g., tetracycline-resistant *Staphylococcus*.

Natural resistance is a stable characteristic of all strains of the same bacterial species to an antibiotic. It is permanent, known, and predictable. Natural resistance may be intrinsic (always expressed in the species) or induced (the genes are naturally occurring in the bacteria but are only expressed to resistance levels after exposure to an antibiotic). Intrinsic resistance may be defined as a trait that is shared universally within a bacterial species, is independent of previous antibiotic exposure and not related to horizontal gene transfer.

The most common bacterial mechanisms involved in intrinsic resistance are reduced permeability of the outer membrane and the natural activity of efflux pumps. Multi-drug efflux pumps are also a common mechanism of induced resistance. Table 9.2 shows some examples of bacteria with intrinsic antimicrobial resistance.

Acquired resistance is a genotype modified by gene mutation or gene acquisition. It is evolutionary, unknown, unpredictable and dependent on the use of the antibiotic. The genetic mechanisms of acquired resistance are mutation, transduction, conjugation and transposition. Acquired resistance may be temporary or permanent, chromosomal or plasmid-mediated. Bacteria have an average mutation rate of 1 for every 10^6 to 10^9 cell divisions. The use of antimicrobials can lead to the selection of drug-resistant mutants in successive generations.

Antibiotics do not induce resistance, but select resistant bacteria by eliminating the susceptible in a heterogeneous population. Known as 'selective pressure', the increase in the frequency of resistant strains is linked to the extensive use of a specific antibiotic.

Multi-drug resistant organisms (MDROs) are defined as microorganisms resistant to one or more classes of antimicrobial agents, e.g., MRSA, VRE and ESBL-producing gram-negative bacilli.

The "90–60 rule" implies that approximately 90% of infections treated based on invitro susceptibility are likely to respond if an "S" drug is selected; yet, up to 60% will respond even if an "R" drug is selected. The most likely situations where the latter is true is if the infection is at a site in which the drug concentration is much higher than that achieved in the test tube, i.e., much higher than the MIC.

Table 9.2 Natural resistance to antimicrobial agents

Bacteria	Resistant to
Gram-negative	
Bacteroides (anaerobes)	Aminoglycosides, quinolones
K. pneumonia	Ampicillin and amoxicillin
Enterobacteriaceae	Macrolides
All gram-negative bacilli	Glycopeptides (vancomycin), lipopeptides (linezolid)
Pseudomonas, Enterobacter and *Serratia*	All β-lactams except third-generation cephalosporins (ceftazidime, cefoperazone)
Gram-positive	
All gram-positive bacteria	Aztreonam
Streptococci	Aminoglycosides
Gram-positive cocci	Colistin
Enterococci	Aminoglycosides, cephalosporins and nalidixic acid

ANTIMICROBIAL RESISTANCE IN CLINICAL THERAPY

Natural resistance and susceptibility of gram-positive cocci

Antimicrobials that do not act on gram-positive bacteria are aztreonam, polymyxin, nalidixic acid and colistin. Antimicrobials effective against gram-positive bacteria are penicillin, cloxacillin, cotrimoxazole, ciprofloxacin, tetracycline, erythromycin, gentamicin, chloramphenicol and clindamycin.

Methicillin-sensitive *Staphylococcus aureus* (MSSA)

MSSA is sensitive to anti-staphylococcal penicillins, e.g., penicillin G, cloxacillin, nafcillin, cefazolin and cephalexin.

Methicillin-resistant *Staphylococcus aureus* (MRSA)

Beta-lactam antibiotics inhibit cell wall synthesis by binding to a cell membrane protein—PBP—essential for cell wall synthesis. The presence of the **mecA** chromosomal gene in *S. aureus* alters the PBP, rendering the bacterium resistant to all β-lactam antibiotics.

Methicillin resistance correlates with the presence of this resistance gene—mecA—that codes for a unique penicillin binding protein—PBP2a, that is not affected by β-lactamase resistant penicillins. There is a reduction in the affinity of PBPs of the staphylococcal cell wall to β-lactam antibiotics.

In India, the incidence of MRSA reported is 30–40% in various geographical regions of the country. MRSA can be community- or hospital-acquired. Community-acquired strains are more virulent than MSSA and hospital-acquired strains are associated with hospital-acquired infections, especially surgical site infection.

As human skin is the source of staphylococci, any carrier of MRSA, be it a patient or a healthcare worker, can be an endogenous or exogenous source of infection. In the event of an MRSA outbreak, HCWs are screened for the presence of MRSA on hands and nasal mucosa and treated with topical mupirocin or chlorhexidine. Hand-washing is the single most important step in the control of the spread of MRSA in a hospital setting. Coagulase-negative staphylococci, though normal commensals on the skin and rare pathogens, can also be sensitive or resistant to methicillin.

Antimicrobials effective against MSSA are penicillin, cefazolin, ciprofloxacin, erythromycin, cotrimoxazole, gentamicin and clindamycin.

MRSA is resistant to all β-lactams including cephalosporins and carbapenems. Hospital-acquired strains are also resistant to erythromycin, clindamycin and tetracycline, in addition to the above-mentioned antimicrobials. Community-acquired strains are resistant to erythromycin in addition to the above-mentioned antimicrobials. Antimicrobials effective against MRSA are trimethoprim-sulfamethoxazole, doxycycline, linezolid, vancomycin, teicoplanin and daptomycin.

Due to the irrational use of vancomycin in the treatment of infections due to MRSA, the MIC of vancomycin is on the rise. Although vancomycin-resistant *S. aureus* (VRSA) is rare in India, vancomycin-intermediate *S. aureus* (VISA) has been reported. The **van A** gene associated with VRSA is acquired from vancomycin-resistant *Enterococcus faecalis* (VRE) by horizontal transfer.

Antimicrobials effective against VRSA are linezolid, daptomycin, telavancin and quinupristin/dalfopristin.

***Streptococcus* spp.** are sensitive to penicillin, erythromycin and clindamycin, while *S. pneumonia* is sensitive to penicillin G, amoxicillin, ceftriaxone, clindamycin, macrolides and fluoroquinolones. Drug resistance in *Streptococcus* spp. is rare, but can be encountered in *S. pneumoniae*. Penicillin resistance in pneumococci is attributed to an altered PBP.

Vancomycin-resistant enterococci (VRE)

Enterococci are naturally resistant to β-lactams, aminoglycosides, sulphonamides and cotrimoxazole, but due to the synergism between β-lactams and aminoglycosides, their combination can be used in treatment. In the event of failure of this combination, vancomycin is indicated.

Due to the increase in the use of vancomycin, the incidence of VRE is on the rise, with India reporting 5–10% enterococcal isolates as VREs. Vancomycin acts by binding to the peptidoglycan in the cell wall of a gram-positive coccus. The presence of the inducible van genes located on transposons alters this binding site for vancomycin and renders the bacterium a VRE. Enterococci, being part of the normal microflora in the intestine, are a likely source of infection and transmission of VRE in hospital-acquired infections. The van gene can also be transmitted to *S. aureus*. VREs are resistant to ampicillin, penicillin, erythromycin, ciprofloxacin, gentamicin and vancomycin. Antibiotics effective against VREs are linezolid, chloramphenicol, tetracycline, daptomycin and quinupristin/dalfopristin.

Antimicrobial resistance in *Neisseria gonorrheae*

Gonococci were naturally sensitive to β-lactams, quinolones and sulphonamides, but have acquired resistance with the extensive use of these drugs. The resistance to β-lactams is largely due to the production of enzyme β-lactamase or penicillinase that is plasmid or chromosomal and mediated by horizontal transfer through conjugation. Third-generation cephalosporin—ceftriaxone—is the recommended drug of choice.

Natural resistance and susceptibility of gram-negative bacilli

Antimicrobials that do not act on gram-negative bacilli are vancomycin, erythromycin, clindamycin, dalfopristin and linezolid. Antimicrobials effective against gram-negative bacteria are ampicillin, amoxicillin-clavulanic acid, cefuroxime, cefotaxime, ceftriaxone, cotrimoxazole, nalidixic acid, ciprofloxacin and amikacin.

Beta-lactam antimicrobials effective against *Pseudomonas* spp. are piperacillin, piperacillin-tazobactam, ceftazidime, cefoperazone, cefepime, aztreonam and imipenem. Other anti-pseudomonal antimicrobials are ciprofloxacin, amikacin and tobramycin.

Beta-lactamases produced by gram-negative bacilli are ESBLs, AmpC, KPCs, MBLs and NDMs depending upon the spectrum of action against β-lactams.

ESBLs are often encoded by genes located on plasmids and confer resistance to third-generation cephalosporins and monobactams but render sensitivity to cephamycin and carbapenems. ESBL-producing isolates are resistant to ampicillin, amoxicillin-clavulanic acid, cefotaxime, cefuroxime, cotrimoxazole, ciprofloxacin and amikacin, and are sensitive to piperacillin-tazobactam, cefotetan, cefepime and imipenem.

Carbapenem resistance in gram-negative bacilli

Carbapenem resistance can be developed either due to acquired or intrinsic resistance mechanisms or both since bacteria have acquired numerous resistance mechanisms including mutations in the target site, efflux pumps and enzymatic inactivation.

The most common resistance mechanism against various β-lactams is the hyperproduction of chromosomal AmpC cephalosporinases, which contributes to carbapenem resistance in bacteria with upregulated efflux pumps, porin loss or both.

KPC-producing bacteria are resistant to all β-lactam antibiotics but sensitive to aminoglycosides, tigecycline, polymyxin B and colistin.

MBLs, e.g., NDM-1, are those that confer resistance to imipenem, broad-spectrum cephalosporins and aminoglycosides but are susceptible to ciprofloxacin, tigecycline and colistin.

Natural resistance and susceptibility of anaerobic bacteria

Antimicrobials effective in anaerobic infections are penicillin, amoxicillin-clavulanic acid, cephalosporins, chloramphenicol, metronidazole, clindamycin and imipenem. Amongst anaerobic bacteria, antimicrobial resistance is on the rise in *Bacteroides* spp.

Multi-drug resistant TB

Drug resistance in *M. tuberculosis* is attributed to the occurrence of mutations in the rpoB or inhA gene. The use of a multi-drug regimen prevents the selective multiplication of the mutant strain. This is achieved by the action of one of the alternative antibiotics to which the mutant strain is sensitive. Multi-drug resistant TB (MDR TB) is resistant to isoniazid and/or rifampicin. Extensively drug-resistant TB is resistant to rifampicin and isoniazid, resistant to fluoroquinolone, and resistant to one or more of the following injectable drugs—amikacin, capreomycin and kanamycin.

Chemoprophylactic use of antimicrobials

Prophylaxis of an infection with an antimicrobial agent can either be pre-exposure or post-exposure. Pre-exposure chemoprophylaxis is recommended either in travellers or in outbreaks or in high-risk individuals, e.g., anti-malarials in travellers, doxycycline for leptospirosis and scrub typhus, ciprofloxacin for meningococcal meningitis, tetracycline for cholera and plague outbreaks, and hydroxychloroquine in HCWs for COVID-19. Post-exposure chemoprophylaxis is recommended by doxycycline for anthrax, isoniazid for TB in the HIV-infected, oseltamivir for influenza in HCWs and nevirapine for infants born to HIV-positive mothers.

Rational antimicrobial therapy

Rational antimicrobial therapy is defined as prescribing the right drug, in the right dose, at the right frequency and for the right duration. The first principle of rational antimicrobial therapy is that antibacterial agents are not indicated in viral infections like the upper respiratory common cold, uncomplicated diarrhea or asymptomatic bacteriuria. In bacterial infections, the rational choice of an antimicrobial depends on the site of infection, the causative agent and the spectrum of activity.

Antimicrobial resistance can be prevented by:

1. Authorised practitioners who prescribe after making a diagnosis and limiting empiric antibiotic therapy
2. Prescribing broad-spectrum antimicrobials only on evidence of resistance
3. De-escalating therapy based on evidence of sensitivity or clinical response
4. Prescribing optimal dose and schedule of the antimicrobial therapy
5. Patient compliance with the therapy as prescribed by practitioner. Self or pharmacy prescribing antimicrobials over the counter contributes to resistance.

Antimicrobial stewardship program (AMSP)

Antimicrobial resistance (AMR) has given rise to multi-drug resistant organisms (MDROs) due to the irrational use of antibiotics in humans, animals and agriculture. An antimicrobial stewardship program (AMSP) involves a multi-disciplinary team in a hospital to strategise, implement and audit an antibiotic policy for that hospital.

An antibiotic policy is a set of rules, where an antibiotic may or may not be prescribed in a defined setting—that is in a community, hospital or nation. It is aimed at reducing the occurrence of antimicrobial resistance in that setting.

Questions

1. What are plasmids? How do they cause antimicrobial resistance? Explain, giving examples.
2. What is a mutation? How does it cause antimicrobial resistance? Explain, giving an example. How can it be overcome?
3. What is a transposon? How does it cause antimicrobial resistance? Explain, giving examples.
4. Define 'minimum inhibitory concentration' and 'minimum bactericidal concentration'. Add a brief note on the methods of antimicrobial susceptibility testing.
5. Classify antimicrobial agents with their mechanism of action and spectrum of activity.

6. What is the clinical relevance of the terms 'susceptible, intermediate and resistant' in antimicrobial susceptibility testing? Explain the need for antimicrobial susceptibility testing.
7. Compare and contrast natural and acquired antimicrobial resistance. How does extensive use of a specific antimicrobial agent cause AMR?
8. Discuss phenotypic mechanisms of antimicrobial resistance, giving examples of each.
9. Discuss natural resistance and susceptibility of gram-positive cocci, giving examples.
10. Discuss natural resistance and susceptibility of gram-negative bacilli, giving examples.
11. Discuss natural resistance and susceptibility of anaerobic bacteria.
12. Explain the relationship between β-lactamases produced by gram-negative bacilli and the effective therapeutic options available.
13. Define MRSA, giving its epidemiology, mechanism and impact on therapeutic options.
14. Define VRE, giving its epidemiology, mechanism and impact on therapeutic options.
15. Define MDR TB, giving its epidemiology, mechanism and impact on therapeutic options.
16. Write a brief note on the use of antimicrobials for chemoprophylaxis.
17. What is 'antibiotic policy'? What practices can be adopted to ensure rational antimicrobial therapy in a clinical setting?

Immunity

Immunology (Latin 'immunitas' meaning 'freedom from') deals with the study of the immune system. Each system in the human anatomy is responsible for a specific physiological function, i.e., the gastrointestinal system for alimentation, the respiratory system for oxygenation, and so on and so forth. Similarly, there exist specific organs and cells that form what is called an 'immune system'. However, as microbial assault can occur anywhere in the body, the immune system too is functional across the body in all tissues irrespective of the anatomical location.

The primary function of the immune system is to discriminate between self and non-self molecules (Fig. 10.1). Therefore, in health, the immune system is not only competent to respond to microbes (non-self) but also does not respond to self (human cells) or contacted foreign molecules. The immune system responds to microbes (normal immune response) when infection occurs, in a bid to prevent disease.

The failure of the immune system to respond to microbial invasion is seen as immunodeficiency. On the other hand, the occurrence of an immune response to self molecules is seen as autoimmunity. An abnormal, exaggerated response to contacted foreign molecules is seen as hypersensitivity.

Further, the immune response is also seen to be activated when normal cells undergo malignant transformation and are seen as abnormal (tumour immunology) or, subsequent to an organ transplant, the recipient's immune system responds to the foreign organ (transplantation immunology).

Immunity is defined as the resistance of a host to injury caused by microorganisms (an infection) and/or their products. Whether or not an infection results in disease would depend on the host's immunity.

Fundamental properties of immunity

The three fundamental properties of immunity are:
1. Immunologic diversity or the presence of highly diverse antigen receptors that help to recognise an infinite range of pathogens, although the response is pathogen-specific.
2. Immune memory is defined by the ability for a rapid recall and the early mount of immune response on re-exposure.
3. Immunologic tolerance is that which helps prevent immune-mediated damage to self-antigens.

Immunity can be classified as

I) Innate/constitutive

II) Acquired/adaptive
- a) Active: Natural, Artificial
- b) Passive: Natural, Artificial

Innate or constitutive immunity is the resistance to infection by virtue of an individual's genetic and constitutional make-up.

Acquired or adaptive immunity is the resistance that an individual acquires during life. It depends on the 'immune response' to an infection. The immune response is mediated via cells (cell-meditated immunity, CMI) and antibodies (humoral immunity).

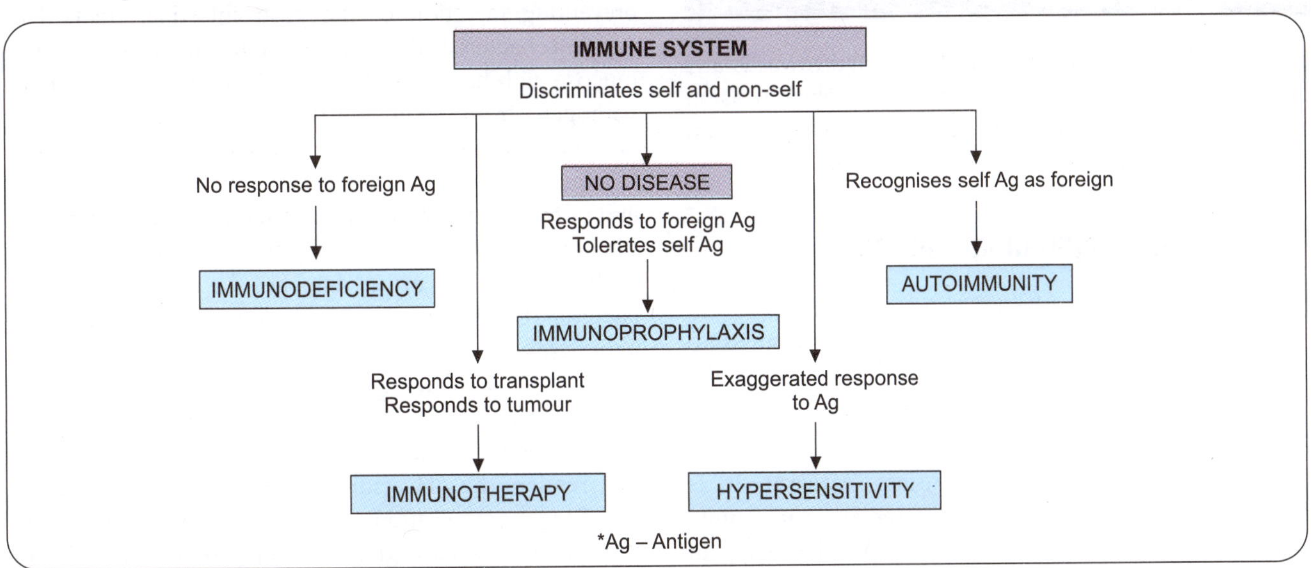

Fig. 10.1 The spectrum of immunity and disease

As innate immunity comes into play against all infections, irrespective of the identity of the microorganism, it lacks specificity. Additionally, it lacks specificity, and, since it is non-specific, it does not have the ability to recognise the microorganism when it reinfects, thus lacking memory (Table 10.1).

As opposed to this, the acquired immune response depends on the identity and antigenic specificity of the microorganism and is therefore highly specific. Subsequently, it remembers the antigenic specificity of the microorganism it once contacts and responds with a memory on re-infection with the same.

The macrophage that has phagocytosed the microorganism triggers the microorganism-specific acquired immune response, involving lymphocytes, resulting in the production of specific antibodies and lymphokines.

Table 10.1 Comparison of innate and adaptive immunity

Property	Innate	Adaptive
Mechanisms Physical barriers	Skin and mucous membrane	None
Soluble factors	Lysozyme, complement factor, acute phase proteins, interferons, antimicrobial peptides	Antibodies Lymphokines
Cells	Macrophages, polymorphonuclear leucocytes, neutrophils, eosinophils, natural killer cells	Macrophages T & B lymphocytes
Properties		
Specificity	No	Yes
Memory	No	Yes

INNATE IMMUNITY

Innate immunity is not specifically directed towards any microbe but is equally applicable to all microbial invasion and infection. It is independent of antigen recognition and therefore lacks a memory to any specific antigen.

Factors influencing innate immunity

Species: All human beings are insusceptible to plant pathogens and vice versa. This is attributed to physiological and biochemical differences between the tissues. For example, tobacco mosaic virus that affects plants does not affect farmers.

Race: (Mongolian, Caucasian, Asian, African) It is difficult to demonstrate marked differences in immunity in the human race. One example is red cell sickling in Africans, that confers immunity to the malarial parasite.

Individual: Factors influencing innate immunity in an individual are:

1. *Age:* Extremes of life (neonate and geriatric) carry a higher susceptibility to infections. In the neonate, the immune system is yet to be fully developed, while in the aged, comorbidities and metabolic disturbances are more likely.
2. *Hormones:* Diabetes mellitus (hyperglycemia), hypothyroidism and adrenal dysfunction (increased corticosteroid levels) enhance susceptibility to infections. The hyperglycemic environment in diabetes favours immune dysfunction and steroids are known to cause immunosuppression.
3. *Nutrition:* Qualitative and quantitative changes in protein, carbohydrates, lipids, vitamins and minerals in the body affect immunity. Malnutrition, especially protein energy malnutrition, is a common cause of immunodeficiency in developing nations like India.
4. *Therapy with immunomodulators and chemotherapy:* Chemotherapeutic agents act on immunocompetent cells, causing neutropenia and leading to immunosuppression.

Mechanisms of innate immunity (Table 10.1)

1. Physical and chemical barriers to the entry of microorganisms

The skin and mucous membranes protect the underlying tissues from microbial invasion. The cilia and mucous at the surface of the respiratory mucous membrane mechanically prevent the entry of microbes. Microbial and other foreign particles trapped within the adhesive mucus are removed by 'mechanical stratagems', such as ciliary movement, coughing and sneezing.

The acidic pH of the stomach kills microbes, thereby preventing infection of the gastrointestinal tract. The resident bacterial flora in the intestine colonises the mucosa with harmless commensals, preventing intestinal pathogens from occupying the mucosal surface in their presence, which is termed 'microbial antagonism'. Commensals prevent infection by competing for adhesion and nutrition, and by inducing a basal immune response. The lysozyme present in all secretions (except CSF), such as sweat and urine, damages the cell wall of many bacteria. Lung surfactant and skin defensins are both antimicrobial.

The vaginal ciliary epithelium and lactobacilli prevent vaginal infection. Lactobacilli metabolise the glycogen secreted by the vaginal epithelium to produce lactic acid that renders the pH acidic, thereby preventing bacterial and fungal colonisation. The use of antimicrobials can disrupt the normal vaginal flora and increase the risk of candidal infection.

2. Internal constitutive defence

Phagocytes belonging to the mononuclear phagocytic system (MPS), previously known as the reticulo-endothelial system (RES), the complement system, properdin system, acute phase proteins, and interferons constitute the innate defence against microbes. Monocytes and macrophages are 'professional phagocytes' while polymorphonuclear cells are 'microphages' that have a non-dividing, short-lived, multi-lobed nucleus with granules. The MPS is present throughout the connective tissue and basal membrane of small blood vessels.

The mononuclear phagocytic system consists of circulating monocytes and tissue-resident sentinel cells—tissue macrophages and dendritic cells—and a phagocyte is the first cell to respond to the infecting microorganism. It plays a role in the innate as well as acquired immune response. Its role in innate immunity is phagocytosis. Circulating macrophages, monocytes and tissue-resident macrophages—Langerhans cells in skin, such as intestinal macrophages in the gut, Kupffer cells in the liver, microglial cells in the brain, osteoclasts in bone, alveolar macrophages in the lung, sinus histiocytes in the spleen or littoral cells and placental Hofbauer cells at the materno-fetal interface are able to phagocytose microorganisms independent of its specificity or antigenic recognition.

Other cells of innate immunity include **natural killer (NK) cells**, dendritic cells, polymorphonuclear leucocytes, neutrophils and eosinophils. A macrophage is a phagocytic mononuclear cell that derives from bone marrow monocytes. Mature monocytes differentiate into tissue-specific macrophages or into dendritic cells (DCs). The DCs in the lymph nodes are called follicular dendritic cells (FDCs).

3. Signals for innate immunity

Pathogen-associated molecular patterns (PAMPs) or microbe-associated molecular patterns (MAMPs) are molecules expressed by microorganisms but not by human cells. Bacterial cell surface peptidoglycans, lipopolysaccharides, teichoic acid, flagellin, and endosomal bacterial DNA and viral RNA are PAMPs. Pattern recognition receptors (PRRs) are membrane proteins in human cells that are non-specific to a pathogen but important in innate immunity. Mannose-binding lectins, C-reactive protein (CRP), heat shock proteins and toll-like receptors (TLRs) are examples of PRRs. Microbial PAMPs that interact with human PRRs lead to the release of cytokines, complement activation by the alternate pathway, and opsonisation and phagocytosis of the microorganism. PAMPs exert an adjuvant-like effect on the immune system by interacting with TLRs and other PRRs.

4. The antiviral state

Interferons, produced in response to viral infection, act on virus-infected and uninfected cells to inhibit viral replication, giving rise to an antiviral state. The antiviral state is the result of a signalling pathway induced by IFN following any viral infection. It leads to the transcription of various cellular antiviral genes that code for host defence proteins.

Inflammation and phagocytosis

WBCs or leucocytes are classified as polymorphonuclear leucocytes (PMNLs) or macrophages—neutrophils, basophils and eosinophils, and mononuclear cells—lymphocytes and monocytes. PMNLs are predominantly inflammatory cells that form part of innate immunity. Lymphocytes are specific and play a role in acquired immunity.

Monocytes, macrophages and granulocytes are cells of innate immunity, their main function being phagocytosis. Macrophages are able to internalise material external to the cell (endocytosis) by means of phagocytosis (ingestion of particulate material) or pinocytosis (ingestion of soluble material). **Phagocytosis** is the process of internalising external material by a cell. The steps in phagocytosis are shown in Fig. 10.2.

Normally, phagocytic cells are not equipped for the highly efficient uptake of particles, unless stimulated by substances called **"opsonins"** (*opsonium* = Latin for 'relish'). Several substances normally serve as opsonins, amongst them are the C3b component of the complement and antibodies.

Therefore, body surfaces (skin and mucous membrane) form the first line of defence against penetration by

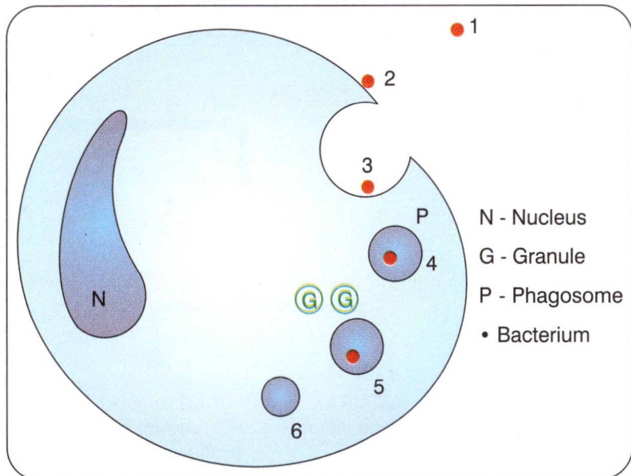

Fig. 10.2 Phagocytosis.

Steps in phagocytosis

1. Chemotaxis of the bacterium towards the macrophage
2. Recognition, adherence and opsonisation of the bacterium by the macrophage
3. Ingestion by formation of a vesicle
4. Formation of the phagolysosome
5. Degranulation and release of lysosomal enzymes into the phagosome, resulting in
6. Bacterial death

microorganisms. When penetration does occur, the invading microorganism encounters internal constitutive defences such as:

Bacterial infections

Internal constitutive defences in bacterial infections include phagocytosis by neutrophils, opsonisation by complement, activation of the alternative complement pathway directly by bacteria and secretion of acute phase proteins, e.g., CRP binds to bacteria and activates the alternative complement pathway.

Viral infections

Natural killer cells are lymphocytes capable of directly binding to and killing virus-infected cells, as they do not bear the T cell receptor (TCR) or surface immunoglobulins. Interferons are cytokines secreted by virus-infected cells that in turn, exhibit antiviral properties.

Leucocyte—endothelial interaction in inflammation

The innate and adaptive immune responses are dependent on the migration of leucocytes across endothelial cells. Diapedesis, in which the leucocyte crawls between tightly apposed endothelial cells, is a unique and complex process. Several molecules concentrated at the junctions of endothelial cells have a role in the emigration of leucocytes. These leucocyte-endothelial–cell adhesion molecules are named the junctional adhesion molecule (JAM) family.

ACQUIRED IMMUNITY

The acquired immune system is unique and has a set of fundamental properties:

1. *Antigen-specific recognition:* To recognise all potential antigens, the immune system has a specific receptor. This allows a focused response to an unwanted antigen.
2. *Diversity of lymphocyte repertoire:* The availability of an enormous spectrum of reactive lymphocytes.
3. *Discrimination of self versus non-self:* This ability is not genetically coded and, therefore, not inherited but achieved by mechanisms of self-tolerance.
4. *Immunologic memory:* Lymphocytes "memorise" prior antigen-specific stimulation, giving the ability to respond faster and more rigorously when re-exposed to an antigen. However, unwanted immune responses can be equally amplified.
5. *Self-limitation:* The antigenic stimulation of T cells not only triggers proliferation but also induces programmed cell death (apoptosis).

Acquired immunity is of two types—active and passive (Fig. 10.3). **Active immunity** is the resistance developed by an individual as a result of an antigenic stimulus. It sets in only after a latent period as it involves the active functioning of the person's immune apparatus but once developed is long-lasting and may last for years. If an individual who has been actively immunised against an antigen contacts the same antigen subsequently, the immune response occurs more quickly and abundantly than during the first encounter. This is known as **secondary response,** attributed to the 'immunological memory' that is a characteristic of active immunity.

In **passive immunity,** the individual's immune system is not involved and it is therefore immediate, without latency but short-lasting as the antibodies neutralise the

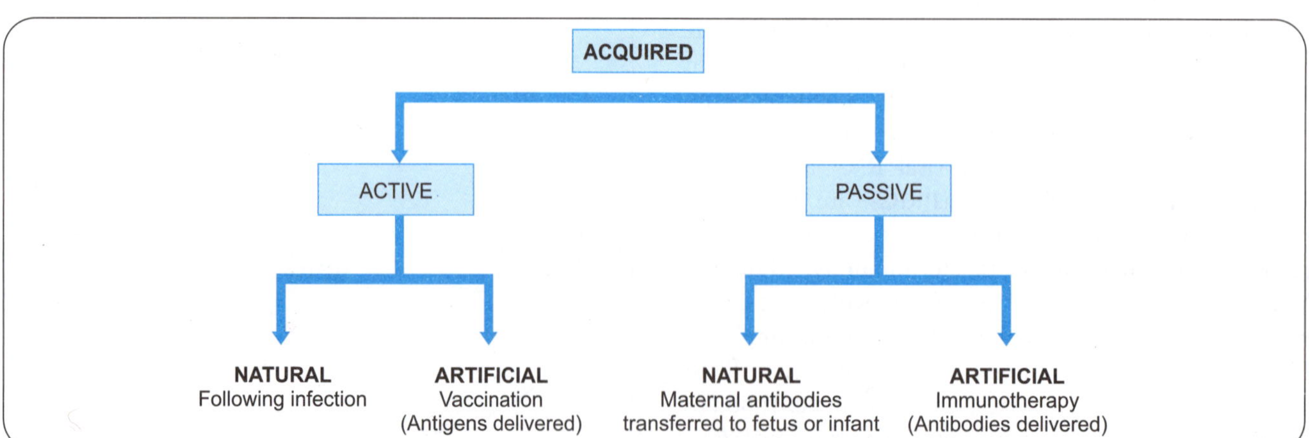

Fig. 10.3 Spectrum of acquired immunity

antigen and are metabolised. On re-administration, the antibodies are eliminated more rapidly than initially—this phenomenon is called 'immune elimination'. In practice, active immunisation is more relevant as immunoprophylaxis (to prevent infections) and passive immunisation as immunotherapy (once infection occurs) (Table 10. 2).

Active immunity comes into play naturally following any infection or exposure to a microbe. It can also be activated artificially by way of delivering microbial antigens as vaccines. Passive immunity comes into play naturally during pregnancy and postpartum, when maternal antibodies are passively transferred through the placenta and breastmilk, respectively, and protect the fetus or infant.

Antibodies are passively transferred from mother to child up to 9–18 months of age. This fact gains importance when diagnosing an infection in a newborn as it is important to differentiate whether the antibodies detected in a newborn are of maternal or neonatal origin. In utero, antibody synthesis in the fetus begins at 20 weeks of pregnancy.

Herd immunity implies the level of immunity in a community. In communicable diseases, when herd immunity is low, the chances of an outbreak/epidemic are high. A vaccine strain may be transmitted in a geographical population by herd immunity, e.g., the 'pulse polio' program with oral polio vaccine in India contributed to the spread of the vaccine strain in children.

Premunition is an infection-immunity that results from repeated antigenic exposure that protects from high numbers of the microorganisms and illness without completely eliminating the microbe. It is seen in malaria, syphilis and relapsing fever.

Table 10.2 Comparison of active and passive immunity

Active	Passive
Participation of host's immune system	Host's immune system does not participate
Induced by infection or immunogens	Conferred by antibodies
Effective after a lag period but long-lasting	Effective immediately but short-lasting
'Immunological memory' effect on subsequent challenge (booster effect)	'Immune elimination' effect on subsequent administration, rendering the subsequent dose less effective
Immunoprophylaxis with live organisms contraindicated in immunodeficient hosts	Immunotherapy applicable in immunodeficient hosts
Examples of immunoprophylaxis—DPT, typhoid, cholera, polio, MMR vaccines	Examples of immunotherapy—tetanus, diphtheria, gas gangrene, HBV antisera

Questions

1. What is the primary function of the immune system? What are its key properties?
2. Define immunity and the types of immunity, citing examples.
3. Compare and contrast innate and acquired immunity.
4. What are the factors affecting innate immunity?
5. Discuss the mechanisms of innate immunity.
6. Enumerate the properties of acquired immunity.
7. Compare and contrast active and passive immunity.
8. Define herd immunity and premunition, citing examples.

Antigen and Vaccine

An immunogen is a molecule that is capable of stimulating an immune response. An antigen is a molecule that specifically reacts with antibodies or T cell receptors, evoked by immunogens. The antibodies that are reacted with may or may not have been induced by the antigen. Therefore, all immunogens are antigens but the reverse may not be true.

Antigens may be cell-associated if they are a part of the microbe structure, either on the surface (bacterial capsule) or core, or secretory (bacterial toxin) if the molecule is secreted by the microbe.

Only a restricted portion or segment of an antigenic molecule is involved in actual binding with antibody combining sites. Such a segment, which determines the specificity of an antigen-antibody reaction, is designated as an '**epitope**'. Antigens are large molecules made up of numerous antigenic 'epitopes'. An epitope or antigenic determinant is defined as a specific 4–6 amino acid sequence or sugar residue that combines with specific antibodies or T cell receptors.

In an antigen-antibody reaction, an epitope from an antigen combines with its corresponding 'idiotype' on the F_{ab} fraction of an immunoglobulin molecule. Likewise, it is the 'paratope' of a receptor that comes in contact with an epitope.

Therefore, it follows that each epitope induces a specific antibody formation. Virtually any region on the exposed surface of an antigen may serve as an epitope. Epitopes are either linear or conformational. A linear epitope has a sequential primary structure of continuous amino acids and comes in contact with antibodies. A conformational epitope has a three-dimensional structure of discontinuous amino acids and comes in contact with a cell receptor (Fig. 11.1).

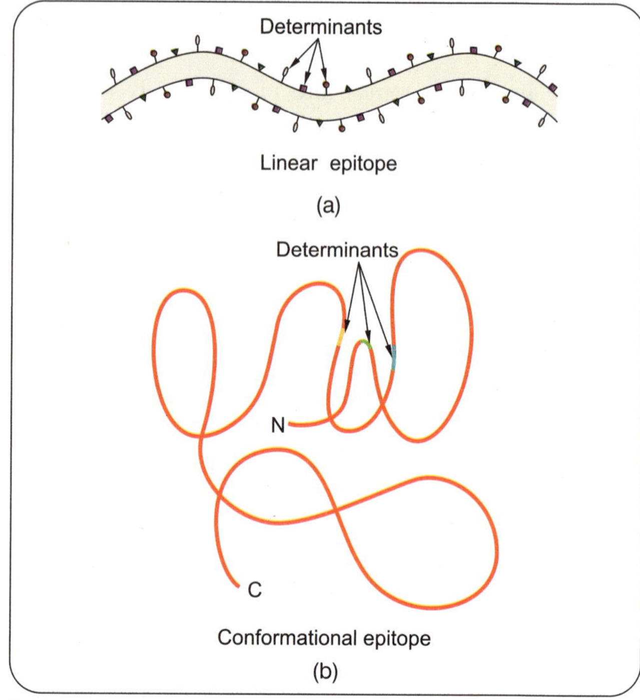

Fig. 11.1 Linear and conformational epitopes

Factors affecting antigenicity or immunogenicity

1. ***Chemical nature***: Biochemically, proteins are more immunogenic than carbohydrates or lipids. Most antigens associated with microorganisms are proteins, e.g., bacterial flagella. Polysaccharides known to be antigenic are bacterial capsule and RBC iso-antigens. The lipopolysaccharide (LPS) cell wall of bacteria is also antigenic. A molecule must also possess a certain degree of chemical complexity to be immunogenic.
2. ***Foreignness***: The immune system is known to discriminate between the 'self' (autoantigens) and 'non-self', so that only molecules that are foreign to the individual are normally immunogenic. In general, the antigenicity of a molecule is related to the degree of its foreignness. Antigens sourced from other individuals of the same species (iso-antigens) are less antigenic than those from other species (species specificity). For example, the Forssman antigen, which is a lipid carbohydsrate complex, is widely distributed in man, animals, birds, plants and bacteria.
3. ***Molecular size***: Extremely small molecules are not immunogenic. The most potent immunogens are macromolecules, with a molecular weight greater than 10,000 daltons. Adjuvants like aluminum hydroxide are used in vaccines along with antigens to increase their molecular size.
4. ***Genetic constitution of the individual***: The ability to respond to a particular antigen is a function of the way the immune response is controlled genetically. In humans, the class II major histocompatibility complex (MHC) genes control the ability of an individual to respond to a particular antigen (epitope). The MHC genes, phenotypically expressed as HLAs, are allelic in nature. An **allele** is a variant form of a gene. MHC genes have a variety of different forms, which are located at the same position, on chromosome 6.

   The class II MHC genes are allelic and therefore class II HLA molecules are polymorphic, i.e., they vary from individual to individual, within the same species. The variation of the antigen binding site among different class II molecules and, hence, the variation in the ability of a given class II molecule to bind a given antigenic epitope predict that only certain class II molecules can present particular antigens. These variations account for the fact that only certain individuals, i.e., those who inherit the appropriate class II genes, can respond to a given antigen.

5. ***Dose and mode of antigen administration***: These determinants are important in the area of immunisation schedules. Vaccines (antigens) are administered orally, intramuscularly, subcutaneously or intradermally.

Nomenclature of antigens

1. An **autoantigen** is a self-antigen, a molecule that is not foreign to the body, e.g., human cell components. An **alloantigen** is species-specific; that is, all individuals of a species possess it, e.g., allotypes of immunoglobulins in humans. An iso-antigen is a variant of antigenic specificity within a species, e.g., human RBC blood group iso-antigens.
2. A **complete antigen** is able to induce antibody formation and produce a specific reaction with the antibody so produced. An **incomplete antigen (hapten)** is incapable of inducing antibody formation by itself but can specifically react with an antibody.

   Such haptens can be conjugated with immunogenic carrier proteins to develop hapten-protein conjugates against which an immune response can be elicited, e.g., in an experimental pregnancy vaccine, when human chorionic gonadotropin (hapten) was conjugated with tetanus toxoid (carrier protein) it could induce an anti-HCG response to prevent pregnancy. However, the drawback in such conjugates is that the immunological memory response is directed only against the carrier molecule (in this case the TT) and not to the hapten (HCG).
3. **Heterophile antigens or cross-reactive antigens**: Two molecules that show some degree of sharing of antigenic specificity or epitopes are called heterophile antigens.

   Heterophile antigens that exhibit cross-reactivity in in vitro antigen-antibody reactions can be used to detect unrelated etiological agents, e.g., the Weil-Felix reaction uses *Proteus* antigens in rickettsial typhus fever, the Paul-Bunnell test uses sheep RBCs in infectious mononucleosis caused by the Epstein–Barr virus, and the cold agglutinin test is used in primary atypical pneumonia caused by *Mycoplasma*.

   When human and microbial antigens cross-react in vivo, it could lead to tissue damage and autoimmune reactions, e.g., rheumatic heart disease. The Guillain–Barré syndrome is a rare but serious autoimmune disorder that follows an infectious disease (gastrointestinal or respiratory) where the immune response attacks the myelin of the peripheral nerves.
4. Based on the mechanism of inducing their corresponding antibody, antigens can be classified as T-dependent antigens, T-independent antigens and **superantigens**. Most antigens are T cell-dependent to produce antibodies; some like bacterial LPS directly produce antibodies by binding to B cell receptors (T independent) while others like staphylococcal and streptococcal exotoxins directly activate T cells without antigen processing by antigen presenting cells (APCs), resulting in a cytokine tsunami or storm and the polyclonal activation of B cells.

VACCINE IMMUNOLOGY

The first vaccine to be developed by Edward Jenner in 1796 was by 'variolation' of cowpox vesicular fluid to prevent smallpox. Since then, numerous vaccines have been developed, which constitute the immunisation schedule in children.

A vaccine (antigen) effects an innate and an adaptive immune response to it. The adaptive immune system provides the antigen-specific immune response with memory—both cellular (T cell) and humoral (B cell). B cells respond to T cell-independent and T cell-dependent antigens. B cells respond to T cell-independent antigens, without the antigens being processed by the T cell, the response being weak with poor memory as opposed to the response to T cell-dependent antigens. T cell-independent antigens are lipopolysaccharides, while proteins are T cell-dependent antigens.

All vaccines elicit a cellular and humoral immune response, the predominant response depending upon the nature of the vaccine. Live attenuated vaccines predominantly elicit cytotoxic T cell responses. Vaccines containing unconjugated capsular polysaccharide elicit a weak, short-lasting, T cell-independent response that becomes T cell-dependent response when conjugated with a protein carrier molecule.

Immune response to vaccines

Dendritic cells (DCs) or antigen presenting cells (APCs) that travel all over the body are activated on encountering the vaccine antigen, leading to their migration to the draining lymph node. At the lymph node, the APCs cause antigen-specific activation of T and B cells.

Live attenuated vaccines (bacterial and viral) disseminate throughout the body similar to natural infection. Thus, DCs at multiple sites are activated and migrate to many lymph nodes, leading to T and B cell activation in multiple lymph nodes. This renders live vaccines more immunogenic, irrespective of their site and route of injection.

Non-live vaccines do not disseminate and produce an immune response only in one group of draining lymph nodes, making the site and route of injection important. The skin and muscles have a large number of DCs, hence, the intradermal or intramuscular route is preferred over subcutaneous or mucosal administration.

T cell-independent antigen (polysaccharide) vaccines activate both T and B cells while T cell-dependent (protein) vaccines activate predominantly B cells.

The immune response to T cell-dependent vaccines is primary and secondary. The primary immune response to T cell-dependent vaccines has two components: extrafollicular reaction and germinal centre reaction.

> **Extrafollicular reaction:** Antigen-specific B cells bind the vaccine antigen to the specific IgM receptor on its surface, leading to activation and proliferation of the B cell clone into plasma cells. This process, called 'extrafollicular reaction', produces low-affinity IgM antibodies that switch to IgG and IgA. These plasma cells die within a few days by apoptosis, therefore, rendering a short-lived response.
>
> **Germinal centre reaction:** Some activated B cells migrate to the outer T cell zone of lymph nodes and encounter T cells activated by the same antigen (presented to it by a DC), to form a single antigen-specific germinal centre. It takes 2 weeks for a germinal centre to develop, but the antibodies produced are high-affinity and longer-lasting. A subset of antigen-specific memory T and B cells is also formed.

Thus, in the **primary immune response** to a vaccine, specific antibodies first appear after 1 week of the first dose but the peak response occurs at 4 weeks, and then the antibody titre rapidly drops. Thus, the booster dose is administered 4–6 months after the first dose. The intensity of this primary humoral response depends on the nature of the vaccine (live or non-live), dose interval or booster dose, and genetic (HLA) makeup of the individual. Antibody response is poor at extremes of age.

The memory B cells migrate to the extrafollicular regions of the spleen and lymph nodes, their prolonged survival depending on the persistence of the antigen. Live attenuated viral vaccines produce memory B cells that survive lifelong without any booster.

The **secondary immune response** occurs when the individual encounters the infection. It is brisk and produces high-affinity antibodies due to activation of the memory B cells. Its intensity is directly proportional to the intensity of the primary immune response to the vaccine antigen.

Capsular polysaccharide antigens of *Haemophilus influenzae, Streptococcus pneumoniae* and *Neisseria meningitidis* elicit a T cell-independent immune response. Low-titre, low-affinity antibodies appear within a week of the first dose but memory B cells are not generated, necessitating the need for booster doses. Conjugating the polysaccharide antigen with carrier proteins results in improved immune response.

T cell response to vaccines

The vaccine antigen is taken up by the DC that migrates to the lymph node and presents it to the CD4 T cell. The CD4 T cell differentiates into the T_H1 and T_H2 subsets depending on the secretion of IL2 and IL4, respectively. The T_H1 cells secrete IFNγ and tumour necrosis factor (TNF)α that kill infected cells directly or by activating CD8 T cells. T_H2 cells activate B cells to produce specific antibodies against extracellular microbes and toxins. Memory T cells are also generated that live lifelong.

Maternal IgG transfer from mother to fetus occurs during the third trimester of pregnancy and this passive antibody titre is the highest at term delivery and subsequently declines, lasting up to 6–9 months of postnatal age. The immune system is immature in neonates, therefore, vaccines administered before 1 year of age need a primary series and booster doses. Maternal antibodies can also interfere with the vaccine antigen—the higher the maternal antibody titre, the lower the antibody response of the infant. When compared with older children, the response to polysaccharide vaccines is poor up to 2 years of age, even after conjugation with carrier proteins.

Active immunisation by vaccines is done by using the antigenic components of microorganisms in order to induce a protective immune response. The various strategies used to obtain potent antigens are (Table 11.1):

1. ***Whole-cell vaccines:*** The whole microorganism is utilised in a way by which the microorganism loses its virulence or pathogenicity but maintains its antigenicity. In whole-cell live vaccines, live microorganisms are treated in a manner to maintain antigenicity and lose their virulence. This process is called **attenuation**. On the other hand, the microorganism can be killed using heat or chemicals to render a killed vaccine.
2. ***Subunit or component vaccines:*** Instead of using the whole cell, an immunogenic part of the microorganism is separated out to make a vaccine. The polysaccharide capsule of the bacteria is used as a vaccine against pneumococcal and meningococcal infections.

Table 11.1 Vaccine strategies

Vaccine strategy	Bacterial	Viral
1. Whole-cell vaccines Live attenuated	Bacille Calmette-Guerin (BCG) for tuberculosis, oral vaccine for typhoid and cholera	Oral (Sabin) for polio, measles, mumps, rubella (MMR), chickenpox
Killed	Typhoid-paratyphoid A and B vaccine (TAB) for typhoid, cholera, pertussis	Salk for polio, rabies, COVID-19
2. Subunit	Vi Ag for typhoid, capsular Ag for pneumococcal and meningococcal infections, Hib for influenza	Hepatitis B virus surface antigen (HBsAg)
3. Toxoids	Diphtheria, tetanus	
4. Modified viral vector		COVID-19

3. ***Antigenic subunits of viruses:*** These can be developed in vitro by using the recombinant DNA technology, e.g., hepatitis B virus surface antigen (HBsAg).
4. ***Bacterial products:*** Extracellular proteins secreted by bacteria, e.g., toxins, are 'toxoided' to retain their antigenicity and destroy their toxigenicity. Gram-positive bacilli causing tetanus and diphtheria secrete exotoxins which are available as toxoid vaccines.
5. ***Synthetic vaccines:*** These are artificially constructed immunogenic peptides (sequences of amino acids).
6. ***Mucosal vaccines:*** These induce local and systemic immunity, are easy to administer, and induce immune response at the site of administration and at distal mucosal sites, e.g., oral polio vaccine
7. ***Nucleic acid vaccines:*** mRNA vaccines teach our cells to make a microbe protein that triggers an immune response, e.g., COVID-19 vaccine.
8. ***Viral vector vaccines:*** They use a modified version of a different harmless virus as a vector to deliver instructions to our cells to produce a virus protein, e.g., COVID-19 vaccine.

Adjuvants

The response to an immunogen can be enhanced if it is administered as a mixture with substances called adjuvants. Adjuvants function by prolonging the retention of the immunogen by its slow release from the site of injection, increasing its effective size and providing additional signals to APCs. The most widely used adjuvant in humans is a suspension of aluminium hydroxide on which the immunogen is adsorbed (alum precipitate).

> There are licensed vaccines available for the following infections:
> **Bacterial**: TB (BCG), diphtheria, pertussis, tetanus, influenza (Hib), pneumococcal and meningococcal infections, typhoid and cholera.
>
> **Viral**: Polio, hepatitis A, hepatitis B, rotavirus infection, mumps, measles, rubella, varicella zoster, influenza, human papillomavirus, Japanese B encephalitis, rabies, COVID-19 and yellow fever.

The adjuvant alum has been shown to generate the T_H2 type of immune response.

Vaccines under development are those for enteroviruses (EV71), dengue virus, chikungunya virus, HCV, HIV, HAV, HEV, RSV and parainfluenza virus.

In 1985, the Govt. of India launched the 'Universal Immunisation Programme' (UIP) with the objectives of eliminating neonatal tetanus and paralytic poliomyelitis. A reliable surveillance system, self-sufficiency in the production of vaccines and the development of a safe cold chain system has led to a decline in the incidence of vaccine preventable diseases and, subsequently, infant mortality.

The Department of Health Research, Ministry of Health & Family Welfare in its 2017 report *India: Health of the Nation's States* mentions diarrheal diseases, lower respiratory tract infections and tuberculosis as the leading causes of communicable diseases. The vaccines included in the National Immunization Schedule (NIS) for Infants, Children and Pregnant Women are depicted in Table 11.2.

Advantages of live vaccines

Live vaccines **"simulate"** an infection and therefore more potently stimulate cell-mediated immunity as well as humoral immune response. Moreover, in the case of intracellular microorganisms, the microorganism is able to provide antigenic stimulation for long periods, therefore reducing the need for 'boosters' or secondary doses of the vaccine. For example, viral vaccines like the MMR and chickenpox vaccines are single-dose vaccines providing lifelong protection.

Disadvantage of live vaccines

They cannot be administered to immunocompromised individuals.

As opposed to live vaccines, killed or subunit vaccines can be administered to immunocompromised individuals, but necessitate the need for booster doses for long-term protection.

Table 11.2 National Immunization Schedule (NIS) for Infants, Children and Pregnant Women

Vaccine	When to give	Dose	Route	Site
For Pregnant Women				
TT-1	Early in pregnancy	0.5 mL	Intra-muscular	Upper Arm
TT-2	4 weeks after TT-1*	0.5 mL	Intra-muscular	Upper Arm
TT- Booster	If received 2 TT doses in a pregnancy within the last 3 yrs*	0.5 mL	Intra-muscular	Upper Arm
For Infants				
BCG	At birth or as early as possible till one year of age	0.1 mL (0.05 mL until 1 month age)	Intra-dermal	Left Upper Arm
Hepatitis B - Birth dose	At birth or as early as possible within 24 hours	0.5 mL	Intra-muscular	Antero-lateral side of mid-thigh
OPV-0	At birth or as early as possible within the first 15 days	2 drops	Oral	Oral
OPV 1, 2 & 3	At 6 weeks, 10 weeks & 14 weeks (OPV can be given till 5 years of age)	2 drops	Oral	Oral
Pentavalent 1, 2 & 3	At 6 weeks, 10 weeks & 14 weeks (can be given till one year of age)	0.5 mL	Intra-muscular	Antero-lateral side of mid-thigh
Rotavirus#	At 6 weeks, 10 weeks & 14 weeks (can be given till one year of age)	5 drops	Oral	Oral
IPV	Two fractional dose at 6 and 14 weeks of age	0.1 mL	Intra dermal two fractional dose	Intra-dermal: Right upper arm
Measles /MR 1st Dose$	9 completed months-12 months (can be given till 5 years of age)	0.5 mL	Sub-cutaneous	Right Upper Arm
JE - 1**	9 completed months-12 months.	0.5 mL	Sub-cutaneous	Left Upper Arm
Vitamin A (1st dose)	At 9 completed months with measles-Rubella	1 mL (1 lakh IU)	Oral	Oral
For Children				
DPT booster-1	16–24 months	0.5 mL	Intra-muscular	Antero-lateral side of mid-thigh
Measles/MR 2nd dose $	16–24 months	0.5 mL	Sub-cutaneous	Right Upper Arm
OPV Booster	16–24 months	2 drops	Oral	Oral
JE-2	16–24 months	0.5 mL	Sub-cutaneous	Left Upper Arm
Vitamin A*** (2nd to 9th dose)	16–18 months. Then one dose every 6 months up to the age of 5 years.	2 mL (2 lakh IU)	Oral	Oral
DPT Booster-2	5–6 years	0.5 mL	Intra-muscular	Upper Arm
TT	10 years & 16 years	0.5 mL	Intra-muscular	Upper Arm

*Give TT-2 or Booster doses before 36 weeks of pregnancy. However, give these even if more than 36 weeks have passed. Give TT to a woman in labour, if she has not previously received TT.

**JE Vaccine is introduced in select endemic districts after the campaign.

*** The 2nd to 9th doses of Vitamin A can be administered to children 1–5 years old during biannual rounds, in collaboration with ICDS.

#Phased introduction, at present in Andhra Pradesh, Haryana, Himachal Pradesh and Orissa from 2016 and expanded in Madhya Pradesh, Assam, Rajasthan, and Tripura in February 2017 and planned in Tamil Nadu and Uttar Pradesh in 2017.

$ Phased introduction, at present in five states namely Karnataka, Tamil Nadu, Goa, Lakshadweep and Puducherry. (As of Feb' 2017)

QUESTIONS

1. Explain the concept of 'immunogen', 'antigen' and 'epitope'.
2. Discuss factors that affect the antigenicity of molecules.
3. What are autoantigens, alloantigens and iso-antigens? Provide examples of each.
4. How does the immune response to complete and incomplete antigens differ?
5. Write a brief note on heterophile antigens, citing examples.
6. Define the mechanism of antibody production in T-dependent antigens, T-independent antigens and superantigens.
7. Elaborate on the immune response to vaccines.
8. Enumerate the strategies of antigens used as vaccines, citing examples.
9. Compare and contrast live and killed vaccines.
10. What is an adjuvant? How does it function?

Chapter 12

Section II: Immunology

Antibody and Complement

Antibodies or immunoglobulins (Ig) are glycoproteins that combine specifically with the antigen that elicited their formation. They constitute 20% of the total plasma proteins. On contact with a specific antigen, B cells differentiate into plasma cells to secrete specific antibodies and form the humoral component of the acquired immune response.

STRUCTURE OF AN IMMUNOGLOBULIN MONOMER

A 'domain' is a globular region of polypeptide chains on an Ig molecule. A 'receptor' is a functional macromolecule present on cells.

A monomer of Ig consists of two light (L) and two heavy (H) polypeptide chains bound together by disulphide bridges (Fig. 12.1). The monomer has either kappa or lambda light chains, never both together.

The variable (V) regions of the H and L chains are responsible for antigen binding and therefore determine the idiotype of the Ig. The constant regions form three domains: C_H1, C_H2 and C_H3, with the functions of isotype/allotype determinants, complement fixation and F_c receptor binding, respectively.

Idiotypes represent the antigenicity of the antigen binding site of an antibody.

Species-specific C_H1 region differences are called isotypes. In humans, five isotypes are seen—G, A, M, D and E. They give rise to five distinct classes of immunoglobulins, i.e., IgG, IgA, IgM, IgD and IgE. Each normal individual expresses all these isotypes characteristic of the species.

The allelic forms of isotypes (genetic variations between individuals of the same species) are called **allotypes**.

IgM and IgA exist as polymers (i.e., more than one basic monomer unit): IgA as a dimer or trimer and IgM as a pentamer (Fig. 12.2). When immunoglobulins exist as polymers, the monomers are held together by a polypeptide chain called the J chain. Besides the J chain, the IgA molecules also consist of a 'secretary component' molecule, giving it the ability to be secreted across a mucosal barrier.

Immunoglobulins can be found in association with cells (B cells) or free in body fluids. The predominant antibody in serum is IgG but the predominant B cell-associated Ig (in the resting state) is IgM. Therefore, on antigenic stimulation, the first antibody to appear is IgM, referred to as 'primary antibody'. Following which, an isotype switch (gene rearrangement) occurs in the B cell to synthesise IgG. Isotype switch can occur from IgM to IgG to IgA but never in the reverse.

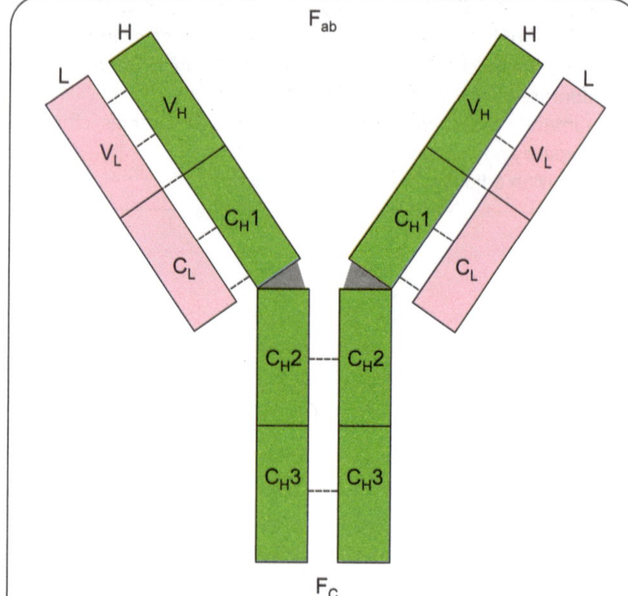

L = light chain (two subtypes kappa κ **or** lambda λ): domain
V_L = variable region of light chain
C_L = constant region of L chain domain
H = heavy chain (G, A, M, D, E)
V = variable, C = constant
V_H domain = variable region of H chain (amino terminal) consists of a hyper-variable region (idiotype) that combines with the antigenic epitope.
C_H1 domain = constant region of H chain responsible for the isotype and allotype specificity of Ig molecules.
C_H2 domain = constant region of H chain site for complement fixing and activation
C_H3 domain: constant region of H chain F_c receptor-binding site (carboxyl terminal)
F_{ab} = antigen-binding fragment of Ig
F_c = F_c receptor-binding fragment of Ig (C = crystallisable) on cold storage.

Fig. 12.1 Structure of an immunoglobulin monomer

Antibody **affinity** refers to the strength with which the epitope binds to an individual paratope (antigen binding site) on the antibody. **Avidity** gives a measure of the overall strength of an antibody-antigen complex. These properties are utilised in antigen-antibody assays. High-affinity antibodies bind quickly to the antigen and permit greater sensitivity in assays. Low-affinity antibodies, in contrast, bind weakly to the antigen and often do not detect the antigen in vivo or in assays.

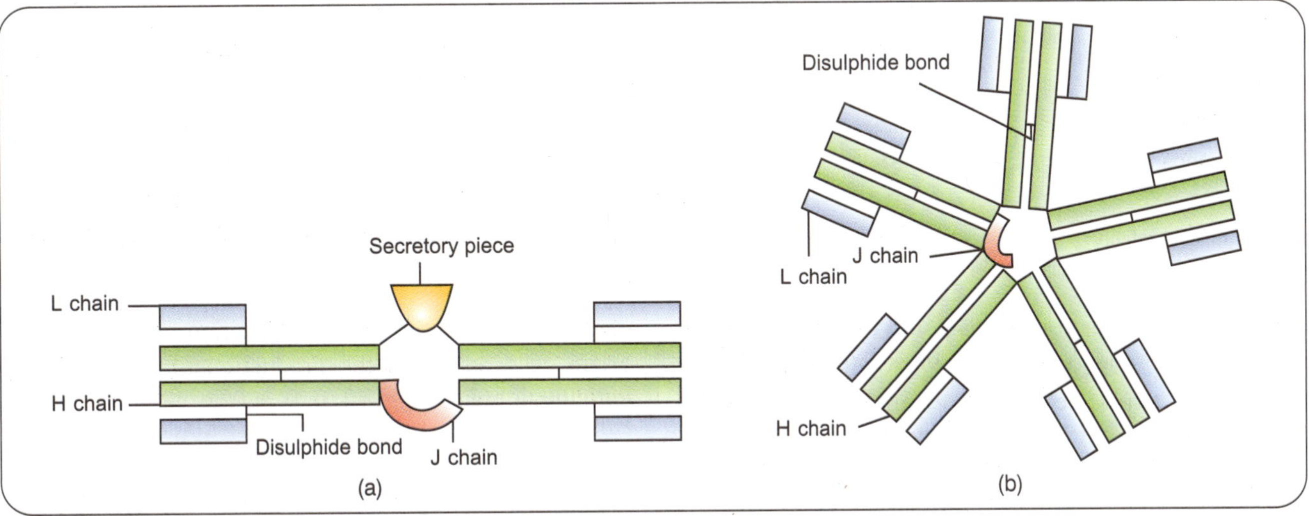

Fig. 12.2 Structure of immunoglobulin isotypes: (a) Secretory IgA molecule, (b) IgM molecule

On specific antigenic stimulation, B cells are transformed to plasma cells that secrete epitope-specific immunoglobulins. These immunoglobulins form the humoral immune response against an infection and help to control the infection in any of the following effector mechanisms:

1. **Neutralisation of free antigens or toxins:** Antibodies have the ability to neutralise free antigens (bacterial exotoxins).
2. **Immunoglobulins can combine with target cell-associated antigens:** This in vivo antigen-antibody reaction activates the classical complement pathway to lyse the target cell.
3. **Opsonisation:** Immunoglobulins function as opsonins and thereby enhance phagocytosis by macrophages. Opsonin is any molecule that enhances phagocytosis by marking an antigen for an immune response. The ability of antibodies and complement molecules to coat antigens for enhanced recognition by antibody or complement receptors on phagocytes is **opsonisation**.
4. **Antibody-dependent cell cytotoxicity (ADCC)** (Fig. 12.3): Non-specific cytotoxic cells like neutrophils, NK cells, macrophages and eosinophils possess F_c receptors for immunoglobulins. Thus, a target cell expressing the reciprocal antigen is rendered amenable to killing by these cells as the antibody on the F_c receptor recognises the antigen.

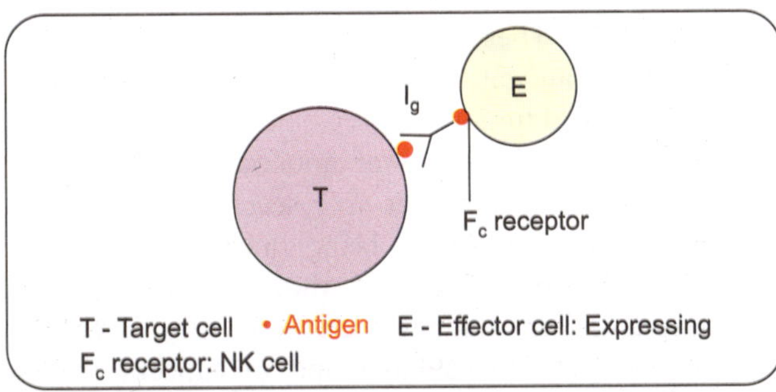

Fig. 12.3 Antibody-dependent cell cytotoxicity (ADCC)

A tumour cell or virus-infected cell expressing an antigen can be coated with a specific antibody and made amenable to killing by non-specific cytotoxic cells expressing F_c receptors.

Antibodies can also result in disease when formed against self-antigens, as seen in autoimmune disorders or due to activation of the complement subsequent to the formation of immune complexes.

Classes of immunoglobulins

The salient properties of the classes of immunoglobulins are depicted in Table 12.1. The role of each class in health and disease is as follows:

IgG: It is the only antibody to cross the placental barrier and is also present in breastmilk. It plays a role in protecting the fetus and newborn from infection. Therefore, the detection of IgG in neonates or infants does not imply infection of the child.

In case of an infection, IgG appears after IgM but persists for a long time, maybe lifelong. Therefore, the presence of IgG may denote either active or past infections. If used to diagnose active infections, it is important to determine the antibody titre (significant or not) or demonstrate a rise in the IgG levels with a paired serum test (to be repeated after 7–10 days).

IgM: In infections, it is the first to appear but is short-lived. Therefore, if present, it denotes active or recent infection. IgM does not cross the placental barrier. Its presence in a neonate denotes infection of the neonate.

IgA: It is secreted across the mucosal barriers (colostrum, saliva, tears), providing localised immunity. It is predominantly found in secretions (respiratory,

Table 12.1 Classes of immunoglobulins

	IgG	IgA	IgM	IgE
Molecular weight	150,000	160,000	900,000	190,000
% in serum	75	15	10	0.004
Serum concentration (mg/dL)	1000	200	120	0.05
Associated cell	B cell	B cell	B cell	Mast cell
Predominant activity	Neutralising	Mucosal	Primary response	Type I hypersensitivity

vaginal, urethral, gastrointestinal, oral) on infection of the mucous membrane. A small amount is found in serum.

IgE: This is a mast cell-associated antibody that plays a role in immediate, IgE-mediated hypersensitivity (atopy, anaphylaxis) and immunologic defence against parasitic infestations.

IgD: It is expressed when the B cell exits from bone marrow and signals the B cells to be activated.

Kinetics of antibody response (Fig. 12.4)

Following an acute infection, the antigen is the first to appear. The antigen disappears as IgM begins to appear. The IgM is the first antibody to appear but is short-lived. The IgG antibody appears after IgM, peaks and then declines, but may persist for life.

Terminology of antibodies

- ***Polyclonal antibodies***

Multi-specific antibodies arising from more than one cell, e.g., antibodies in convalescent sera or those raised in an in vivo animal model (rabbit, horse).

- ***Monoclonal antibodies***

Monospecific antibodies arising from a single cell (clone) have been developed in vitro by the hybridoma technique.

Hybridoma technique: Specifically sensitised B cells are fused with myeloma cell lines and grown in tissue culture medium. The continuous multiplication of the myeloma origin cells results in the production of mono-specific immunoglobulins (Fig. 12.5).

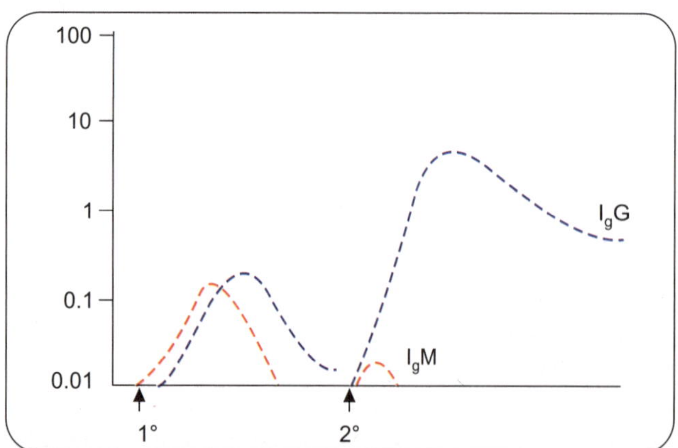

Fig. 12.4 Kinetics of antibody response

- ***Humanised/chimera antibodies***

They are immunoglobulins designed with F_{ab} fragments from mice, and the F_c portion from humans. The F_{ab} fragment can have a desired antigenic specificity.

- ***Bence Jones proteins***

Monoclonal light chains present in the urine of patients with paraproteinemic disorders like multiple myeloma.

Nomenclature

The nomenclature of monoclonal antibodies is generic and defined by the WHO. All monoclonal antibody names end with the stem '*mab*', preceded by *sub-stems*. The sub-stem is derived from the source of the *mab*—mouse (*o*), chimeric (*xi*), humanised (*zu*), chimeric humanised (*xizu*) or human (*u*) and the target of the *mab,* e.g., lymphocyte (*li*), nervous system (*ner*), circulatory system (*cir*) or musculoskeletal system (*mu*). The chimeric antibodies are 34% mouse and 64% human while the humanised ones are 90% human.

The sub-stems have a prefix that has no meaning but specifies the drug and may have an additional suffix if another substance is attached to the *mab*—either polyethylene glycol, a cytotoxic agent (to bind to cancer cells), or a chelator (to bind to radioisotope). Thus, "tocilizumab" comes from "toci-li-zu-mab", and is an anti-lymphocyte, humanised monoclonal antibody.

Applications of monoclonal antibodies

1. ***Diagnostic***: These antibodies are used as specific antibody conjugates to detect specific antigens in infections and tumours.
2. ***Therapeutic***: The use of monoclonal antibodies as therapeutic molecules is researched extensively, with newer molecules being increasingly added to its repertoire. In general, they are used in organ transplantation (kidney rejection), in infectious diseases (anti-interleukins) and in cancers (breast cancer, lymphoma) as immunomodulators.

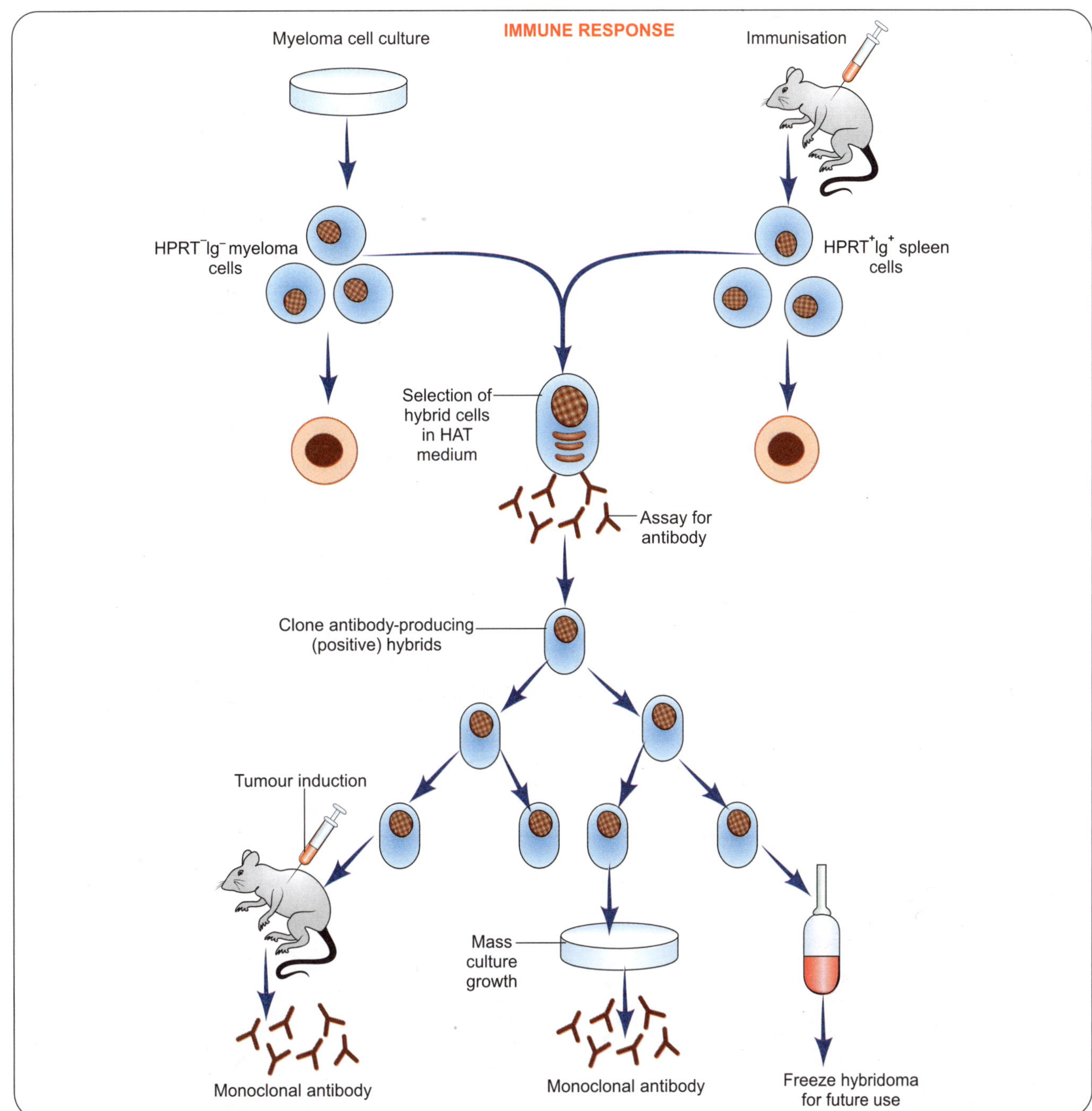

Fig. 12.5 Monoclonal antibody production by hybridoma technique

3. ***Research***: Specific monoclonal antibodies are used to characterise specific surface antigens on immunoreactive cells.

THE COMPLEMENT SYSTEM

The meaning of the word "humoral" is "molecules in solution in a body fluid". Humoral effector mechanisms, though activated by or dependent on the formation of antibodies, exist inertly in blood. The humoral effector systems are: the complement system, the kinin system and the properdin system.

Jules Bordet discovered a heat-labile component of serum and named it 'complement' for its ability to complement the antimicrobial properties of antibodies. The complement system comprises 30 protein molecules present in the circulation in an inactive form. These proteins are secreted by the liver, spleen and intestinal epithelium. The complement system is constitutive and in the immunological sense non-specific, i.e., it does not vary with antigenic specificity. It has a role to play in innate as well as adaptive/acquired immunity.

Nomenclature of the complement system

These protein molecules are named numerically from C1 to C9 with alphabetically denoted break-up molecules. For e.g., C3 is cleaved into C3a and C3b. These molecules interact with specified cells via the presence of complement receptors on the cells, e.g., CR2 on B cells.

Activation of the complement system can occur in three ways (Fig. 12.6)

1. The **'classical' pathway** that is dependent on a specific antigen-antibody interaction and therefore seen predominantly in acquired immune response as it is activated by the presence of antibodies.
2. The **'alternative' pathway** that can be directly activated by substances like the bacterial lipopolysaccharide and other bacterial cell surface molecules in the absence of antibodies.
3. The **mannose lectin binding pathway,** wherein mannan-binding lectin (MBL), a protein produced by the liver, initiates the complement cascade by binding to pathogen surface sugars that are components of microbes like *Salmonella, Neisseria, Listeria,* HIV and *Cryptococcus.*

Activation of the complement system results in cleaving and activation of the complement molecules as a cascade. Irrespective of whether the complement cascade is activated by the classical or the alternative pathway, both finally result in the formation of a **'membrane attack complex'.** This membrane attack complex has the ability to pierce the bacterial cell wall and cause cell lysis. This mechanism of killing bacteria is seen predominantly in capsulated bacteria like *Neisseria* and enveloped viruses that are otherwise known to resist phagocytosis.

Functions of the complement

1. Complement causes lysis of the microbial cell (especially of capsulated bacteria and enveloped viruses) by the membrane attack complex.
2. Complement molecules can act as 'opsonins' and enhance phagocytosis by macrophages. Opsonin is any molecule that enhances phagocytosis by marking an antigen for an immune response. The ability of antibodies and complement molecules to coat antigens for enhanced recognition by antibodies or complement receptors on phagocytes is "opsonisation".
3. The C5a component is chemotactic for neutrophils and monocytes, while C3a is a potent anaphylatoxin.

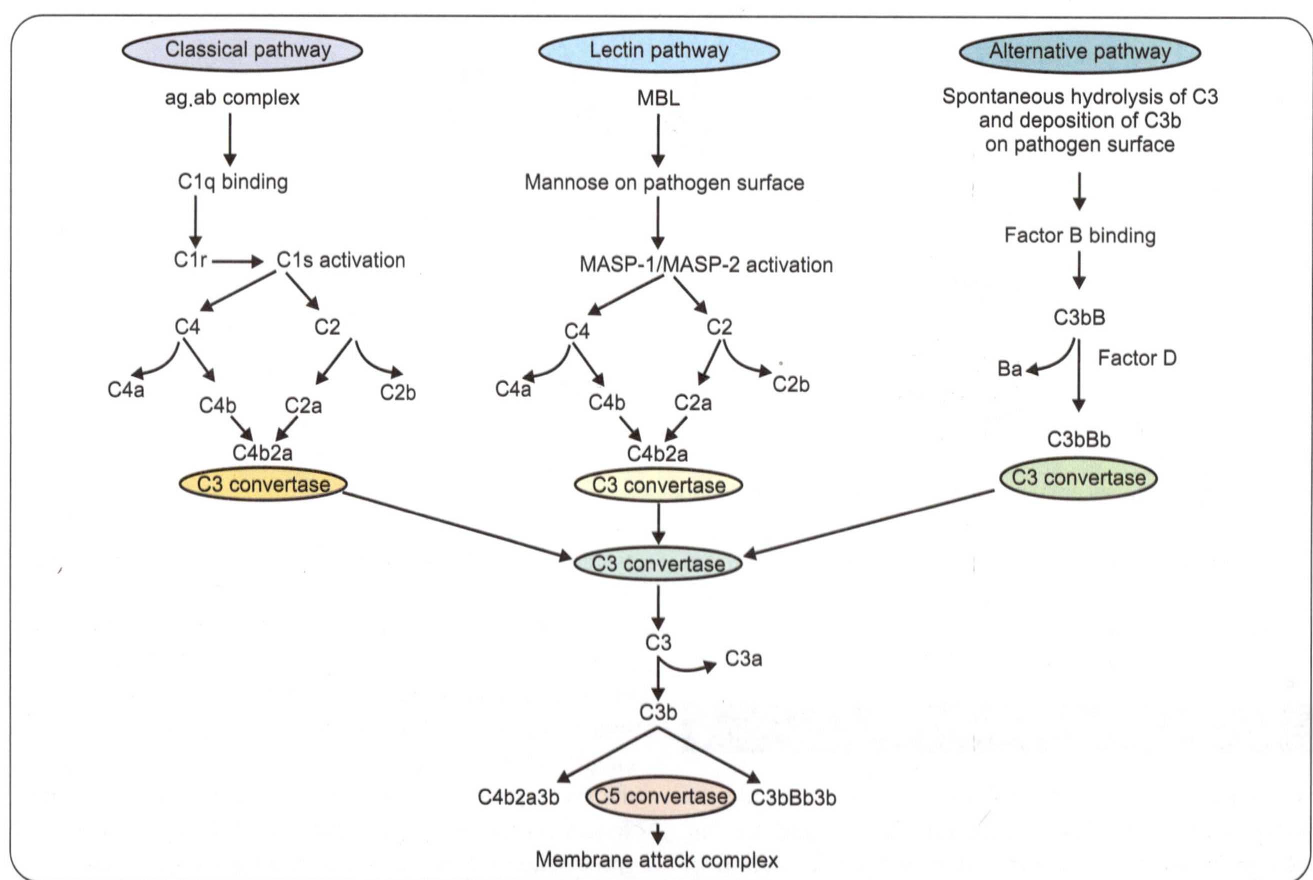

Fig. 12.6 Activation of the complement system can occur in three ways

4. It promotes the inflammatory response and increases the vascular permeability at the site of infection.

Complement component deficiencies

1. Increased susceptibility to pyogenic infections, especially fulminant infections with *Neisseria meningitidis,* is associated with deficiency of C3.
2. Immune complex syndromes, including systemic lupus erythematosus, glomerulonephritis and vasculitis, are associated with deficiency of C1.

The complement system is programmed to destroy and is therefore strictly regulated by serum regulatory proteins that inhibit the components of both the pathways.

Misregulation of the complement cascade results in

1. Recurrent pyogenic infections
2. Hereditary angioneurotic edema
3. Paroxysmal nocturnal hematuria

Complement levels in blood can be measured by complement assays

1. **CH_{50} (complement hemolytic) assay:** The hemolytic activity (amount of complement that lyses 50% of RBCs) is determined and interpolated with the concentration of the complement.
2. **Immunoassays:** ELISA can measure the level of specific complement components.

Questions

1. Draw, label and explain the structure of an immunoglobulin monomer, listing the function of its domains.
2. Define "affinity" and "avidity" of an antibody.
3. What are the effector mechanisms by which antibodies play a role in the immune response?
4. Explain the functions of each class of immunoglobulins.
5. Why is IgM the first antibody to appear? Explain the rationale behind the kinetics of an immune response.
6. What are monoclonal antibodies? Cite their applications.
7. State the mechanisms of activating the complement system.
8. What is the complement system and how does it function?
9. Write a brief note on complement deficiency and its misregulation.

The Immune System

The immune system has central lymphoid organs, peripheral lymphoid organs and cells that are functional in immune response.

> **The organs of the immune system are:**
> **Central (Primary):** Bone marrow, thymus
> **Peripheral (Secondary):** Lymph nodes, mucosa-associated lymphoid tissue (MALT) and spleen.

All immunocompetent cells arise from their progenitor stem cells in the fetal liver and the bone marrow of adults. The thymus is responsible for maturation and education of T lymphocytes, also referred to as "thymocytes". In embryonic life, self-reactive T cells are either destroyed or rendered non-reactive in the thymus before being released into circulation, and it is this central tolerance that prevents autoimmunity.

The immune system is organised as lymphoid follicles in its peripheral organs. The lymph nodes consist of primary lymphoid follicles of resting B cells prior to any antigenic challenge. On antigenic challenge, these B cells proliferate into secondary lymphoid follicles with a germinal centre. The spleen contains T and B cells and is a site of destruction of microbes. Splenectomised individuals or splenic defects result in an increased propensity towards infection with capsulated bacteria and need to be vaccinated. The mucosa-associated lymphoid tissue (MALT) is found in the intestines (gut-associated lymphoid tissue, GALT), the bronchus (BALT) and the tonsils.

> The mucosal immune system consists of:
> 1. Lymphoid follicles situated in the mucosa (MALT).
> 2. Mucosal B cells that differentiate into IgA plasma cells. IgA is the predominant immunoglobulin to be secreted across the mucosal barrier. IgA is also found in circulation.
> 3. Specialised epithelial cells called M cells that are capable of endocytosis of enteric pathogens, such as *Salmonella, Shigella, Vibrio* and poliovirus.
> 4. Mast cells with IgE receptors mediating immediate hypersensitivity on contact with allergen at the mucosa.
> 5. Non-immunologic factors, such as resident bacterial flora, presence of a ciliated epithelium, mucus secretion and gastric acidity that prevents infection.

CELLS OF THE IMMUNE SYSTEM

1. Lymphocytes

Both T and B cells are differentiated from the common lymphocyte progenitor which in turn arises from hematopoietic stem cells. There exist a total of 2×10^{12} lymphocytes distributed in tissues and circulation (only 1%). In the peripheral blood, 70–80% are T cells, 10–15% are B cells and 10–15% are NK cells.

Depending on their immune function and presence or absence of specific receptors, the lymphocyte lineage is further divided into types and subsets. Those lymphocytes that develop in the thymus are called **T lymphocytes** or **T cells.** The other major lymphocyte population—the B cells—are so named because, in birds, their development takes place in an organ called the bursa (B) of Fabricius. As humans do not possess this bursa, B cell development probably occurs in hemopoietic tissue (bone marrow) and in secondary lymphoid organs.

A **clone** is a lineage of cells arising from a single cell. A **'receptor'** is a functional macromolecule present on the cell surface. The functional macromolecule (receptor) expressed on these cell surfaces (surface marker) is denoted as "CD" (cluster of differentiation) along with a requisite numerical. The development of T cells begins in the bone marrow. The pre-T cell enters the thymus for differentiation and leaves the thymus to reside in the lymph node. In the thymus T cell, selection occurs in such a way that positive selection ensures MHC restriction and negative selection ensures self-tolerance.

The diversity of T and B cells is the consequence of genetic recombination that occurs in the development of these cells in the thymus and bone marrow, respectively. Through a process known as V(D)J recombination, the variable (V), junction (J) and diverse (D) genes are rearranged in a random manner to produce a repertoire of antigen-binding immunoglobulins and T cell receptors (TCRs) for all pathogens.

Dendritic cells bring the antigen to the lymph node. On contacting an antigenic epitope of a specific microbe, there occurs clonal selection of a specific lymphocyte followed by clonal expansion by mitotic division of the cells.

The types of lymphocytes are T cells, B cells and natural killer (NK) cells. The subsets of T cells are helper, cytotoxic, regulatory and memory T cells.

T cells

All mature T cells possess a T cell receptor (TCR) called CD3. Helper T cells (T4) bear the CD4 molecule whereas cytotoxic T cells (CTL) bear the CD8 molecule (Table 13.1). T cells are able to recognise an antigenic epitope only when complexed with the MHC molecule. The subsets of T cells are:

Helper T cells/CD4+ cells

T helper cells (T_H) recognise antigenic epitopes when complexed with MHC II molecules by antigen presenting cells (APCs) (Table 13.1). T_H, on activation, generates either T_H1 or T_H2 response. The T_H1 response involves interleukin (IL)2 and interferon (IFN)γ and promotes cell-mediated immunity by killing intracellular pathogens. The T_H2 response involves IL4, IL5 and IL6 and stimulates B cells to produce immunoglobulins.

Cytotoxic T cells/CD8+ cells

Cytotoxic T cells (T_c) recognise antigenic epitopes when complexed with MHC I molecules by any nucleated cells. They destroy virus-infected cells and tumour cells (Table 13.1).

Regulatory T cells (TREGs or TREG)/CD4+ CD25+

Regulatory T cells (TREGs or T_{REG}), formerly called T suppressor cells, are a specialised sub-population of T cells that suppress immune response and prevent autoimmunity by peripheral tolerance.

Γδ T cells/CD4– CD8–

They are a part of innate immunity and found in the gut mucosa. Their exact function is unknown.

Memory T cells

T_m or T memory cells exist in both CD4 and CD8 cells.

B cells

B cells express the F_c receptor and thus possess an immunoglobulin molecule on their surface (Table 13.1). With T cell help, B cells differentiate into plasma cells that secrete immunoglobulins. They also differentiate into memory B cells.

NK cells

Within the lymphocyte there exists a sub-population that has the capacity for spontaneous cytotoxicity (independent of antigen recognition) towards various targets cells, especially virus-infected cells and tumour cells (Table 13.1) These lymphocytes possess CD16 and CD56 and are termed natural killer (NK) cells. They are a part of the innate or constitutive defence mechanisms. NK cell activity is augmented by interferons. CD16 is an F_c receptor for IgG that confers the NK cell its ability to kill IgG-coated target cells. This phenomenon is called **antibody-dependent cell cytotoxicity** (ADCC).

2. Macrophages

Cells of the mononuclear phagocytic system (MPS) play a role in constitutive and in acquired immunity. Their role in innate/constitutive immunity is phagocytosis (Table 13.1). In acquired immunity, macrophages function as APCs, as they possess the class II MHC. Macrophages may be circulating (monocytes) or fixed (tissue macrophages). Activated macrophages also secrete a variety of cytokines (IL1, IL6, TNFα, IFNα and β and colony stimulating factors).

Table 13.1 The cells of the immune system

Cell		Surface marker	Function
Lymphocytes	T cell	CD3 (TCR)	Cell-mediated immunity
	T_H	CD4	Recognises antigen with class II MHC. Promotes differentiation of B cells and T_c. Activates macrophages.
	T_c	CD8	Recognises antigens with class I MHC Kills antigen-expressing cells
	B cell	F_c receptor, Ig Class II MHC	Recognises antigen directly Differentiates into plasma cell Antigen presentation
	NK cell	CD16, CD 56	Non-specific target cell killing, ADCC*
Macrophages		Class II MHC F_c receptor	Phagocytosis (Innate) Antigen presenting cells (Acquired) ADCC
Dendritic cells		Class II MHC	Antigen presenting cells
Mast cell		IgE receptor	Allergic response
Neutrophils		F_c receptor	Phagocytosis of bacteria, fungi, ADCC
Eosinophils		F_c receptor	Phagocytosis and elimination of parasites, ADCC
Basophils		IgE receptors	Allergic response

*Antibody-dependent cell cytotoxicity

3. Dendritic cells

They are located in all tissues that are in contact with the external environment, like the skin (Langerhans cells) and nasal, respiratory and gastrointestinal mucosa. All dendritic cells possess class II MHC and function as the most potent antigen presenting cells (APCs) (Table 13.1). They capture the antigen at one location and present it to T cells at another location, thereby acting as peripheral sentinels.

Interstitial dendritic cells are found in visceral organs and interdigitating dendritic cells in the thymus. Follicular dendritic cells are found in lymphoid follicles within lymph nodes and interact with B cells. They do not express class II MHC but can recognise an antigen-antibody complex.

4. Polymorphonuclear cells

Neutrophils are capable of phagocytosis, eosinophils play a role in the elimination of parasites and basophils, along with mast cells, possess receptors for IgE and play a role in allergic responses (Table 13.1).

MAJOR HISTOCOMPATIBILITY COMPLEX (MHC) AND HUMAN LEUCOCYTE ANTIGEN (HLA)

Human leucocyte antigen (HLA) was first described by Snell, Dassuat and Benacerraf on the surface of leucocytes, due to which it got its name. The researchers were awarded the Nobel Prize in 1980 for this discovery. HLA molecules are the cell surface products expressed as a result of MHC genes found on the short arm of chromosome 6. Various allelic forms of MHC genes are known to exist, giving rise to polymorphism of HLA molecules between individuals within the human species.

There exist three classes of HLA molecules:

1. Class I, denoted as A, B and C with numerals

The HLA class I molecule has the ability to present an antigenic epitope to a T_8 cell. Recognition of an epitope by cytotoxic T_8 cells occurs only and only when the epitope exists in combination with an HLA class I molecule. They are present on ALL nucleated cells, therefore, no human cell can escape cytotoxic T_8 killing.

2. Class II, denoted as DP, DQ and DR with numerals

The HLA class II molecule has the ability to present an antigenic epitope to a T_4 cell. T_4 cells can recognise the epitope only and only when the epitope exists in combination with an HLA class II molecule. They are present on all APCs (macrophages, dendritic cells), B cells and activated T cells, therefore, antigen presentation to the T_4 MHC is restricted and controlled.

Nature has designed the genetic control of the immune response such that, if infected, all cells of the body are amenable to cytotoxic T cell killing, thereby ensuring protection. On the other hand, only APCs and B cells can present antigens, ensuring control of the immune response against self-antigens. It therefore follows that antigen recognition by T cells is MHC-restricted and depends on an individual's HLA.

3. Class III that codes for the certain properdin and complement system molecules

Role of HLA class I and II molecules in health and disease

1. In the normal immune response:
   Class I: Killing of target cells by cytotoxic T_8 following recognition of the (epitope + class I) complex on the target cell surface.
   Class II: Induction of immune response following recognition of the (epitope + class II) complex on APCs.
2. In transplantation immunology:
   - Donor class I molecules are recognised by host T_8 as foreign, causing graft rejection.
   - Donor T_4 recognises host class II molecules as foreign, causing graft versus host disease.
3. Disease association:
   Certain diseases (especially autoimmune disorders) show an increased incidence of occurrence in individuals with certain HLA molecules. In the presence of certain HLA types in an individual, the relative risk of certain diseases may be increased or decreased. Some examples of positive associations are:
   - Spondylo-arthropathies and HLA B27
   - Insulin-dependent diabetes mellitus (IDDM) and HLA DR3
   - Rheumatoid arthritis and HLA DR4
   - Multiple sclerosis and HLA DR2
   - Myasthenia gravis and HLA B8
   - Systemic lupus erythematosus (SLE) and HLA DR3, DR2

   An example of a strong negative association is between type 1 DM and HLA DQB1.
4. HLA allelic diversity increases the likelihood that a new pathogen will be recognised by some HLA molecule, maintaining host immunity. As a consequence of this diversity, some alleles become capable of recognising 'innocent bystander' molecules, including drugs or environmental molecules, to develop hypersensitivity,

e.g., clindamycin-induced cutaneous drug reaction and HLA B5101, dapsone hypersensitivity in leprosy and HLA B1301.

Role of macrophages in immune response

The macrophage plays an important role in both innate and acquired immunity. In the innate processes, macrophages are capable of non-specific (independent of antigen recognition) phagocytosis. In acquired immunity, the macrophage plays a myriad of roles.

Macrophage effector functions in the immune response

1. **Microbicidal effect:** The antimicrobial spectrum of macrophages exceeds that of neutrophils, although neutrophils kill more rapidly than macrophages. In the presence of T_H cells, macrophages become actively phagocytic cells organised into granulomas that are the hallmark of delayed hypersensitivity.
2. **Antigen presenting cells (APCs):** A large proportion of the antigens ingested by macrophages are degraded and rendered non-immunologic. Some are presented to the T_H cell in conjunction with surface class II MHC molecules. Unlike B cells, T cells can recognise antigens only when presented by macrophages. This recognition, besides the class II MHC molecule, also requires the presence of certain co-stimulatory signals between the T_H cell and the macrophage.
3. **Tumouricidal effect:** Macrophages can effect direct killing of tumour cells. They also secrete a cytokine called tumour necrosis factor (TNF), which is responsible for tumour necrosis.
4. **Immunomodulatory effect:** Macrophages secrete a number of cytokines that are known to exert specific (stimulatory or inhibitory) actions on other immunocompetent cells.

T cell interactions with antigen presenting cells

The main role of MHC is in mediating cell-cell communication. MHC molecules "hold" foreign antigens for recognition by appropriate corresponding T cells.

Class II MHC molecules are expressed on APCs, i.e., macrophages, dendritic cells and B cells. The T_H cell cannot recognise an antigen unless presented to it by an APC with a class II MHC molecule. Thus, the T cell recognition of an antigen is MHC-restricted. The APCs presenting antigens to T_H cells intracellularly follow the endocytic pathway within the cell.

Class I MHC molecules are expressed on all nucleated cells, including RBCs and platelets. Similarly, T_c cells cannot recognise antigens unless complexed with a class I MHC molecule. The intracellular pathway followed for antigen recognition by T_c cells is the cytosolic pathway.

Though phagocytosis is a constitutive defence mechanism independent of antigen recognition, macrophages play the important role of APCs in acquired immunity.

Since the MHC gene exists in allelic forms, the ability of an individual to process a particular antigenic epitope depends on the presence of a specific MHC. Unlike T cells, B cells can directly recognise antigens without the need for an MHC molecule but require T cell help to form antibodies.

Lymphocyte development and circulation

Lymphocytes are derived from the pluripotent stem cells in the fetal liver and the bone marrow of adults. In the T lymphocyte pathway, the cells enter the thymus. Thymic education allows these T cells to express receptors for antigens and to select appropriate specificities. Upon exiting the thymus, they travel to the peripheral lymphoid organ. Removal of the thymus at birth prevents the development of T cells.

From their peripheral bases, lymphocytes patrol the body continuously, re-circulating between the blood and lymph. This re-circulation promotes the contact of lymphocytes with antigens and ensures that information about a localised antigenic insult is dispersed throughout the body. Thus, systemic immunity is produced.

Consider a microbe that has penetrated the skin. Through the inflammatory response, the microbe will be swept into the afferent lymphatic present in all tissues (except the CNS and placenta). The microbe will be ingested by the MPS and this triggers the acquired immune response. The lymph exiting the node via the efferent vessel and ultimately emptying into the blood is microbe-free.

QUESTIONS

1. Enumerate the organs of the immune system and the role of each organ.
2. Explain the innate and acquired mechanisms of defence at the mucosa.
3. Elaborate on the T cell types and subsets, and their functions.
4. Compare and contrast T and B lymphocytes.
5. What is the role of dendritic cells in the immune response?
6. Explain how the genetic constitution of an individual affects the immune response.
7. What is the role of HLA in health and disease?
8. Discuss lymphocyte development, circulation and function.
9. Discuss the role of macrophages in normal immune response.

Chapter 14

Section II: Immunology

The Immune Response

The normal acquired immune response entails a cell-mediated and an antibody-mediated humoral response that are interdependent (Fig. 14.1).

Cell-mediated response

A macrophage or dendritic cell that has innately phagocytosed the microbe functions as an antigen presenting cell (APC). The antigen-processing pathways in the APCs are either endocytic or cytosolic. In the endocytic pathway, exogenous antigens are presented to CD4 cells along with class II MHC. In the cytosolic pathway, endogenous antigens are presented to CD8 cells along with class I MHC. The most important APCs are dendritic cells, macrophages and B cells.

Complexed with the class II MHC molecule, the antigen is presented to the appropriate T_H cell via the TCR or CD3 for recognition (signal 1). An additional signal 2 is provided by co-stimulatory molecules binding to CD28 on T cells and IL1 secreted by macrophages. Only on receiving both signals is the T_H cell activated to secrete cytokines that activate T_c and B cells. Following activation of T_H cells, three immunologic consequences occur:

1. Activation of T_c (cytotoxic T lymphocyte, CTL): T_c activation requires the CD8-MHC I signal, the CD28-B7 costimulatory signal and IL2. The T_c cell thus activated has the ability to kill the target cell on recognition of the specific antigen class I MHC molecule on the target cell. The activated T_c produces the enzymes perforins and granzymes to cause target cell lysis.
2. Though B cells can directly interact with antigens, T_H help is rendered to the B cell (T and B cell collaboration via chemokine receptors) for its conversion to plasma cell and to produce specific antibodies. Specific antibodies bind to the corresponding antigen and clear it.
3. Some T_H cells differentiate in T_H memory cells. These memory cells are responsible for secondary immune response when the same antigen is encountered much later in time.

In the event of occurrence of signal 1 in the absence of signal 2, the T cell either dies (apoptosis) or becomes

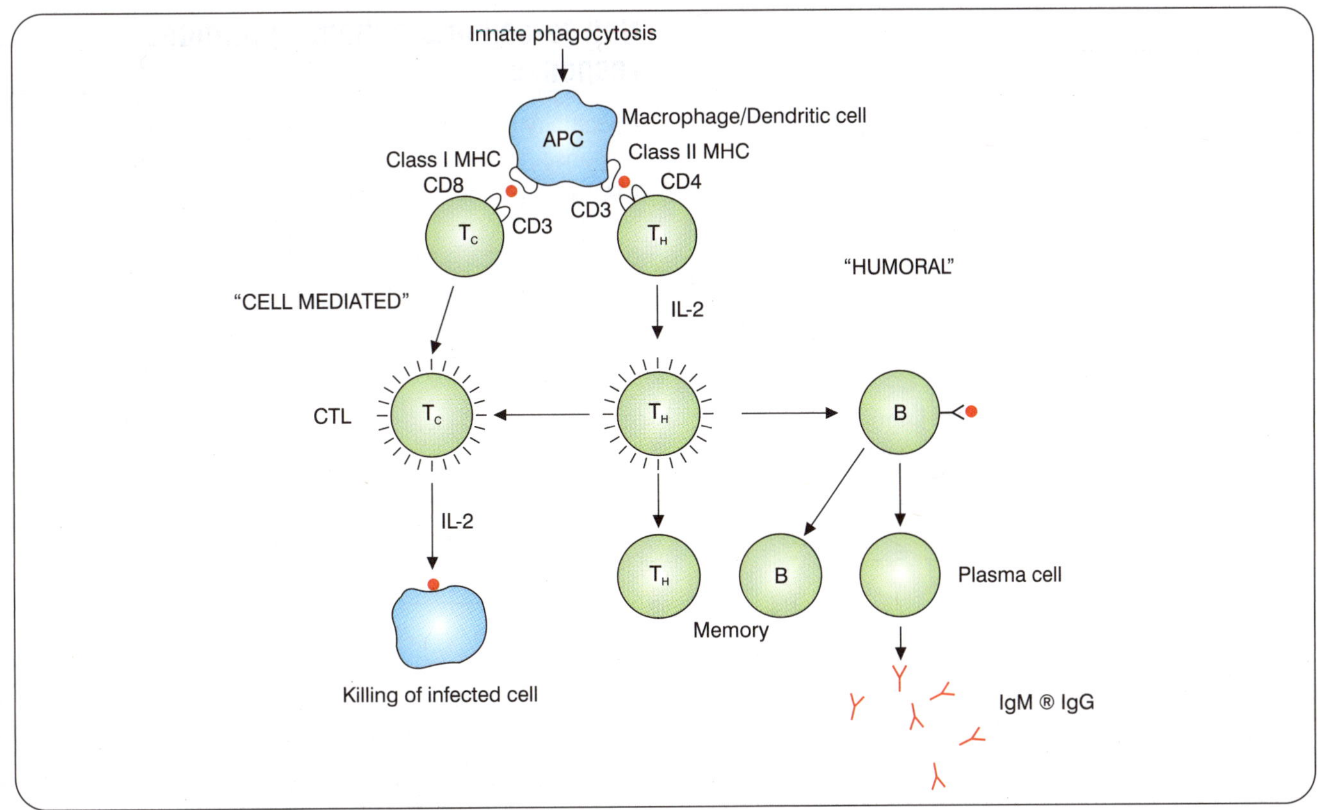

Fig. 14.1 The acquired immune response

anergic or non-responsive to that antigen.

Humoral response

B cells can act as APCs and recognise antigens directly but require T_H help in converting to plasma cells. As the predominant B cell surface immunoglobulin is IgM, the first antibody to appear is IgM followed by IgG and IgA due to isotype switching. A subset of B cells differentiate into memory B cells, awaiting antigenic challenge later in life.

In general, cell-mediated immunity plays a larger role against intracellular pathogens, especially viruses, whereas antibodies are directed at extracellular, free microbes and cell surface antigens.

Besides cytotoxic T cell lysis, target cells can be lysed by two other mechanisms:

1. NK cell-mediated cytotoxicity, which is a part of innate immunity, especially against virus-infected cells.
2. Antibody-dependent cell cytotoxicity (ADCC) (Fig. 14.2) exhibited by NK cells, macrophages, neutrophils and eosinophils as they express F_c receptors on their surface. On coming in contact with a target cell coated with antibody, these cells bind to the F_c fragment of that antibody and release cytotoxic enzymes that lyse the target cell. ADCC plays a role in eosinophil/IgE-mediated immune response in helminths and NK cell/IgG-mediated response to tumours and virus-infected cells. It is independent of the complement system.

Regulation of the immune response

The immune response to foreign antigens, once initiated, must be regulated. The major ways in which it is down-regulated are:

1. The functional life-span of effector cells is short and therefore, self-limiting. This self-killing of the effector cell is known as apoptosis or programmed cell death (PCD).
2. An efficient immune response eliminates the antigen, in turn subsiding the immune response.
3. Idiotypic network of antibodies: The binding site of the epitope to the Ig molecule is the idiotypic determinant in the hypervariable region of the immunoglobulin. This idiotypic determinant too can be antigenic and result in the formation of an anti-idiotype antibody that plays a role in the down-regulation of the first antibody.
4. T_H2 response: On the recognition of the antigen, the T_H cell, via cytokines, is activated. Depending on the cytokines (IL2 or IL5), there exist two pathways of the T_H cell—T_H1 and T_H2. The T_H1 pathway leads to the CTL killing of target cells whereas the T_H2 pathway results either in apoptosis or host tissue damage and delayed hypersensitivity to the antigen. This T_H2 pathway is also believed to play a role in the down-regulation of the immune response.

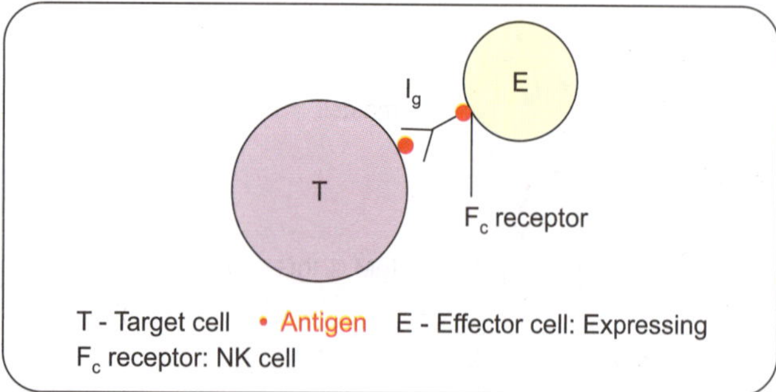

Fig. 14.2 Antibody-dependent cell cytotoxicity

Kinetics of immune response (Fig. 14.3)

The reaction evoked the first time an antigen is encountered is called the primary immune response, and that caused by subsequent encounters with the same antigen is called the secondary or anamnestic or memory response. The secondary response occurs after a shorter lag period and is of a greater intensity and longer duration. Immunologic memory is due to the fact that memory lymphocytes (T_H and B) are long-lived and can persist for years.

Cell-to-cell interactions in immune response

Specificity is the central principle of the acquired immune response. The immune response is not only directed towards specific antigens, it also shows specificity in

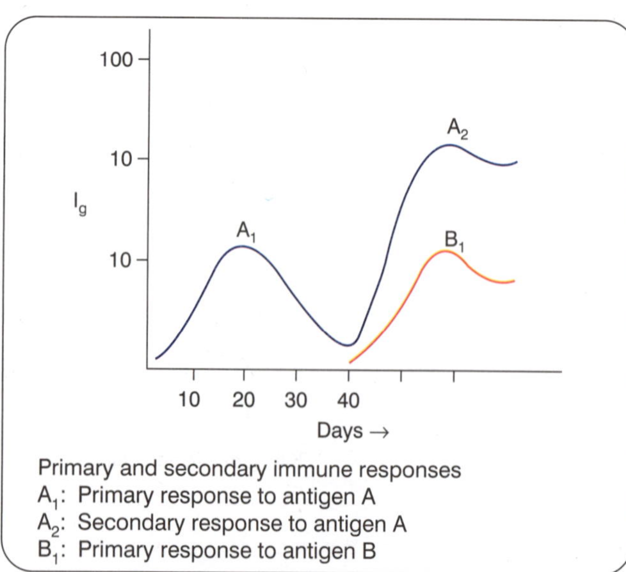

Fig. 14.3 Kinetics of immune response

cellular interactions. Besides interacting via specific cell surface receptors and co-stimulatory molecules, there are other means of cellular interactions.

An example is lymphocyte adhesion with endothelial cells or peripheral lymph nodes, or diversion of gut-derived plasma cells to breast connective tissue for the transfer of passive immunity from the mother to child. These cellular interactions take place via the presence of certain **'homing' receptors** present on cell surfaces ensuring correct trafficking of the cells. Cellular interactions occur by Toll-like receptors (TLRs), cell-adhesion molecules and high endothelial venules (HEVs), and the secretion of cytokines.

Toll-like receptors (TLRs) are a class of proteins that play a key role in the innate immune system and are usually expressed on macrophages and dendritic cells that recognise molecules from microbes. Once these microbes have breached physical barriers such as the skin or intestinal tract mucosa, they are recognised by TLRs, which activate immune cell responses. They are also expressed on non-immune cells, such as epithelial and endothelial cells, and fibroblasts.

Cell-to-cell interactions involve multiple ligands and **cell-adhesion molecules** (CAMs), which are a diverse group of integral membrane proteins. CAMs fall into five main classes—the cadherins, immunoglobulin-superfamily CAMs, selectins, mucins and integrins.

High endothelial venules (HEVs) are specialised endothelial cells found in the venules of lymph nodes and MALT, that enable the lymphocytes circulating in blood to directly enter a lymph node.

Cytokines (Table 14.1), secreted by lymphocytes and macrophages, are peptide mediators of the immunologic and inflammatory response. They are not usually present in the serum and are found only on stimulation of the immunocompetent cells by an antigen. They act in an autocrine (self-cell) or a paracrine (neighbouring cell) manner but never in an endocrine manner. They are specific in their action. For example, IFNγ secreted by the T_H1 cell activates macrophages while IL5 secreted by T_H2 inhibits macrophage activation.

> There exist various cytokines secreted by specific cells and having a specific function in the immune response. They are classified as:
> - Interleukin 1–16 (IL1 to IL16)
> - Granulocyte monocyte colony stimulating factors (GM-CSF)
> - Tumour necrosis factors α and β (TNFα and β)
> - Interferons α, β and γ (IFN α, β and γ) and others

Table 14.1 Cytokines

Cytokine	Main sources	Major functions
A. INTERLEUKINS		
IL1 (α and β)	Macrophages and other cell types	Proliferation and differentiation of T, B and other cells; pyrogenic; induce acute phase proteins; bone marrow cell proliferation
IL2	T cells	Promote growth and differentiation of T and B cells, cytotoxicity of T and NK cells, secretion of other lymphokines
IL3	T cells	Multi-CSF
IL4	T_h cells	Proliferation of B and cytotoxic T cells; increase IgG and IgE production; enhance MHC Class II and IgE receptors
IL5	T_h cells	Proliferation of eosinophils, stimulate IgA and IgM production
IL6	T_h, macrophages, fibroblasts	Promote B cell differentiation; IgG production, acute phase proteins
IL7	Spleen, bone marrow stromal cells	B and T cell growth factor
IL8	Macrophages, others	Neutrophil chemotactic factor
IL9	T cell	T cell growth and proliferation
IL10	T, B cells, macrophages	Inhibit IFN production and mononuclear cell functions
IL11	Bone marrow stromal cells	Induce acute phase proteins
IL12	T cells	Activate NK cells
IL13	T cells	Inhibit mononuclear cell functions
IL17	T_h 17 cells	Proinflammatory marker
B. COLONY-STIMULATING FACTORS		
GM-CSF	T cells, macrophages, fibroblasts	T cell and macrophage growth stimulation
G-CSF	Fibroblasts, endothelium	Granulocyte growth stimulation
M-CSF	Fibroblasts, endothelium	Macrophage growth stimulation

(Continued...

...continues)

Cytokine	Main sources	Major functions
C. TUMOUR NECROSIS FACTORS		
TNFα	Macrophages, monocytes	Tumour cytotoxicity, lipolysis, wasting, acute phase proteins, phagocytic cell activation, antiviral and antiparasitic effects, endotoxic shock
TNFβ	T cells	Induce other cytokines
D. INTERFERONS		
IFNα	Leucocytes	Macrophage activation, augmentation of monocyte and neutrophil function and anti-tumour activity
IFNβ	Fibroblasts	Antiviral activity
IFNγ	T cells	Antiviral, macrophage activation; MHC Class I and II expression on cells
E. OTHERS		
TGFβ	T and B cells	Inhibit T and B cell proliferation and hematopoiesis; promote wound healing
LIF	T cells	Proliferation of stem cells; eosinophil chemotaxis

Questions

1. Draw and label a diagram and explain the events in an acquired immune response.
2. Explain the three mechanisms of cytotoxic cell lysis.
3. How is the immune response down-regulated?
4. Explain the mechanisms by which specificity is maintained in cell-to-cell interactions in the immune response.

Hypersensitivity

The term **hypersensitivity reactions** is used for immune responses that occur in an exaggerated or inappropriate manner. They occur in individuals who have been previously sensitised (priming dose) to an antigen and are therefore secondary responses to a shocking dose of the same antigen. Hypersensitivity can occur to exogenous environmental antigens or endogenous self-antigens. They are often associated with the inheritance of particular susceptibility genes (HLA and non-HLA). The mechanisms of hypersensitivity follow the same effector mechanisms as the normal immune response. In 1963, based on the speed of the reaction and the immunologic pathway, Coombs and Gell classified hypersensitivity into four types. Types I, II and III are antibody-mediated whereas type IV is T cell-mediated.

Type I or immediate hypersensitivity

Clinical Case 1

A 30-year-old lady on a forest trip suddenly develops sneezing and congestion of the eyes and nasal mucosa, with itching all over the body. On arrival to the emergency room, she has tachycardia, tachypnea and a systolic blood pressure of 90 mm Hg.

1. What is the most likely diagnosis?
2. Explain the immunologic basis of this condition.
3. Provide three other examples of localised disease with the same immunologic basis.
4. What is the drug of choice to relieve her distress?

Type I hypersensitivity is also called an allergic reaction, the antigens involved being initiator 'allergens', and occurs within minutes of exposure to the antigen. First exposure to environmental antigens stimulates T_H2 responses and IgE production in genetically susceptible individuals. This IgE binds to the F_c receptors on the mast cells in sensitised individuals. The IgE sensitises the mast cells for degranulation. Re-exposure to a minute quantity of the same antigen results in the antigen binding to the IgE antibody on the surface of a mast cell and triggers the release of histamine following degranulation of the mast cell (Fig. 15.1).

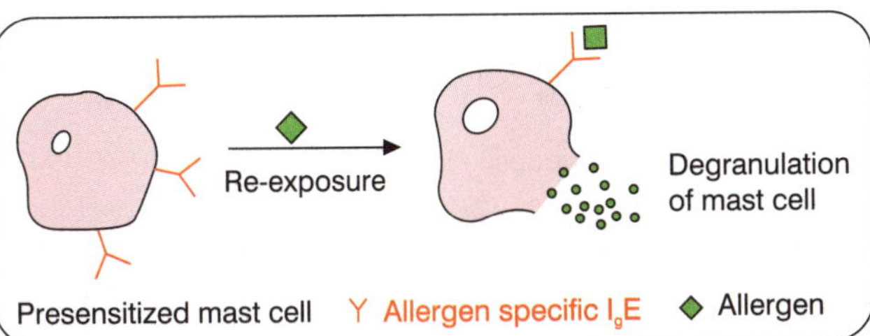

Fig. 15.1 Type I hypersensitivity: IgE-mediated

This results in the release of several pharmacologic mediators with various effects. Primary mediators such as histamine and serotonin increase vascular permeability and smooth muscle contraction. Secondary mediators such as prostaglandins, leukotrienes and bradykinin cause vasodilation, as well as an increase in vascular permeability and smooth muscle contraction. Cytokines cause systemic anaphylaxis, characterised by vascular shock, widespread edema and difficulty in breathing. The short time lag is due to the presence of preformed mediators in mast cells (Table 15.1).

Examples:

1. **Atopy:** This refers to an inherited/familial propensity to respond to naturally occurring allergens (inhaled or ingested), e.g., allergic rhinitis, allergic asthma, food allergies and atopic dermatitis usually involve a specific target site/organ. Atopy can be recurrent but mostly is not fatal. Allergy testing by skin tests or the detection of serum IgE levels can be done.
2. **Systemic anaphylaxis:** Non-atopic (lacking a genetic predisposition or specific target organ), this reaction occurs in multiple organs, leading to anaphylactic shock, e.g., drug allergy. It is characterised by bronchospasm due to smooth muscle contraction and hypotension due to vasodilation. It can be fatal if not treated immediately with epinephrine, that counters the effects of the mediators of anaphylaxis.
3. **Urticaria angio-edema:** This is a cutaneous form of anaphylaxis.
4. **Casoni's test:** Intradermal test for the diagnosis of hydatid disease.

Anaphylactoid reaction is a non-immune, non-IgE-mediated direct release of mast cell mediators without prior sensitisation, e.g., radio contrast media reaction.

Type II or antibody-mediated cytotoxicity

Clinical Case 2

A 40-year-old man admitted to a trauma unit following a road traffic accident has massive blood loss. The doctor on duty requests for the patient's A+ group for blood transfusion. On receiving the transfusion, the patient develops fever, chills and hematuria.

1. *What is the most likely diagnosis?*
2. *Explain the immunologic basis of this condition.*
3. *Provide three other examples of localised disease with the same immunologic basis.*
4. *How could the above condition have been prevented?*

Antibodies that react with cell surface or extracellular antigens cause disease by destroying the cells and triggering inflammation (Table 15.1).

The mechanisms by which antibodies (IgG and IgM) mediate injury are:

1. Activation of the complement pathway, e.g., erythroblastosis fetalis (Rh incompatibility) and lysis of RBCs or platelets in autoimmune hemolytic anemia or autoimmune thrombocytopenic purpura, respectively. Erythroblastosis fetalis can be prevented by identifying Rh-negative mothers in early pregnancy and giving RhoGAM that can destroy anti-Rh antibodies that enter her circulation following an Rh-positive delivery. This helps protect future Rh-positive babies.
2. Opsonisation and phagocytosis.
3. Antibody-mediated damage of specific cell receptors, resulting in cellular dysfunction (antibody-dependent cellular dysfunction, ADCD). For example, the TSH receptor in Grave's disease results in hyperthyroidism and the ACh receptor in myasthenia gravis. This mechanism is, by some, referred to separately as type V hypersensitivity.
4. Antibody-dependent cell cytotoxicity (ADCC) (Fig. 15.2). IgG antibodies bind the F_{ab} end to cell surface antigens on the target cell and F_c end to the F_c receptor on effector cells—NK cells, macrophages, neutrophils or eosinophils.

Type III or immune complex-mediated hypersensitivity

When an antigen and antibody combine, an immune complex is formed that activates the complement system, resulting in tissue damage by eliciting inflammation at the sites of deposition (Table 15.1).

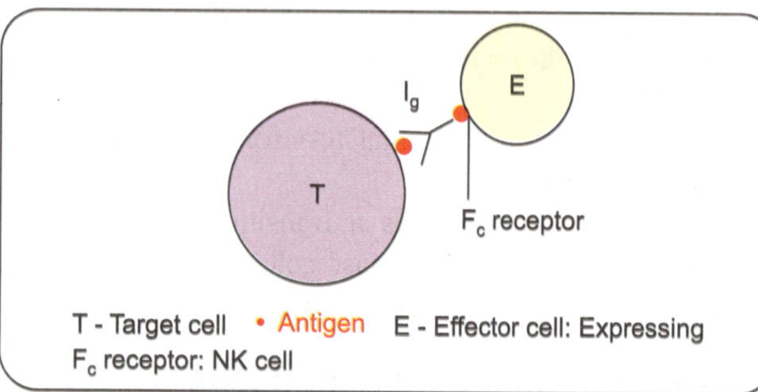

Fig. 15.2 Type II hypersensitivity: antibody-dependent cell cytotoxicity

Clinical Case 3

A 17-year-old boy presents to the emergency room with dark-coloured urine, anasarca (generalised edema) and a decreased urine output. He gives a history of cough with sore throat 3 weeks back that was treated with oral antibiotics. On examination, his systolic blood pressure is 160 mm Hg; he has pedal edema and his renal function tests are deranged.

1. *What is the most likely diagnosis?*
2. *Explain the immunologic basis of this condition.*
3. *What is the causative organism in this case?*
4. *What laboratory test will confirm it?*

This is seen in the event of an excess in circulating immune complexes. The site of deposition of the immune complex dictates the type of pathology (Fig. 15.3).

Examples:

1. Systemic immune complex disease: Acute serum sickness following passive immunotherapy with anti-tetanus or anti-diphtheritic serum of foreign (horse) origin, characterised by arthritis, vasculitis and nephritis.
2. Local immune complex disease: Post-streptococcal glomerulonephritis following streptococcal cell wall antigen deposition in the glomerular basement membrane and arthritis due to immune complex deposition in hepatitis B, hepatitis C and dengue. Arthus reaction is an induced localised immune complex-mediated hypersensitivity in experimental animals.

Type IV cell-mediated reaction or delayed-type hypersensitivity (DTH)

Clinical Case 4

A 60-year-old man with chronic cough and weight loss is given the tuberculin test. After 48 hours, he develops induration of 20 mm at the site of injection.

1. *Interpret the test result.*
2. *Explain the immunologic basis of this test.*
3. *Provide one example of localised disease with the same immunologic basis.*
4. *Provide one example of another infection test with the same immunologic basis.*

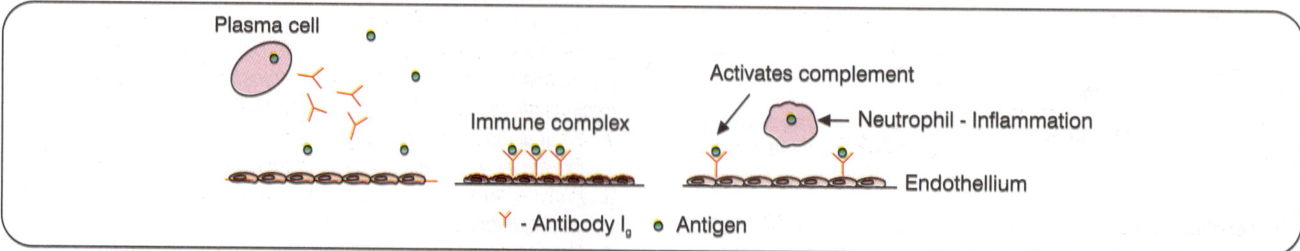

Fig. 15.3 Type III hypersensitivity: immune complex-mediated

Delayed hypersensitivity occurs 48–72 hours after antigenic exposure and is caused by activation of sensitised T_{DTH} cells, release of cytokines and influx of macrophages. The individual gets sensitised to the antigen, resulting in the formation of T_{DTH} cells (Table 15.1).

Subsequently, on exposure to the same antigen, there occurs a DTH response with influx of T_{DTH} cells and the release of cytokines that attract macrophages to the site. This cellular inflammatory response is seen as a localised induration at the site (Fig. 15.4).

Examples:

1. Inflammation-mediated tissue injury in rheumatoid arthritis, multiple sclerosis, type 1 diabetes mellitus and allergic contact dermatitis.
2. Tuberculin test to detect exposure to tubercle bacilli.
3. DTH is often a characteristic of infections by intracellular, slow-growing pathogens—tuberculosis, leprosy and histoplasmosis. Cell-mediated immune reactions in these infections result in a lesion called a granuloma—a densely packed collection of macrophages which fuse to produce characteristic giant cells, surrounded by epithelial cells and lymphocytes. Granulomas are slow-progressing but active lesions. As the host attempts to contain the infection, the microorganisms continue to grow intracellularly at slow rates. Whether the lesion progresses or resolves depends on many immunological factors. Especially the release of specific cytokines that determine the activation of the T_H1 or T_H2 pathway.

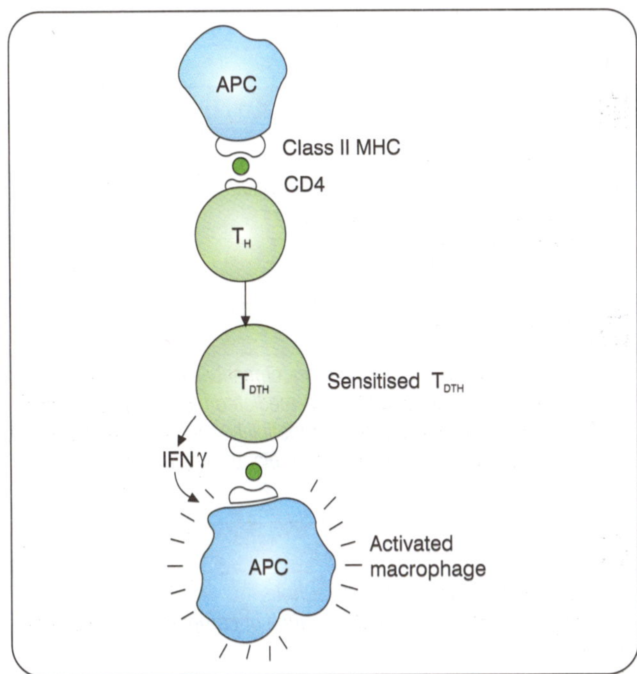

Fig. 15.4 Type IV hypersensitivity: T cell-mediated

Hypersensitivity skin tests

Both immediate and delayed hypersensitivity can be tested in vivo in any individual in a controlled manner on the skin of the forearm, as the skin provides all the immunological cells and factors required for a reaction to occur, if any. Examples of hypersensitivity tests on the skin are provided in Table 15.2.

Table 15.1 Coombs and Gell classification of hypersensitivity

Type	Mechanism	Examples
I/Immediate (minutes)	IgE–Mast cell → Degranulation of mast cell releasing histamine	• Atopy: allergic rhinitis allergic asthma • Systemic anaphylaxis • Cutaneous anaphylaxis (urticaria angioedema) • Casoni's test
II/Cytotoxic	Cell surface Ag + Ig → Opsonisation and phagocytosis → Complement-mediated cytolysis	Hemolytic anemia
III/Immune Complex	Ag + Ab → deposited immune complex → Neutrophil inflammation	• Serum sickness (after anti-diphtheritic and anti-tetanus serum therapy) • Persistent microbes like *Streptococcus pyogenes*
IV/Delayed (24–48 hr)	Activation of sensitised T cells → release of cytokines → influx of macrophages	• Tuberculin test • Lepromin test • Allergic contact dermatitis • Granuloma formation in tuberculosis, leprosy, histoplasmosis

Table 15.2 Hypersensitivity tests on the skin

	Immediate	Delayed
Example	Casoni's test in hydatid disease	Tuberculin test in tuberculosis
Mechanism	IgE–mast cell–histamine	T cell–cytokines–macrophage
Passive transfer	Possible	Not possible
Time to read	20 min	24–48 hours
Reaction seen	Erythema, wheal	Erythema, induration

Answers

Clinical Case 1

1. *Systemic anaphylaxis*
2. *Mast cell—IgE-mediated type I immediate hypersensitivity*
3. *Asthma, allergic rhinitis, drug allergy*
4. *Epinephrine (adrenalin)*

Clinical Case 2

1. *Transfusion reaction due to ABO incompatibility*
2. *Antibody-dependent complement-mediated type II hypersensitivity*
3. *Erythroblastosis fetalis (Rh incompatibility), autoimmune hemolytic anemia, idiopathic thrombocytopenic purpura*
4. *Ensure ABO and Rh compatibility in blood transfusion*

Clinical Case 3

1. *Post-streptococcal glomerulonephritis*
2. *Immune complex-mediated activation of complement causing tissue injury in type III hypersensitivity*
3. *Streptococcus pyogenes*
4. *Raised ASO titre*

Clinical Case 4

1. *The individual is tuberculin-positive as the induration measures 20 mm.*
2. *T cell-mediated delayed type IV hypersensitivity*
3. *Contact dermatitis*
4. *Lepromin test*

Questions

1. Compare and contrast the immunologic basis of immediate and delayed hypersensitivity.
2. Outline the immunologic reaction in the event of a mismatched blood transfusion or erythroblastosis fetalis. How can the latter be prevented?
3. Explain the immunologic basis of induration in the tuberculin test.
4. Explain the immunologic basis of post-streptococcal glomerulonephritis.
5. What is the difference between 'anaphylaxis' and 'atopy'?
6. Compare and contrast immediate and delayed hypersensitivity skin tests.

Autoimmunity and Immunodeficiency

AUTOIMMUNITY

Clinical Case 1

A 50-year-old lady presents with generalised purpuric spots on the skin. On examination, there are no significant findings other than a platelet count of 50,000 cells/mm³. Her bone marrow biopsy shows a normal picture.

1. *What is the most likely cause of an isolated thrombocytopenia?*
2. *What is the underlying cause of this condition?*
3. *Define self-tolerance.*
4. *Mention two mechanisms of tolerance.*

The central principle of the immune system is to distinguish between the self and the non-self. Immune responses are not always protective and can be associated with tissue destruction and disease. Pathologic immune processes in autoimmune disease use immune cells and mediators similar to those in protective immune reactions.

Autoimmunity implies the presence of antibodies or T cells that react with self-antigens as an immune aberration, not necessarily with pathogenic consequences. Autoimmune disease implies a breakdown of the immune system's ability to discriminate between the self and the non-self, which is the central dogma of the immune system.

The immune system is constantly exposed to self-antigens without inducing lymphocyte stimulation, in a phenomenon called **self-tolerance**. It is not an inherited feature but somatically acquired. How does the immune system distinguish the self from non-self? The precise mechanism of tolerating self-antigens is not known but there are two possibilities:

Mechanisms of tolerance

1. *Central tolerance*

The formation of the receptor repertoire of T cells is a random process and is followed by a selection procedure in the thymus (thymic education) that sorts lymphocytes into those with self-reactivity and those lacking self-reactivity. Since self-tolerance is somatically acquired, it is subject to failure.

Clones of lymphocytes potentially reactive to a particular antigen (self) may be eliminated. For the T cell lineage, this **clonal deletion** occurs in the thymus and forbidden self-reactive clones are eliminated. Clonal deletion has been observed in B cells. This elimination of self-reactive clones in the thymus is independent of antigen recognition and occurs by a process of positive and negative selection via receptors on thymocytes.

Tolerance, like immune recognition, is antigen-specific and the central major mechanisms clonally delete or anergise unwanted lymphocytes (self-reactive lymphocytes). Clonal deletion in the thymus is irreversible. Clonal anergy may be a reversible process as the cells continue to be available but are functionally paralysed.

The antibody responses facilitated by B cells generally depend on the help provided by T lymphocytes. This central role of T cells in the immune response makes them an ideal target for tolerance induction.

Tolerance to self-proteins is mainly controlled through the elimination or the functional impairment of autoreactive T cells. However, there is evidence that clonal deletion and clonal energy exist for B cells, the site probably being the bone marrow.

The specificity and memory of acquired immunity are mediated by lymphocytes and their receptors through a process of clonal selection. The clonal selection hypothesis suggests that there are millions of different lymphocytes, each with specific receptors for a particular antigen. These lymphocyte surface receptors are an immunoglobulin on B cells and the TCR on the T cell. When an antigen binds to its specific receptor on a lymphocyte, that particular lymphocyte proliferates and differentiates. The result is the clonal expansion of the lymphocytes including memory lymphocytes of that specificity.

2. *Peripheral tolerance*

Peripheral T cell non-responsiveness can be achieved by induction of T cell anergy. Specific (self-reactive) clones may be present but may be rendered non-responsive to self-antigens by certain negative signals. This functional inactivation is termed clonal anergy. There exist certain costimulatory signals between T cells and antigen presenting cells that are also essential for triggering the immune response.

To undergo complete activation, T cells require two signals. The first is provided by the TCR binding to the antigen. The second signal derives from the interaction of co-stimulatory molecules expressed on the T cell surface and the antigen presenting cells. If T cells recognise antigens in the absence of appropriate co-stimulatory signals, they enter a state of anergy characterised by survival but inability to proliferate when re-exposed to the antigen. Recently, more emphasis has been put on the microenvironment (inflammation, cell types, cytokines) in which antigen exposure occurs.

3. *Induced/acquired tolerance*

Immune tolerance can be artificially and specifically induced **by gradually increasing the amount of antigen to a threshold**. Specifically induced tolerance could represent an avenue for therapy of autoimmune diseases, allergic conditions and allograft rejection.

Principles and features of autoimmunity

Due to a yet unknown antigenic stimulus, prior tolerant lymphocytes break the state of tolerance and initiate an immune response, resulting in autoimmunity. Almost any organ system of the body can be affected by a syndrome classified as an **autoimmune disease**. Despite this heterogeneity, autoimmune syndromes share certain features:

1. Autoimmunity is not necessarily a sign of disease. Autoantibodies in low titres are frequently detected in normal individuals, e.g., rheumatoid factors. Rheumatoid factors are predominantly IgM antibodies directed against the F_c fragment of IgG present in high titres in the serum of patients with rheumatoid arthritis and other rheumatoid diseases. Rheumatoid factors may be seen in normal individuals and also in non-autoimmune diseases like leprosy and tuberculosis.
2. Autoimmunity has a tendency to occur in women more than men.
3. More than one immune abnormality is required to induce autoimmune disease.
4. Autoimmune inflammation includes multiple different cell types.
5. The pathogenesis of autoimmunity involves multiple inherited risk factors (familial) combined with environmental contributions.

Despite intense research, the exact pathogenesis of autoimmunity has not been clarified. The likely etiologic mechanisms are a combination of genetic factors (HLA association) and environmental factors:

1. Release of sequestered antigens like the ocular lens and corneal proteins following infection, spermatozoa-associated antigens following trauma or vasectomy, myelin basic protein following infection and cardiac muscle antigen following infarction.
2. Failure of thymic education, leading to escape of autoreactive T cells from the thymus.
3. Breakdown of peripheral tolerance, that prevents activation of self-reactive T cells that have escaped the thymus. Inflammation may provide ideal conditions for breaking peripheral tolerance.
4. Polyclonal stimulation of lymphocytes by "superantigens" derived from bacteria, e.g., staphylococcal protein A. This is seen especially in systemic autoimmune disorders characterised by production of a wide array of autoantibodies.
5. Molecular mimicry: An immunogenic peptide (epitope) in a foreign antigen exhibits sequence similarity to a self-peptide. Such cross-reactivity results in an autoimmune response against the self-peptide. Examples are the M protein of *Streptococcus pyogenes* and glycoproteins of the heart leading to RHD, type 1 diabetes mellitus (IDDM) and multiple sclerosis following viral infections as well as Reiter's syndrome and the Guillain–Barré syndrome following bacterial gastrointestinal infections.
6. Abnormalities in lymphocyte interactions, especially the T_H1 and T_H2 subsets, leading to functional impairment.
7. Host responsiveness: Besides professional antigen presenting cells, many different cell types can acquire the ability to aberrantly function as antigen presenting cells.

It has been suggested that the immune system has a major function in surveying tissue integrity, to remove necrotic tissue or help in tissue repair. Autoimmune disease could result from dysregulation of such cleaning and repair mechanisms.

Tissue infection, ischemia, malignancy and necrosis attract immune cells into all organs. Neoantigens are formed under these circumstances and tolerance against such antigens has not been established.

Genetic susceptibility in autoimmune diseases

Autoimmune disease genes in the sense of a mutated allele are non-existent. Patients inherit multiple genetic risk factors, each of which is normal if considered alone. Autoimmune syndromes run in families, suggesting clustering of susceptibility factors in families. Autoimmunity is associated with HLA genes. HLA disease association means that the frequency of a defined HLA allele is significantly increased in patients with a certain disease when compared with ethnically matched controls. For example:

HLA DR 4 — Rheumatoid arthritis
HLA DR 2 — Multiple sclerosis
HLA DR 3 — Insulin-dependent diabetes mellitus
HLA B 27 — Ankylosing spondylitis

Autoimmune diseases are divided into two clinical types

1. **Organ-specific autoimmune disease:** This is seen when the autoimmune response targets a specific organ or tissue. Examples include Graves' disease,

Hashimoto's thyroiditis, type 1 diabetes mellitus, insulin-resistant diabetes mellitus, vitiligo, autoimmune hemolytic anemia, autoimmune thrombocytopenic purpura, pernicious anemia, myasthenia gravis, multiple sclerosis, the Guillain–Barré syndrome and acute rheumatic fever.

2. **Systemic autoimmune disease:** This is sometimes called 'collagen vascular disease', as it targets connective tissue and the vascular system. Examples include systemic lupus erythematosus (SLE) and rheumatoid arthritis

Rheumatoid arthritis is a chronic inflammatory joint disease with systemic involvement. Immunologic findings include rheumatoid factors, immune complexes and antinuclear antibodies.

Systemic lupus erythematosus is a chronic, inflammatory multi-organ disorder that predominantly affects women of child-bearing age. Immunologic findings include lupus erythematosus (LE) cells, antibodies to DNA and other autoantibodies. LE cells are seen when nuclear material is opsonised by anti-DNA antibody or complement and then phagocytised by neutrophils.

Acute disseminated encephalomyelitis may occur following vaccination (e.g., rabies vaccine) or viral infection (e.g., measles and influenza). The disease represents a cell-mediated response to the basic protein of myelin.

Guillain–Barré syndrome (acute idiopathic polyneuritis) occurs after a viral infection (measles, hepatitis) or vaccination (influenza) and affects all age groups. It is characterised by T cell-mediated damage to the peripheral nerves.

Multiple sclerosis is a relapsing neuromuscular disease with exacerbations and remissions.

Hashimoto's thyroiditis demonstrates antibody to thyroid-specific antigens, while *Graves' disease* (hyperthyroidism) demonstrates antibody to thyrotropin receptors.

Laboratory diagnosis of autoimmune diseases

1. **General tests:** Tests to detect elevated serum gamma globulin levels, the presence of immune complexes in serum and deposited in tissues, depressed levels of serum complement and the presence of various autoantibodies.
2. **Specific tests:** Tests to detect antibodies to the particular antigen involved in the disease, e.g., anti-nuclear antibodies (ANA) in SLE, rheumatoid (RA) factor and anti-cyclic citrullinated peptide (anti-CCP) in rheumatoid arthritis and anti-platelet antibodies in immune/idiopathic thrombocytopenic purpura.

IMMUNODEFICIENCY

Clinical Case 2

A 2-year-old male child from a refugee camp presents with fever and chronic diarrhea. On examination, the child is underweight, with wasting of muscles. Microscopic examination of the stool revealed cysts of Cryptosporidium spp. with no laboratory evidence of HIV infection.

1. *What is the most likely cause of immunodeficiency?*
2. *Name three other causes of acquired immunodeficiency.*
3. *What is the normal leucocyte, lymphocyte, CD4 and CD8 count in blood?*
4. *What are the normal immunoglobulin values in serum?*

The four major components of the immune system are **T cells, B cells, phagocytes** and the **complement molecules.** A deficiency in any of these systems may be congenital or acquired. The type of the infection (infective agent) that occurs as a result of the immunodeficiency provides clues to the type of the immunodeficiency present. For example, T cell defects result in viral infections and complement deficiency in infections due to capsulated bacteria.

General features of immunodeficiency

1. Chronic, recurrent infection with usual agents, infection with an unusual agent, incomplete resolution of an infection following adequate therapy, or reactivation of latent infections.
2. Clinical features: Skin rash, chronic diarrhea, loss of weight, growth failure/failure to thrive and hepatosplenomegaly are all pointers towards an immunodeficiency.
3. Live immunisation is contraindicated in immunodeficient individuals.

Immunodeficiency can be classified as **primary,** i.e., those that are congenital or present at birth, or **secondary,** i.e., those acquired during life.

Congenital/primary immunodeficiencies can be defects in:

1. **T cells:** Congenital thymic aplasia and chronic mucocutaneous candidiasis
2. **B cells:** Congenital hypogammaglobulinemia
3. **T and B cells:** Severe combined immunodeficiency (SCID)
4. **Phagocytes:** Chronic granulomatous disease of childhood and enzyme deficiencies, (Chediak–Higashi syndrome)
5. **Complement:** Component or receptor deficiency

Factors resulting in acquired/secondary immunodeficiencies are:

1. Malnutrition: In India, malnutrition is the commonest cause of immunodeficiency. Inadequate intake of protein, protein energy malnutrition or loss of protein due to renal or intestinal disease can cause immunodeficiency, especially in children.

2. Iatrogenic: Radiation therapy for malignancy and the use of immunosuppressive agents (steroids) or cytotoxic drugs.
3. Neoplastic disorders of the lymphocytes.
4. Following viral infections like measles, a transient immune-deficient state is known to occur.
5. Acquired immunodeficiency syndrome (AIDS), where the HIV targets the CD4 T lymphocytes. Other viruses infecting the immune system itself are the Epstein–Barr virus (EBV), cytomegalovirus (CMV), human herpes virus type 6 (HHV-6) and human T lymphotropic virus type 1 (HTLV-1).

Evaluation of cell-mediated immunity

Isolated T cell defects are rare as T cell defects invariably lead to B cell dysfunction. The level of cell-mediated immunity in an individual to a particular antigen can be studied by performing an intradermal skin test and observing its results. Following injection of the antigen, a controlled immune response occurs (inflammation) at the site of the infection resulting in the influx of lymphocytes that can be measured in 'mm' as induration. Local vasodilatation results in erythema at the local dermal site. The size of the induration denotes the cell-mediated immune status of that individual to that antigen.

> The normal values of cell counts (in peripheral blood) are:
> Total leucocytes: 4000–10,000/μL
> Total lymphocytes: 800–5,000/μL
> CD4 lymphocytes: 500–1,200/μL
> CD8 lymphocytes: 150–1,000/μL

However, mere normal values cannot denote the functional integrity of lymphocytes. The functional competence of lymphocytes is measured by the intradermal skin test (Table 16.1).

This skin test may show:

No reaction at the site or complete non-responsiveness to the antigenic challenge. This could mean either that the individual has not been exposed (T cells not primed) to the antigen or there exists severe immunodeficiency, such that the cells are rendered non-reactive or anergic.

Any reaction at the site needs to be measured and interpreted accordingly. Reaction to the specific antigen proves priming of cells to that antigen, i.e., prior exposure to the antigen by infection, if cross-reaction with other antigens is ruled out. The degree of reactivity to the antigen determines the cellular immunity vis-à-vis the delayed hypersensitivity status of the individual. For example, the tuberculin test and lepromin test.

Evaluation of humoral immunity

In blood, 20% of the total plasma proteins consists of immunoglobulins (gammaglobulins): The normal concentrations of these are IgG, 1000 mg/dL; IgA, 200 mg/dL and IgM, 120 mg/dL. They can be quantitatively measured (Table 16.1).

The Schick test is a skin test that demonstrates the presence of circulating diphtheria antitoxin. In an individual who has been immunised by the diphtheria toxoid, the antitoxin level should be high (unless the individual is suffering from an immunodeficiency) and therefore would react negatively. A positive Schick test in an immunised person denotes immunodeficiency.

Similarly, the isohemagglutinin titres (anti-A or anti-B) measure 1:4 by 1 year of age. Like the Schick test, any specific antibody level can be measured following immunisation with specific antigens, e.g., hepatitis B vaccination should lead to protective levels of anti-HBs in the (HBsAg) immunised person.

Evaluation of phagocyte function

A qualitative estimation of circulating monocytes, neutrophils and eosinophils is feasible but is unable to detect the functional competence of these cells. The functional competence is detected by an in vitro test called the nitro blue tetrazolium test (Table 16.1).

For the evaluation of complement, complement molecule levels can be estimated by CH_{50} assays or ELISA.

Table 16.1 Evaluation/screening protocol for immunodeficiency

	Immunodeficiency	Test
1.	T cell	In vitro: Total count, differential count, T subset count (CD4, CD8) In vitro: Intradermal skin test
2.	B cell	Quantitative Ig levels: IgG, IgM, IgA Schick test following diphtheria immunisation Antibody titres following specific immunisation
3.	Phagocyte	Neutrophil, monocyte count (Functional) nitro blue tetrazolium test
4.	Complement	Level assays by ELISA

Answers

Clinical Case 1

1. *Immune or idiopathic thrombocytopenic purpura*
2. *Autoimmune disease, where anti-platelet antibodies destroy platelets*
3. *The immune system is constantly exposed to self-antigens without inducing lymphocyte stimulation, in a phenomenon called self-tolerance*
4. *Deletion of self-reactive clones in the thymus (central) or non-responsiveness/anergy of self-reactive clones (peripheral)*

Clinical Case 2

1. *Malnutrition*
2. *Viral infections of the immune system, e.g., HIV, iatrogenic, e.g., steroid therapy, neoplasia of lymphocytes*
3. *Total leucocytes: 4000–10,000/μL, total lymphocytes: 800–5,000/μL, CD4 lymphocytes: 500–1,200/μL, CD8 lymphocytes: 150–1,000/μL*
4. *IgG: 1,000 mg/dL, IgA: 200 mg/dL, IgM: 120 mg/dL*

Questions

1. What is the central principle of the immune system? Define and explain the mechanisms of self-tolerance.
2. Explain the mechanisms of pathogenesis in autoimmunity, citing examples.
3. What is rheumatoid factor? Write a brief note on the genetic association in autoimmune diseases.
4. Outline the laboratory diagnosis of autoimmune disorders.
5. Mention the general features of immunodeficiency.
6. Classify congenital immunodeficiency, citing examples.
7. List the factors resulting in acquired immunodeficiency.
8. Outline screening tests for immunodeficiency.

Transplantation and Tumour Immunology

Transplantation of kidney, bone marrow, liver and heart is being carried out for a variety of indications, like renal failure, SCID, aplastic anemia, leukemias and others. For successful transplantation, certain immunologic investigations are necessary and precautions need to be taken, that involve the donor and the recipient. Allograft survival is highest when the donor and recipient are HLA identical.

Pre-transplantation evaluation

1. ABO blood grouping
2. Histocompatibility testing/HLA typing (class I and II) to detect the HLA and type of the recipient and donor to determine histocompatibility. This test is also used to determine the presence of any preformed antibodies (due to pre-sensitisation) in the recipient-to-donor HLA.

Post-transplantation immunosuppression

This is achieved by one or more of the following modalities:

1. Drugs used: Steroids, cyclosporine, azathioprine and cyclophosphamide, with monitoring for toxicity.
2. Immunotherapy: Use of anti-lymphocyte globulins or monoclonal antibodies directed against specific T cell subsets.

The outcome of the transplantation depends on the degree of histocompatibility between the donor and recipient. In spite of compatibility and immunosuppression, two serious immunologically mediated outcomes may occur:

1. Graft rejection and
2. Graft versus host disease (GVHD)

Graft rejection: The recipient's immunocompetent cells recognise graft HLA as foreign and mount an immune response directed against the graft. It results in rejection of the graft by cytotoxic T (CD8) cells; recipient CD8 cells found in graft tissue.

GVHD: The donor's (graft) immunocompetent CD8 cells mount an immune attack against recipient antigens; donor CD8 cells are found in all recipient tissues. The reaction is systemic, resulting in severe immunodeficiency in the recipient.

TUMOUR IMMUNOLOGY

Malignancy is the autonomous growth and multiplication of transformed, monoclonal cells. The transformation may be spontaneous (mutational) or induced by carcinogens or oncogenic viruses. Tumour cells (transformed) show antigenic differences from normal (non-transformed) cells.

Antigens expressed on tumour cells

1. *Tumour-specific antigens:* Unique antigens expressed only on tumour cells but have not yet been characterised in humans.
2. *Tumour-associated antigens:* Antigens expressed on tumour cells and normal cells but differing qualitatively and quantitatively.

Examples:

Oncofetal antigens like carcinoembryonic antigen (CEA) and α-fetoprotein (AFP). CEA is found on the fetal gut during embryogenesis and on colonic carcinoma cells but not on normal adult colon cells. Alpha fetoprotein in serum is secreted by fetal liver and transformed hepatocytes but not by normal hepatocytes. Common acute lymphoblastic leukemia antigen (CALLA) is expressed only on leukemic T cells and prostate specific antigen (PSA) in prostate cancer.

As specific tumour-associated antigens can be identified, tumour-reactive monoclonal antibodies can be used in the diagnosis (conjugated with fluoro-isotope enzymes) and immunotherapy of tumours (conjugated with cytotoxic drugs).

Tumour markers are molecules produced by the cancer tissue or by the immune response to cancer. They can be detected in clinical samples to diagnose/stage cancer, monitor response to therapy or detect recurrence and prognosis. Recently, they have been used for early detection and guide targeted treatment in certain cancers. They provide useful information but also have limitations in interpretation and are employed in conjunction with histopathologic examination. Examples of tumour markers in cancers are: α-fetoprotein in liver, beta-2 microglobulin in multiple myelomas, CA 15-3, estrogen receptors in breast, CA-125 in ovaries, CEA in colon, chromogranin A in neuroendocrine tumours, PSA in prostate and thyroglobulin in thyroid.

Tumour antigens, using specific monoclonal or polyclonal antibodies, can also be detected by immunohistochemistry. Immunohistochemistry of cancer biopsy/specimen can differentiate benign and malignant tumours, stage and classify the cancer, determine the primary origin of metastatic tumour of unknown origin and detect micrometastasis. Examples of diseases for which tumour markers are used in immunohistochemistry are breast cancer, lymphomas, epithelial cancers like squamous cell carcinoma and adenocarcinomas.

Immunologic effector mechanisms against tumour cells comprise of

1. Direct killing of tumour cells by T_c (CD8) cells
2. Secretion of antibodies against tumour antigens by B cells
3. Natural killer (NK) cell activity against tumour cells, including antibody-dependent cell cytotoxicity
4. Secretion of tumour necrosis factors α and β by activated macrophages

Immune surveillance

In all normal individuals, the immune response actively looks for tumour cells and destroys them wherever transformation of cells occurs. Failure in this surveillance results in the occurrence of tumours. Immune surveillance is of importance in tumours induced by oncogenic viruses (HBV, HSV, EBV, HHV-6).

Questions

1. Give an account of the immunological basis of graft rejection and graft versus host disease and list the strategies to minimise these events.
2. Give an account of tumour antigens and the immunologic effector mechanisms against tumours.

Chapter 18

Abscess, Cellulitis, Myositis, Tetanus, Infective Arthritis and Osteomyelitis

The skin is the largest organ of the body, covering the entire exterior surface. It consists of the outer epidermis and underlying dermis containing sebaceous and sweat glands. Beneath the skin is the subcutaneous connective tissue and adipose tissue (Fig. 18.1). Superficial infections (e.g., fungal infections) affect only the dermis, while deeper infections can involve the underlying soft tissues and muscles as in gas gangrene. Being a large organ exposed to the environment, clinical manifestations of skin infections can be caused by bacteria, viruses, fungi and parasites, including ectoparasites. The risk factors for skin infection include malnutrition, trauma, medical devices, occlusion of blood supply and diabetes. Microorganisms commonly infecting the skin and soft tissues are listed in Table 18.1.

SKIN AND SOFT TISSUE BACTERIAL INFECTIONS

Endogenous skin and soft tissue bacterial infections

The skin is colonised by bacterial flora up to 10^4 CFU/mm^2 (colony forming units). This bacterial flora is either transient or resident. The transient bacteria are acquired by contact with the environment and survive for a short period of time, the commonest being *Staphylococcus aureus* and *Streptococcus pyogenes*. Resident flora multiply on the skin and are regularly present, the commonest being *Propionibacterium acnes*, *Staphylococcus epidermidis* and *Acinetobacter baumanii*. Resident bacteria can give rise to endogenous infections during the use of percutaneous intravenous catheters, following intravenous catheterisation. Intracardiac catheterisation can give rise to bloodstream infection (BSI) and central line-associated bloodstream infection (CLABSI). Endogenous medical device-related hospital-acquired infections include those related to the use of central venous lines, peritoneal dialysis catheters, tubes to drain body cavities, chemotherapy lines, parenteral nutrition lines and temporary pacemaker lines.

Endogenous skin infections or secondary infection of the skin with a primary focus elsewhere

Primary infection can be osteomyelitis, septic arthritis, lymphadenitis (tuberculous, atypical mycobacterial or pyogenic), actinomycotic jaw or meningococcal bacteremia. Pharyngitis caused by *S. pyogenes* results in the secretion of an exotoxin called the **erythrogenic factor** that can spread through the bloodstream and cause scarlet fever, characterised by a red rash and desquamation of the extremities. Staphylococcal exotoxin can also cause scalded skin syndrome and toxic shock

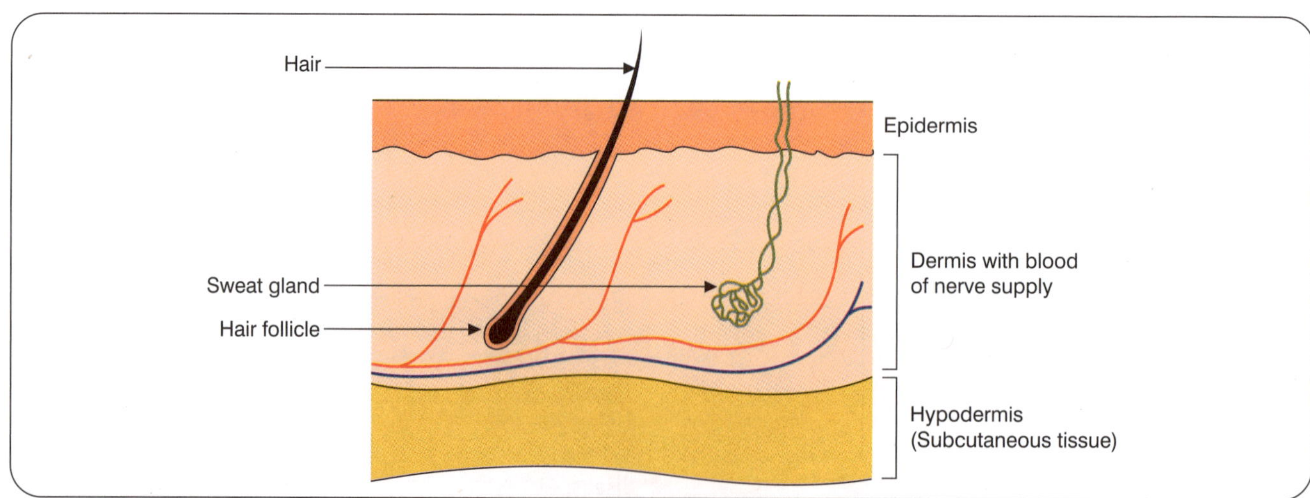

Fig. 18.1 Diagrammatic representation of the microscopic appearance of skin

Table 18.1 Microorganisms commonly infecting the skin and soft tissues

Group	Microorganism	Disease
Bacteria	*Staphylococcus aureus* *Streptococcus pyogenes* *Non-sporing anaerobes* *Clostridium perfringens* *Clostridium tetani* *Mycobacterium leprae*	Abscess, folliculitis Cellulitis, fasciitis Abscess, cellulitis, fasciitis Gas gangrene Tetanus Leprosy
Fungi	*Candida albicans and other non-albicans Candida* spp., *Dermatophytes,* *Actinomyces, Nocardia* *Sporothrix,* chromoblastomycosis, phaeohyphomycosis	Dermal candidiasis Tinea infection Mycetomas Subcutaneous mycosis
Virus	Measles, rubella, chickenpox, parvovirus B19 Herpes simplex virus 1 and 2 Human papillomavirus Molluscum contagiosum virus	Exanthems (childhood fevers) Genital herpes Warts Nodules
Parasite	*Dracunculus medinensis* *Trichinella spiralis* *Taenia solium* *Leishmania donovani* *Leishmania tropica* *Ancylostoma duodenale* *Strongyloides stercoralis* Filariasis	Blisters Cysts in muscle Cysticercus cellulosae Post kala-azar dermal leishmanoid Dermal leishmaniasis Cutaneous larva migrans Cutaneous larva currens Lymphangitis, elephantiasis

syndrome. Decubitus (pressure) ulcers, diabetic ulcers, bite infections, burn infections and injection site abscess are secondary skin infections due to trauma or underlying disease.

Exogenous skin and soft tissue bacterial infections

Folliculitis, furuncles, carbuncles and abscesses are mostly caused by *S. aureus* while erysipelas, lymphangitis, cellulitis and fasciitis are caused predominantly by *S. pyogenes*. Impetigo can be caused either by streptococci or staphylococci and fasciitis by streptococci, enterococci and anaerobes.

S. aureus infections tend to localise while *S. pyogenes* tends to spread through tissues. This is due to the presence of the enzyme coagulase in *S. aureus* infection, that is responsible for laying a fibrin barrier around the site of infection, and causing abscesses known as **folliculitis**, **furuncles** or **carbuncles**. On the other hand, *S. pyogenes* secretes hyaluronidase that has the ability to break down tissue causing locally spreading infections like impetigo (epidermis), erysipelas (lymphatics) and cellulitis (subcutaneous fat). Necrotising infections of the skin include fasciitis and gas gangrene caused by anaerobic clostridia. Cellulitis refers to an acute inflammation of the subcutaneous tissue that spreads rapidly (24–48 hours) wherein the borders are not clearly demarcated from unaffected tissues, as in erysipelas. Pyoderma refers to skin infections caused by pyogenic bacteria, mostly *S. aureus* and *Streptococcus* genera.

STAPHYLOCOCCUS

Clinical Case 1

An 18-year-old boy presents with a history of high fever for 3 days and is delirious at presentation. He gives a history of glass injury to his right ankle during a hike. On examination, the site of injury is erythematous, tender and purulent (Fig. 18.2). The laboratory report received is as follows: Methicillin-resistant Staphylococcus aureus *(MRSA) grown, sensitive to linezolid, vancomycin, teicoplanin, cotrimoxazole and daptomycin.*

1. *Name the source of the above bacterium.*
2. *What is your course of action?*
3. *What microbiological investigation would you request for?*
4. *Name two antimicrobials recommended for empirical therapy.*
5. *If untreated, what complications may occur?*
6. *What is your choice of antimicrobial in this case?*

Staphylococci are gram-positive cocci arranged in clusters measuring 1 µm in diameter (Fig. 18.3). *S. aureus* produces the enzyme coagulase while coagulase-negative staphylococci (CoNS) do not produce it. Skin is the natural habitat of *S. aureus* and carriers are known to exist. The anterior nares, perineum, axillae and toe webs are the sites where it is likely to colonise and contribute to disease. CoNS are also carried on the human skin but are largely non-pathogenic and cause disease under specific

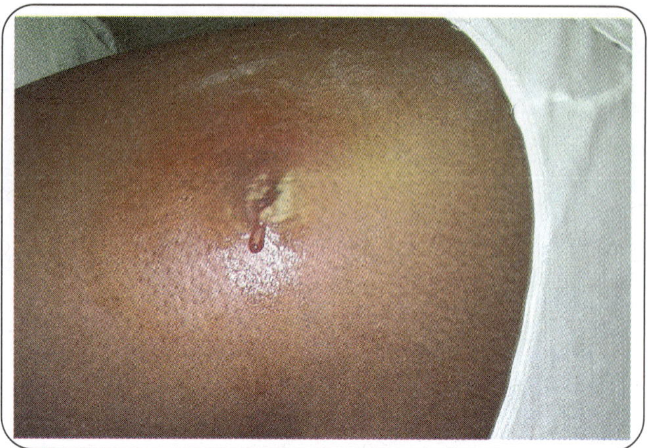

Fig. 18.2 Cutaneous abscess

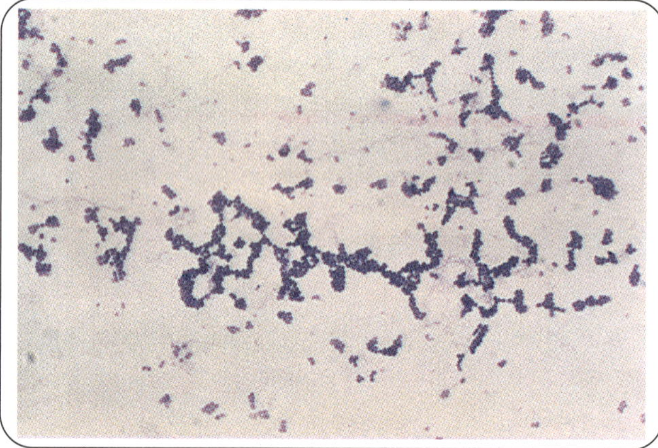

Fig. 18.3 Primary smear showing gram-positive cocci in clusters, Gram stain, 100X, e.g., *Staphylococcus aureus* (Source: Public Health Image Library, ID 2296 / Centers for Disease Control and Prevention [CDC] / Dr. Richard Facklam)

circumstances, especially in the presence of long-standing medical devices and among neonates.

The virulence factors of *S. aureus* include its cell wall, hemolysins, leucocidins (Panton–Valentine leucocidin toxin), exfoliative toxin, enterotoxins, toxic shock syndrome (TSS) toxin and the enzyme coagulase. In the cell wall, peptidoglycan mediates inflammation, teichoic acid mediates adhesion and protein A inhibits phagocytosis. The hemolysins and leucocidins are hemolytic and leucocidal. Exfoliative toxin causes staphylococcal scalded skin syndrome (SSSS), and pre-formed enterotoxin causes food poisoning. The enzyme coagulase combines with a plasma protein to activate prothrombin that converts fibrinogen to fibrin. Other staphylococcal enzymes include hyaluronidase, DNase and lipase.

The toxic shock syndrome toxin (TSST) is an exotoxin that enters the bloodstream and acts as a superantigen. As a result, it causes non-specific activation of T cells leading to a cytokine storm with fatal multi-system damage. Initially associated with the use of tampons, it is also a complication of other localised staphylococcal pyogenic infections. Toxic shock syndrome presents with fever, desquamation and multi-organ failure. It is diagnosed by detection of the TSST using a serologic or molecular method.

S. aureus can cause pyogenic infections, toxin-mediated disease, hematogenous spread and hospital-acquired infection. It can also produce β lactamases—enzymes that confer resistance to most β-lactam antibiotics. Methicillin-resistant *Staphylococcus aureus* (MRSA) is now encountered in both hospital and community settings.

Skin and soft tissue infections caused by *S. aureus* are folliculitis, furuncle, carbuncle, impetigo, necrotising fasciitis, cutaneous abscess, botryomycosis (mycetoma), pyomyositis, toxic shock syndrome and staphylococcal scalded skin syndrome. *S. aureus* is the commonest bacterium causing osteomyelitis. It can also cause pneumonia, empyema, UTI, infective endocarditis and sepsis.

CoNS implicated in human disease are *S. epidermidis* and *Staphylococcus saprophyticus*. *S. epidermidis* is a commensal on the skin that possesses surface adhesins and secretes an extracellular polysaccharide or slime. These facilitate its pathogenic role in infections due to medical devices—intra-cardiac catheters, IV lines, valvular prosthesis and ventricular peritoneal shunts. *S. epidermidis* has also been implicated in cases of neonatal sepsis and *S. saprophyticus* in urinary tract infection as it has the ability to bind to the urinary tract epithelium.

STREPTOCOCCUS

Clinical Case 2

A 50-year-old diabetic lady presents with sudden onset pain and diffuse erythema and swelling of the skin over the right leg. On examination, the skin is tender and the erythema not well-demarcated (Fig. 18.4).

1. *Which layers of the skin and underlying soft tissues are likely to be infected?*
2. *Enumerate the bacteria likely to cause this infection and the most common bacterium.*
3. *What are the risk factors for this condition?*
4. *Which life-threatening infective condition do you need to rule out?*

Streptococci are non-motile, capsulated, gram-positive cocci in short chains, each measuring 1 μm in diameter (Fig. 18.5).

Depending upon their ability to lyse RBCs, they are classified as:

1. α hemolytic (causes partial hemolysis): Examples are *Streptococcus viridans* group, *Streptococcus pneumoniae* (in pairs).

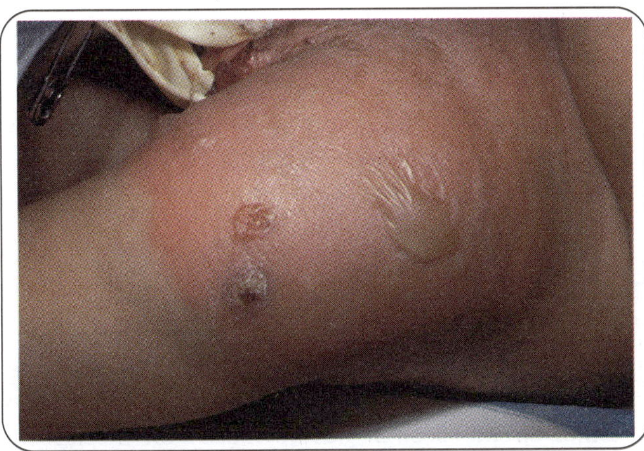

Fig. 18.4 Cellulitis

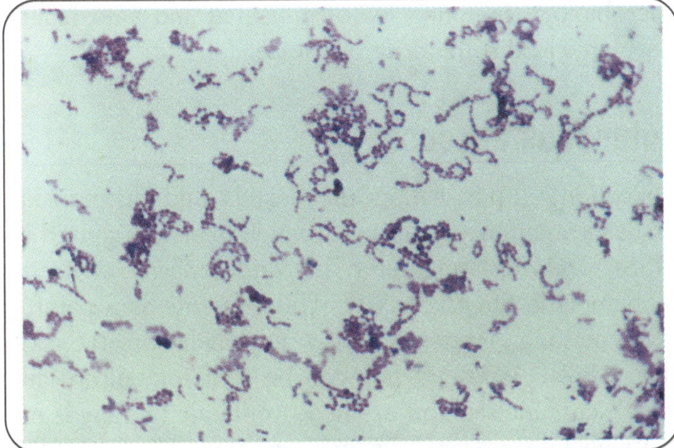

Fig. 18.5 Primary smear showing gram-positive cocci in pairs and short chains, Gram stain, 100X, e.g., *Streptococcus pyogenes*

2. β hemolytic (causes complete hemolysis): These are further classified by the Lancefield grouping based on the C carbohydrate in the cell wall into 20 serotypes, i.e., A to V (except I and J). The most important pathogens are Group A *Streptococcus* (GAS or Streptococcus pyogenes) and Group B *Streptococcus* (GBS or *Streptococcus agalactiae*).
3. γ hemolytic or non-hemolytic: An example is Enterococcus spp. (in pairs).

The *S. viridans* group causes subacute bacterial endocarditis, and *S. pneumoniae* causes pneumonia and meningitis. *S. agalactiae* is carried in the female genital tract and can cause neonatal meningitis or neonatal sepsis. Streptococcal gangrene caused by Group A, C or G organisms is rare and restricted to the superficial fascia, sparing the deep muscle fascia, but can evolve into streptococcal shock syndrome.

Group B streptococci (GBS) or *S. agalactiae* that colonises the female genital tract can cause early-onset neonatal sepsis as well as skin and soft tissue infections. Neonatal sepsis follows normal vaginal delivery in female carriers of GBS.

Enterococcus spp. are commensals in the intestines and can cause UTI, endocarditis and intra-abdominal soft tissue infections.

S. pyogenes is a commensal in the upper respiratory tract and can cause pharyngitis (sore throat) in children and a wide variety of skin and soft tissue infections. Recurrent infections can lead to post-streptococcal non-suppurative complications of rheumatic fever and glomerulonephritis. The pathogenicity of *S. pyogenes* is due to the virulence of its cell wall antigens, exotoxins and enzymes.

The cell wall antigens include peptidoglycan (thrombolytic), lipoteichoic acid (adhesion) and outer M protein, besides a hyaluronic acid capsule (adhesion and antiphagocytic) in some strains. The M protein is an important virulence factor as it inhibits phagocytosis and binds to fibrinogen, causing streptococcal toxic shock syndrome. It cross reacts with myocardial antigen, therefore repeated attacks of streptococcal sore throat may lead to rheumatic fever.

The exotoxins secreted by *S. pyogenes* include hemolysins—streptolysin O (oxygen labile) and S (oxygen stable)—and streptococcal pyrogenic exotoxin. The streptococcal pyrogenic toxin induces fever and causes scarlet fever, necrotising fasciitis and toxic shock syndrome.

Streptolysin O is cytotoxic for cardiac tissue and highly antigenic. Antibodies to streptolysin O, termed anti-streptolysin O (ASO), are measurable in blood for the diagnosis of rheumatic heart disease. The ASO titres are not raised in streptococcal skin diseases and the bacteria largely continue to be sensitive to penicillin. Post-streptococcal glomerulonephritis may be a complication of streptococcal skin infections while rheumatic fever is not.

The enzymes secreted by *S. pyogenes* are streptokinase (fibrinolysin), streptodornase (DNase) and hyaluronidase. The enzymes streptokinase (fibrinolysin) and streptodornase (DNase) have the ability to break down fibrin and DNA, respectively. Fibrinolysis helps in spreading the infection and extending it locally to surrounding tissues. Streptokinase is also used in the treatment of thromboembolic episodes of the heart and brain. Hyaluronidase breaks down hyaluronic acid in tissues to spread the infection.

S. pyogenes and other β hemolytic streptococci are susceptible to penicillin and resistant to tetracycline and cotrimoxazole, with variable resistance to clindamycin and macrolides. Therefore, it becomes important to establish the identity of *Staphylococcus* and *Streptococcus* in order to prescribe the right antibiotic and avoid the emergence of drug resistance.

SKIN INFECTIONS CAUSED BY STAPHYLOCOCCI AND STREPTOCOCCI

Impetigo

Impetigo is a contagious superficial (involves epidermis) skin infection that can be bullous (blister filled with serum) or non-bullous. Predominantly non-bullous, it is more common in children and spreads through close contact with affected persons, shared personal items being fomites. Handwashing and maintenance of good personal hygiene prevent the spread of impetigo. Bullous impetigo is caused by *S. aureus*, due to its locally acting exfoliative toxins (ET-A and ET-B) whereas in staphylococcal scalded skin syndrome (SSSS), the same exfoliative toxins are disseminated hematogenously. The diagnosis is clinical; pus culture can be done for establishing the causative bacteria, i.e., *Staphylococcus* or *Streptococcus*.

In patients with limited disease, topical mupirocin or neomycin are effective only in staphylococcal impetigo, including MRSA. Topical fusidic acid and retapamulin are effective in staphylococcal and streptococcal impetigo, including MRSA.

Oral antibiotics recommended are cloxacillin, amoxicillin-clavulanic acid, cephalexin, erythromycin, cotrimoxazole, clindamycin and linezolid depending upon the identity of the causative agent—methicillin susceptible *S. aureus* (MSSA) or MRSA or *Streptococcus* and its antimicrobial susceptibility test result. In recurrent infections, staphylococcal carriage needs to be ruled out and treated with mupirocin.

Staphylococcal scalded skin syndrome

The exfoliative toxins of *S. aureus,* when acting locally, cause bullous impetigo and when disseminated by blood cause the generalised staphylococcal scalded skin syndrome (SSSS). This is a whole-body generalised exfoliative skin infection characterised by erythema and necrotic debridement of the superficial epidermis. More common in children, it generally follows a localised staphylococcal disease and presents with fever and well-demarcated erythematous tender patches leading to blisters and epidermal exfoliation, finally resulting in painful, moist, raw scalding. Diagnosis is clinical as skin swab and blood culture may be negative. Early administration of anti-staphylococcal antibiotics results in a good prognosis, without scarring.

Ecthyma

Ecthyma is a cutaneous bacterial infection that extends into the dermis and heals with scarring. More common in children, it is caused by staphylococci, streptococci or both in combination. It begins with an erythematous plaque, progressing to vesicles, pustules, ulceration and induration (fixed hardness in the skin due to inflammation). Diagnosis is clinical with antimicrobial therapy effective against both causative agents.

Erysipelas

Erysipelas is a bacterial infection of the dermis and upper subcutaneous tissue, a subtype of cellulitis. Predominantly seen in adults, it is caused by *S. pyogenes* or other streptococci. It occurs mostly on the legs or face as a single erythematous, well-demarcated, raised "orange peel" appearance of the skin. Diagnosis is clinical, microbiology sampling is not indicated and amoxicillin-clavulanic acid is the drug of choice.

Cutaneous abscess

Collection of pus within the dermis and deeper skin tissues is a cutaneous abscess, which is mostly caused by MSSA or MRSA and rarely by *Streptococcus anginosus* (α hemolytic *Streptococcus*). This presents as a painful, erythematous, tender swelling anywhere on the skin (Fig. 18.2). Aspirated pus can be sent for culture and the abscess incised and drained. Antibiotic therapy is not routinely indicated and if considered should be for MSSA. However, in case of MRSA, appropriate antibiotics need to be prescribed. If untreated, the bacterium from the localised abscess can be seeded in blood to cause bacteremia and septicemia.

Bacterial folliculitis (infection of the hair follicle) and furuncle (abscess involving hair follicle)—commonly called "boil"—are mostly caused by *S. aureus*. Folliculitis is a pyoderma located within hair follicles and the apocrine regions. A furuncle is a deep inflammatory nodule extending into subcutaneous tissue that develops from preceding folliculitis. A carbuncle is a larger, deeper, indurated lesion, usually on the neck, back or thighs. Carbuncle is a deep, multiple, contiguous hair follicle and surrounding subcutaneous tissue infection caused by *S. aureus*. Pus can be cultured and anti-staphylococcal antibiotics administered.

Cellulitis

The term "cellulitis" refers to inflammation of the loose connective tissue that may or may not be of infective origin. Cellulitis is a spreading infection of the dermis and the subcutaneous tissue predominantly caused by streptococci (Fig. 18.4). *S. aureus* is associated with purulent cellulitis and *Haemophilus influenzae* with facial cellulitis sequelae to otitis media. Risk factors include lymphedema, obesity and diabetes. This disease presents with acute onset pain, erythema, and swelling of the affected area of the skin with poor demarcation. Diagnosis is clinical but there is a need to rule out necrotising fasciitis, which is a life-threatening condition requiring immediate surgical intervention. Cellulitis is empirically treated with antibiotics covering *S. pyogenes* and MSSA unless proven to be caused by other bacteria. The antimicrobials indicated are cefazolin, cephalexin or amoxicillin-clavulanic acid with or without clindamycin.

Other infections

Other dermal infections include acute paronychia, botryomycosis and toxic shock syndrome caused by *S. aureus* and scarlet fever caused by *S. pyogenes*.

Acute paronychia is a painful, tender, purulent, erythematous swelling of the nailfold. Botryomycosis is a granulomatous and purulent subcutaneous lesion like mycetoma with sinus tracts draining to the surface. The colour of the 'sulphur' granules being golden yellow in *S. aureus* mycetomas. Toxic shock syndrome is mediated by toxic shock syndrome toxin-1 (TSST-1), which is a pyrogenic toxin superantigen. It is a serious disease with fever, rash, desquamation, hypotension, multi-organ involvement needing hospitalisation and intravenous antibiotics. *S. pyogenes* can also rarely cause toxic shock syndrome.

Scarlet fever caused by the pyrogenic exotoxin of *S. pyogenes* affects mostly the lower extremities in children. The bacterium enters by the pharynx or a skin wound and presents with fever, rash and vomiting after 5 days. The rash desquamates after a week and the oral mucosa and tongue appear red and smooth. Penicillin or erythromycin (in penicillin allergy) is the drug of choice.

Laboratory diagnosis of skin infections caused by staphylococci and streptococci

1. Gram stain examination of the primary/direct smear from the pus or affected lesions. Staphylococci are gram-positive cocci 1 μ in size, arranged in clusters. Streptococci are gram-positive cocci arranged in chains.
2. Culture of pus before antibiotic is administered followed by identification of the bacterium. The presence of the enzyme coagulase is important to distinguish *S. aureus* from CoNS.
3. Antimicrobial susceptibility spectrum:

MSSA — cloxacillin, cefazolin, amoxicillin-clavulanic acid, cotrimoxazole, ciprofloxacin, erythromycin, clindamycin, doxycycline

MRSA — linezolid, teicoplanin, vancomycin, daptomycin, cotrimoxazole, clindamycin, doxycycline

Streptococci — penicillin, cefazolin, erythromycin, clindamycin, ciprofloxacin, cotrimoxazole, tetracycline

National Treatment Guidelines (NTGs) recommend ceftriaxone or amoxicillin-clavulanic acid for cellulitis caused by *S. pyogenes*, abscess caused by *S. aureus* and necrotising fasciitis caused by both.

Other gram-positive bacteria that can infect the skin are *Corynebacterium minutissimum, Actinomycetes* spp. and *Nocardia* spp. Gram-negative bacteria that have dermal manifestations include *Neisseria meningitidis, Pseudomonas aeruginosa, Yersinia, Salmonella, H. influenzae, Rickettsiae, Vibrio vulnificus, Brucella, Borrelia, Leptospira, Treponema* and non-tuberculous mycobacteria (Table 18.2).

Table 18.2 Other bacteria causing skin and soft tissue infections

Bacterium	Primary disease	Dermal manifestation
Neisseria meningitidis	Meningococcemia	Non-blanching purpuric rash over trunk and extremities
Pseudomonas aeruginosa	Septicemia in the immunocompromised	Ecthyma gangrenosum—necrotic eschar
Yersinia pestis	Plague	Purpuric rash
Salmonella spp.	Enteric fever	Pink, blanching maculopapular exanthema (rose spots)
Haemophilus influenzae	Cellulitis	Blue cellulitis in children
Rickettsia	Scrub typhus	Maculopapular rash on trunk and extremities
Vibrio vulnificus	Acute gastroenteritis	Skin blisters
Bacillus anthracis	Cutaneous anthrax	Malignant pustule—necrotic eschar with localised edema
Non-tuberculous *mycobacteria*	Granuloma Abscess	Swimming pool granulomas (*Mycobacterium marinum*), injection abscess (*Mycobacterium fortuitum*)
Corynebacterium minutissimum	Erythrasma	Pruritic, erythematous, macular lesions in genitocrural area
Francisella tularensis	Tularemia	Ulceration with lymphadenitis
Burkholderia pseudomallei	Meliodosis	Abscess and ulceration

INFECTIOUS GANGRENE/GANGRENOUS CELLULITIS

It is a rapidly progressive cellulitis with extensive necrosis of subcutaneous tissues and the overlying skin. The infection mostly develops following the introduction of a microbe at the disease site. Streptococcal gangrene caused by Group A, C or G streptococci is rare and restricted to the superficial fascia, sparing the deep muscle fascia, but can evolve into streptococcal shock syndrome.

Clostridial anaerobic cellulitis is a necrotising clostridial infection of the subcutaneous tissue, sparing the deep fascia and muscle. Gas formation is common and extensive. *Clostridium perfringens* and some other *Clostridium* spp. are introduced into the tissue either on traumatic or surgical injury. The incubation period is several days, longer than the 2 days for clostridial myositis/myonecrosis. Clostridial cellulitis and clostridial myositis can be differentiated only at surgery by the viability and appearance of the muscle.

Non-clostridial anaerobic cellulitis caused by non-sporing anaerobes (*Bacteroides,* peptococci, peptostreptococci) and facultative anaerobes (*E. coli, Klebsiella, Aeromonas*) is similar to clostridial cellulitis and can be differentiated on Gram stain examination.

Necrotising fasciitis is an uncommon severe infection of the superficial and deep fascia, following surgical or other trauma. It can be type 1 or type 2 depending on the causative organism. It commonly occurs on the lower extremities but can affect any part of the body, such as the abdominal wall, the perianal or groin areas or post-operative wounds. In type 1, at least one strictly anaerobic species is involved with other facultative anaerobes, as opposed to type 2 caused by GAS with others, most commonly *S. aureus*. Piperacillin-tazobactam with clindamycin is recommended. A form of necrotising fasciitis occurring in the male genitalia is known as Fournier's gangrene, involving the scrotum, perineum and abdominal wall.

Clinical Case 3

A 14-year-old boy is operated for a recurrent retropharyngeal abscess. The pus drained is foul-smelling and sent for microbiological examination. The laboratory report received on day 1 states:
Gram stain of the pus smear shows the presence of pus cells with gram-negative bacilli. Culture: No growth on routine aerobic culture (Fig. 18.6).

1. *Which group of bacteria is most likely to be the cause of the above infection? Why?*
2. *Name four genera of gram-negative bacilli belonging to this group.*
3. *What is the source of infection in the above case?*
4. *Name two antimicrobial agents of choice to treat so that the abscess does not recur*

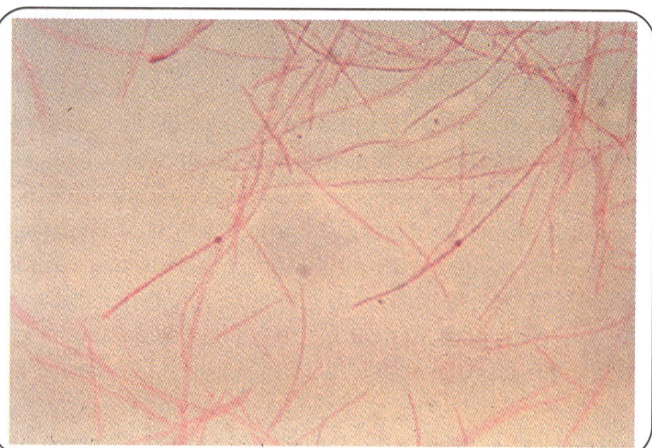

Fig. 18.6 Anaerobic non-sporing gram-negative bacilli, Gram stain, 100X, e.g., *Fusobacterium* spp. (*Source:* Public Health Image Library, ID 3044 / CDC / Dr. V.R. Dowell. Jr)

NON-SPORING ANAEROBES

Non-sporing anaerobic bacteria are normal commensals in the skin, mouth, intestine and vagina. They can cause a variety of pyogenic skin and soft tissue infections, especially abscess in the mouth, head, neck, brain, lung, abdomen, pelvic and bone.

Based on their morphology, non-sporing anaerobic bacteria are classified as:

1. Gram-positive cocci: *Peptococcus, Peptostreptococcus*
2. Gram-negative cocci: *Veillonella*
3. Gram-negative bacilli: *Bacteroides, Prevotella, Porphyromonas, Fusobacterium*
4. Gram-positive bacilli: *Bifidobacterium, Eubacterium, Propionibacterium, Lactobacillus, Actinomyces*

Of these, *Veillonella, Bifidobacterium, Eubacterium, Propionibacterium* and *Lactobacilli* are commensals and non-pathogenic while *Peptococcus, Peptostreptococcus, Bacteroides, Prevotella, Porphyromonas, Fusobacterium* and *Actinomyces* are known to cause disease.

Bacteroides fragilis is the predominant commensal in the intestine, far outnumbering the facultative anaerobe *E. coli.* Amongst non-sporing anaerobes, it is also the commonest cause of human infections. *Fusobacterium* spp. are commensals present in the oral cavity (Fig. 18.6) and are known to cause a painful ulceration in the pharynx called Vincent's angina in association with the spirochaete *Borrelia vincentii*.

A clinical diagnosis of non-sporing anaerobic infection should be suspected:

1. If the site of infection is close to a mucosal surface or deeper.
2. Following oral, abdominal or pelvic surgical procedures, trauma or perforation.
3. If foul-smelling discharge, aspirate or pus is present.

4. If routine aerobic culture is negative, especially with Gram stain showing bacteria.
5. If infection recurs following failure of antimicrobial therapy.

Clinical specimens are to be collected and processed, avoiding exposure to oxygen by anaerobic methods. Surgical debridement or drainage of pus is important along with antibiotics. Clindamycin, amoxicillin-clavulanic acid, ampicillin-sulbactam or other β lactam + β lactamase inhibitor combinations are the NTG-recommended drugs of choice. Alternatively, vancomycin is used for resistant gram-positive anaerobes and carbapenems are used for gram-negative anaerobes.

MYOSITIS AND MYONECROSIS

Classification of infectious myositis

1. **Pyogenic and localised**: Pyomyositis is an acute, pyogenic and predominantly localised skeletal muscle infection caused by *S. aureus* that may or may not follow blunt trauma. It is rarely caused by Group A *Streptococcus, S. pneumoniae, Fusobacterium, Clostridia, Mycobacterium tuberculosis, Mycobacterium avium-intracellulare*, and *Cryptococcus neoformans*.

   Gas gangrene is largely caused by *C. perfringens* and non-clostridial gangrene by *S. pyogenes** and *Peptostreptococcus*. Psoas abscess is usually caused by *S. aureus*, gram-negative bacilli or *M. tuberculosis*.
2. **Non-pyogenic and generalised**: Myalgias are generally caused by viral fevers. Muscle contractions and spasm are caused by *C. tetani*. Myalgia with cysts and eosinophilia can be caused by parasites *Trichinella spiralis* and *Taenia solium*.

Both streptococcal necrotising fasciitis and myositis or clostridial myositis (with gas and crepitation) are surgical emergencies necessitating exploration and debridement of necrotic tissue. A Gram stain examination of pus, serous fluid or tissue can confirm the causative bacterium for appropriate antimicrobial therapy.

The current NTG recommends piperacillin-tazobactam or cefoperazone with clindamycin as empirical therapy following surgical intervention.

**Streptococcus pyogenes* necrotising myositis or necrotising fasciitis is associated with streptococcal toxic shock syndrome, with a rapid course, and intense pain and swelling in the affected area, with fever. It most commonly affects the extremities. The overlying skin may not be affected or, if affected, may be erythematous or violaceous with petechiae or bullae.

GAS GANGRENE/CLOSTRIDIAL MYONECROSIS

Clinical Case 4

A 24-year-old man presents with tender swelling of the right foot, crepitant on touch, with a serosanguinous foul-smelling discharge (Fig. 18.7). He gives a history of trauma at the site following an accident 3 days back that he ignored.

1. *What laboratory test will you order and why?*
2. *The microbiology laboratory reports gram-positive bacilli without spores. What is your diagnosis?*
3. *Name the bacterium most likely to cause this condition and explain its pathogenesis in this case.*
4. *Name other infections caused by this bacterium.*

Clostridia

Clostridia are gram-positive, spore-forming bacilli that are saprophytes in the soil and organic matter and commensals in the human intestine. Human pathogens are *C. perfringens* causing gas gangrene, *C. tetani* causing tetanus, *C. botulinum* causing food poisoning and *C. difficile* causing pseudomembranous colitis. The spores of *Clostridia* are larger than the bacillary body, thus producing a bulge in the bacillary body. This is a feature that distinguishes *Clostridia* from another gram-positive bacillus—*Bacillus anthracis*. The shape (spherical or oval) and the position of the spore—central, sub-terminal or terminal—provides clues to the species. Clostridia are obligate aerobes, with some being aerotolerant or microaerophilic, and are cultivated in anaerobic culture media.

Gas gangrene is a rapidly progressive myonecrosis predominantly caused by *C. perfringens* following muscle trauma and wound contamination. The mere presence of gram-positive bacilli resembling *C. perfringens* in wounds, without myonecrosis, does not imply gas gangrene. Non-traumatic gas gangrene can occur due to *Clostridium septicum* arising from endogenous gastrointestinal mucosal lesions.

As *C. perfringens* is a soil saprophyte and a commensal in the human intestine, gas gangrene can

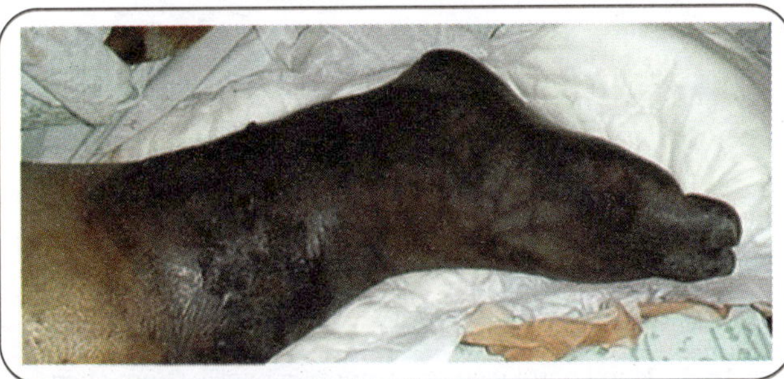

Fig. 18.7 Gas gangrene

occur following accidental trauma, penetrative injuries, abdominal surgery wounds and septic abortion. Spores of the histotoxic species of *Clostridia* contaminate the wounds, and then multiply and secrete exotoxins that cause myonecrosis and hypotension. Crush injury causes tissue anoxia, conducive to the growth of *C. perfringens* contaminating the wound. Poor oxygenation and impaired tissue perfusion reduce the redox potential causing the spores to germinate and multiply. Exotoxins, lecithinase and collagenase secreted by the bacterium further the local destruction of tissue.

The incubation period is 2–3 days and may be as short as 6 hours, acute onset pain being the first symptom. The rapid course is because the generation time of *C. perfringens* is in minutes, causing rapid multiplication in the contaminated tissue. Locally, the muscle is swollen, tender and crepitant, with a serosanguinous, foul-smelling discharge (Fig. 18.7). Hypotension, tachycardia, shock and renal failure rapidly progress, with a fatal outcome, if not aggressively treated.

Gram-stained smear of the wound exudate reveals thick, gram-positive bacilli measuring 4 μm × 1.5 μm (Fig. 18.8). *C. perfringens* spores are oval and subterminal but usually absent in the sample from gas gangrene. *C. perfringens* is a microaerophilic anaerobe. The sample is cultured in an anaerobic liquid medium, such as Robertson's cooked meat medium (RCM) and examined by Gram stain after incubation, for the presence of gram-positive bacilli. Besides *C. perfringens*, other clostridial species have been implicated rarely, along with other facultative anaerobes like *E. coli* and *Enterococcus* spp., causing a polymicrobial mixed infection due to variable contamination of the wound.

Emergency surgical exploration is mandated to distinguish between gas gangrene (visibly unhealthy appearance of the muscle) and anaerobic crepitant cellulitis and for debridement of necrotic tissue. Subsequently, antimicrobial therapy with penicillin and clindamycin, along with *C. perfringens* antitoxin and hyperbaric oxygen is indicated.

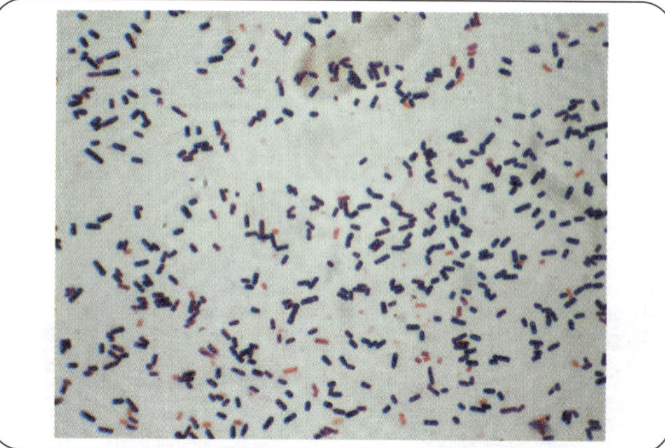

Fig. 18.8 Non-sporulating gram-positive bacilli, Gram stain, 100X, e.g., *Clostridium perfringens*

Non-clostridial myositis can occur due to anaerobic streptococci, *Aeromonas hydrophila* (following trauma in freshwater or associated with marine animals) or infection secondary to arterial insufficiency. Streptococci do not cause crepitus because they are non-gas producing bacteria.

C. perfringens being a commensal in the intestine is also known to cause other soft tissue infections like gangrenous appendicitis, necrotising enteritis and postsurgical intra-abdominal infections, especially in the biliary tract.

TETANUS

Clinical Case 5

A 50-year-old lady engaged in farming presents with a locked jaw (Fig. 18.9), dysphagia and stiffness of the neck. She gives a history of abrasions on the leg 2 weeks earlier while working in the farm that healed spontaneously.

1. *What is the most likely clinical diagnosis and the causative agent?*
2. *Trace the course of events and pathogenesis leading to this condition.*
3. *Could it have been prevented? How?*
4. *Can further progression of her disease be arrested? How?*

Clostridium tetani

Clostridium tetani is an obligate anaerobic gram-positive bacillus with spherical, terminal spores transmitted percutaneously through wounds—either traumatic or at childbirth (neonatal tetanus) (Fig. 18.10). Neonatal tetanus has an infant mortality rate of 80–90%. In August 2015, the WHO declared India to be free of

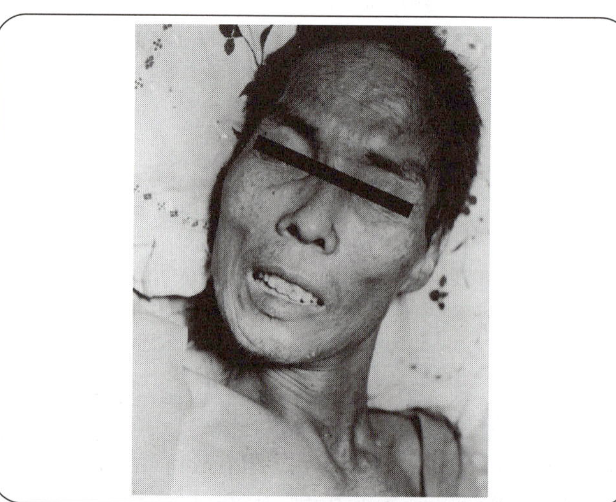

Fig. 18.9 Lockjaw (*Source*: Public Health Image Library, ID 1675 / CDC / Armed Forces Institute of Pathology (AFIP)/ C. Farmer)

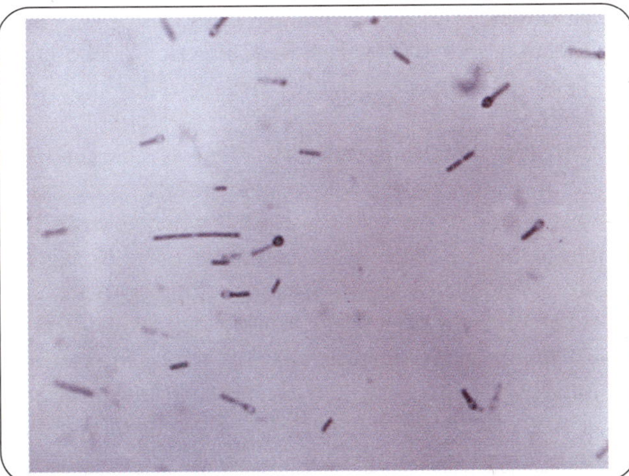

Fig. 18.10 Gram-positive bacilli with spherical terminal spores, Gram stain, 100X, e.g., *Clostridium tetani* (*Source:* Public Health Image Library, ID 12056 / CDC / Dr. Holdeman)

maternal and neonatal tetanus due to extensive antenatal vaccination. The reservoir of *C. tetani* are herbivorous animals—the bacterium resides in their intestine. Their excreta contaminates the soil and disseminates in the environment.

Pathogenesis and clinical features

Following the percutaneous contamination of wounded tissue, *C. tetani* produces two exotoxins—tetanolysin and tetanospasmin. Tetanospasmin is a neurotoxin that is responsible for tetanus. It is antigenic, can be toxoided, and neutralised by an antitoxin. It binds to and enters motor nerve terminals and prevents the presynaptic release of gamma-aminobutyric acid (GABA) and glycine, leading to spastic muscle contraction.

The bacilli enter through a wound caused by injury, surgery or obstetric delivery which is contaminated by the bacilli. Therefore, it is important to take the tetanus toxoid (TT) vaccine following injury and follow strict asepsis in surgical and obstetric practices. Neonatal tetanus was earlier seen in India due to unhygienic practices in home deliveries.

After an incubation period of 10 days, the effect of tetanospasmin is first seen to appear in the muscles of the jaw, causing a 'lockjaw'. Following pain and spasm of the jaw, the disease progresses to dysphagia, neck stiffness and a descending spastic paralysis, involving the back and extremities. If untreated, it will result in sustained spasm of the facial muscles, abnormal posture or respiratory spasm. Hence, it should be treated immediately.

The diagnosis is largely clinical as a wound may or may not be present. If present, the wound material can be examined by Gram stain for the presence of *C. tetani* and cultured anaerobically.

Prevention and treatment

The disease can be prevented by strict asepsis practices in surgery and obstetrics, proper wound toilet in case of injuries, and administering TT following injuries and in pregnancy. If disease occurs, the shorter the incubation period the graver is the prognosis. The disease progression can be arrested by passive immunisation with a specific antitoxin immediately. Human tetanus immunoglobulin (HTIG) is preferred over anti tetanus serum (ATS) of equine origin as it is devoid of hypersensitivity reactions and its effect lasts for 30 days. In non-vaccinated persons, passive immunisation is combined with active immunisation by administering the first single dose of TT on the other arm. The antibiotic indicated is penicillin or metronidazole, especially if the causative wound still exists.

Active immunisation contains a formol toxoid adsorbed on alum, given IM, with the protective antitoxin titre being > 0.01 unit/mL. In India, seven doses of primary immunisation are given to children at age 6, 10 and 14 weeks, 16–24 weeks, and 5, 10 and 16 years. If primary immunisation has not been given, an adult can be immunised with four doses of TT at 0, 1 month, 1 year and 6 years. At the antenatal care (ANC) visit, two doses of TT are indicated to the mother at 16–20 weeks and 20–24 weeks, to prevent neonatal tetanus. If the mother has a history of previous ANC vaccination in the last 5 years, only a single dose is indicated.

LYMPHADENITIS AND LYMPHANGITIS

Lymphadenitis is an acute or chronic inflammation of the lymph node that can be localised, regional or generalised. Histopathologically, it can be suppurative, granulomatous, non-suppurative or necrotising, depending on the microorganism involved.

Acute lymphadenitis with palpable regional nodes may occur due to localised infection—cutaneous, pharyngeal, dental, or of the extremities. Acute suppurative lymphadenitis caused by *S. aureus* or *S. pyogenes* is common in children. Other microorganisms rarely implicated are *Klebsiella, Salmonella, Pseudomonas* and *Nocardia*. In immunodeficiency, opportunistic pathogens may cause lymphadenitis, including *Histoplasma, Cryptococcus* and *Candida.*

Granulomatous lymphadenitis is most often caused by *M. tuberculosis* and rarely by *M. avium-intracellulare, Chlamydia trachomatis* (lymphogranuloma venereum [LGV]) and *Bartonella henselae* (cat scratch disease).

Inguinal lymphadenopathy is associated with sexually transmitted infections like LGV, primary genital herpes, or the non-STD *Bacillus anthracis* infection (bubonic plague). Generalised lymphadenopathy is a feature of secondary syphilis, HIV, infectious mononucleosis, acute toxoplasmosis, leptospirosis, military tuberculosis and filariasis.

Lymphangitis is an inflammation of the lymph channels, usually in the subcutaneous tissue. Acute lymphangitis is of bacterial origin caused by *S. pyogenes*, rarely by *S. aureus* or *Pasteurella multocida*, or due to filariasis. Chronic lymphangitis is rare and attributed to sporotrichosis, atypical mycobacteria or filariasis.

INFECTIVE ARTHRITIS

Infectious arthritis of single or multiple joints may be caused by bacteria, mycobacteria, viruses or fungi. **Bacterial arthritis**, also called **suppurative, pyogenic** or **septic arthritis,** is the most common. **Viral arthritis** involves multiple joints as part of a systemic infection. Arthritis due to mycobacteria or non-*Candida* fungi is usually chronic, slowly progressive and monoarticular. In addition to arthritis caused by direct infection of the joint by microbes, reactive or sterile arthritis is associated with systemic or local infection at a site away from the joint.

Infectious arthritis can be classified as:

1. **Acute bacterial arthritis:** Some bacteria exhibit tropism for the synovial membrane and possess surface components that adhere to it, while others produce toxins that can damage the joint, but in most cases it is the inflammatory response to the infection that promotes joint injury. Factors that predispose one to bacterial arthritis are rheumatoid arthritis or osteoarthritis, diabetes, chronic renal failure, chronic liver disease, immunosuppression and arthroscopic or surgical interventions. The most common causative agent is *Staphylococcus aureus*, followed by *Streptococcus pyogenes, Streptococcus pneumoniae, Haemophilus influenzae* and *Neisseria gonorrhoeae*. Gonococcal arthritis is a presentation of disseminated gonococcal infection, secondary to gonococcal mucosal infections. The disease is usually monoarticular, with the knee joint being the most commonly affected. Non-specific laboratory findings are elevated leucocytes, ESR and CRP. Specific findings are raised cell count, positive culture or PCR in synovial fluid. It is advisable to culture synovial fluid in blood culture bottles.
2. **Viral arthritis:** Or arthropathy due to viruses is a part of systemic viral fevers, either directly infecting the joint or through host mediated immune response, and is usually polyarticular. Viral infections that can involve the joints are chikungunya or other arbovirus infections, rubella, hepatitis B, hepatitis C, HIV infection and parvovirus B19 infection. Laboratory diagnosis is by detection of virus-specific antibodies or PCR.
3. **Chronic infectious arthritis:** This is characterised by insidious onset, slow course and progressive destruction, that is usually monoarticular or oligoarticular. The most common causative agents are *Mycobacterium tuberculosis* and other non-tubercular mycobacteria. Worldwide, about 10% of extrapulmonary tuberculosis cases involve the bones or joints. Other bacteria that can cause chronic infectious arthritis are *Borrelia burgdorferi, Treponema pallidum* and *Nocardia* or fungi *Candida* and *Cryptococcus*. The laboratory diagnosis is by microbe-specific serology, culture or PCR.
4. **Reactive arthritis:** Formerly referred to as Reiter's syndrome, this is arthritis along with conjunctivitis and urethritis. The disease affects different organs, and symptoms may or may not appear at the same time. It is mostly seen in HIV-infected individuals following concomitant non-gonococcal urethritis or shigella gastroenteritis, and is linked to HLA B27.

OSTEOMYELITIS

Infection of the bone termed 'osteomyelitis' may arise exogenously following trauma or endogenously from infected soft tissue (in diabetic foot ulcer), joints or blood (hematogenous). The most common causative agents are *S. aureus*, followed by *Salmonella typhi, S. pyogenes, S. pneumoniae, H. influenzae* and *Enterococcus* spp., and rarely *M. tuberculosis*. Factors predisposing osteomyelitis are listed in Table 18.3. The common sites of disease are the long bones and vertebrae in adults. Blood culture as well as biopsy culture of the infected site, if feasible, is indicated with intravenous antimicrobial therapy.

Osteomyelitis can also occur as postoperative infection in orthopedic surgery, including the placement of a prosthesis or fixation device in contaminated open

Table 18.3 Factors predisposing osteomyelitis

Predisposing factor	Bacterium
Infants	Group B streptococci
Children	*Haemophilus influenzae*
Sickle cell disease	*Salmonella* spp.
Immunosuppression	*Pseudomonas, Nocardia*
Trauma to the jaw	*Actinomyces*
Zoonotic	*Brucella* spp.
EPTB (with or without primary disease in lungs)	*Mycobacterium tuberculosis*

fracture. Vascular insufficiency and nerve damage in diabetics predisposes osteomyelitis subsequent to diabetic ulcer on the feet, caused by mixed bacterial flora (Table 18.3). Non-specific laboratory findings are elevated leucocytes, ESR and CRP with microbe-specific serology, culture or PCR.

ANSWERS

Clinical Case 1

1. *Staphylococcus aureus is a skin commensal that contaminated the wound and caused a pyogenic abscess at the local site*
2. *Aspiration and drainage of the pus, wound toilet under aseptic conditions with local application of an antibacterial ointment and oral antimicrobial therapy*
3. *Gram stain smear, culture and antimicrobial sensitivity test of the pus*
4. *Amoxicillin-clavulanic acid or cefazolin*
5. *Bacteremia, septicemia*
6. *Linezolid or daptomycin*

Clinical Case 2

1. *Dermis and subcutaneous tissue*
2. *Streptococcus pyogenes, Staphylococcus aureus and Haemophilus influenzae, the most common being S. pyogenes.*
3. *Risk factors include lymphedema, obesity and diabetes*
4. *Necrotising fasciitis or gangrenous cellulitis*
5. *Ceftriaxone or amoxicillin-clavulanic acid*

Clinical Case 3

1. *Non-sporing anaerobic bacteria, because the abscess is close to the pharyngeal mucosa, pus is foul-smelling and culture is negative, though gram-negative bacilli are seen*
2. *Bacteroides, Prevotella, Porphyromonas, Fusobacterium*
3. *Oral cavity, as Fusobacterium is a commensal in the mouth*
4. *Amoxicillin-clavulanic acid, ampicillin-sulbactam*

Clinical Case 4

1. *Gram stain and anaerobic culture of the discharge*
2. *Anaerobic myositis or gas gangrene*
3. *Clostridium perfringens, an anaerobic saprophytic, sporulating gram-positive bacillus. The wound, due to trauma, was contaminated with spores of the causative agent. The bacteria multiply and secrete exotoxins that cause myonecrosis and hypotension. Crush injury causes tissue anoxia, conducive to the growth of C. perfringens contaminating the wound. Poor oxygenation and impaired tissue perfusion reduce the redox potential, causing the spores to germinate and multiply. Exotoxins, lecithinase and collagenase secreted by the bacterium further the local destruction of tissue*
4. *Gangrenous appendicitis, necrotising enteritis and post-surgical intra-abdominal infections, especially in the biliary tract*

Clinical Case 5

1. *Tetanus caused by Clostridium tetani*
2. *Wound contamination and multiplication of C. tetani that produces exotoxin tetanospasmin, a neurotoxin. It binds to and enters motor nerve terminals and prevents the presynaptic release of gamma-aminobutyric acid (GABA) and glycine, leading to spastic muscle contraction. After an incubation period of 10 days, the effect of the tetanospasmin is first seen to appear in the muscles of the jaw, causing a 'lockjaw'.*
3. *If primary immunisation has not been given, an adult can be immunised with four doses of TT at 0, 1 month, 1 year and 6 years to prevent tetanus, followed by a single dose every 5 years.*
4. *Disease progression can be arrested by immediate passive immunisation with human tetanus immunoglobulin (HTIG). In non-vaccinated persons, passive immunisation is combined with active immunisation by administering the first single dose of tetanus toxoid on the other arm.*

Questions

1. Provide a brief account of the superficial skin infections caused by *Staphylococcus aureus* and *Streptococcus pyogenes*.
2. Compare and contrast the virulence factors of *Staphylococcus aureus* and *Streptococcus pyogenes*, and elaborate on their role in subcutaneous infections.
3. How does staphylococcal toxin-mediated disease differ from pyogenic staphylococcal infections?
4. Elucidate the importance of establishing the identity of the causative bacterium in a case of cellulitis.
5. Explain the difference in the pathologic process of cellulitis, non-clostridial anaerobic cellulitis and clostridial cellulitis.
6. How does clostridial cellulitis differ from clostridial myositis?
7. Elaborate on the role of non-sporing anaerobic bacteria in causing soft tissue infections.
8. How can mortality due to tetanus be prevented in a case presenting with early signs of the disease?
9. Compare and contrast acute bacterial, viral and chronic infective arthritis.
10. Enumerate the microorganisms causing osteomyelitis, its source of infection and risk factors.

Leprosy and Cutaneous Mycobacterial Infections

Mycobacterial infections of the skin include infection by ***Mycobacterium leprae***, *Mycobacterium tuberculosis* and non-tubercular mycobacteria (NTM).

LEPROSY

Clinical Case

A 24-year-old man presents with a hypopigmented macule and loss of sensation on the right lower back (Fig. 19.1).

1. *What will you clinically examine this case for?*
2. *How will you proceed to confirm your clinical diagnosis?*
3. *Explain the features based on which you would classify the disease.*
4. *What advice would you offer him on confirmation of the diagnosis?*

Leprosy is a chronic, granulomatous disease of the skin, nasal mucosa and peripheral nerves caused by *M. leprae,* a slow-growing acid-fast bacillus (AFB) that has not been cultured in vitro.

Epidemiology

Humans are the only source of this infection, which is transmitted predominantly by nasal droplets and rarely by prolonged close skin contact. The transmissibility being low, its incidence is decreasing worldwide. In India, the incidence of leprosy is highest in Bihar and Chhattisgarh, with the national prevalence in 2011 being less than 0.69 per 10,000 population. Two non-human sources that *M. leprae* can multiply in are the armadillo and foot pad of mice. It has not yet been cultivated in vitro, except for the Indian Cancer Research Centre (ICRC) bacillus, whose relation to *M. leprae* is uncertain. In 1962, the ICRC, Mumbai, isolated an AFB from leprosy patients in human fetal spinal ganglion culture.

Mycobacterium leprae

The bacterium has a capsule, cell wall, cell membrane and cytoplasm. The capsule contains phenolic glycolipid-1 (PGL-1) that is highly immunogenic. The cell wall contains peptidoglycans, mycolic acid and PGL-1. PGL-1 interacts with the laminin of Schwann cells (neural macrophages). The cytoplasm and cell membrane consist of polypeptides and proteins.

The generation time of the bacterium is 12 days, rendering an incubation period of 6 years due to slow multiplication. The bacterium enters through the nasal mucosa, is phagocytosed by macrophages and travels to cooler parts of the body—skin and peripheral nerves (Schwann cells)—and multiplies intracellularly within the macrophages.

Pathogenesis and clinical features

Depending on the immune resistance to infection, the clinical spectrum varies from paucibacillary (PB) or tuberculoid (TL) to multibacillary (MB) or lepromatous (LL) leprosy. Between these two poles rest the borderline forms of leprosy. The immunologically unstable borderline group has three categories: **borderline tuberculoid** (BT), **borderline** (BB) and **borderline lepromatous** (BL), BB being the most unstable form of leprosy. Additionally, indeterminate leprosy (IL) is an early and transient presentation in children that mostly self-heals and rarely moves to tuberculoid leprosy. The WHO classifies PB and MB disease based on the number of lesions in the patient: more than five lesions qualify as multibacillary disease and up to five lesions (including one nerve) as paucibacillary disease.

Tuberculoid leprosy is seen in patients with good cell-mediated immunity (CMI). It is characterised by anesthetic, dry, hairless skin lesions (less than five in number) with thickening of the nerve in the vicinity. Slit skin examination does not show AFB and the lepromin reaction is strongly positive.

In borderline leprosy, the immunity is unstable and patients may shift within the spectrum—upgrade or downgrade.

Lepromatous leprosy is seen in patients with low CMI. In the early phase, the patient develops diffuse brown,

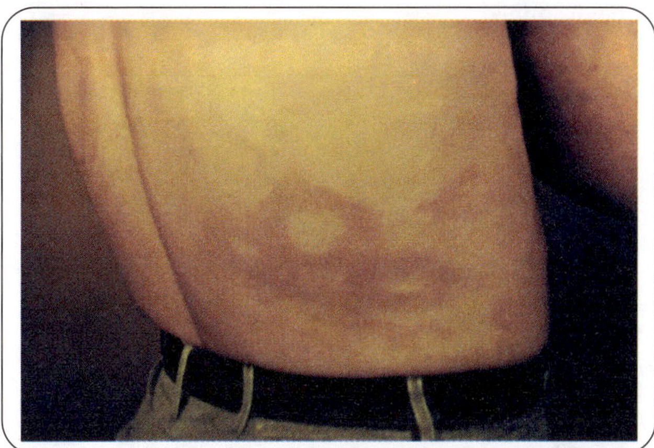

Fig. 19.1 Hypopigmented macule in lepromatous leprosy (*Source*: Public Health Image Library, ID 18971 / Centers for Disease Control and Prevention [CDC])

hypopigmented or erythematous macules (Fig. 19.1). If untreated, LL advances to thick ear lobes, collapse of bridge of the nose, loss of eyebrows and thickening of forehead skin. Glove and stocking anesthesia with skin papules, plaques and nodules occurs. Slit-skin examination shows plenty of AFB and the lepromin reaction is negative (Fig. 19.2).

The diagnosis of leprosy is based upon three cardinal signs, the presence of any one of which is sufficient: hypopigmented or erythematous patch with sensory impairment, enlarged peripheral nerve trunk/s or presence of AFB on slit-skin examination. The peripheral nerves commonly affected are the posterior tibial, peroneal, radial cutaneous, ulnar, median and great auricular nerves.

Laboratory diagnosis

Early diagnosis is important. Clinical examination alone may suffice, but the classification and monitoring of treatment needs laboratory help. The following tests are used for the laboratory diagnosis of leprosy:

1. Slit-skin smear (SSS)

The smear is prepared and stained by Ziehl–Neelsen (ZN), using 5% H_2SO_4 for decolourisation. The bacilli are 4 × 0.4 microns in size seen singly or in groups, free or intracellular, or bound together in masses known as 'globi' (Fig. 19.2).

The sites for demonstration of *M. leprae* are ear lobes, nasal mucosa, forehead skin lesions, the biopsy of nodular lesions and thickened nerves, or the puncture of lymph nodes. The morphological index is expressed as the percentage of uniformly stained bacilli of the total number of stained bacilli counted. The bacteriological index is calculated by the total number of pluses (+s) scored in all the smears and dividing this by the number of smears. For calculating the bacteriological index, a minimum of four skin lesions, a nasal swab and both the ear lobes have to be examined.

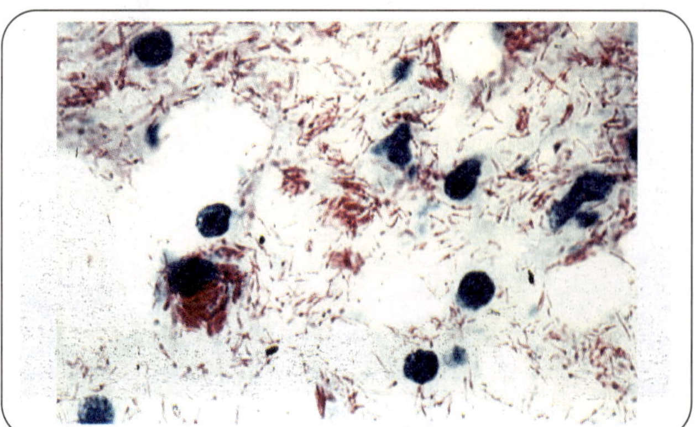

Fig. 19.2 Slit-skin smear showing AFB-*Mycobacterium leprae*, ZN stain, 100X

The smears are graded based on the number of bacilli between 1+ (1–10 bacilli in 100 fields) and 6+ (more than 100 bacilli in every field).

2. Skin biopsy

Histopathologic examination by H&E or tissue ZN called Fite-Faraco staining.

Tuberculoid leprosy is characterised by a granuloma of epithelioid cells, giant cells and lymphocytes with the absence of AFB. Lepromatous leprosy shows a macrophage granuloma, few lymphocytes and innumerable AFB.

3. Lepromin test

This is a delayed hypersensitivity skin test to assess the cell-mediated immunity to lepra bacilli and has **no diagnostic value.** It may be done to classify leprosy and monitor the therapeutic response. It also acts as a marker for candidate leprosy vaccines (lepromin seroconversion).

4. Other tests

These include serological tests using PGL-1 antibody to detect leprosy antigen in lepromatous cases and PCR assays in smear-negative cases. However, these tests are not routinely employed.

The WHO regimen for leprosy includes rifampicin, dapsone and clofazimine. Dapsone resistance and MDR leprosy have been reported, causing relapse of the disease; newer drug regimens are used in such cases. Vaccines with BCG + killed *M. leprae*/*Mycobacterium vaccae*/ ICRC bacillus have been tried with no success.

Immune reactions in leprosy

Reactions in leprosy are immunologically mediated episodes of inflammation affecting the skin, nerves or mucous membranes during the course of the disease.

The type 1 reaction (**lepra reaction**) is associated with sudden alteration of the CMI associated with a shift in the patient's position in the leprosy spectrum. This is usually seen in borderline cases. It is a delayed hypersensitivity reaction that may develop without anti-leprosy treatment and is localised to skin and nerves without constitutional symptoms.

The type 2 reaction (erythema nodosum leprosum) is associated with T_H2 humoral response leading to antigen-antibody complexes, usually in LL. It is an immune complex-mediated hypersensitivity that develops 3–6 months after treatment, with multi-tissue involvement and constitutional symptoms.

Management of leprosy

Like in TB, single drug therapy leads to drug resistance, hence, multi-drug therapy is recommended. The WHO regimen for leprosy includes dapsone and rifampicin in PB leprosy for 6 months. In MB leprosy, it is dapsone, rifampicin and clofazimine for 12 months. In case of drug resistance, the WHO has approved modifications to the above regimens.

Prevention

Prevention is by early detection and prompt treatment. The household contacts of patients with LL should be monitored annually for 5 years following diagnosis. Vaccines trials have been conducted with BCG, BCG + killed *M. leprae* or *M. vaccae* or ICRC bacillus.

CUTANEOUS TUBERCULOSIS (CTB)

Tuberculosis is predominantly a pulmonary disease with 10% being extrapulmonary and 1% showing skin involvement. Cutaneous tuberculosis is caused by *M. tuberculosis* and rarely by *Mycobacterium bovis*. It can be exogenous, by direct inoculation of the bacilli on the skin, or endogenous, from tuberculosis elsewhere in the body. The common forms of endogenous cutaneous tuberculosis are **scrofuloderma, lupus vulgaris** and **tuberculids**. Scrofuloderma results from a contiguous spread of underlying primary lymph node tuberculosis. This disease is more common in children, usually affecting the cervical lymph nodes, either due to *M. tuberculosis* or *M. bovis* ingestion by drinking unpasteurised milk. Lupus vulgaris is CTB secondary to a primary focus disseminated to skin via hematogenous or lymphatic spread. Tuberculids are a cutaneous immunologic reaction to *M. tuberculosis* in a patient with an immune response. The laboratory diagnosis of CTB is by demonstration of the AFB in ZN smear, culture on solid (Lowenstein–Jensen [LJ]) or liquid media (MGIT) and the detection of *M. tuberculosis* DNA by CBNAAT.

CUTANEOUS NON-TUBERCULOUS MYCOBACTERIAL INFECTIONS

Non-tuberculous mycobacteria, earlier referred to as 'atypical' or mycobacteria other than *M. tuberculosis* (MOTT), are species other than the *Mycobacterium tuberculosis* complex (*M. tuberculosis, Mycobacterium africanum, M. bovis*) and *M. leprae*. Based on the time taken to grow on culture, they are classified as **rapid growers** and **slow growers**. Rapid growers *Mycobacterium fortuitum*, *Mycobacterium chelonae* and *Mycobacterium abscessus,* and slow growers *Mycobacterium marinum* and *Mycobacterium ulcerans* commonly cause cutaneous infection. NTMs are ubiquitous in the environment, present in soil or water. Predisposing factors are wound contamination by environmental NTM, an immunocompromised patient or iatrogenic infection. Iatrogenic infection follows injections, endoscopic or surgical intervention with contaminated medical instruments or solutions. A cutaneous lesion caused by NTM may be a solitary localised skin infection (verrucous or ulcerative) or spread to the lymph nodes or bloodstream in immunocompromised patients. The sample collected from the site of infection is examined for AFB by ZN smear and cultured on LJ medium.

ANSWERS

Clinical Case

1. *Palpate for thickening of peripheral nerve and sensory loss*
2. *Slit-skin smear to demonstrate M. leprae using 5% ZN staining*
3. *The clinical spectrum varies from paucibacillary (tuberculoid) to multibacillary (lepromatous) leprosy. Tuberculoid leprosy presents with anesthetic, dry, hairless skin lesions (less than five in number) with thickening of the nerves in the vicinity. Slit-skin examination does not show AFB and the lepromin reaction is strongly positive. In lepromatous leprosy, the patient develops diffuse brown, hypopigmented or erythematous macules. If untreated, the disease advances to thick ear lobes, collapse of bridge of the nose, loss of eyebrows and thickening of forehead skin. Glove and stocking anesthesia with skin papules, plaques and nodules occurs. Slit-skin examination shows plenty of AFB and the lepromin reaction is negative.*

QUESTIONS

1. Elaborate on the pathogenesis of leprosy from infection to disease. What is its incubation period and why is it so?
2. Compare and contrast lepromatous and tuberculoid leprosy.
3. Give an account of the role of MTB and NTM in cutaneous disease.

Viral Infections With Dermal Manifestations

Viruses can cause a variety of cutaneous morphologic manifestations, like macular, maculopapular, papular and vesicular rashes, warts, etc. Some viruses primarily affect the skin, others secondarily, and some establish latent infections. Diagnosis of exanthematous dermatologic viral infections is largely based on clinical examination of the skin, as most are self-limiting and do not require additional laboratory tests or specific treatment.

Based on the morphology of the skin lesions, the commonly infecting viruses are:
- Macular — Rubella virus
- Maculopapular — Measles virus, parvovirus
- Maculopapular, petechial — Viral hemorrhagic fevers, e.g., dengue
- Vesicular — Herpes simplex virus, coxsackievirus
- Vesiculopapular — Varicella zoster virus
- Papulovesiculopustular — Vaccinia virus, variola virus, cowpox virus
- Papular (warts) — Molluscum contagiosum virus, HPV

CHILDHOOD FEVERS

A childhood disease in which a skin eruption or rash is the predominant manifestation is often referred to as an **exanthematous disease** or a childhood **exanthematous fever**. It may or may not be accompanied by mucosal manifestations (enanthems). It is epidemic and contagious, transmitted by respiratory droplets. In India, the common childhood fevers are measles, rubella, chickenpox and parvovirus infection, amongst which rubella and chickenpox are teratogenic viruses during pregnancy. The dermal manifestations of viruses infecting the skin are shown in Table 20.1.

Table 20.1 The dermal manifestations of viruses infecting the skin

	Disease/Virus	Dermal manifestation
1.	Measles	Maculopapular rash
2.	Rubella/German measles	Maculopapular rash
3.	Chickenpox/ Varicella zoster Herpes zoster	Vesicles become purulent followed by crusted lesions Shingles characterised by painful vesicles over specific dermatomes
4.	Parvovirus B19	Erythematous plaques on the malar eminences on cheeks
5.	Herpes simplex virus (HSV)	Vesicles on lip and genital area

Measles (Rubeola)

Clinical Case 1

A 10-year-old boy presents with an erythematous maculopapular rash on the forehead, behind the ears, on the neck and extremities (Fig. 20.1). On examination, the pharynx shows lesions as seen in Fig. 20.2.

1. *What is your clinical diagnosis?*
2. *What other history would you enquire about from the parent?*
3. *What advice and counsel would you offer to the parent?*
4. *How can this disease be prevented?*
5. *Name the four other childhood fevers.*

Epidemiology

It is a global disease with epidemic potential. In India, it is the third most common cause of death among those under the age of 5 years. Since the implementation of immunisation, we find an average of 25,000 cases reported annually in India. In 2012, it was found that 14% of the total global mortality due to measles was from India. The WHO has called for the global eradication of measles by 2020, a goal which is being actively pursued in India with the measles-rubella vaccination campaign.

Virus: Paramyxovirus, single-stranded RNA, enveloped virus

Pathogenesis and clinical features

Measles has a worldwide distribution and its incidence is declining due to vaccination. The infection is highly contagious. It is transmitted in humans from the upper

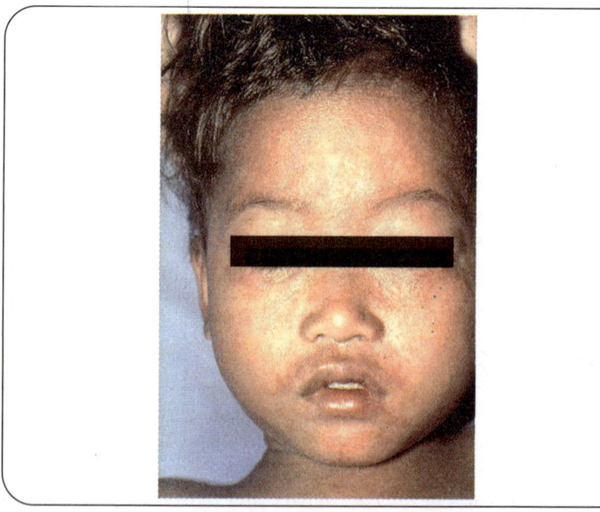

Fig. 20.1 Erythematous maculopapular rash in measles

respiratory tract, with the patient being infectious for 2 days before the onset of symptoms to 4 days after the appearance of rash. The infection occurs in the mononuclear phagocytic system followed by viremia that causes fever, with deposition of the viral antigens in the capillaries. The cell-mediated immune response to these antigens leads to the characteristic macular rash 4 days after the onset of fever.

Measles is characterised by a transient decrease in cell-mediated immunity for a period of 1–2 months that could lead to infection of the brain, causing sub-acute sclerosing panencephalitis (SSPE) as a late sequela. Natural infection confers lifelong immunity.

The incubation period of 10 days is followed by prodromal symptoms of fever, malaise, conjunctivitis, upper respiratory involvement and appearance of Koplik spots (Fig. 20.2). Koplik spots, 1–2 mm in size, are bluish-white spots surrounded by bright red macules on the buccal mucous membrane.

Subsequently, the rash appears in 3 days and disappears spontaneously in 4 days with residual desquamation. It is characterised by erythematous macules and papules starting from the forehead and behind the ears and spreading to the neck, trunk and extremities (Fig. 20.1). This uncomplicated, self-limiting disease lasts for 10–12 days and is diagnosed clinically without laboratory aids. Infection can spread to cause otitis media, laryngotracheobronchitis, hepatitis or thrombocytopenia.

Complications of pneumonia, encephalitis and death are rare. They are seen in malnourished and immunocompromised patients and need laboratory confirmation by virus-specific antigen, antibody detection or histopathologic examination. One in 10,000 people with measles can develop SSPE due to a mutated form of the virus, especially if primary measles occurs before the age of 2 years. In these cases, the virus mutates and is able to survive in the neurons and the glial cells. SSPE develops 7–10 years after primary measles. It is characterised by convulsions and progressive deterioration of cognitive and motor function. The condition is diagnosed by detecting measles virus antibodies in the CSF.

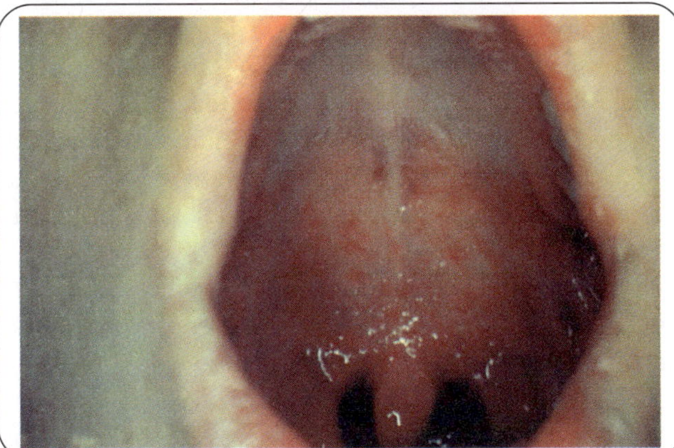

Fig. 20.2 Koplik spots in measles (*Source:* Public Health Image Library, ID 3187 / Centers for Disease Control and Prevention [CDC] / Heinz F. Eichenwald, MD)

A single-dose, live attenuated, cell-cultured vaccine provides lifelong protection in 95% of individuals. A single dose vaccine is given between 9 and 12 months of age along with vaccines for mumps and rubella. Presently, a second dose is being provided in the selected states of Karnataka, Tamil Nadu (TN), Goa, Puducherry and Lakshadweep.

Rubella (German measles)

Clinical Case 2

A 22-year-old married lady presents with a rash, of discrete pink macules that become confluent as a diffuse erythema, on the face, neck and chest (Fig. 20.3). She gives a history of fever, malaise, myalgia, cough and sore throat for 3 days before the appearance of rash.

1. *What is your clinical diagnosis?*
2. *What important family history would you like to elicit in her case?*
3. *What subsequent advice would you like to give? And why?*
4. *What is the national immunisation program to prevent this infection?*

Virus: Non-arthropod borne single-stranded RNA togavirus

Pathogenesis and clinical features

Infection is transmitted by the respiratory route by nasopharyngeal droplet secretions 7 days prior to the onset of symptoms and until 14 days after, and provides lifelong immunity.

Rubella affects children, adolescents and young adults, with prodromal symptoms of fever, malaise, myalgia, cough and sore throat for 3 days, followed by the appearance of rash. The soft palate may show the presence of red macules or petechiae in the prodromal phase. Generalised lymphadenopathy involving the suboccipital, postauricular and cervical glands occurs 1 week before the rash. The rash first appears on the face and spreads downwards, consisting of discrete pink macules that become confluent as a diffuse erythema that subsequently fades (Fig. 20.3). After the rash subsides, rarely, arthritis may develop in women, involving small and large joints.

Rubella infection during pregnancy runs the risk of causing congenital rubella syndrome, as follows:

1. Between 1 and 11 weeks of pregnancy there is 85% risk of developing multiple defects (cardiac and cataract)

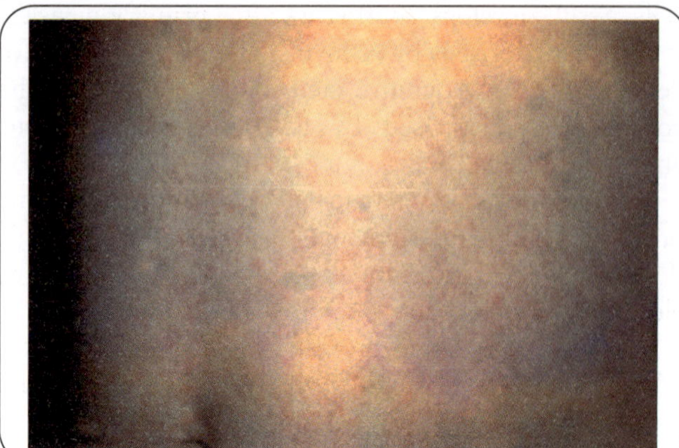

Fig. 20.3 Discrete erythematous pink macules in rubella (*Source*: Public Health Image Library, ID 4514 / CDC)

2. Between 12 and 16 weeks of pregnancy there is 35% risk of developing nerve deafness and mental retardation
3. After 16 weeks of pregnancy there is no risk to the fetus

Maternal rubella is associated with abortions and stillbirths. In the event of a live birth, the severity and multiplicity of the defects are increased the earlier in pregnancy the infection occurs. The neonate is likely to have hepatosplenomegaly, thrombocytopenic purpura, jaundice, anemia and low birthweight. Laboratory confirmation is by the presence of IgM antibodies in the patient and in the neonate, if infected.

A single-dose, live attenuated, cell-cultured Wistar strain vaccine provides lifelong protection in 95% of individuals. The vaccine is given at age 9–12 months along with the vaccines for mumps and rubella (MMR). Vaccination is contraindicated 1 month before or during pregnancy and in immunocompromised individuals.

Chickenpox

There are eight pathogenic human herpesviruses: herpes simplex virus (HSV) type 1, HSV type 2, varicella zoster (VZ) virus, cytomegalovirus, Epstein–Barr virus (EBV), human herpes virus (HHV)-6, HHV-7 and HHV-8.

Clinical Case 3

A 12-year-old girl presents with fever and a pruritic rash crop from macules to papules to vesicles which initially started on her trunk and later spread to the extremities (Fig. 20.4). She lives in a family of four with her parents and younger brother.

1. *What is your diagnosis?*
2. *What factors would influence the likelihood of her family being infected?*
3. *What are the complications of this disease?*
4. *Where does the virus reside in latency? What can it cause?*

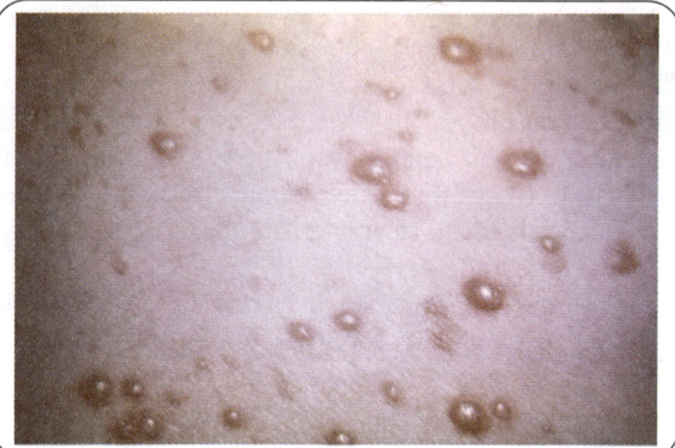

Fig. 20.4 Vesicular rash in chickenpox

Epidemiology

A worldwide disease in endemic and epidemic form, it is most common in children aged 5–9 years. Its incidence has been declining since the introduction of the vaccine. Chickenpox is transmitted by droplet infection and the transplacental route.

Virus: Varicella (chickenpox) and **zoster** (shingles) are diseases due to the VZ virus—a double-stranded DNA herpesvirus. Varicella is the primary illness and zoster the reactivation of the infection.

Pathogenesis and clinical features

Chickenpox is transmitted by the respiratory route via droplet infection acquired by contact with either varicella or (less commonly) zoster cases. The incubation period is 14 days and infectivity period is from 48 hours before the onset of rash until all skin lesions have fully crusted.

Following entry, the virus replicates in the lymph nodes and spills into a primary viremia with seeding into the MPS. Subsequently, secondary viremia occurs that persists through the rash, which is self-limiting.

The prodromal symptoms include fever, malaise, pharyngitis and loss of appetite. The rash is itchy and comes in successive crops, moving from macules to papules to vesicles (with an erythematous base) to crusts in 1 week, with a centripetal distribution on the body (Fig. 20.4). Secondary bacterial infection of varicella skin lesions with *Staphylococcus aureus* or *Staphylococcus pyogenes* may occur.

A single episode confers lifelong immunity to varicella, but latency of the virus in the affected sensory dorsal root ganglion may reactivate to cause zoster.

Chickenpox is a benign illness in children while in adults it may be associated with pneumonitis, encephalitis, hepatitis or other organ disease leading to death. Chickenpox in pregnancy contracted before gestational age 20 weeks is associated with microcephaly, limb hypoplasia, skeletal defects, deafness, cataract and chorioretinitis. If contracted after 20 weeks in pregnancy, it is associated with congenital varicella in the newborn, characterised by vesicles on the skin and CNS involvement.

Laboratory diagnosis

Diagnosis is largely clinical, based on the morphologic appearance of the skin lesions and does not depend on the laboratory. If necessitated, in the case of pulmonary or CNS involvement, VZ antigen detection in specimens with immunofluorescence or immunoperoxidase, or PCR for VZ DNA can be done. **Tzanck smear**—scraping from the ulcer base to look for Tzanck cells (acantholytic cells, multi-nucleated giant cells)—is useful, but unable to distinguish between VZ and HSV infection.

Prophylaxis

The live attenuated, cell-cultured Oka strain vaccine is 80% efficacious in preventing varicella infection and 100% efficacious in preventing severe varicella. It is recommended as double doses at 15 months and 4–6 years. Unimmunised adults and adolescents require two doses 1–2 months apart. Its use is contraindicated 1 month before or during pregnancy and in immunocompromised individuals.

Passive immunisation with varicella zoster immunoglobulin (VZIG) is indicated in neonates, immunocompromised children and pregnant women with close exposure to a patient of varicella or zoster. Acyclovir is an effective alternative to passive immunisation.

Herpes zoster

Clinical Case 4

An 85-year-old lady presents with intensely painful lesions since 2 days (Fig. 20.5).

1. *What is your diagnosis?*
2. *Explain the pathogenesis of these lesions.*
3. *What would you advise the patient regarding the future course of the disease?*
4. *Is this an indication for antiviral medication?*

Herpes zoster (HZ) occurs following reactivation of the latent VZ virus in the affected sensory root ganglion. It is seen in those older than 60 years of age and in the immunocompromised. HZ can cause primary varicella by the respiratory route in non-immune persons.

The latent VZ virus is reactivated due to any decline in immunity and destroys the ganglion, and travels along the sensory nerves causing painful varicella skin lesions along the dermatome (Fig. 20.5). The lesions are commonly found in the thoracic dermatome and show multinucleated giant cells with intranuclear inclusion bodies.

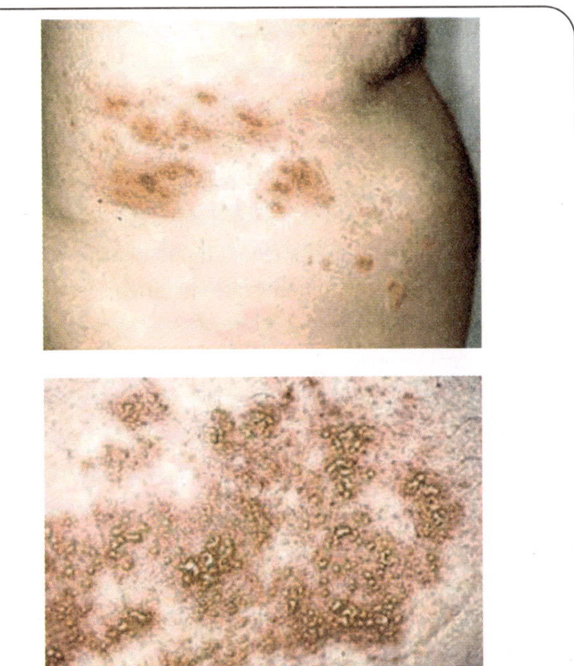

Fig. 20.5 Rash along the dermatome in herpes zoster (*Source*: CDC)

Acute herpetic neuralgia—the pain that accompanies the rash—is stinging, agonising and lasts for 30 days. If it persists for 30–120 days, it is considered **subacute neuralgia** and beyond 120 days **postherpetic neuralgia** (PHN).

Complications are neurologic, such as PHN, encephalitis, transverse myelitis, the Guillain–Barré syndrome and peripheral motor neuropathy, and ocular, such as keratitis, iridocyclitis and panophthalmitis. Typical HZ is clinically diagnosed without laboratory tests and treated with antiviral acyclovir.

Parvovirus B19

Parvovirus B19 is known to cause erythema infectiosum, a disease that can occur at all ages, across the world and throughout the year.

Pathogenesis and clinical features

Parvovirus is transmitted by the inhalation of droplets, patients being infective before the appearance of the rash. The incubation period is 10 days and asymptomatic. The rash begins on the cheeks as erythematous plaques on the malar eminences. It fades and is followed by the appearance of erythematous macules and papules on the neck, trunk and extensor aspect of the extremities, lasting for 10 days or rarely longer.

The rash may be accompanied by acute symmetric polyarthropathy, flu-like symptoms or pruritis. The virus is known to infect RBCs, giving rise to a transient aplastic anemia as a complication.

Parvovirus B19 is also associated with papular-purpuric glove and sock syndrome. Fetal infection, though rare, may cause spontaneous abortion, hydrops fetalis, congenital anemia or fetal death. Currently, neither specific treatment nor a vaccine is available for parvovirus infection.

Herpes simplex virus

Virology

HSV is a dsDNA virus that causes a large spectrum of diseases, the primary lesion being a vesicle. There are two types of HSV that are related but can be differentiated. HSV type 1 is common and mostly causes oro-facial lesions but can infect the genitals. HSV type 2 is the main cause of genital herpes.

HSV prevalence is worldwide, and virtually all humans get infected at some time in life. Infection with HSV-1 is common in children due to close personal contact. HSV-1 is also spread by kissing and HSV-2 by sexual intercourse. Both types display asymptomatic latency after disease, that can be a source of infection. Disease due to the viruses can be **primary**—when the virus is first encountered—or reactivation due to the latent virus.

Primary infections are mostly asymptomatic. If symptomatic, they can present as vesicles inside the mouth and gums (gingivostomatitis), conjunctivitis and keratitis, or vesicular eruptions on the genitals. Rarely, primary infection may lead to acute necrotising encephalitis. Primary genital infection during pregnancy can cause neonatal infection, characterised by vesicular lesions on the skin, jaundice and thrombocytopenia. During primary infection, HSV-1 travels via the nerves, from the mouth to the trigeminal and other cranial and cervical ganglia and HSV-2 to the sacral ganglia.

Reactivation of the virus is provoked by stress, common cold, menstruation, sunlight or organ transplant immunosuppression. Reactivation happens sporadically and at times recurrently throughout life, manifested by vesicles around the mucocutaneous junction of the nose and mouth, genitals, or keratitis (Fig. 20.6).

Laboratory diagnosis

Diagnosis is by demonstration of HSV-1 or 2-specific antigens or antibodies by ELISA, or DNA by PCR only in the case of complications. In these cases, antiviral therapy with acyclovir is recommended.

Other viral infections with dermal manifestations

1. **Viral hemorrhagic fevers** (VHFs)
   a) Chikungunya: maculopapular rash on the trunk and limbs during disease and residual facial pigmentation after recovery.
   b) Dengue: Maculopapular generalised rash during febrile illness and petechiae in the case of hemorrhagic manifestation.
   c) Other VHFs may show petechiae as a hemorrhagic manifestation.
2. **Coxsackievirus and echoviruses** cause hand, foot and mouth disease characterised by oral vesicular eruptions.
3. **Herpesviridae group of viruses:**
   a) Cytomegalovirus infection can present with a wide variety of skin lesions.
   b) Epstein–Barr virus causes infectious mononucleosis, characterised by fever, lymphadenopathy and pharyngitis. The rash is erythematous maculopapular on the trunk and extremities with palatal petechiae. EBV is associated with oral hairy leukoplakia, cutaneous T cell lymphoma and other uncommon cutaneous disorders.
   c) Human herpes virus 6 (HHV-6) causes roseola infantum or exanthema subitum in children, characterised by erythematous maculopapular rash on the neck and trunk.
   d) Human herpes virus 8 (HHV-8) is associated with Kaposi's sarcoma, a systemic neoplasm affecting the skin and viscera.
4. **Sexually transmitted infections:**
   a) Hepatitis B rarely manifests an erythematous macular rash.
   b) Hepatitis C is associated with a variety of cutaneous manifestations, like vasculitis, lichen planus, etc.

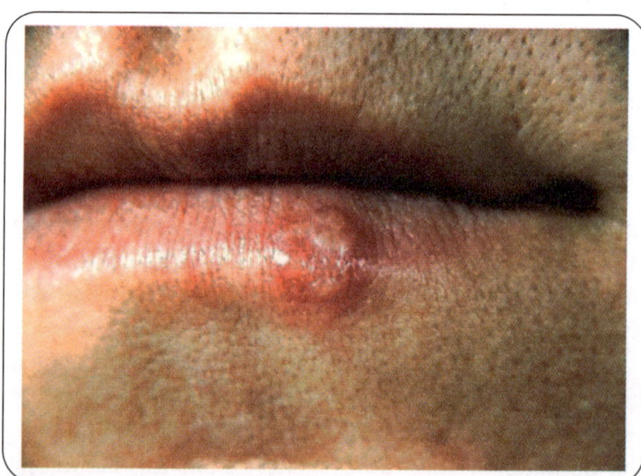

Fig. 20.6 HSV infection (*Source*: Public Health Image Library, ID 1573 / CDC / Dr. Herrmann)

c) HIV infection is associated with Kaposi's sarcoma and opportunistic skin infections of tuberculosis, candidiasis, dermatophytosis, herpes simplex, herpes zoster, molluscum contagiosum, etc.

5. **Pox group of viruses:**
   a) Smallpox caused by variola has been eradicated since 1980 with effective vaccination. Cowpox is seen in Europe, while camelpox and buffalopox have been encountered in India.
   b) Molluscum contagiosum is a benign infection transmitted by mechanical trauma in children and sexually in adults. The characteristic lesions are single or multiple pink, umbilicated papules 3 mm in diameter. In children, the lesions are on the face, chest or extremities while in sexually active adults they are limited to the perineum, genitalia or buttocks. Histopathologic examination shows large, eosinophilic, intracytoplasmic inclusion bodies.

WARTS

Warts or verruca are benign proliferations of skin and mucosa caused by human papillomavirus (HPV). Warts are common non-genital or sexually transmitted anogenital warts. Transmission of non-genital warts occurs by non-sexual, close person-to-person contact or fomites.

Virology

HPV is a dsDNA virus that replicates in keratinocytes, with many numerical serotypes.

Pathogenesis

HPV enters and multiplies in keratinocytes, largely causing benign cutaneous warts, or benign mucosal anogenital warts or laryngeal papillomas. However, HPV has also been associated with malignant lesions, the foremost being cervical cancer, showing 90–100% causality. Cervical infection with HPV is a precursor to cervical cancer along with other cofactors that contribute to it. HPVs 16 and 18 are the most carcinogenic types, associated with invasive cervical carcinoma, the others being HPVs 31, 33, 35, 39, 45, 51, 52 and 58. Other malignancies associated with HPV are genital mucosal carcinoma, oropharyngeal carcinoma, laryngeal carcinoma and esophageal carcinoma.

Current HPV vaccines include genetically engineered type-specific viral capsid protein L1 and are bivalent (types 16 and 18) or quadrivalent (types 6, 11, 16 and 18). Both are recommended intramuscularly, scheduled at 0, 2 and 6 months in three doses for women aged 9–26 years.

ANSWERS

Clinical Case 1

1. *Measles*
2. *Whether the patient has siblings and if yes, their age and symptoms, if any*
3. *To avoid close contact with siblings until 4 days after the appearance of the rash*
4. *A single-dose, live attenuated vaccine provides lifelong protection in 95% of individuals. It is given at 9–12 months along with the vaccines for mumps and rubella*
5. *Mumps, rubella, chickenpox and parvovirus B19*

Clinical Case 2

1. *Rubella or German measles*
2. *Whether she is pregnant, as rubella infection during pregnancy runs the risk of causing congenital rubella syndrome*
3. *To avoid pregnancy until recovery*
4. *A single-dose, live attenuated vaccine provides lifelong protection in 95% of individuals. It is administered at 9–12 months along with the vaccines for mumps and rubella (MMR). Its use is contraindicated 1 month before or during pregnancy and in immunocompromised individuals*

Clinical Case 3

1. *Chickenpox*
2. *The patient is infective till all lesions are crusted*
3. *Secondary bacterial infection in children. In adults, it may be associated with pneumonitis, encephalitis, hepatitis or other organ disease leading to death*
4. *The virus resides in the affected sensory dorsal root ganglion and may reactivate to cause herpes zoster*

Clinical Case 4

1. *Herpes zoster*
2. *The latent VZ virus is reactivated due to any decline in immunity and destroys the ganglion, and travels along the sensory nerves causing painful varicella skin lesions along the dermatome*
3. *The painful rash is likely to last for 30 days and is infectious. It can cause chickenpox in non-immune persons*
4. *Yes. Treated with acyclovir*

Questions

1. Describe the distinguishing clinical features of skin rash among childhood exanthematous fevers that aid in diagnosis.
2. Enumerate childhood viral diseases and discuss the late sequelae presentations of each, if any.
3. What is the likelihood of congenital rubella syndrome if rubella occurs during pregnancy? How can rubella be prevented?
4. Give a brief account of non-sexual HSV infection.

Superficial and Subcutaneous Mycotic Infections

Cutaneous mycosis can be superficial (confined to the stratum corneum), subcutaneous (by exogenous inoculation) or the dermal manifestation of a systemic fungal infection. The common superficial fungal infections are dermatophytosis, pityriasis versicolor and cutaneous candidiasis. The sources of infection are spores from the soil, animals or humans, through contagious contact.

DERMATOPHYTOSIS

Dermatophytes are fungi that have the ability to breakdown and digest keratin in skin, hair and nails. The human dermatophytes (asexual forms) are of three genera: *Microsporum*, *Trichophyton* and *Epidermophyton*. These genera can be identified by the microscopic appearance of fungal growth and not from the microscopic examination of skin scrapings. Predisposing factors are trauma, occlusion, marching, sweaty socks, moisture and maceration.

Clinical Case 1

A 30-year-old man presents with lesions on the trunk that appear as red annular patches or plaques with active borders showing scaling and papules that tend to spread peripherally and clear centrally (Fig. 21.1). The microscopic examination of the fungal culture on lactophenol cotton blue (LPCB) mount shows (Fig. 21.2).

1. *Describe the morphologic appearance in the above LPCB mount.*
2. *What is your clinical and microbiologic diagnosis?*
3. *Which layer of the skin does it affect?*
4. *Name three genera of dermatophytes.*

According to the site of infection, they are classified as **tinea corporis** (body, except palms, soles and groin), **tinea barbae** (beard), **tinea faciei** (face), **tinea capitis** (scalp), **tinea pedis** (athlete's foot), **tinea manuum** (hands), **tinea cruris** (groins, perianal, perineal) and **tinea unguium** (nails [onychomycosis]). Of these, tinea0 corporis, tinea cruris and tinea capitis are commonly seen.

The body (tinea corporis) lesions appear as pink-to-red annular patches or plaques with raised borders that tend to spread peripherally and clear centrally, and are commonly caused by *Trichophyton rubrum*, *Trichophyton mentagrophytes* or *Microsporum canis* (Fig. 21.1). The scalp (tinea capitis) lesions are round patches of dry scaling or alopecia that is contagious (Fig. 21.3). The groin infection (tinea cruris) shows pruritic, erythematous ring lesions, usually bilateral and extending to the inner thighs (Fig. 21.4). It is commonly caused by *T. rubrum* or *T. mentagrophytes*.

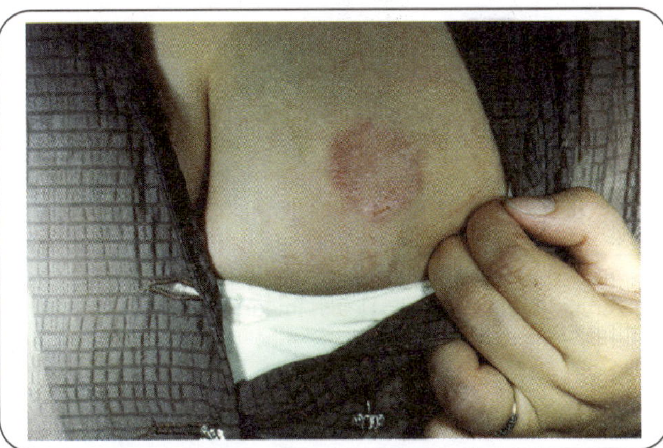

Fig. 21.1 Red annular patches with clear centre and active borders in tinea corporis (*Source:* Public Health Image Library, ID 22271 / Centers for Disease Control and Prevention [CDC] / Dr. Lucille K. Georg)

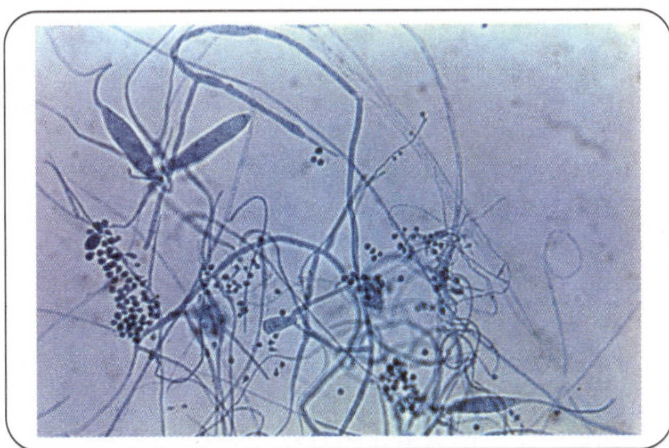

Fig. 21.2 *Trichophyton mentagrophytes*, LPCB mount, 40X (*Source:* Public Health Image Library, ID 22031 / CDC / Dr. Lucille K. Georg)

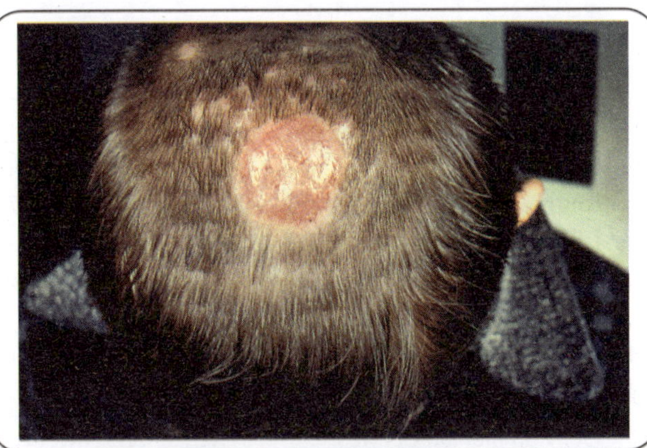

Fig. 21.3 Round patches of dry scaling with alopecia in tinea capitis (*Source:* Public Health Image Library, ID 22272 / CDC / Dr. Lucille K. Georg)

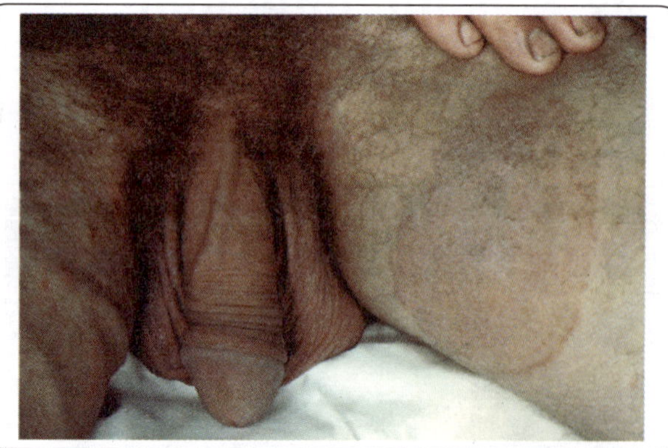

Fig. 21.4 Erythematous, annular ring lesions of the groin in tinea cruris (*Source:* Public Health Image Library, ID 21482 / CDC / Dr. Lucille K. Georg)

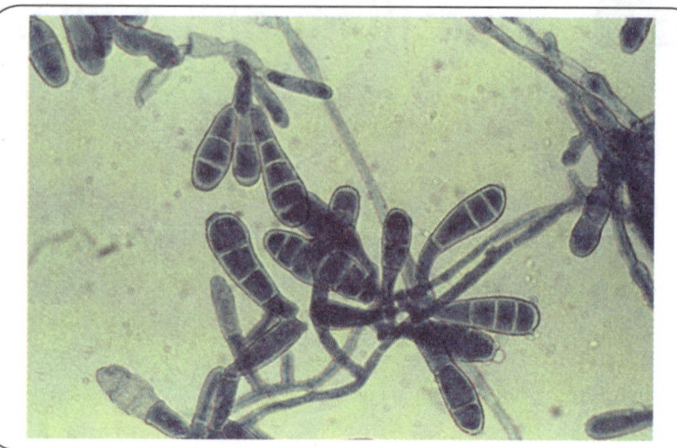

Fig. 21.5 *Epidermophyton* spp. microscopic appearance on LPCB, 40X (*Source:* Public Health Image Library, ID 14588 / CDC / Dr. Lucille K. Georg)

Laboratory diagnosis of dermatophyte infection

1. **Microscopy:** Samples, such as those from skin scraping, nail clippings or hair close to the scalp are to be collected in a dry folded piece of paper. The same is mounted on a glass slide with 10–30% KOH and examined under 40X. Nails need to be kept in KOH for a longer time, maybe overnight. On microscopic examination of KOH mount under 40X, the dermatophyte appears as thin, septate, branched, hyaline hyphae.
2. **Culture of the sample on Sabouraud's dextrose agar (SDA):** Depending on the genera and species, the dermatophyte will take 7–10 days for visible growth that can be identified by macroscopic (surface, texture, colour) and microscopic (microconidia, macroconidia) examination (Figs. 21.2, 21.5 and 21.6) in an LPCB mount as shown in Table 21.1.

Treatment of dermatophytes is specific to the site of infection and involves topical and systemic use of antifungal agents. The systemic agents used are griseofulvin, azoles and terbinafine. The topical agents being azoles, terbinafine, tolnaftate, ciclopirox and amorolfine.

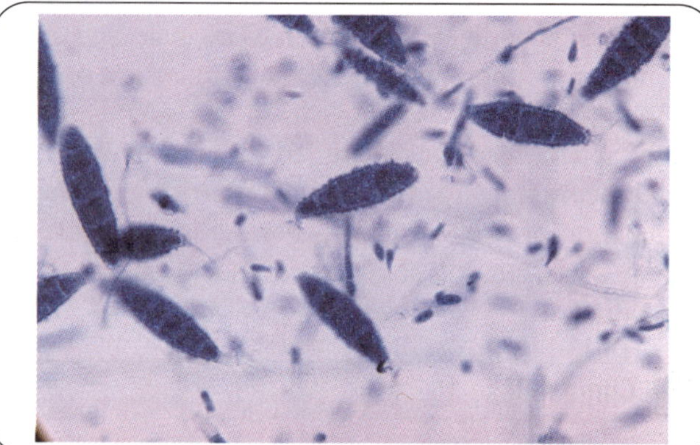

Fig. 21.6 *Microsporum* spp. microscopic appearance on LPCB, 40X (*Source:* Public Health Image Library, ID 21580 / CDC / Dr. Libero Ajello)

Pityriasis versicolor (Tinea versicolor)

The infection of the stratum corneum caused by *Malassezia* spp. is characterised by hypopigmented or hyperpigmented macular lesions on the trunk (Fig. 21.7). The LPCB mount from culture shows a combination of hyphae, 4 μ wide, and spherical yeast cells 8 μ in size.

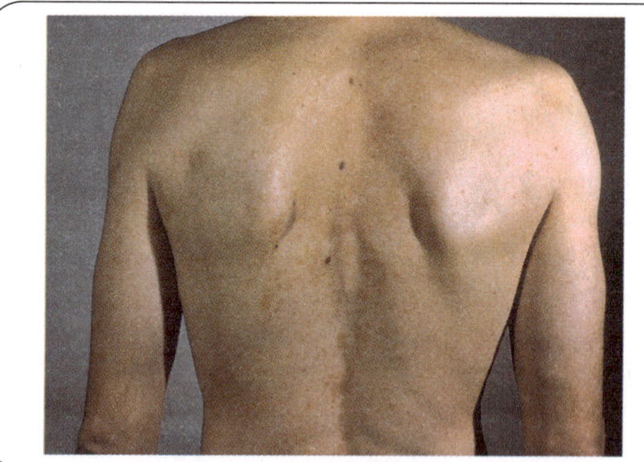

Fig. 21.7 Well-defined hypopigmented macular lesions in tinea versicolor (*Source*: Public Health Image Library, ID 12534 / CDC / Dr. Lucille K. Georg)

Table 21.1 Macroscopic and microscopic morphology of dermatophytes

Dermatophyte	Macroscopic	Microscopic morphology
Trichophyton mentagrophytes	White-tan, powdery	Microconidia numerous, round/pyriform-shaped
Epidermophyton floccosum	Yellow-green, powdery	Macroconidia numerous, club-shaped
Microsporum audouinii	Brown, velvety	Macroconidia numerous, spindle-shaped

Treatment agents are selenium sulphide, ketoconazole and terbinafine.

Cutaneous candidiasis

Candidiasis is the most common opportunistic fungal disease, causing superficial and deep infections. *Candida* is a yeast-like fungus with pseudohyphae, the common species being *Candida albicans*. The predisposing factors are extremes of age, pregnancy, low immunity, antibiotics and diabetes.

Pathogenicity and clinical features

Candida possesses adhesins, enzymes, toxins and the ability to produce pseudohyphae. The presence of pseudohyphae in lesions implies active infection and not its mere presence as a commensal. The diseases caused by *Candida* are:

1. Mucosal — oropharyngeal infection (oral thrush), vulvovaginitis, ocular infection, esophageal candidiasis, stomatitis and balanitis
2. Cutaneous — diaper rash, paronychia (nails), interdigital candidiasis
3. Deep/invasive/systemic — urinary tract infection, meningitis, pulmonary, endocarditis

Superficial candidiasis includes candida infection of the skin, nail and mucosae (oral, urethral, vaginal) (Fig. 21.8). Infection of the skin and nail can be diagnosed by microscopic examination and culture identification. The specimen in KOH mount (400X) or Gram stain shows budding yeast cells sized around 7–10 μm with pseudohyphae (Fig. 21.9). Samples can be cultured on SDA and identified. Treatment is by the systemic use of azole compounds topically or orally. Onychomycosis or fungal infection of the nail is common and presents with nail discoloration (Fig. 21.10). It can be caused either by dermatophytes or by *Candida*.

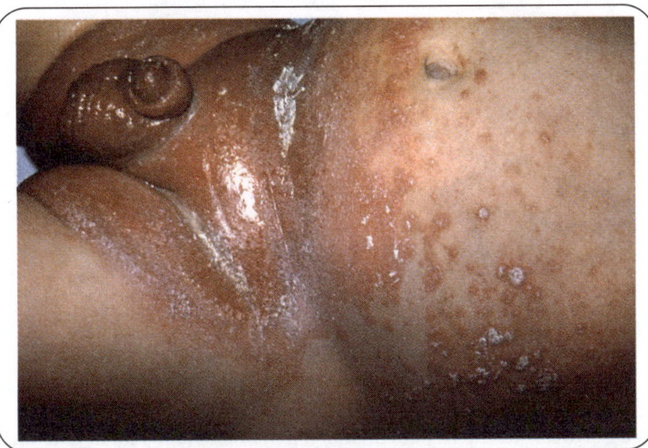

Fig. 21.8 *Candida albicans* diaper rash (*Source:* Public Health Image Library, ID 5143 / CDC)

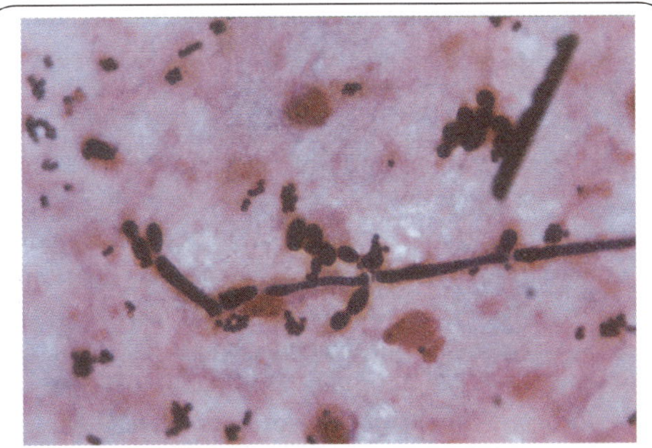

Fig. 21.9 Gram-positive budding yeast cells with pseudohyphae—*Candida albicans,* Gram stain, 100X

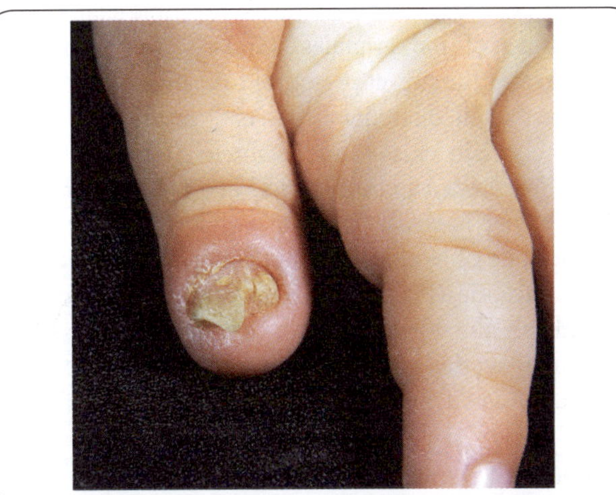

Fig. 21.10 Candidal onychomycosis (*Source:* Public Health Image Library, ID 15669 / CDC / Mr. Gust)

SUBCUTANEOUS MYCOSES

Subcutaneous mycoses are chronic infections occurring from exogenous inoculation of saprophytic fungi following trauma. It is usually seen in rural populations exposed to soil, like farmers. The subcutaneous mycoses seen in India include mycetoma, sporotrichosis, chromoblastomycosis, phaeohyphomycosis and zycomycosis. Rhinosporidiosis is now considered to be a protozoan infection.

Mycetoma

Clinical Case 2

A 50-year-old farmer presents with a localised swelling on the right foot as seen in the picture, gradually manifesting over 1 month (Fig. 21.11).

1. *What is your clinical diagnosis?*
2. *How will you proceed to diagnose in the laboratory?*
3. *What is the source of this infection?*
4. *Name two bacterial and two fungal causative agents.*

This is a chronic, localised, subcutaneous infection of exposed sites characterised by the sinuses discharging pus and granules to the skin surface (Fig. 21.11). Mycetoma is endemic across India. It is seen in TN and is also called **Madura foot** or **maduramycosis**. The microorganisms are mostly soil saprophytes that enter following trauma and burrow into the subcutaneous tissues but rarely affect the underlying muscle and bone. Mycetomas are caused either by fungi (eumycotic —*Trematosphaeria grisea, Madurella mycetomatis, Streptomyces* spp.) or bacteria—*Actinomyces* spp. or *Nocardia* spp. (actinomycotic) and *Staphylococcus aureus* (botryomycosis). The colour of the granules, which are the microcolonies of the causative agent, varies with the agent. The granules can be white, yellow, red, brown or black.

Laboratory diagnosis is by the microscopic examination and culture of the granules. Gram stain examination demonstrates the presence of gram-positive, filamentous, branched bacilli of *Actinomyces* or *Nocardia,* or gram-positive cocci in clusters of *S. aureus* (Figs. 21.12 and 21.13). *Nocardia* can be distinguished from *Actinomyces* as the former is weakly acid-fast while the latter is not. The presence of hyphal elements confirms the mycetoma to be eumycotic (Fig. 21.14). It is of utmost importance to establish the identity of the causative agent of the mycetoma as the antimicrobial therapy (antibiotics or antifungal) will depend on it.

Actinomyces

Actinomyces are thin, filamentous, branching, anaerobic gram-positive bacilli that are soil saprophytes and commensals in the oral cavity and intestine (Fig. 21.12). Actinomycosis can be endogenous, as in cervicofacial **'lumpy jaw'** or post-surgical abdominal infection, or exogenous, as in mycetomas with granules. Rarely, it can cause brain, lung or liver abscess. Lumpy jaw is a slow-growing, painless mass with fistula. Specimens from the

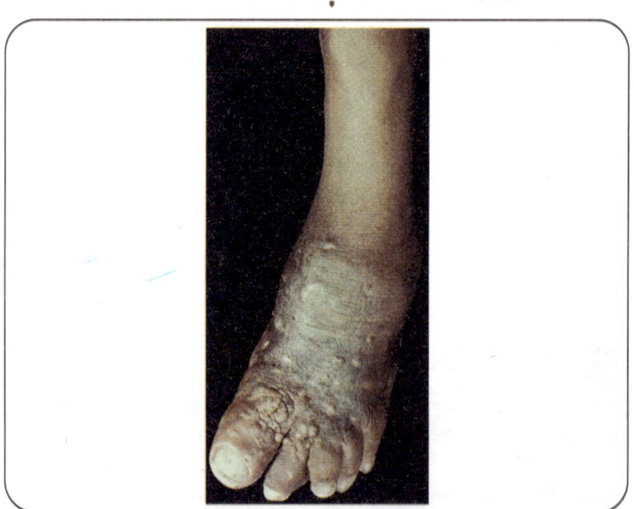

Fig. 21.11 Localised swelling of the right foot with sinus tracts in mycetoma (*Source:* Public Health Image Library, ID 14816 / CDC / Dr. Victoria (Mexico); Dr. Lucille K. Georg)

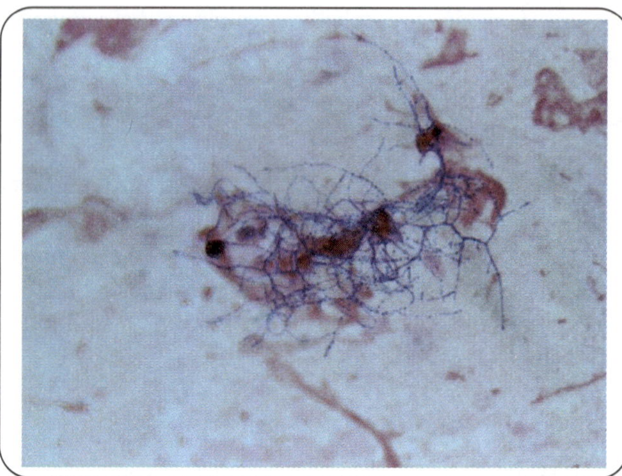

Fig. 21.12 Gram-positive, filamentous, branched bacilli of *Actinomyces*, Gram stain, 100X

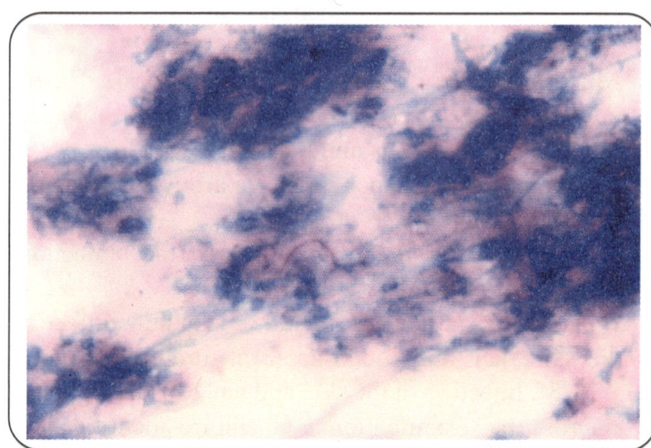

Fig. 21.13 Acid-fast, filamentous, branched bacilli of *Nocardia*, ZN stain, 100X (*Source:* Public Health Image Library, ID 2797 / CDC / Dr. Leanor Haley)

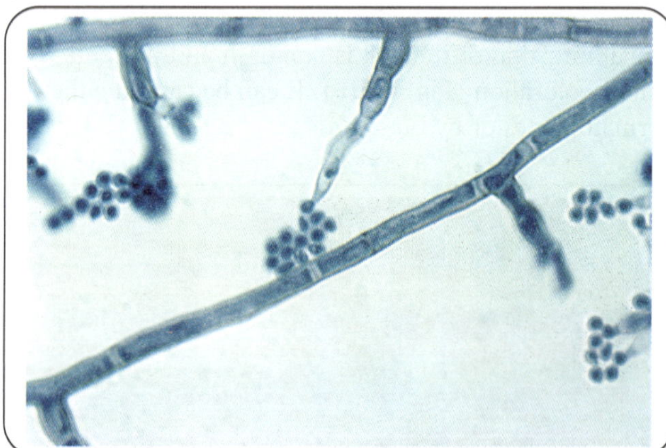

Fig. 21.14 *Madurella mycetomatis*, LPCB mount, 100X (*Source:* Public Health Image Library, ID 4034 / CDC / Dr. Lucille K. George)

site of disease can be examined by Gram stain and culture or histopathologically to demonstrate the presence of *Actinomyces* spp. This bacterium is sensitive to penicillin.

Nocardia

Nocardia are thin, filamentous, branching, aerobic, acid-fast, gram-positive bacilli that are soil saprophytes (Fig. 21.13). Nocardiosis is an opportunistic disease presenting as lobar pneumonia that may disseminate via blood to the brain, skin, muscle, kidney or bone. Specimen from the site of disease can be examined by 1% ZN and culture or histopathologically to demonstrate the presence of *Nocardia*. The bacterium is sensitive to cotrimoxazole.

SUBCUTANEOUS MYCOSES OTHER THAN MYCETOMA

Sporotrichosis

This is a chronic, subcutaneous pyogranulomatous infection by *Sporothrix schenckii,* involving skin, subcutaneous tissue and lymphatics. The fungus enters through a minor skin injury or thorn/splinter prick. It can be diagnosed by a KOH mount, histopathology, culture on SDA and detection of specific antibodies using a latex agglutination test (LAT) with mannan cell wall antigen. The histopathology reveals cigar-shaped asteroid bodies that appear as a central basophilic fungus surrounded by eosinophilic immune complexes called the **Splendore–Hoeppli phenomenon**.

Chromoblastomycosis

This is an infection caused by dematiaceous or pigmented fungi found in the soil. Chromoblastomycosis manifests as wart-like cutaneous nodules on the lower extremities, resembling cauliflower heads. It is caused by traumatic implantation in plantation workers. It can be demonstrated on KOH mount, histopathology and culture on SDA. Phaeohyphomycosis caused by pigmented soil fungi is seen in the immunocompromised to cause abscess in the brain, muscle or subcutaneous tissue.

Phaeohyphomycosis

These are chronic subcutaneous lesions caused by dematiaceous (pigmented) mycelial fungi found in the soil, other than those causing chromoblastomycosis. The causative fungi include *Alternaria* spp., *Curvularia* spp., *Bipolaris* spp. and *Exophiala* spp.

Answers

Clinical Case 1

1. *Thin, septate, branched, hyaline hyphae with numerous round-to-pyriform microconidia*
2. *Tinea corporis caused by Trichophyton mentagrophytes*
3. *Stratum corneum*
4. *Trichophyton, Epidermophyton, Microsporum*

Clinical Case 2

1. *Mycetoma*
2. *Macroscopically look for coloured granules in the sinus tracts. Microscopic examination by Gram stain, ZN stain and KOH to demonstrate gram-positive bacilli (Actinomyces), AFB (Nocardia) or hyphal elements (subcutaneous fungi) and culture of granules.*
3. *Soil saprophytes introduced by trauma*
4. *Bacteria are Staphylococcus aureus, Actinomyces or Nocardia, and fungi are Madurella or Streptomyces.*

Questions

1. Differentiate between tinea versicolor and tinea corporis.
2. Which conditions predispose to infection by *Candida albicans*? Give an account of its pathogenicity and laboratory diagnosis.
3. How will you distinguish between eumycotic and actinomycotic mycetoma? Why is this distinction important?

Parasitic Infections With Dermal Manifestations

The anatomical location of most parasites commonly infecting humans is the intestine, blood or lymphatics. With the exception of a few cases, the primary disease caused by these parasites is not in the skin, and cutaneous manifestations are secondary or rare.

Parasitic diseases affecting the skin, mucosa or subcutaneous tissue are:

1. Subcutaneous infection — *Dracunculus medinensis*
2. Skeletal muscle — *Trichinella spiralis*
3. Post kala-azar dermal leishmanoid (PKDL) — *Leishmania donovani*
4. Oriental sore — *Leishmania tropica*
5. Cutaneous larva migrans — *Ancylostoma duodenale*
6. Cutaneous larva currens — *Strongyloides stercoralis*
7. Acute adeno-dermato-lymphangitis in filariasis — *Wuchereria bancrofti, Brugia malayi*
8. Nasal polyps, rhinosporidiosis — *Rhinosporidium seeberi*
9. Nocturnal enuresis — *Enterobius vermicularis*
10. Muscle (cysticercus cellulosae) — *Taenia solium*
11. Urticaria — *Echinococcus granulosus* (hydatid disease)
12. Scabies — ectoparasite *Sarcoptes scabiei*

DRACUNCULIASIS (GUINEA WORM DISEASE)

India was declared free of dracunculiasis in 2000, but recently a few cases have been reported from Rajasthan. The disease has been largely eradicated globally, with a few foci in Africa.

Man is the definitive host for *Dracunculus medinensis* and water-residing cyclops the intermediate host (Fig. 22.1). Infection is transmitted by ingestion of water carrying the infected cyclops. The larva released

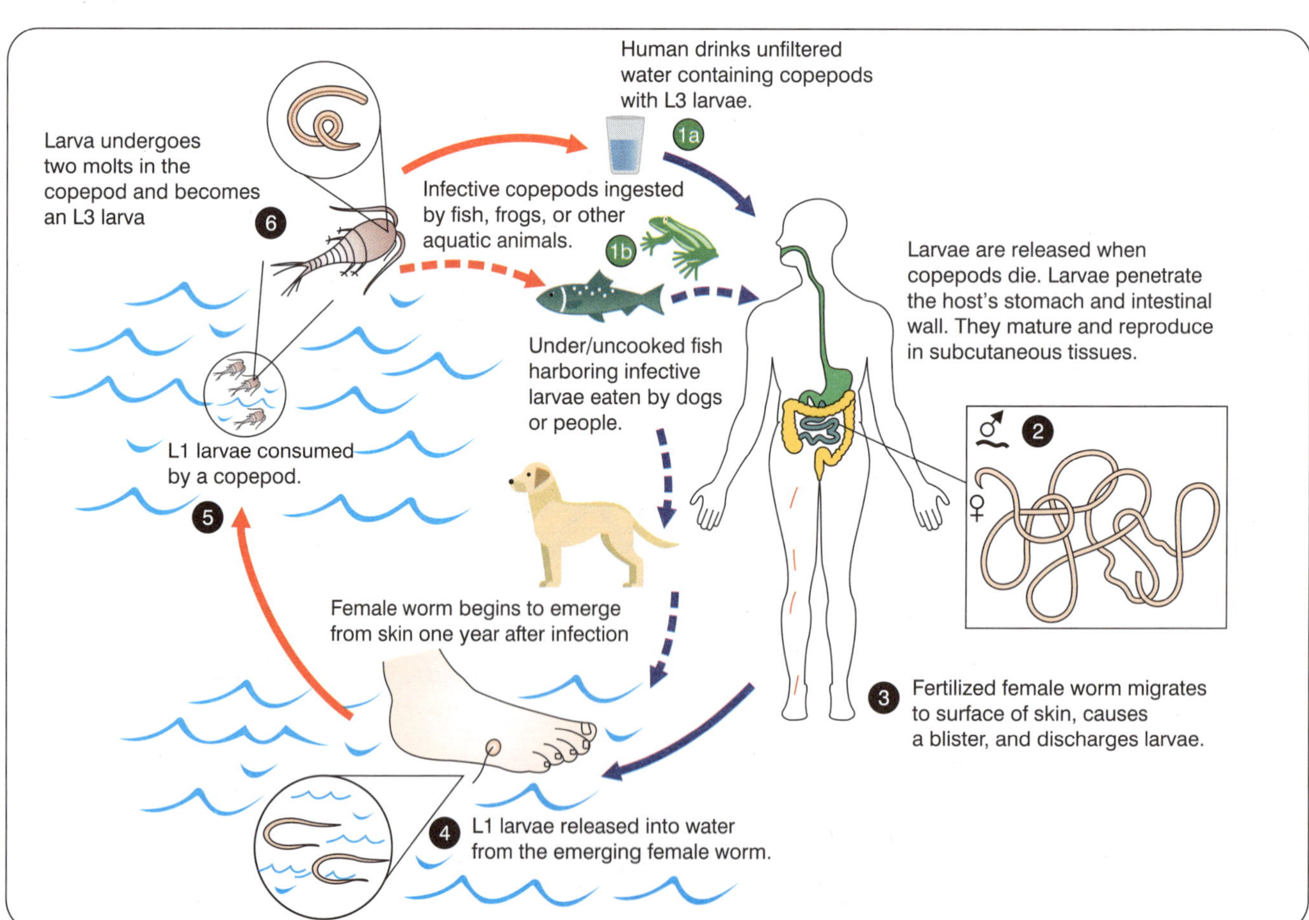

Fig. 22.1 Life cycle of *Dracunculus medinensis* (*Source:* Adapted from CDC's Parasites and Health page, at https://www.cdc.gov/parasites/guineaworm/biology.html)

in the intestine penetrates the intestinal wall and matures in the retroperitoneal space. After mating, the male dies while the female grows and migrates to the lower extremities. A localised swelling and an ulcer are seen where the adult finally resides, from where millions of larvae are discharged upon contact with water. These infect cyclops in the water and develop to become infective to man, and the life cycle continues. Peripheral blood eosinophilia occurs during the infection. Eventually, the adult worm is resorbed or calcified.

The cutaneous lesion is seen only at the time when the subcutaneous adult residing in the lower extremities discharges larvae into water. The cutaneous lesion begins as a papule, progresses to a vesicle, and leads to ulceration that is associated with pruritus. Diagnosis is clinical and the adult parasite can be palpated under the skin and mechanically removed.

TRICHINOSIS

Trichinella spiralis, a nematode infection, is rare in India. Humans are **dead end** or **accidental hosts**, infected by consumption of infected meat. The *T. spiralis* larvae present in infected meat reach the duodenum and penetrate the submucosa to grow into adults. After mating, more larvae are produced that invade and travel to the skeletal muscle where they are encysted and infective and survive for years. However, in man, the cycle ends and the adults die in a lifespan of 2 months. The condition is diagnosed by histopathological examination of muscle biopsy, demonstrating cysts in the muscle (Fig. 22.2). Azole compounds are effective only prior to encystation of the larvae in muscle.

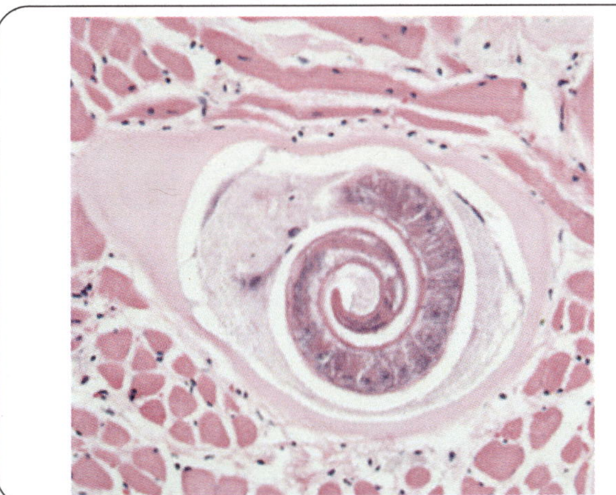

Fig. 22.2 *Trichinella spiralis* larva, H&E stain, 100X (*Source:* CDC, DPDx)

LEISHMANIASIS

The protozoan *L. donovani* infects the mononuclear phagocytic system and causes the systemic disease **kala-azar** (visceral leishmaniasis). In 10% of kala-azar cases, following 1–2 years of successful treatment of visceral leishmaniasis, dermal manifestations may appear. This is termed as **post kala-azar dermal leishmaniasis** (PKDL), in which the lesions are characterised by scattered maculopapular rash, butterfly erythema on the face and soft, painless, granulomatous nodules, without ulceration. The condition can be diagnosed by demonstration of the amastigote forms of *L. donovani* in the lesion biopsy.

Completely separate from PKDL, the parasite *Leishmania tropica* does not cause visceral leishmaniasis but exclusively infects the Langerhans cells of the skin causing cutaneous leishmaniasis also known as **oriental sore, tropical sore** or **Delhi boil**. The morphologic forms, life cycle and laboratory diagnosis of cutaneous and visceral leishmaniasis do not differ except that, besides man, dogs too serve as a reservoir of infection. Unlike PKDL lesions, the cutaneous leishmaniasis lesions are chronic ulcerative granulomas that occur on the face and extremities. The amastigote forms of the parasite can be demonstrated in the margins of the ulcer. Additionally, unlike visceral leishmaniasis, cutaneous leishmaniasis is associated with lifelong immunity following a single episode.

CUTANEOUS LARVA MIGRANS

It is also known as **creeping eruptions**, as the lesions creep or migrate as they are due to the presence of parasites moving in the skin. It is caused by *Ancylostoma duodenale* larvae and a number of other animal hookworms (*Ancylostoma braziliens, Ancylostoma caninum, Gnathostoma spinigerum*) accidentally infecting humans.

The larva penetrates the skin and presents with intensely pruritic papules to begin with, that start spreading or migrating locally with linear eruptions. Pruritic maculopapular dermatitis that occurs at the site of entry of the larvae is called 'ground itch'. Classical lesions with a history of contact with soil or sand are diagnostic. The condition is treated with ivermectin and may be associated with pulmonary eosinophilia.

CUTANEOUS LARVA CURRENS

This is caused by the wandering larvae of *Strongyloides stercoralis*. Infection is characterised by pruritis and urticaria in the perianal skin and buttocks.

FILARIASIS

Clinical Case

An elderly man limps into the OPD with a swollen left lower limb associated with thickening of the skin into folds, with black pigmentation, warty growth and non-healing ulcers (Fig. 22.3).

1. *What is the pathogenesis of this condition? Name the parasite.*
2. *Name a parasite that resides in the subcutaneous tissue of the lower extremity.*
3. *Name two parasites causing creeping eruptions in the skin.*

Lower limb lymphedema is a chronic feature of lymphatic filariasis caused by *Wuchereria bancrofti* or *Brugia malayi*, that leads to thickening of the skin into folds, with black pigmentation, warty growth and non-healing ulcers (Fig. 22.3).

Other dermal manifestations of parasites include perianal pruritis (nocturnal enuresis) in *Enterobius vermicularis* infection, muscle cysts (cysticercus cellulosae) in *Taenia solium* infection and urticaria in *Echinococcus granulosus* (hydatid disease) infection.

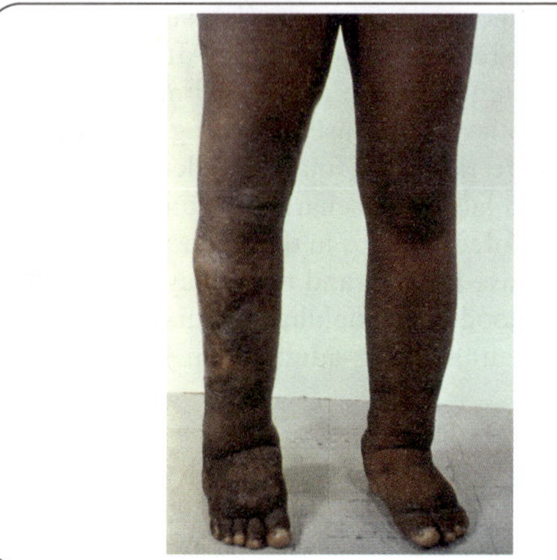

Fig. 22.3 Elephantiasis (*Source*: Public Health Image Library, ID 16536 / CDC / Dr. Robert S. Craig)

SCABIES

Scabies is caused by the ectoparasite *Sarcoptes scabiei* (itch mite) transmitted in humans by skin-to-skin contact within a family or through sharing of towels, clothing or bedding. The mite burrows in the superficial stratum corneum layer of the epidermis in the skin of the finger web, or fold of the wrist, elbow or knees and causes intense nocturnal pruritis. The burrows are seen on the affected areas and the mite can be demonstrated microscopically. Oral ivermectin with local application of benzyl benzoate or permethrin is recommended along with decontamination of clothing material in hot water. People in close contact with the patient also need to be examined and treated if required.

ANSWERS

Clinical Case

1. *Repeated attacks of lymphangitis cause lymphatic blocks, leading to chronic lymphedema that presents as thickening of the skin into folds. Wuchereria bancrofti or Brugia malayi.*
2. *Dracunculus medinensis*
3. *Ancylostoma duodenale, Strongyloides stercoralis*

QUESTIONS

1. Enumerate the parasites infecting the skin, subcutaneous tissue or skeletal muscle.
2. Distinguish between post-kala-azar dermal leishmaniasis and cutaneous leishmaniasis.
3. Elaborate on how wandering parasitic larvae can cause disease.

SECTION SUMMARY

Section III: Musculoskeletal, Skin and Soft Tissue Infections

Infection/pathogen	Virulence/pathogenesis	Laboratory diagnosis	Vaccine/recommended antimicrobial
Bacteria **Abscess** *Staphylococcus aureus*	Coagulase, exotoxins	Smear Gram stain, Culture	Cefazolin Amoxicillin-clavulanic acid
Cellulitis *Streptococcus pyogenes*	Streptolysins	Smear Gram stain, Culture	Penicillin amoxicillin-clavulanic acid Ceftriaxone
Abscess Non-sporing anaerobes	Streptokinase	Smear Gram stain, Culture	Amoxicillin-clavulanic acid Ampicillin+sulbactam
Gas gangrene *Clostridium perfringens*	Exotoxins, lecithinase, collagenase	Smear Gram stain, Culture	Penicillin Clindamycin Antitoxin
Tetanus *Clostridium tetani*	Exotoxin tetanospasmin		Toxoid Antitoxin
Leprosy *Mycobacterium leprae*	Resists phagocytosis, cell wall interacts with Schwann cells	ZN stain Lepromin test	Dapsone, rifampicin, clofazimine
Viral exanthems (Childhood fevers)			Live attenuated
Measles	MPS infection, rash and Koplik's spot		Live attenuated
Rubella	URTI, lymphadenopathy Rash		Live attenuated
Chickenpox	Vesicular rash	ELISA, PCR	Live attenuated
Varicella zoster	Herpes zoster in dermatome		
Parvovirus B19	Rash on cheek	PCR	
Genital herpes Herpes simplex virus 1 and 2	Kissing disease Asymptomatic latency		
Warts Human papillomavirus	Multiplies in keratinocytes, oncogenic		Viral capsid protein Bivalent (type 16, 18) Quadrivalent (type 6, 11, 16, 18)
Nodules Molluscum contagiosum	Pearly pink lesions	Intracytoplasmic, eosinophilic inclusion bodies	
Fungi *Candida albicans* and other non-albicans *Candida* spp.	Skin, nail, mucosa (diaper rash, onychomycosis, oral thrush, vulvovaginitis)	KOH mount, Smear Gram stain, Culture	Clotrimazole
Dermatophytes	Tinea capitis/corporis/ cruris	KOH mount, Culture	Griseofulvin
Mycetomas, Subcutaneous mycosis *Actinomyces, Nocardia*	Following trauma, localised	Bacterial Gram smear, culture	Antibacterial for bacterial-cotrimoxazole
Sporothrix, chromoblastomycosis, phaeohyphomycosis	Swelling, sinus tracts, granules	Fungal KOH mount, culture	Antifungal for fungi-itraconazole
Parasites **Blisters** *Dracunculus medinensis*	Ingestion of larvae-intestinal wall to subcutaneous tissue-Localised ulcer on extremities	Removal of adult	

Infection/pathogen	Virulence/pathogenesis	Laboratory diagnosis	Vaccine/recommended antimicrobial
Cysts in muscle *Trichinella spiralis*	Ingestion of infected meat-duodenum-skeletal muscle-encysted	HPE of muscle	Albendazole
Cysticercus cellulosae *Taenia solium*	In skeletal muscle	HPE of muscle	Albendazole
Post Kala-azar dermal leishmanoid *Leishmania donovani*	Following visceral leishmaniasis, no ulceration	Biopsy amastigote forms	Sodium stibogluconate
Dermal leishmaniasis *Leishmania tropica*	Bite of vector, ulceration seen Lifelong immunity	Biopsy amastigote forms	Sodium stibogluconate
Cutaneous larva migrans *Ankylostoma duodenale*	Larvae penetrate skin, pruritic papules	Ova in stool	Albendazole
Cutaneous larva currens *Strongyloides stercoralis*	Larvae penetrate skin, pruritic papules, perianal region	Larvae in stool	Ivermectin
Lymphangitis, elephantiasis Filariasis	Bite of *Culex*-chronic lymphangitis, lymphatic obstruction	Adult in lymph node	Diethylcarbamazine
Osteomyelitis *Staphylococcus aureus*	Endogenous or exogenous	Smear Gram stain, Culture	Cloxacillin, cefazolin

Infective Endocarditis, Rheumatic Fever, Bacterial Sepsis, FUO, Anemia, Transfusion and Transplantation-Associated Infections

INFECTIVE ENDOCARDITIS

Blood is a sterile compartment but transient carriage of microorganisms in blood can occur. Some infections also specifically target the RBC or vascular compartment. However, rarely, bacteria may infect the endothelial surface of an artery (endarteritis), the heart (endocarditis) or a vein (phlebitis). Infective phlebitis commonly occurs due to intravenous catheters.

Clinical Case 1

A 35-year-old man with a known history of valvular heart disease presents with fever, malaise and dyspnea. On examination, he has splenomegaly, skin petechiae and a recent onset murmur.

1. *What is your most likely diagnosis? Name two common etiologic agents?*
2. *What is HACEK?*
3. *What steps will you take, in this case, to isolate the causative agent in the laboratory?*
4. *Name a fungal cause of the above condition.*

Pathogenesis

Infective endocarditis can be sub-acute, caused by non-pyogenic bacteria on abnormal valves, or acute, caused by pyogenic bacteria involving abnormal or normal valves.

Bacteria causing infective endocarditis include *Streptococcus viridans, Streptococcus pneumoniae*, oral streptococci (*Streptococcus sanguis, Streptococcus mutans*), Group A, B and G streptococci, *Enterococcus faecalis, Staphylococcus aureus*, CoNS, non-typhoidal *Salmonella, Pseudomonas aeruginosa*, and the HACEK group. The HACEK group of bacteria includes *Haemophilus* spp., *Actinobacillus, Cardiobacterium, Eikenella* and *Kingella*. *Candida* spp. are also known to cause infective endocarditis.

Transient bacteremia is common, but endocarditis is rare. It is more often seen on aortic and mitral valves, as the normal vascular endothelium is resistant to infection. The bacteria adhere to thrombotic vegetations or the platelet fibrin aggregates resulting from cardiac abnormalities. To cause endocarditis, bacteria need to enter the blood, survive host defences, adhere to the endocardium (of valve/vegetations) and multiply on the endocardium. This involves interactions between the bacteria, host defence, endothelium, host protein and coagulation components.

Streptococci causing endocarditis produce extracellular dextran to adhere to cardiac valves and vegetations. Fibronectin is an extracellular glycoprotein in human tissues that can bind to other receptors and proteins. *Staphylococcus aureus* and *E. faecalis* bind to fibronectin and promote adherence to normal valves. CoNS, with its polysaccharide capsule and slime production, adheres to prosthetic valves.

A vegetation is a cardinal pathologic feature of infective endocarditis. It is a mass of fibrin, platelets and bacteria varying from 1 mm to 1 cm in size and may be single/multiple. Bacteria are continuously shed in blood and diagnosed by blood culture. Infective endocarditis is also associated with fever, heart murmur, embolic events, circulatory immune complexes, splenomegaly and glomerulonephritis. Risk factors include known rheumatic heart disease (RHD) or valvular malformation or patients with implanted prosthetic valves. Other pathologic conditions that can contribute to infective endocarditis are abscess, infarction or glomerulonephritis in the kidneys, mycotic aneurysms, cerebral or pulmonary emboli, splenic infarction, skin petechiae or retinal hemorrhage.

The clinical presentation of infective endocarditis is myriad, featuring fever with chills, weakness, dyspnea, anorexia and malaise. Diagnosis depends on the pathologic criteria of microbial culture and pathologic heart lesions and clinical evidence of endocardial involvement (follows Major and Minor criteria). If not diagnosed and treated in time, it can lead to mortality.

Laboratory diagnosis

1. *Non-specific* — Abnormal hematologic parameters, anemia, raised erythrocyte sedimentation rate (ESR) and proteinuria.
2. *Specific* — Blood culture is of utmost importance. A minimum of three blood cultures over 24 hours performed at least 1 hour apart with 10 mL of blood drawn at each venepuncture is recommended. Though

a variety of bacteria can cause infective endocarditis, 90% of the cases are attributed to streptococci and staphylococci, which are easy to identify in the laboratory by Gram staining and culture.

3. *Other specific tests* are used if fastidious bacteria or *Candida* spp. are suspected.

Management of infective endocarditis requires consideration of the choice, dose and duration, and monitoring of antimicrobial therapy. The National Treatment Guidelines recommend exclusion of candidemia before antibacterial therapy. In the event of streptococci or enterococci, ampicillin-sulbactam is recommended for 4–6 weeks. In the event of *S. aureus* with a risk of gram-negative sepsis, vancomycin with meropenem is recommended.

Infective endocarditis may be prevented by the correction of predisposing cardiac conditions, reducing skin and oropharyngeal colonisation by relevant bacteria, and the administration of prophylactic antibiotics in dental, diagnostic and therapeutic procedures.

α-hemolytic streptococci

Also known as the viridans group, α-hemolytic streptococci are gram-positive cocci in chains. They are resident flora in the upper respiratory tract and oral cavity and rarely cause endocarditis either in pre-existing cardiac lesions or following dental procedures. Sub-acute bacterial endocarditis is diagnosed by isolating the bacterium in blood culture. The antimicrobial susceptibility to penicillin and other agents is similar to that in *S. pyogenes*.

Genus *Enterococcus*

Enterococci are gram-positive cocci in pairs arranged at an acute angle to each other (Fig. 23.1).

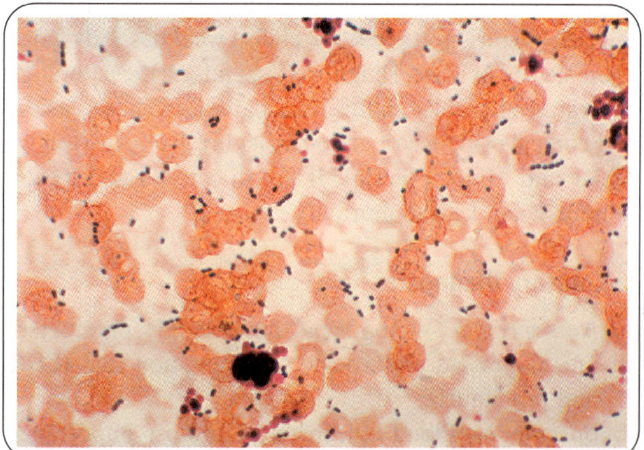

Fig. 23.1 Gram-positive cocci in pairs—*Enterococcus* spp., Gram stain, 100X (*Source*: Public Health Image Library, ID 2899 / CDC / Dr. Mike Miller)

They are non-hemolytic and present as commensals in the intestine, genital tract and saliva. They can cause urinary tract infection, wound infection especially surgical site infection, endocarditis, septicemia and intra-abdominal infections. *E. faecalis* and *Enterococcus faecium* are the common species. Enterococci are intrinsically resistant to cephalosporins and, to some extent, aminoglycosides too. It is important to test the susceptibility to ampicillin, gentamicin (high level), norfloxacin, nitrofurantoin, vancomycin, linezolid, teicoplanin and tetracycline. In life-threatening disease, vancomycin has been the drug of choice but **vancomycin-resistant enterococci (VRE)** strains have been reported.

RHEUMATIC FEVER

Clinical Case 2

A 10-year-old boy presents with a history of intermittent fever, fatigue and migratory joint pains. On examination, he is febrile, with elevated ESR and CRP values. There are palpable subcutaneous, non-tender nodules on his elbow.

1. *What is your most likely diagnosis?*
2. *What is the etiologic agent?*
3. *Name two complications of this condition.*
4. *Name the drug of choice.*

Acute rheumatic fever (ARF) is a disease characterised by non-suppurative inflammatory lesions involving primarily the heart, joints, subcutaneous tissue and CNS. Classical ARF is acute, febrile and self-limiting. Its clinical manifestations are so variable that it's a syndrome with no specific diagnostic test.

Epidemiology

It is seen in children aged 5–15 years and is related to *S. pyogenes* pharyngitis, a rise in anti-streptolysin O (ASO) titres and prolonged convalescent streptococcal carriage.

Pathogenesis

It is a delayed non-suppurative sequela of repeated upper respiratory infection caused by *S. pyogenes*. The strains of *S. pyogenes* vary in their rheumatogenic potential, and certain M serotypes exhibit M protein molecules that share antigenicity with human tissue, especially cardiac muscle protein myosin. However, the theories suggested are: damage due to streptolysins, immune complexes or an autoimmune response. The clinical presentation has a variety of signs and symptoms classified as major and minor manifestations. The major manifestations include carditis, polyarthritis, chorea, erythema marginatum and subcutaneous nodules. The minor manifestations include arthralgia, fever, elevated CRP and ESR and a prolonged PR interval. Supportive evidence includes antecedent positive streptococcal throat culture or antigen and rising ASO titres.

Laboratory diagnosis

1. Throat swab culture for *S. pyogenes,* although differentiating between infection and colonisation is a problem.
2. *S. pyogenes* antigen detection from throat swab by rapid immunochromatography.
3. ASO titre measured by the latex agglutination test. A titre > 200 Todd units/mL is considered to be significant. If the test is done within 2 months of onset, 80% patients will have a significant ASO titre. Although a raised ASO titre is not diagnostic of ARF, failure to demonstrate a rise in titre makes the diagnosis of ARF doubtful.
4. The enzyme streptodornase (DNase) secreted by *S. pyogenes* is also antigenic, giving rise to anti-DNase B antibodies that can be measured. When used together with the ASO test, elevated or rising titres, 90% of past streptococcal infections can be correctly identified.

Prevention

A single injection of penicillin G benzathine is recommended every 4 weeks to prevent the recurrence of ARF.

BACTERIAL SEPSIS

Bacteremia is defined as bacteria multiplying in blood while septicemia means bacterial multiplication with bacterial lysis and the release of toxins. Clinical sepsis is characterised by fever or hypothermia, tachycardia, hypotension, and a decreased urine output due to increased cardiac output and decreased peripheral vascular resistance. Besides bacteremia, it may also be seen in candidemia, malaria, tuberculosis and some viral infections. It occurs due to the release of cytokines by macrophages stimulated by microorganisms or their toxins. The host mononuclear phagocytic system interacts with specific microbial products, resulting in hemodynamic and multi-organ malfunction.

The bacteria implicated in bacteremia and septicemia are

Gram-positive cocci: Streptococci and staphylococci from wounds, puerperal fever and abortion.

Gram-negative bacilli: Salmonella causing enteric fever, *Escherichia coli* from intra-abdominal infection, intestinal perforation or rupture.

Gram-negative cocci: Neisseria meningitidis following pyogenic meningitis.

Rarely, *Brucella, Bartonella, Borrelia, Chlamydia, Legionella* and *Coxiella.*

Causes of sepsis

1. Localised pyogenic infection or septic foci either in a wound or in an organ. Pyogenic means 'pus forming'. Pus is a thick white-yellow fluid collection that consists of white blood cells in the immune response to infection.
2. Infection of a foreign body — iatrogenic, e.g., intravenous catheters. Catheter-related bloodstream infection (CRBSI) is defined as an intravascular infection wherein the source of infection is proven to be the central venous catheter (CVC). The source of bacteria for CVC colonisation can be the skin of the patient or that of medical personnel or a contaminated device/intravenous fluid or another focus of infection in the patient.
3. Immune defects — granulocyte, complement or splenic defects, AIDS.
4. Intravenous drug abuse.

Mediation of sepsis

Specific bacterial products interact with the mononuclear phagocyte system (MPS) to produce sepsis. These products include gram-positive cell wall peptidoglycan and teichoic or lipoteichoic acid and gram-negative cell wall lipopolysaccharide (LPS or endotoxin), specifically its lipid A. LPS binds to the CD14 receptor on macrophages and leads to the production of TNFα. A variety of immunologic mediators may finally result in hemodynamic abnormalities leading to disseminated intravascular coagulation.

Laboratory diagnosis

1. Two blood cultures are done from two separate sites of collection at two different times within 24 hours. Ideally, the blood culture should be done before starting antimicrobial therapy or, if started, before the next dose of antibiotic. Automated blood culture reduces the turnaround time for results when compared to manual processing.
2. Tests for CRP and procalcitonin levels are done in bacterial sepsis:

The normal CRP level is < 0.2 mg/dL and can be markedly raised up to > 10 mg/dL in the event of acute bacterial infections. It can be detected by latex agglutination assay or ELISA.

A procalcitonin level of 0.10–0.49 ng/mL is considered insignificant or indicative of a local infection, 0.50–1.99

ng/mL indicates a moderate risk of progression to sepsis, 2.00–9.99 ng/mL indicates a high risk of progression to sepsis and > 10 ng/mL implies severe sepsis or septic shock.

Bacterial sepsis is treated empirically with broad-spectrum antimicrobials and treated definitively following the availability of the culture and antimicrobial susceptibility results.

FEVER OF UNKNOWN ORIGIN (FUO)

Definition

Fever of unknown origin (FUO) or pyrexia of unknown origin (PUO) is defined as a temperature > 38.3°C for a duration > 3 weeks in an immunocompetent patient, without an established etiology despite intensive evaluation and diagnostic testing.

FUO can be classified as:

1. Classic FUO — as defined
2. Nosocomial FUO — when the definition is applied to a hospitalised patient
3. Neutropenic FUO (WBCs < 500/µL)
4. FUO associated with HIV infection

Approach for classic FUO

In the event of classic FUO (in the absence of HIV infection, neutropenia or hospitalisation), history of travel, immunisation and exposure is important, followed by a physical examination (every 2 days). It is advisable to stop antimicrobial therapy, if any, for a thorough evaluation of the case.

Obligatory investigations

Complete blood count (CBC) with malarial parasite X 2, blood culture X 2, urine microscopy, ESR and CRP

Potential diagnostic clue (PDC)

1. Non-invasive (CT thorax/abdomen, ultrasonography [USG] abdomen, positron emission tomography [PET] scan)
2. Invasive (lymph node biopsy or bone marrow biopsy, splenic aspirate)

Etiology of classic FUO in India

In India, the etiology of FUO is infection (50%, predominant tuberculosis), neoplasm (20%, predominant non-Hodgkin's lymphoma), non-infective inflammatory disease (20%, predominant systemic lupus erythematosus) and miscellaneous (5%).

Some cases remain undiagnosed (5%); these are cases where no underlying source is identified for > 6 months and the prognosis is generally good.

Infections causing FUO in India

Amongst FUOs, of the 50% attributed to infections, the etiology is as follows: TB (60%), melioidosis (*Burkholderia pseudomallei*) (10%), infective endocarditis (4%), visceral abscess (4%), and others—disseminated histoplasmosis, fungal endocarditis, enteric fever, visceral leishmaniasis, lepra reaction and brucellosis.

ANEMIA

Anemia can be caused by several mechanisms, ranging from nutritional deficiencies to malignancies. When anemia is due to an acute or chronic infection or infestation, the underlying mechanism is varied.

In India, infestation with *Ancylostoma duodenale*, *Necator americanus*, *Trichuris trichiura* and *Diphyllobothrium latum* results in anemia subsequent to the attachment of the helminth to the intestinal epithelium. *A. duodenale* and *N. americanus* are known to actively ingest blood. *Trichuris trichiura* causes blood loss from parasite-induced lesions in the intestinal mucosa. *D. latum* dissociates the vitamin B12–intrinsic factor complex in the ileum resulting in decreased absorption of vitamin B12 and megaloblastic anemia. *Schistosoma haematobium,* seen in Ratnagiri, Maharashtra, causes anemia due to chronic hematuria. *Plasmodium* species, especially *Plasmodium falciparum,* causes hemolytic anemia due to the destruction of RBCs and *Leishmania donovani* due to bone marrow dysfunction.

Parvovirus B19 binds to RBC blood group antigens on erythroid progenitor cells in the bone marrow and causes anemia. In the event of underlying hemolytic anemia, parvovirus infection results in severe anemia or transient aplastic crisis. In the event of underlying immunosuppression, parvovirus infection results in anemia. Anemia in chronic infections is due to cytokines of the host immune response inducing anemia by complex pathophysiologic mechanisms.

TRANSFUSION AND TRANSPLANTATION-ASSOCIATED INFECTIONS

Blood transfusion is one of the methods of transmitting infectious diseases. The predominant microbial agents that can be potentially transmitted by donor blood transfusion are:

1. Bacteria — *Treponema pallidum*

2. Viruses — HIV, HBV, HCV, HDV, CMV, HTLVs and parvovirus B19
3. Parasite — *Plasmodium* spp.

To ensure the safety of blood transfusion, donor blood is screened prior to transfusing in a recipient. In India, the FDA donor blood safety regulation includes VDRL test for syphilis, HBsAg by ICT/ELISA for HBV, anti-HIV and anti-HCV antibodies by ICT/ELISA.

Besides HIV, HBV and HCV, the risk of transmission in donor organ transplantation includes CMV, EBV and *Toxoplasma gondii*. Allografts from HIV-, HBV- and HCV-positive donors are contraindicated. Donor seropositivity for CMV, EBV and *T. gondii* is not a contraindication but provides information on the reactivation of these infections post transplantation in the immunosuppressed recipient.

Answers

Clinical Case 1

1. *Infective endocarditis, Streptococcus viridans, Staphylococcus aureus*
2. *The HACEK group of bacteria includes Haemophilus spp., Actinobacillus, Cardiobacterium, Eikenella and Kingella that can cause infective endocarditis.*
3. *A minimum of three blood cultures over 24 hours performed at least 1 hour apart with 10 mL of blood drawn at each venepuncture.*
4. *Candida albicans*

Clinical Case 2

1. *Acute rheumatic fever*
2. *Streptococcus pyogenes pharyngitis*
3. *Post-streptococcal glomerulonephritis, rheumatic heart disease*
4. *Penicillin G benzathine*

Questions

1. Outline the pathogenesis and laboratory diagnosis of infective endocarditis.
2. Outline the pathogenesis, laboratory diagnosis and prophylaxis of rheumatic heart disease.
3. Enumerate the bacteria causing sepsis and describe briefly the mediation of sepsis.
4. Define fever of unknown origin. What is the role of infections in FUO? How will you rule out infections?
5. What role do microbial agents play in causing anemia.
6. What are the microbes that can be transmitted by blood transfusion and organ transplantation? How can this risk be minimised?

Chapter 24

HIV and Viral Hemorrhagic Fevers

The pathogenesis of all viral infections involves multiplication of the virus in the blood following which the virus specifically targets the cell that it affects. For example, exanthematous viruses multiply in blood before affecting the skin with dermal manifestations. However, some viruses not only multiply in the bloodstream but specifically target the cells of the blood, e.g., HIV multiplies in CD4 T lymphocytes and some arboviruses can lead to hemorrhagic fevers.

HIV INFECTION AND ACQUIRED IMMUNODEFICIENCY SYNDROME

Clinical Case 1

A 30-year-old homosexual male presents with a history of unexplained weight loss, persistent fatigue and diarrhea. On examination, the lymph nodes are palpable and the stool shows presence of Strongyloides stercoralis larvae.

1. *What infection will you suspect? How will you confirm your diagnosis?*
2. *What advise will you offer to this patient?*
3. *In India, which is the commonest opportunistic infection encountered in this scenario?*
4. *Name other parasites that can cause opportunistic diarrhea.*

Retroviral infections have evolved from their origin in primates to the human species. The pathogenic human retroviruses include the lenti-viruses (lent=slow), human immunodeficiency viruses HIV-1 and HIV-2, and the oncoviruses human T lymphocyte viruses HTLV-I and HTLV-II. HIV-1 has nine subtypes A, B, C, D, F, G, H, J and K amongst which subtype C is most prevalent in India. HIV-2 has five subtypes A to E amongst which subtype A is most prevalent in India. Co-infection or superinfection with HIV-1 and HIV-2 is rare but known.

The retroviruses possess an RNA-dependent DNA polymerase known as 'reverse transcriptase' that can transcribe the viral RNA into linear dsDNA for subsequent integration into the host cell.

HTLV-I is associated with T cell leukemia, an aggressive T cell lymphoma, tropical spastic paraparesis or HTLV-I-associated myelopathy. HTLV-II is a serologic variant that is 50% homologous with HTLV-I. HIV-2 is a serologic variant of HIV-1 but closer to simian immunodeficiency virus (SIV) than HIV-1 is to SIV.

Epidemiology

The virus is transmitted by sexual contact, blood transfusion, mother to child, and other contact with body fluids like intravenous (IV) drug abuse. The viral load is high in blood, genital fluids and CSF, low in breastmilk and saliva, and absent in urine, sweat, feces, tears, sputum and vomitus, unless contaminated with blood. On blood spillage, there occurs a 90% reduction in the viral load within hours and on drying the virus remains detectable up to 3 days. In the blood bag, it can survive for 3 weeks at 40°C and for years if frozen. People living with HIV and AIDS (PLHA) are the only reservoirs of infection, with groups such as commercial sex workers (CSWs), men having sex with men (MSM), sex with multiple partners, transgenders, healthcare workers (HCWs), intravenous drug abusers and blood/blood product recipients being at a higher risk than the general population.

The global prevalence of HIV infection in adults is 0.8%. In India, it is 0.3% in general, 2.2% in female CSWs, 4.2% in MSMs, 7.5% in transgenders and 9.9% in IV drug abusers.

In India, the first HIV infection was documented in a CSW in Chennai in 1986 and the first AIDS case in a CSW in Mumbai in 1987. As per the National AIDS Control Organisation (NACO), in 2016, the HIV prevalence in India was 0.3%, the total number of people living with HIV (PLHIV) being 2.1 million, with 62,000 AIDS-related deaths. The states with a high prevalence are Maharashtra, TN, Karnataka, Andhra Pradesh (AP), Manipur and Nagaland.

Structure of HIV (Fig. 24.1)

The viral envelope, formed from the host cell membrane, bears two envelope proteins—gp41 and gp120. The matrix between the envelope and the core consists predominantly of Gag protein p17. The core contains two molecules of ssRNA tightly bound to nucleoprotein p7,

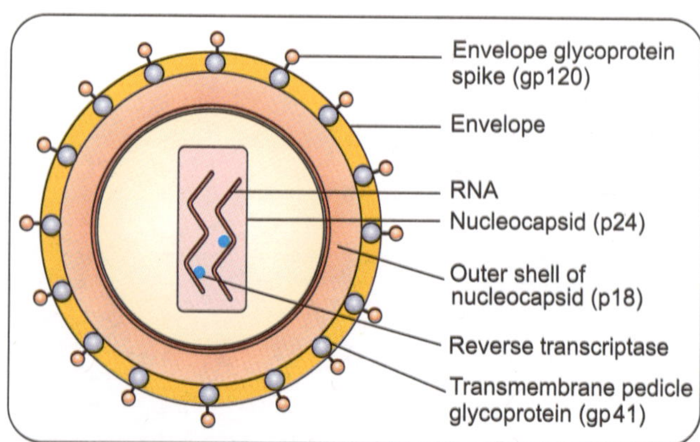

Fig. 24.1 Structure of HIV

major capsid Gag protein p24, and the enzymes—reverse transcriptase, integrase and protease.

Genomic organisation of HIV

The HIV genome consists of the following genes:

1. The gag gene that codes for the p7, p17 and p24 proteins amongst some other proteins.
2. The pol gene that codes for the enzymes—reverse transcriptase, integrase and protease.
3. The env gene that codes for envelope proteins—gp41 and gp120.
4. Amongst the other known genes, the 'vpu' codes for viral protein 'u' in HIV-1 and 'vpx' for viral protein 'x' in HIV-2. The 'vpu' exists only in HIV-1 and 'vpx' only in HIV-2.

Depending on the env gene, HIV has two serotypes—HIV-1 and HIV-2—each consisting of subtypes or clades A-J and A-H, respectively. In India, subtype C in HIV-1 and subtype A in HIV-2 are predominant.

Pathogenesis and clinical manifestations

1. *At the cellular level* (Fig. 24.2)

On entry into the host, the gp120 protein of the envelope binds to the CD4 receptor on T helper cells, macrophages and microglial cells. Besides the CD4 molecule, the specific cell tropism of HIV also requires the co-receptor CXCR4 to gain entry into T helper cells, and CCR5 to gain entry into macrophages and microglial cells. Mutations in the CCR5 receptor in individuals is linked to resistance to HIV infection, or if infected, results in delayed progression to AIDS. Another receptor designated 'integrin $\alpha 4\beta 7$' plays an important role in the transmission of HIV at mucosal surfaces. It is expressed on subsets of lymphocytes

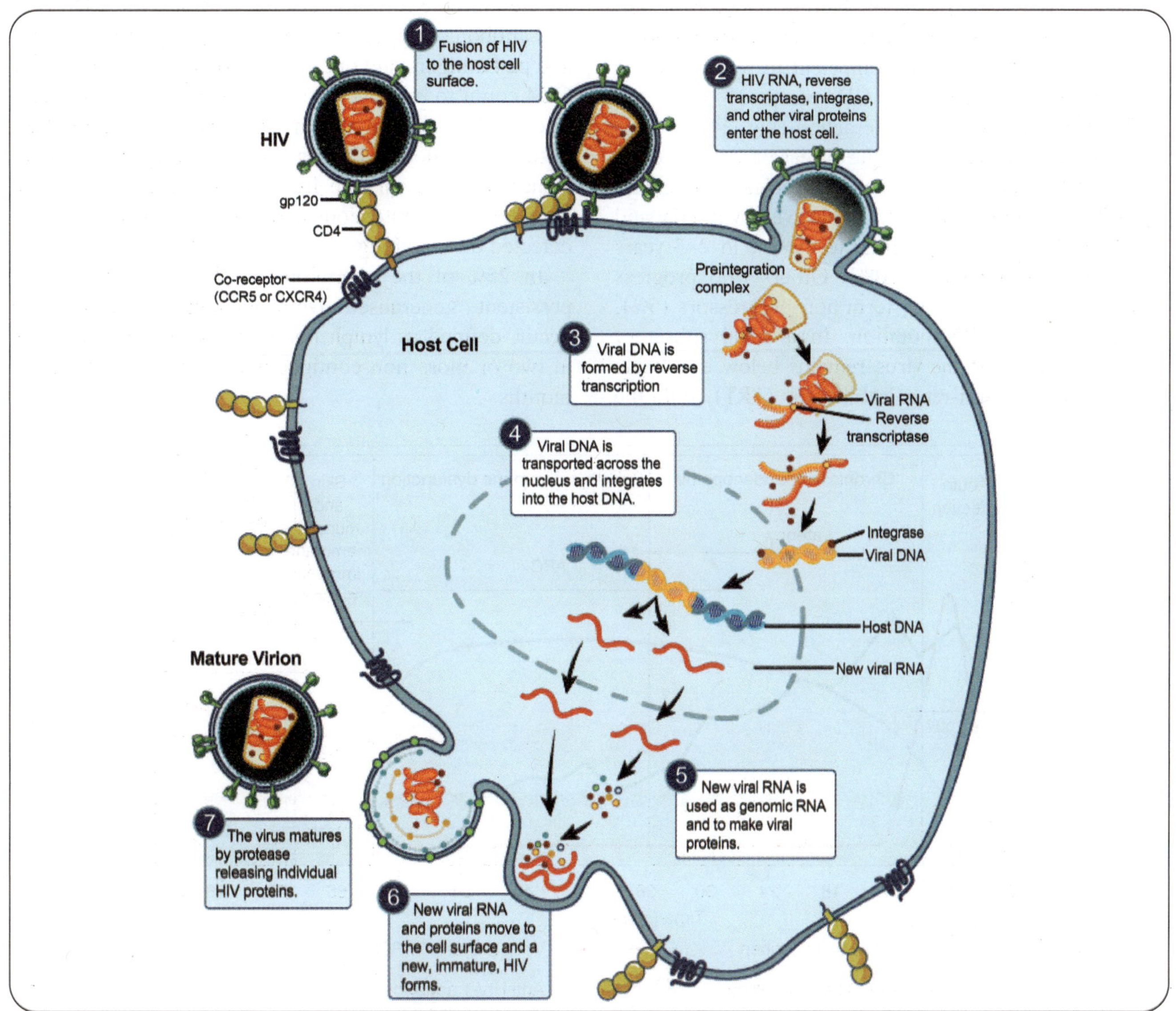

Fig. 24.2 HIV multiplication in host cell

and binds efficiently to mucosal surfaces during sexual transmission of the virus.

On binding to specific cell surface receptors, the envelope of the virion fuses with the host cell membrane and releases only the virion core into the cytoplasm. Reverse transcription begins in the cytoplasm and results in the formation of viral dsDNA, which in turn is integrated into the host cell DNA as a 'provirus'. Integration of viral DNA within the host cell DNA is linear and the replication of the virus then occurs along with the cell replication. Integration lasts till the life of the host cell, with synthesis of new viral genomes and proteins.

The new viral genomes and proteins are assembled in the host cell cytoplasm to form a mature virion. A mature virion contains the core and envelope of the virus and moves towards the host cell membrane. It buds out of the host cell into the surrounding after completing its life cycle within the cell. During budding, it also acquires host cell membrane proteins in its envelope. On release, the virion is competent to re-infect other uninfected cells bearing the CD4 receptor.

2. The virus-host systemic interaction (Fig. 24.3)

The natural course of the disease from infection to AIDS is on average 10 years (typical progressors –80%), without treatment. People living with HIV and AIDS (PLHA) who progress to disease in 2–3 years are rapid progressors (10%). Others who progress in 10–30 years are long-term non-progressors (5%), due to the CCR5 mutation. In a small subset of non-progressors, the virus remains below detectable levels without anti-retroviral therapy (ART). With the advent of ART, life expectancy has increased and HIV infection is now considered a manageable disease. In the absence of ART, the course of HIV from infection to disease is as follows:

Acute HIV Disease

Viral multiplication in the CD4 cells begins soon after infection, resulting in primary viremia and a corresponding p24 antigen rise, with its peak between 3 and 6 weeks. The CD4 levels begin to drop due to CD4 lysis by the virus, and the virus infects and multiplies in the CD4 cells in the lymph nodes. During this time, the patient is usually asymptomatic but highly infectious. In due course, there occurs persistent generalised lymphadenopathy (PGL) and a fall in the CD4 count.

Asymptomatic clinical latency

As the immune response occurs, HIV-specific CD8 cells and antibodies appear and result in a drop in the viremia and p24 antigen, and one sees a corresponding rise in the CD4 count to normalcy. The immune response is unable to clear the infection and the virus continues to multiply in the lymph nodes. During this asymptomatic period of clinical latency, variable from 10–20 years, the patient continues to be infectious and antibodies to HIV can be detected on screening.

In 25% of the asymptomatic clinical latency cases, persistent generalised lymphadenopathy (PGL) can occur, defined as lymph nodes larger than 1 cm in size in two or more non-contiguous locations, for at least 3 months.

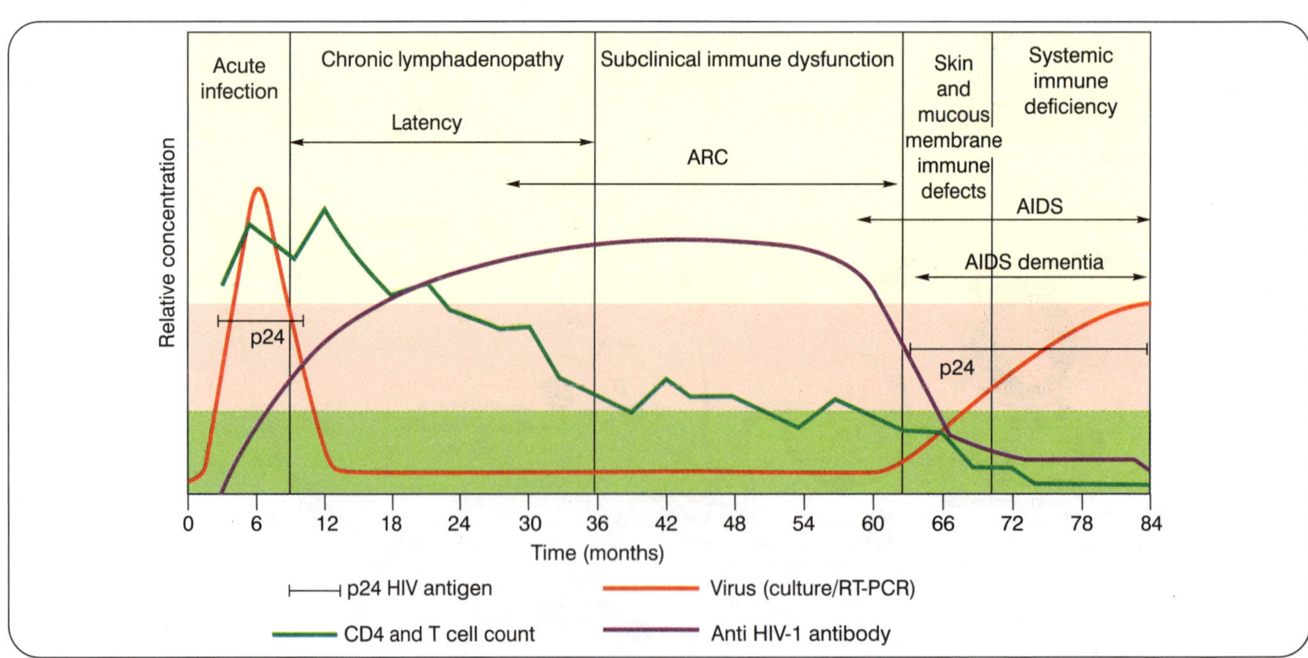

Fig. 24.3 Natural course of HIV infection

Symptomatic HIV infection (AIDS-related complex)

Following the period of clinical latency, the patient eventually develops symptoms of unexplained weight loss, fatigue, persistent diarrhea, or other opportunistic infections. The viremia and p24 antigen begin to reappear, and the CD4 count begins to fall.

Acquired immunodeficiency syndrome (AIDS)

This end stage of the disease sees uncontrolled multiplication of the virus in blood and lymphatics, resulting in a high viral load and a low CD4 count (< 200 cells/μm). Severe immunodeficiency can cause opportunistic infections, neoplasia or encephalopathy.

Globally and in India, the commonest opportunistic infection in HIV is tuberculosis (TB), pulmonary or extrapulmonary. Other opportunistic infections include those caused by viruses HSV and CMV, protozoans *Toxoplasma, Cryptosporidium, Isospora* and *Microsporidium*, and fungi *Candida, Cryptococcus* and *Pneumocystis*. HIV-associated neoplasias include Kaposi's sarcoma, lymphoma and cervical cancer.

Clinically and globally, HIV disease can be classified by two existing systems—the Centers for Disease Control and Prevention, USA (CDC) and the World Health Organisation (WHO). The former includes symptoms and CD4 count while the latter includes only the clinical presentation.

Laboratory diagnosis and NACO guidelines

The laboratory diagnosis and management of HIV patients in India is regulated through the National AIDS Control Organisation (NACO) and respective State AIDS Control Society (SACS) via Integrated Counselling and Testing Centres (ICTCs), State Reference Laboratories (SRLs) and National Reference Laboratories (NRLs). Unlike all other laboratory tests that have an 'implied consent', the HIV test mandates an 'informed consent' wherein pre-test and post-test counselling is provided, with the test results being strictly confidential.

The pre-test counselling provides the person with information on the disease, the right to informed consent, assurance on confidentiality, need for spouse/sex partner testing, safe practices and assessment for other sexually transmitted infections (STIs) and TB.

The post-test counselling is provided to those who test HIV-positive, HIV-negative and HIV-indeterminate. This includes information on prevention practices, demonstration of condom use and provision of condoms, and linkage to ART, TB, STI and/or antenatal care (ANC) centres, as necessitated in individual cases.

A diagrammatic representation of the appearance of viral markers following HIV infection is presented in Fig. 24.3.

Entry of the virus is followed by a transient high level of viremia and p24 antigen, during which the antibodies are absent. Screening tests for HIV antibodies being negative, this 6–8 week period is also called the **'window period'**—the time between HIV infection and the appearance of antibodies. The viremia and p24 antigen are at a low level during the course of infection and reappear only when AIDS sets in later years.

Anti-HIV antibodies to core (p24) and envelope (gp120) proteins are detected and persist throughout the infection. Though all classes of antibodies are formed—IgM, IgG and IgA—the assays used detect IgG. As opposed to the antigen, HIV antibodies can be detected throughout the course of infection except during the window period and after severe immunodeficiency sets in.

Tests for HIV infection

There are three specific tests for the diagnosis of HIV: antibody detection which is the most commonly used, antigen detection in the absence of antibodies, and viral genome detection to estimate the viral load; the most common non-specific marker is the CD4 count.

1. **Antibody detection by immunochromatography (ICT):** Most commonly used for diagnosis, the antibodies appear in 4–8 weeks following infection and persist throughout the infection. The test is negative during the window period and in AIDS. As per NACO guidelines, there exist strategy I, IIA, IIB and III for HIV testing. Of these, strategy I is for ensuring safety by screening blood and blood products for HIV antibodies by ELISA. Strategy IIA is meant for surveillance and IIB for symptomatic or high risk individuals. To detect HIV infection in asymptomatic individuals presenting to the ICTC or PPTCT, strategy III is adopted. In strategy III, HIV infection is diagnosed by three rapid ICTs for antibody detection where different antigens and different principles are used in each ICT. The first ICT is very sensitive, and the other two ICTs are more specific, thereby reducing the risk of false negative and false positive results (Fig. 24.4).
2. **Antigen detection by ELISA**: The p24 (core) antigen is the earliest to appear, i.e., within 1–2 weeks following infection and disappears on appearance of the antibodies. It helps in the diagnosis of HIV infection during the window period, neonatal infection and AIDS.
3. **Polymerase chain reaction (PCR):** This is the most sensitive and specific test.
   i) DNA PCR — pro-viral DNA detected
   ii) RNA PCR — for diagnosis and for monitoring the HIV viral load

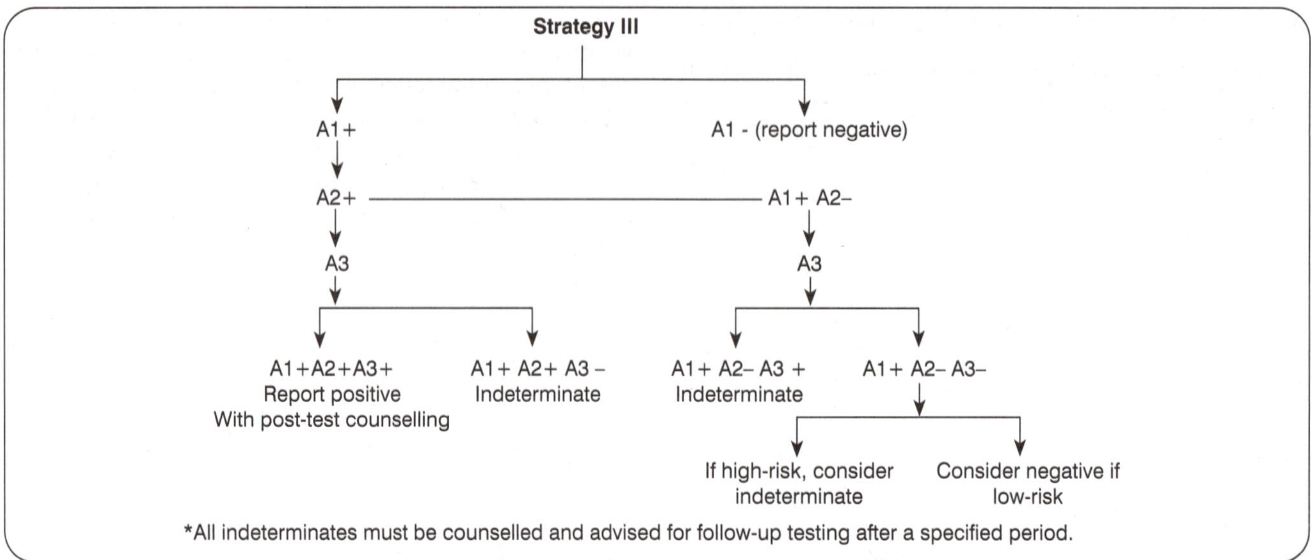

Fig. 24.4 Algorithm for diagnosing HIV infection

Polymerase chain reaction is useful in the diagnosis of neonatal HIV infection during the window period and for early diagnosis following a needle stick injury. As maternal IgG is passively transferred transplacentally to the child, antibody detection cannot be a diagnostic tool in neonates. Dried blood sample (DBS) is collected on a filter paper by a heel prick and transported to a reference centre for PCR. The first sample of the neonate is to be collected at 45 days of age for PCR and, if negative, repeated at 6, 12 and 18 months by a rapid antibody test to be considered HIV-negative.

4. **Non-specific tests:** Total count (TC) and differential count (DC) show leukopenia and lymphopenia, a CD4 cell count < 200/mm^3, T4: T8 ratio reversal, thrombocytopenia, raised non-specific IgG, and diminished CMI seen as skin test anergy.

Neonatal HIV infection

The transmission of HIV infection from a seropositive mother to her child can occur during pregnancy, delivery, or breastfeeding. During pregnancy, transplacental infection can occur as early as 8 weeks. Under the PPTCT program, antenatal cases are screened for HIV antibodies at their first visit anytime after 14 weeks and until delivery. A single dose of nevirapine is given at birth and once daily for 6 weeks to the neonate born to a seropositive mother to prevent HIV infection.

Maternal IgG can persist up to 18 months in the neonate, necessitating the detection of IgA after 3 months by Western blotting, or IgM by ELISA after 6 months. However, the recommended test is HIV RNA by PCR that is detectable within 48 hours of birth in 15% of the cases and by 1 month in 95% of the cases. The neonate is screened by PCR at 45 days and if negative tested every 6 months up to 18 months before being termed as HIV negative.

Prevention and prophylaxis

Safe sexual practices and hospital infection control practices including the use of personal protective equipment considering every patient as infectious irrespective of HIV serostatus prevents exposure to HIV. Safe practices are the use of contraceptives, and avoiding blood donation and the sharing of needles, blades, etc.

Antenatal screening for HIV in pregnancy is mandated to prevent transmission to the child. Post-exposure prophylaxis with single-dose nevirapine is given at birth and daily for 6 weeks. Post-exposure prophylaxis is also provided following occupational exposure in the case of healthcare workers after assessing the risk of seroconversion.

Prevention of TB-HIV includes intensified case finding wherein all TB cases are tested for HIV and all HIV-positives screened for TB. Isoniazid preventive therapy is provided at ART centres. An effort is being made by the Revised National TB Control Programme to identify and treat all TB cases to break the chain of transmission. As other STIs and genital ulcer diseases, especially syphilis, facilitate the transmission of HIV, HIV-positives are also tested for VDRL and referred to STI clinics for assessment.

Opportunistic infections in HIV

Infections that do not occur in immunocompetent individuals but are seen in an immunocompromised host are termed opportunistic infections. HIV infects the CD4 T cells resulting in opportunistic infections.

The diseases arising due to the immunodeficiency in HIV are tuberculosis, mucosal candidiasis, diarrhea (caused by *Strongyloides stercoralis*, *Cryptosporidium*, *Isospora*), pneumonia (caused by *Pneumocystis*), meningitis (caused by *Cryptococcus neoformans*), or toxoplasmosis.

The patient may present with the clinical features of a specific opportunistic pathogen and the laboratory diagnosis will depend on the causative agent.

ARBOVIRAL INFECTIONS

A group of RNA viruses are transmitted by insect vectors to man and are known as arthropod borne viruses or arboviruses. They belong to different families but share common presentations—either fever with rash or arthritis or hemorrhage or encephalitis.

Epidemiology

They are endemic in animals too, and transmitted by mosquitoes, ticks and sandflies. The geographical distribution of arboviral infections varies depending on environmental conditions and the presence of the requisite vector. In India, dengue, chikungunya and Japanese B encephalitis are endemic and frequently cause epidemic outbreaks. Other rarer arboviral infections encountered in India are provided in Table 24.1.

Chikungunya

Chikungunya is an acute fever with arthritis involving multiple joints. It is transmitted by the day feeding mosquito—*Aedes aegypti*. "Chikungunya" means "bent" due to the posture caused by severe joint pain. In this case, humans are infected by the bite of the mosquito. After an incubation period of 5 days, the patient develops fever with arthralgia due to arthritis at multiple joints, myalgia, headache and skin rash. The virus increases intraepidermal melanin retention, rarely presenting as hyperpigmentation on the nose. Most cases recover in a week, and residual arthralgia may persist for weeks, months or a year. Since 2005, the disease has re-emerged due to a new mutant of chikungunya virus and a new vector—*Aedes albopictus*. It is now endemic in Karnataka, Maharashtra, Gujarat, TN, AP and West Bengal (WB) and has four genotypes—West African, East African, Asian and Indian Ocean Lineage. Chikungunya is diagnosed in the laboratory by detecting IgM after 4 days or IgG after 2 weeks of fever in serum either by ELISA or rapid ICT or the viral genome in blood by RT-PCR. Presently, vaccines are not yet available.

Japanese B encephalitis (JE)

Japanese B encephalitis presents with fever accompanied by disorientation, delirium or coma, and is transmitted by the mosquito *Culex tritaeniorhynchus or C. vishnui*. Humans are infected by the bite of the mosquito. After an incubation period of 5–15 days, the patient develops fever with altered sensorium. It is commoner in children, and cases may recover with residual neurodeficit, the case fatality being 20–40%. It is endemic in Uttar Pradesh (UP, Gorakhpur district) Assam, Manipur, TN, Tripura, Bihar, Odisha and WB. JE can be diagnosed in the laboratory by detecting IgM in serum by ELISA or the viral envelope E gene in blood by RT-PCR. A live vaccine is given to children between 1 and 15 years of age in the endemic districts of UP, WB and Assam. A single dose is given subcutaneously followed by a booster dose after 1 year. An inactivated vaccine of Beijing P3 strain, prepared at Central Research Institute, Kasauli, India, is also available.

Table 24.1 Arboviral infections seen in India

Virus	Vector	Reservoir	Location	Disease
Togaviridae Chikungunya virus	*Aedes aegypti*	Monkeys	India	Fever, arthritis
Flaviviridae Japanese B encephalitis virus	Culex	Pigs, birds	India	Encephalitis
Dengue virus	*Aedes aegypti*	Unidentified	India	Hemorrhagic fever
Kyasanur forest disease	Tick	Monkeys, rats	Kyasanur, Karnataka	Hemorrhagic fever
Bunyaviridae Sandfly fever virus	Sandfly	Mammals	–	Fever, myalgia
Ganjam virus	Tick	Mammals	–	Fever
Rhabdoviridae Chandipura virus	Sandfly	Unidentified	–	Encephalitis

HEMORRHAGIC FEVER GROUP

Dengue

Clinical Case 2

A 14-year-old boy presents with fever, myalgia and retro-orbital pain for 4 days. On examination, there is a maculopapular rash on the upper extremities, and no hepatosplenomegaly or lymphadenopathy. The CBC shows mild leukopenia, and thrombocytopenia with a normal erythrocyte count.

1. *What is your probable diagnosis? What test will you ask for?*
2. *What is the role of rapid tests in this disease?*
3. *If the test is positive, what warning signs will you look for?*
4. *Explain the pathogenesis of severity in this disease.*

Epidemiology

Dengue (pronounced den'gee) is one of the most important worldwide endemic diseases, with the sole exception of Europe. In the case of India, it is endemic in WB, TN, Punjab, Kerala, Delhi, Karnataka and Maharashtra, with a seasonal trend of increase in the months of July–November every year. It is a complex relationship between the host (man), agent (virus) and the environment (mosquito breeding, rainfall and water storage or stagnation) that determines the endemicity.

It is transmitted by the bite of the female *Aedes aegypti* and other species of the *Aedes* mosquito from man to man with no other host. Man and mosquito are the reservoirs of infection. The virus is an ssRNA virus with four serotypes—DEN-1, DEN-2, DEN-3 and DEN-4 that are antigenically related with type-specific immunity but also have differences. DEN-5 has been reported in Bangkok. The serotypes are cross-protective only for a few months but any one serotype provides lifelong immunity to re-infection with that particular serotype. DEN-1 and DEN-2 are predominant in India.

Pathogenesis and clinical features

Primary or first infection with any serotype in non-immune persons causes dengue fever (DF), which is self-limiting and renders lifelong immunity to re-infection with that serotype. The incubation period is 3–14 days. Secondary infection with another serotype causes more severe illness like dengue hemorrhagic fever (DHF) or dengue shock syndrome (DSS).

Dengue fever (DF)

On primary infection, the virus multiplies in the blood. DF is characterised by fever for 2–7 days, with two or more of the following symptoms: headache, retro-orbital pain, myalgia, arthralgia, a maculopapular rash or hemorrhagic manifestations.

This infection produces neutralising and non-neutralising antibodies against the infecting serotype. The neutralising antibodies provide lifelong immunity against the infecting serotype, and immunity for a few months against other serotypes. The non-neutralising antibodies too last lifelong but can bind to the secondary infecting serotype and cause the 'antibody-dependent enhancement' immunologic phenomenon. This binding results in a cytokine tsunami that increases vascular permeability and subsequent plasma leakage leading to shock (DSS) and bleeding (DHF) with multi-organ involvement. Dengue virus-infected monocytes and endothelial cells, in the presence of enhancing antibodies produce multiple cytokines, including TNFα that induces plasma leakage. Dengue haemorrhagic fever is associated with high levels of the cytokines TNFα and IL6, and serum complement.

Complications of dengue hemorrhagic fever due to severe thrombocytopenia, with DSS due to hypersensitivity may occur in previously sensitised patients with a history of prior exposure to other serotypes of the virus.

Laboratory diagnosis

This is done by one or more of the following tests:

1. Demonstration of IgM by ELISA after 5–7 days of fever.
2. Demonstration of a four-fold or greater rise in IgG antibody titres to one or more dengue virus antigen in paired sera samples.
3. Demonstration of dengue virus NS1 antigen in serum by ELISA in the first 5 to 7 days of the onset of fever.
4. Detection of viral genomic sequences in serum or CSF sample by PCR.
5. Other hematologic parameters — hematocrit (increased by 20%), thrombocytopenia (< 1,00,000 cells/mm^3)

Prevention of dengue is by control of mosquito breeding and mosquito bites. There are several dengue vaccines under development, e.g., the 'Dengvaxia' is a live attenuated tetravalent chimeric vaccine that has been approved with 56% efficacy in Asians. The schedule is 0, 6 and 12 months subcutaneously for those aged 9 to 45 years.

Kyasanur forest disease (KFD)

Kyasanur is a forest in the Shimoga district of Karnataka in which the KFD virus originated. It originated from monkeys, rodents and squirrels and is transmitted through the bite of hard ticks. It is endemic in Shimoga, Chikmagalur, Udupi, and North and South Karnataka but its incidence has been on the decline ever since a vaccine became available.

Following an incubation period of 1 week, the disease manifests with fever and bleeding from the nose, throat, gums and gastrointestinal tract. Later, after about 15–20 days, meningoencephalitis may be seen with abnormal hematologic and CSF laboratory findings. KFD is diagnosed by specific IgM ELISA or RT-PCR. A killed vaccine from Haffkine Institute, Mumbai, is recommended in the endemic areas.

Ebola virus

Ebola virus causes a severe viral hemorrhagic fever with myalgia, maculopapular rash and hemorrhagic manifestations, with a 50% case fatality rate and ability to transmit from man to man on exposure to blood and body fluids of an infected patient. The virus is an RNA virus with a long filamentous morphology, circulating in blood. It has the potential to cause outbreaks. In India, a single case was reported in 2014 in a man from Liberia.

Answers

Clinical Case 1

1. *HIV. Test for HIV antibodies.*
2. *Post-test counselling regarding the availability of antiretroviral therapy at an ART centre, refraining from blood transfusion and unprotected sex, contact tracing and testing of known contacts, if feasible.*
3. *Tuberculosis*
4. *Cryptosporidium, Isospora, Microsporidium*

Clinical Case 2

1. *Viral hemorrhagic fever, the commonest being dengue. Test for NS1 antigen in blood by ELISA.*
2. *Rapid tests (ICT) have poor sensitivity and specificity and therefore the results need to confirmed by ELISA.*
3. *Vomiting, abdominal pain or tenderness, mucosal bleeding, hepatomegaly > 2 cm, raised haematocrit with decreased platelets.*
4. *In secondary dengue infection, non-neutralising cross-reactive antibodies produced during the primary infection bind to the secondary virus and protect it from the immune response. This 'antibody dependent enhancement' promotes mononuclear cells to release a tsunami of cytokines, leading to severe dengue.*

Questions

1. Enumerate the risk factors for HIV infection with prevention and prophylactic measures for each.
2. Describe the structure of the virus and its relation to its multiplication in a host cell.
3. Co-relate the antigenic structure of the HIV virus with its natural course of disease and the immune response.
4. How will you diagnose HIV infection in a neonate? Can it be prevented?
5. Enumerate the tests for the laboratory diagnosis of HIV with the role of each in the NACO strategies.
6. How will you rule out HIV infection in a child born to a HIV-positive mother?
7. Enumerate arboviral infections seen in India and write a note on chikungunya.
8. Enumerate arboviral infections seen in India and write a note on Japanese B encephalitis.
9. Enumerate arboviral infections seen in India and write a note on the Kyasanur forest disease.
10. Explain the immunopathogenesis of DHF and DSS and their clinical manifestations.
11. What are considered as warning signs for the progression of DF into severe dengue?
12. What are the current national guidelines for the laboratory diagnosis of dengue?

Kala-Azar, Malaria and Filaria

Parasites demonstrable in peripheral blood are the protozoans *Plasmodium* spp., *Leishmania* spp., *Babesia* spp. and *Trypanosoma* spp. Of these, malaria and kala-azar are seen in India. Amongst the tissue nematodes, *Wuchereria bancrofti, Brugia malayi, Loa loa* and *Mansonella ozzardi*, the first two causing filariasis are seen in India.

LEISHMANIASIS

Clinical Case 1

A 30-year-old man presents to a tertiary care hospital in Bihar, with a history of intermittent fever, malaise and emaciation due to unexplained weight loss since 6 months. On examination, the patient is febrile and has hepatomegaly with massive splenomegaly. Laboratory investigations reveal pancytopenia with normal liver enzymes. Tests for malaria, enteric fever and HIV are negative.

1. *What is the most likely infection in the above scenario?*
2. *Name the causative agent and transmission of disease.*
3. *How will you proceed to confirm your diagnosis?*

Leishmaniasis can be visceral (kala-azar) or cutaneous. Visceral leishmaniasis is caused by *Leishmania donovani* that infects macrophages of the mononuclear phagocytic system. Cutaneous leishmaniasis, caused by *Leishmania tropica,* infects the Langerhans cells of the skin. Both are vector-borne and transmitted by the bite of a sandfly (*Phlebotomous argentipes*).

Visceral leishmaniasis

Epidemiology

Leishmaniases are considered zoonotic infections as the parasite is maintained in dogs, rodents and other mammals and transmitted to humans by the bite of an infected sandfly. It is endemic in India in Bihar, WB, UP, Odisha, Assam and TN.

Morphology: amastigote and promastigote

The amastigote (non-flagellate form) is always intracellular and found in the macrophages of the mononuclear phagocytic system, in man. It is an egg-shaped cell measuring around 3 μm, with a nucleus, extranuclear DNA called 'kinetoplast', a vacuole and a rudiment of flagella called 'axoneme' (Fig. 25.1).

The promastigote form (flagellate form) is seen in culture and in the sandfly, and not in man. It is a spindle-shaped cell, measuring around 20 µm, with a nucleus, kinetoplast, a vacuole and an axoneme from which arises a flagellum outside the cell. The flagellum is free, not attached to the cell body and also measures around 20 µm.

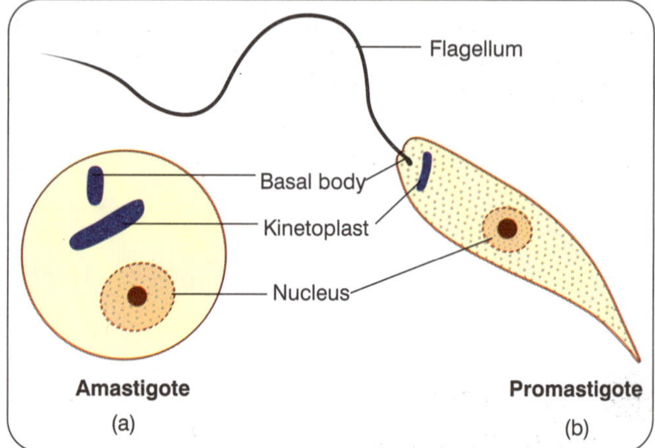

Fig. 25.1 Morphologic forms—amastigote and promastigote—of *Leishmania donovani*

Life cycle

Man is infected by the bite of a sandfly (*Phlebotomous argentipes*). The promastigote forms of the parasite enter the macrophages, are transformed into amastigote forms and begin multiplying by binary fission. Progressively, all cells of the mononuclear phagocytic system are infected as the macrophages teem with amastigotes, lyse and infect more macrophages.

During a blood meal, the sandfly ingests the amastigote forms. In the sandfly, they develop into promastigote forms and multiply by binary fission, the mature forms infecting man (Fig. 25.2). In India, man is the only host or reservoir of infection.

Pathogenesis and clinical features

Cell surface antigens of the amastigote forms of *L. donovani* promote binding, entry and survival within macrophages. Subsequently, it is the T_H1 or T_H2 response of the host that determines the course of the disease. The T_H1 response is beneficial to the host, resulting in a positive leishmanin delayed hypersensitivity skin test. On the other hand, the T_H2 response results in visceral leishmaniasis, hypergammaglobulinemia and a negative leishmanin skin test. The hypergammaglobulinemia causes a reversal of the albumin-globulin ratio. The incubation period is generally 6 months or longer for fever and hepatosplenomegaly to appear. Long-standing infection causes darkening of the

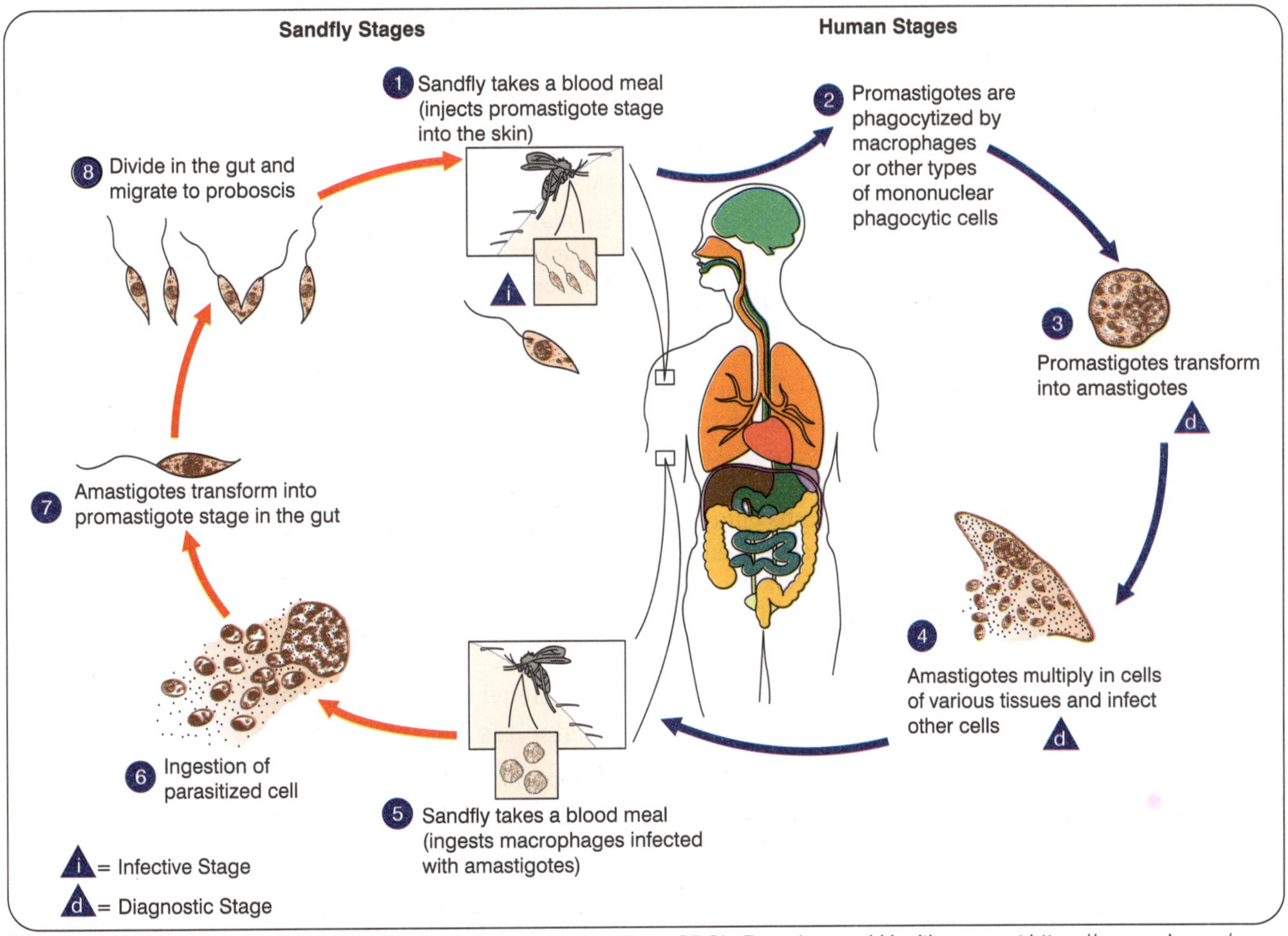

Sandfly Stages

Human Stages

1 Sandfly takes a blood meal (injects promastigote stage into the skin)

2 Promastigotes are phagocytized by macrophages or other types of mononuclear phagocytic cells

3 Promastigotes transform into amastigotes

4 Amastigotes multiply in cells of various tissues and infect other cells

5 Sandfly takes a blood meal (ingests macrophages infected with amastigotes)

6 Ingestion of parasitized cell

7 Amastigotes transform into promastigote stage in the gut

8 Divide in the gut and migrate to proboscis

i = Infective Stage

d = Diagnostic Stage

Fig. 25.2 Life cycle of *Leishmania donovani* (*Source:* Adapted from CDC's Parasites and Health page, at https://www.cdc.gov/parasites/leishmaniasis/biology.html)

skin and emaciation of body due to persistent stimulation of the macrophages releasing TNFα. If untreated, most patients die within a period of 2 years.

Laboratory diagnosis

1. Demonstration of the amastigote forms of the parasite in peripheral blood smear—smear of bone marrow biopsy or biopsy of other viscera, particularly splenic aspirate by Romanowsky stains (Fig. 25.3).
2. Histopathologic examination of liver, spleen for amastigote forms of the parasite.
3. In vitro culture of peripheral blood, bone marrow biopsy, lymph node aspirate or biopsy of other viscera on Novy Nicolle and McNeal (NNN) medium. The method for blood culture is to collect 2 mL blood to 10 mL citrated saline solution and send to the laboratory. The parasite takes 1–4 weeks to grow and appears as promastigote forms.
4. Serological tests to detect antibodies include the direct agglutination test (DAT) using whole or lysed promastigote antigens, ELISA using a specific rK39

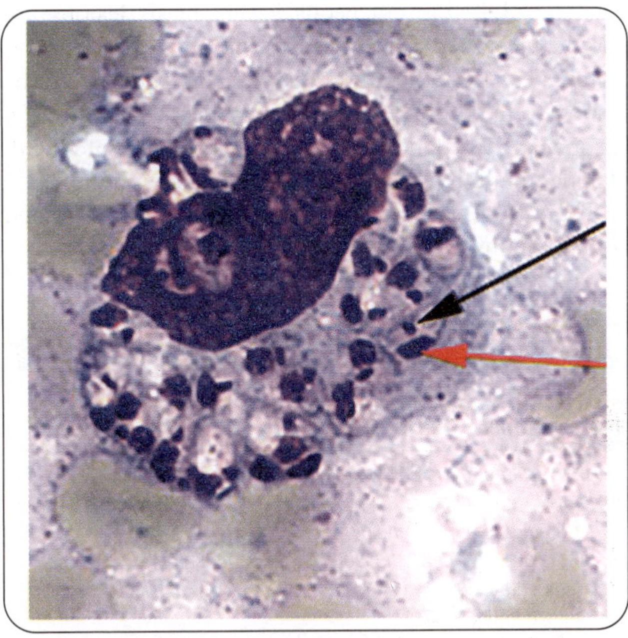

Fig. 25.3 Bone marrow macrophages showing amastigote forms of *Leishmania donovani*, Leishman's stain, 100X (*Source:* CDC, DPDx)

antigen and the indirect fluorescent antibody test. The DAT cannot distinguish from past disease, while the ELISA can distinguish visceral and cutaneous leishmaniasis. However, none are commercially available.
5. The peripheral blood picture shows leukopenia with decreased granulocytes, and non-specific hypergammaglobulinemia.

Dermal leishmanoid (post kala-azar dermal leishmaniasis—PKDL)

In 10% of cases of kala-azar, following 1–2 years of successful treatment of visceral leishmaniasis, dermal manifestation may appear. Termed as post-kala-azar dermal leishmaniasis (PKDL), the lesions are characterised by scattered maculopapular rash, butterfly erythema on the face, and soft, painless, granulomatous nodules, without ulceration. The condition can be diagnosed by demonstration of the amastigote forms of *L. donovani* in the lesion biopsy.

Cutaneous leishmaniasis

Completely separate from PKDL, the parasite *Leishmania tropica* does not cause visceral leishmaniasis but exclusively infects the Langerhans cells of the skin, causing cutaneous leishmaniasis also known as 'oriental sore', 'tropical sore' or 'Delhi boil'. The morphologic forms, life cycle and laboratory diagnosis of cutaneous and visceral leishmaniasis do not differ except that besides man, dogs too serve as a reservoir of infection. Unlike PKDL, the cutaneous leishmaniasis lesions are chronic ulcerative granulomas that occur on the face and extremities. The amastigote forms of the parasite can be demonstrated in the margins of the ulcer. Unlike visceral leishmaniasis, cutaneous leishmaniasis is associated with lifelong immunity following a single episode.

There is no vaccine for leishmaniasis and it is treated with antimony compounds.

MALARIA

Clinical Case 2

A 16-year-old girl presents with sudden onset intermittent fever with chills and rigors since 3 days. On examination, she appears anemic with no other abnormal finding.

1. *What laboratory investigation will you request for?*
2. *Name parasites causing anemia. What is the likely cause of anemia in this case?*
3. *If the parasite is absent on microscopy, what alternatives do we have to confirm the diagnosis?*
4. *Enumerate the complications of this disease.*

Amongst protozoans, sporozoans reproduce asexually (schizogony) and sexually (syngamy) either in the same host or another host. Schizogony is a form of asexual reproduction by which one schizont can produce many merozoites. In man, sporozoans either infect blood, e.g., *Plasmodium* and *Babesia,* or the intestines, e.g., *Toxoplasma* and *Isospora*.

The malarial parasites of man are sporozoans found in blood—*Plasmodium vivax, Plasmodium falciparum, Plasmodium malariae, Plasmodium ovale.*

Epidemiology

Malaria is endemic in India, and more common in the North Eastern states. The number of cases last reported in 2003 was 16.5 lakh and the number of deaths was 943. In the same year, malaria was included under the National Vector Borne Disease Control Programme (NVBDCP). Rainfall, humidity, breeding water sources and sleeping habits contribute to the incidence of malaria.

Morphology: Merozoite, trophozoite, schizont, gametocytes and sporozoite

All morphologic forms are seen in man except the sporozoite, which is the infective form introduced by the bite of a female Anopheles (Figs. 25.4a–25.4e).

Merozoites are seen in the hepatocytes in pre-erythrocytic schizogony and in the RBCs following erythrocytic schizogony. They consist of a fragment of chromatin with little cytoplasm, and no pigment. On lysis of the RBC, the released merozoites infect other healthy RBCs, thus continuing the process of erythrocytic schizogony. During erythrocytic schizogony, the parasite changes form from trophozoite to schizont to merozoite, over a period of 2 days, and 10 days after the infected mosquito bite.

The trophozoite has a red nucleus with a blue ring of cytoplasm that is thinner near the nucleus and measures around 3 μm. It lasts for 10 hours and is actively motile due to pseudopodia.

The immature schizont appears after 36 hours and is a mature, non-motile trophozoite ready to divide, and measures around 10 μm (larger than the RBC itself). It contains scattered pigment granules within the cytoplasm. Nuclear and cytoplasmic division of the schizont is completed in 16 hours giving rise to a mature schizont with merozoites. The RBC bursts releasing the merozoites to infect healthy RBCs, and coincides with the tertian (every third day) appearance of fever.

Some schizonts differentiate into male and others into female gametocytes. Gametocytes contain cytoplasm and nucleus and measure around 10 μm (larger than the RBC itself). They are ingested by the feeding mosquito, and

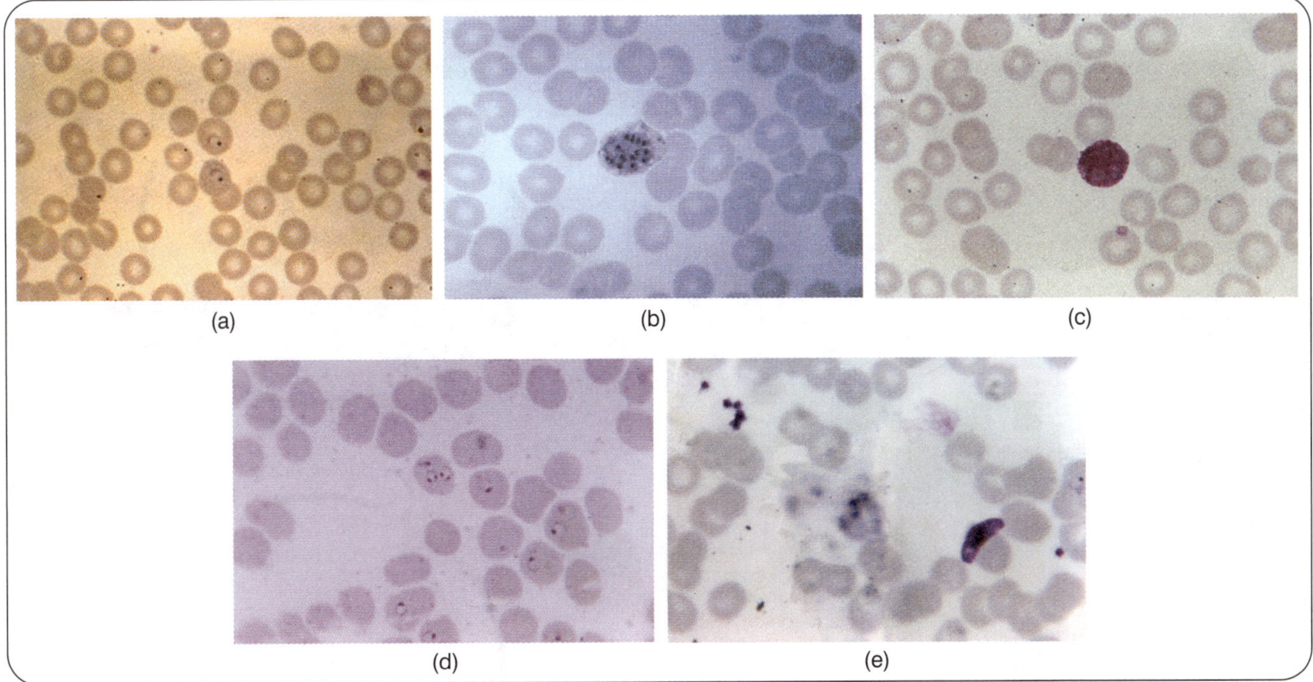

Fig. 25.4 (a) Ring form of *Plasmodium vivax* trophozoite (*Source*: Public Health Image Library, ID 5927 / CDC / Steven Glenn, Laboratory & Consultation Division), (b) Schizont of *Plasmodium vivax* (*Source*: Public Health Image Library, ID 3684 / CDC / Dr. Mae Melvin), (c) Gametocyte of *Plasmodium vivax* (*Source*: Public Health Image Library, ID 3475 / CDC / Dr. Mae Melvin), (d) Multiple ring forms of *Plasmodium falciparum* trophozoite (*Source*: Public Health Image Library, ID 22811 / CDC / Dr. Mae Melvin) and (e) Gametocyte of *Plasmodium falciparum* (*Source*: Public Health Image Library, ID 20374 / CDC / Dr. Mae Melvin)

undergo fertilisation in the mosquito to form oocyst and sporozoites.

Sporozoites are thread-like, curved cells with a central nucleus and no pigment, measuring around 10 μm. They are not seen in man.

Life cycle of a malarial parasite—Asexual life cycle in man (intermediate host)

Day 0 — Mosquito bite releases sporozoites that enter the hepatocytes and undergo pre-erythrocytic schizogony. The **pre-erythrocytic schizogony** results in the formation of merozoites of two kinds:

- Micromerozoites invade RBCs to subsequently undergo erythrocytic schizogony.
- Macromerozoites re-enter the hepatocytes to undergo **exoerythrocytic schizogony** and are responsible for relapse (in 3 years) in *P. vivax*. Exoerythrocytic schizogony is not seen in *P. falciparum*, therefore, there is no relapse in *P. falciparum*. However, *P. falciparum* infection persists in blood giving rise to recurrence of infection (in 1 year) called 'recrudescence' (Fig. 25.5).

Day 10 — Erythrocytic schizogony in RBCs transforming merozoites to trophozoite to schizont to merozoites is completed causing RBC lysis (clinical malaria) by day 10. The difference between pre/exo-erythrocytic schizogony and erythrocytic schizogony is the absence of pigment in the former and presence in the latter.

The released merozoites re-infect new RBCs which, in turn, lyse, releasing merozoites. This results in tertian or quartan fever associated with erythrocytic schizogony and the RBC lysis cycle. Some schizonts undergo gametogony to form M and F gametes in the RBCs. Gametogony occurs in the capillaries of the spleen and bone marrow and only mature gametocytes are found circulating in peripheral blood. They are ingested by the feeding mosquito and undergo fertilisation in the mosquito to form oocyst and sporozoites. Thus, the female *Anopheles*, in whom the sexual cycle takes place, is the definitive host in malaria. Persons with gametocytes in peripheral blood are carriers of malaria.

There exist variations in the morphologic appearances and time for development between the four different species of plasmodia (Table 25.1). In India, *P. vivax* is the most prevalent, followed by *P. falciparum*, the other species being rare if not absent.

Pathogenesis and clinical features

Schizogony destroys RBCs as parasites use the globin of hemoglobin and the pigment hematin gets deposited in the fixed macrophages of the liver and spleen. Clinically,

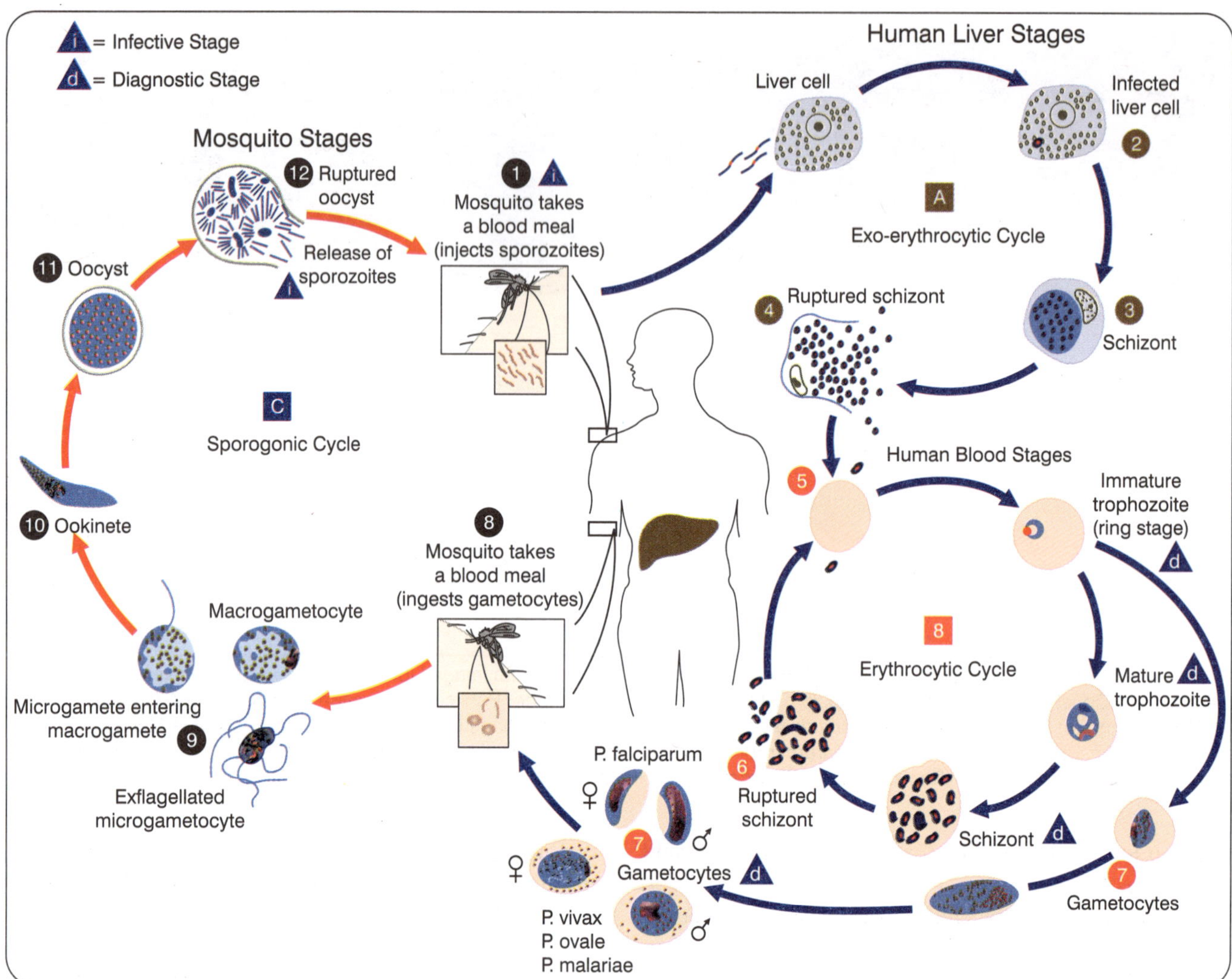

Fig. 25.5 Life cycle of a malarial parasite (*Source:* Adapted from CDC's Parasites and Health page, at https://www.cdc.gov/malaria/about/biology/index.html)

Table 25.1 Differences in *P. vivax* and *P. falciparum*

	Feature	*P. vivax*	*P. falciparum*
1.	Exo-erythrocytic schizogony and relapse	Seen	Not seen
2.	Pre-erythrocytic schizogony	8 days	6 days
3.	Erythrocytic schizogony (trophozoite)	1 in 1 RBC	2–4 in 1 RBC
4.	Erythrocytic schizogony	In circulation	In capillaries of organs
5.	Mature schizont	Larger, 16 merozoites	Smaller, 20 merozoites
6.	Schizont pigment	Yellow-brown	Black
7.	Gametocytes on exposure	Spherical, 16 days	Crescent, 21 days

this results in paroxysmal tertian fever, anemia, due to the erythrocytic schizogony, and splenomegaly. Anemia is of the normocytic or microcytic normochromic type.

Hemoglobin gets converted to hematin (porphyrin in feces), hemotoidin (stercobilinogen, urobilinogen) and hemosiderin. Visceral pathology exhibits pigmentation, hyperplasia, parasitised RBCs, and no fibrosis in the liver or spleen.

Laboratory diagnosis

1. *Light microscopy:* Peripheral blood smears (PBSs) are stained by Romanowsky stains (Giemsa, Wright's, Leishman's, Jenner modifications). Thick smear examination is done to screen for the presence of parasite followed by thin smear examination to identify the *Plasmodium* spp. The features of a thin and thick smear are listed in Table 25.2.

Table 25.2 Comparison between thick and thin blood smear for malaria

Feature	Thin smear	Thick smear
Area on slide	250–450 mm^2	50–90 mm^2
Blood volume	1 μL	5 μL
Mean thickness	0.0025 mm	0.09 mm
Vol. in 100 OI fields	0.007 μL	0.25 μL
Examination time	200 fields/20 min	100 fields/5 min
RBCs	Fixed	Hemolysed
Parasite morphology	Intact	Distorted

Thick smears are used to screen larger amounts of blood, which increases the possibility of detecting light infections. Red cells are lysed before staining, and only WBCs, platelets and parasites, when present, are visible. The morphology of parasites is distorted. Thin smears display parasitised RBCs and parasites clearly and provide definite species identification. The only disadvantage is that a very small volume of blood can be screened. Therefore, examination of a thick smear is more sensitive in picking up infection and can be used to screen patients for malaria. If positive, the parasite can be identified by a thin smear examination.

Parasite density is directly proportional to the severity, duration and outcome of the infection, especially in *P. falciparum*. Quantification of the parasite by the percentage of RBCs infected, or per microscopic field is also used to grade the response to therapy and monitor resistance to antimalarials.

2. *Fluorescent microscopy (FM):* PBS stained by acridine orange stains the plasmodial DNA that fluoresces.
3. *Quantitative buffy coat (QBC) examination*: Blood in a capillary tube is centrifuged to separate the blood components and stained by acridine orange. Parasites get concentrated in the buffy coat, and can be easily visualised under a fluorescent microscope. It is a very sensitive technique as it concentrates the parasite in blood for detection.
4. *Antigen detection:* By immunochromatography, detecting the histidine rich protein (HRP 2) of *P. falciparum* and parasite lactate dehydrogenase (pLDH) of all plasmodia.

   The Pf HRP2 is synthesised only by the asexual forms of *P. falciparum*. Its synthesis begins with immature trophozoites and it is found in the infected RBC cytoplasm and cell membrane. Only 5% is found free in human plasma; the concentration is even lower in the CSF in cerebral malaria. Therefore, whole blood is the preferred sample to detect Pf HRP2. It takes almost 1 month for HRP2 to completely disappear after acute infection, hence, this test cannot be used for monitoring therapy.

   The pLDH does not cross-react with human LDH. It is produced by all four species of *Plasmodium* in the asexual and sexual stages, and can distinguish between *P. vivax* and *P. falciparum* infection. It is cleared rapidly from blood within days of onset of treatment, making it useful in monitoring therapy and the detection of drug resistance.
5. *Antibody detection* is not employed as the presence of antibodies is not diagnostic of malaria.
6. *PCR* to detect plasmodial DNA, RNA and the enzyme Qβ replicase has been developed but is not used routinely.

The performance of the tests available for the laboratory diagnosis of malaria is depicted in Table 25.3.

Complications of malaria

1. Black water fever may occur due to inadequate treatment with quinine, leading to massive intravascular hemolysis, methalbuminemia and hemoglobinuria (dark coloured urine) and renal failure.

Table 25.3 Comparison of laboratory diagnostic tests in malaria

Test	Asexual parasite/μL	Sensitivity %	Specificity %	Total time
Giemsa	4	–	–	80 min
AO (400X)	100	90	90	10 min
QBC	84	90	90	10 min
PfHRP2	30	90	90	16 min
pLDH	60	80	90	16 min
PCR	0.3	100	100	4 hours

2. Pernicious malaria occurs due to the block of capillary vessels of internal organs by parasitised RBCs and may result in
   i) Cerebral malaria, characterised by hyperpyrexia, coma, and motor paralysis.
   ii) Algid malaria, characterised by cold, clammy skin with vascular collapse, and diarrhea.
   iii) Septicemic malaria, characterised by high temperature, pneumonia and cardiac syncope.
3. Hypoglycemia is more common in children and pregnant women.
4. Severe anemia due to RBC hemolysis is seen.

Prevention and treatment

Repeated infection with plasmodia results in the development of an immunity called 'premunition' wherein the parasite is not cleared but the immune response protects against high parasitemia and disease. There is no vaccine available as yet and prevention rests with preventing mosquito breeding and mosquito bites. The circumsporozoite protein (CSP) of *P. falciparum* (*spf*) vaccine, developed specifically for children in Africa to be administered in infancy, offers 36% protection and has not been approved by the WHO.

Malaria is treated with chloroquine, primaquine and artemisinin compounds. Anti-malarial drug resistance is known to exist and can be studied by the parasitic index (density) in RBCs before and after medication.

FILARIASIS

Clinical Case 3

A 40-year-old housewife presents with a history of left upper limb swelling with nodules in the left axilla. There is no history of fever or other complaints. On examination, left axillary lymph nodes are palpable but not tender. An FNAC of the lymph node reveals the presence of motile larvae.

1. *What is your most likely diagnosis?*
2. *Describe the morphologic features of the causative agent*
3. *What laboratory investigations are indicated in this case?*
4. *Name the drug of choice in treating her.*

Filarial nematodes include *Wuchereria bancrofti, Brugia malayi* and *Brugia timori* that reside in the lymphatics, and *Loa loa, Onchocerca volvulus* and *Mansonella* spp. that reside in the subcutaneous tissue or serous cavity. Of these, *Wuchereria bancrofti* and *Brugia malayi* are found in India, causing filariasis.

Epidemiology

W. bancrofti is globally distributed and accounts for 99% of filariasis cases in India, the remainder being due to *B. malayi* in Kerala. Filariasis is highly endemic in UP, Jharkhand, Bihar and WB and has a low prevalence in Punjab, Northeast India, and Jammu and Kashmir. *B. malayi* has been found in Assam, WB, Odisha and Kerala. It is transmitted by the bite of a mosquito—*Culex* spp.

India contributes to one-third of filariasis cases, globally, with 6 million acute annual attacks amongst 20 million chronic cases. We have 7.5 million lymphedema cases, 13 million hydrocele cases and 31.6 million microfilaria carriers.

Morphology

The adult resides in the lymph nodes and lymph vessels and is long (4 cm M, 8 cm F), slender and white with tapering ends (Fig. 25.6). *B. malayi* adults are smaller in size. Being viviparous, the larvae called 'microfilaria' are found in peripheral blood nocturnally. The larvae are 250 μ long, thin, slender, colourless and sheathed, with a blunt head and tail tip free of nuclei. The larva moves within the hyaline sheath. When stained, it is pink with violet nuclei present all through the body except at the head and tail end (Fig. 25.7). *B. malayi* microfilariae have few differences including that nuclei are present in the tail end unlike in *W. bancrofti*.

Life cycle

Man is the definitive host and the only reservoir, and *Culex* mosquito the vector (Fig. 25.8). The microfilariae are ingested from man by mosquito bites during a blood meal and they develop into infective filariform larvae in the mosquito. These infective filariform larvae are deposited in the skin during an infected mosquito bite (Day 0). From

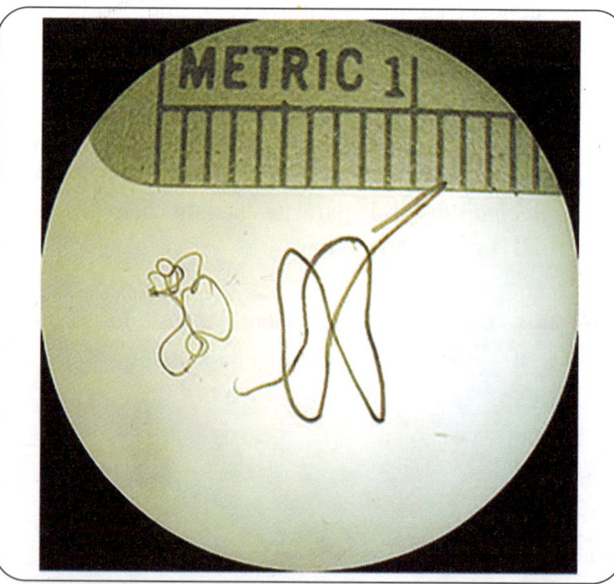

Fig. 25.6 Adult of filarial parasite (*Source:* CDC, DPDx)

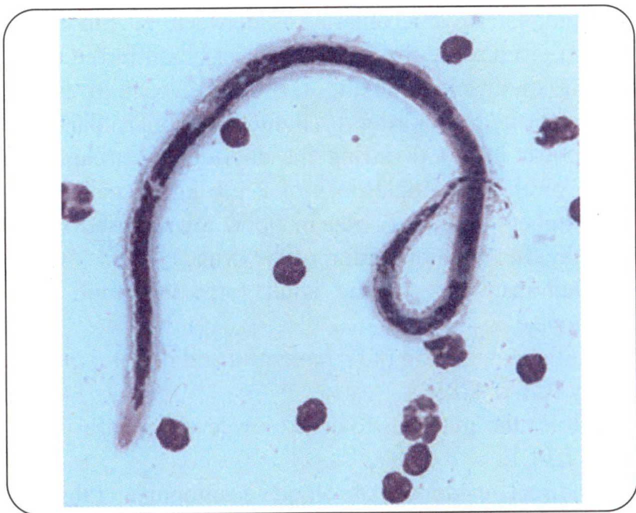

Fig. 25.7 Peripheral smear showing microfilaria of *Wuchereria bancrofti*, Leishman's stain, 40X (*Source:* CDC, DPDx)

the skin, the larvae enter the lymphatic and reach the lymph nodes where they develop into male and female adults. After mating, the male dies while the female can reside up to 20 years discharging microfilariae in blood which, in turn, are ingested by a *Culex* mosquito. The microfilariae first appear within 3–4 months after the infective bite and can survive up to 1 year in peripheral blood.

Pathogenesis and clinical features

The incubation period from the bite of the infected mosquito to classical lymphatic filariasis is about 1 year. The pathogenicity in classical lymphatic filariasis is largely related to the presence of the adult form in the lymph nodes and lymphatics and the response to it. The microfilariae play a role in occult filariasis rather than in classical lymphatic filariasis. The living adult and its resultant acute inflammation cause damage to the lymphatic vessels in the form of dilatation and thickening of the walls. Mechanical irritation, release of metabolites by the living adult and/or superadded bacterial infection presents as recurrent acute lymphangitis and lymphadenitis.

The lymphatics of the testes and epididymis are most commonly affected followed by those of the lower and upper extremities. During an inflammatory episode,

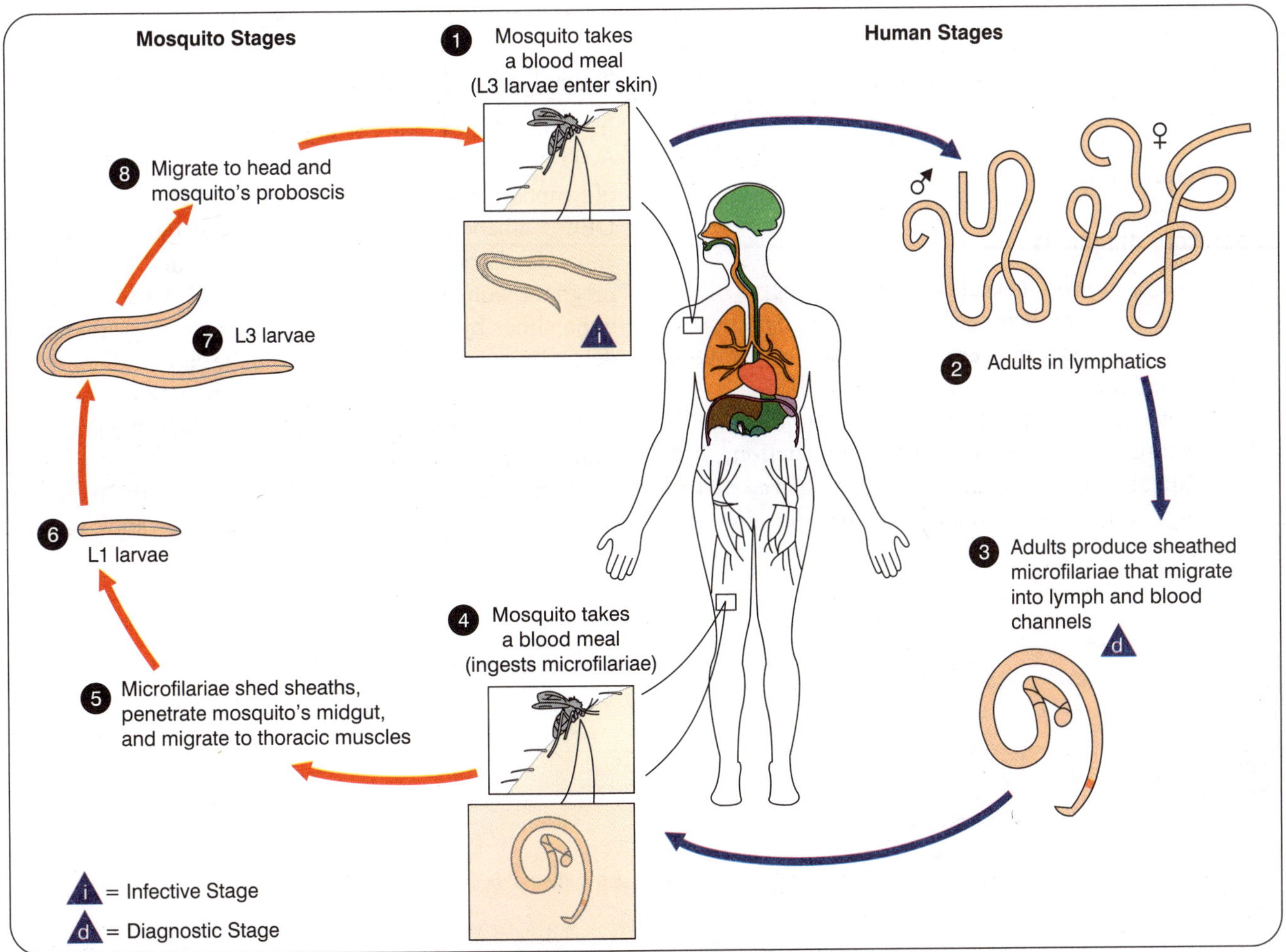

Mosquito Stages
Human Stages
1 Mosquito takes a blood meal (L3 larvae enter skin)
2 Adults in lymphatics
3 Adults produce sheathed microfilariae that migrate into lymph and blood channels
4 Mosquito takes a blood meal (ingests microfilariae)
5 Microfilariae shed sheaths, penetrate mosquito's midgut, and migrate to thoracic muscles
6 L1 larvae
7 L3 larvae
8 Migrate to head and mosquito's proboscis
i = Infective Stage
d = Diagnostic Stage

Fig. 25.8 Life cycle of *Wuchereria bancrofti* (*Source:* Adapted from CDC's Parasites and Health page, at https://www.cdc.gov/parasites/lymphaticfilariasis/biology_w_bancrofti.html)

the lymphatics are enlarged, erythematous, painful and tender. These inflammatory episodes are associated with fever and followed by an epithelioid granuloma around the adult, and tend to recur on a monthly basis. The microfilariae can be demonstrated in peripheral blood. Lymphadenitis of regional nodes usually precedes an attack of lymphangitis, but is painless and not tender.

As the female adult can survive up to 20 years, classical filariasis can develop into a chronic disease after 10 years. The death of the adult leads to fibrosis, endolymphangitis and lymphatic obstruction. Lymphatic obstruction over time causes varicosity of lymphatic vessels, lymphedema and hypertrophy of the affected part that is drained by that vessel (e.g., hydrocele, elephantiasis, chyluria). Hydrocele occurs due to accumulation of fluid in the tunica vaginalis, elephantiasis of the extremities or breast or vulva due to alterations in the skin and subcutaneous tissue, and chyluria due to rupture of lymph vessels at the urinary bladder mucosa.

Occult filariasis or tropical pulmonary eosinophilia is seen in some individuals. The microfilariae are not found in the peripheral blood but are filtered and destroyed in the lungs, giving rise to eosinophilic hypersensitivity reactions and IgE response. The disease presents with fever, nocturnal cough or wheeze and blood eosinophilia (absolute eosinophil count > 3000/μL). Besides the lungs, the microfilariae may also be filtered and destroyed in the spleen or liver.

Laboratory diagnosis

1. Demonstration of microfilariae in peripheral blood, chylous urine or hydrocele fluid either by thick or thin film stained by Leishman's stain or by QBC examination.

   It is difficult to demonstrate microfilariae in elephantiasis due to lymphatic obstruction after an attack of lymphangitis, occult filariasis and during early allergic manifestations. As they are nocturnal in periodicity, the ideal time period to demonstrate microfilariae is between 10 pm and 4 am. To obviate the need to collect blood in the night, the diethylcarbamazine (DEC) provocation test is performed. In order to induce the microfilariae to appear in blood during the day, diethylcarbamazine is given in a single dose of 2 mg/kg. Usually, peak numbers of microfilariae in blood are reported 15–20 min after administration of the drug.
2. Demonstration of the adult form in lymph node biopsy.
3. Antigen detection of *W. bancrofti*, and not of *B. malayi* by ICT or ELISA.
4. Molecular methods to detect low levels of parasites by RT-PCR.
5. Indirect evidence of blood eosinophilia (absolute eosinophil count > 3000/μL), raised serum IgE (> 1000 ng/mL).

   Antibody detection has a role to play in epidemiologic studies in endemic areas and cannot be used for the diagnosis of filariasis except in non-endemic populations. Rapid antibody (IgG) tests and ELISA for *B. malayi* are used for epidemiologic surveillance in the absence of antigen detection.

Prevention and treatment

In India, vector control and mass drug administration (MDA) through the NVBDCP exist for the elimination of lymphatic filariasis. MDA comprises of a single dose of DEC + albendazole given annually to all individuals in the endemic districts except children less than 2 years of age, pregnant women and the sick. Post-MDA surveillance is done either by examination of blood for microfilariae or detection of antigen or antibody in the case of *W. bancrofti* or *B. malayi* infection, respectively. In 2018, Assam, TN, Puducherry, Goa, Daman and Diu achieved elimination status in India.

Diethylcarbamazine with albendazole is the treatment of choice in filariasis as it is effective on the adult parasite and microfilariae.

Answers

Clinical Case 1

1. *Kala-azar or visceral leishmaniasis*
2. *Leishmania donovani, transmitted by the bite of sandfly*
3. *Bone marrow biopsy examination for amastigote forms of the parasite and culture of bone marrow aspirate.*

Clinical Case 2

1. *Peripheral blood smear examination for malarial parasite*
2. *Plasmodium spp., Ancylostoma duodenale and Diphyllobothrium latum*
3. *Detection of HRP2 and pLDH antigen by immunochromatography*
4. *Cerebral, algid or septicemic malaria, and severe anemia*

Clinical Case 3

1. *Filariasis*
2. *Wuchereria bancrofti microfilariae are 250 µ long, thin, slender, colourless and sheathed, with a blunt head and tail tip free of nuclei.*
3. *Examination of peripheral blood smear for microfilariae and detection of filarial antigen in blood, with or without eosinophilia.*
4. *Diethylcarbamazine (DEC)*

Questions

1. Name three parasite genera that are seen in blood and common in India. Compare and contrast thin and thick smear for the diagnosis of parasites in blood.
2. Draw the life cycle and morphologic form of *L. donovani* seen in tissue.
3. How will you diagnose kala-azar in the laboratory?
4. Compare and contrast dermal leishmanoid and cutaneous leishmaniasis.
5. Draw the life cycle and morphologic forms of *P. falciparum* seen in peripheral blood.
6. Enumerate the differences between *P. vivax* and *P. falciparum*.
7. Outline the laboratory diagnosis of malaria.
8. Compare and contrast HRP2 and pLDH detection in the diagnosis of malaria.
9. Discuss the life cycle, pathogenesis and clinical features of filariasis.
10. What is the hetrazan provocation test?

SECTION SUMMARY

Section IV: Cardiovascular System and Blood Infections

Infection/pathogen	Virulence/pathogenesis	Laboratory diagnosis	Vaccine/recommended antimicrobial
Infective endocarditis *Streptococcus viridans*, enterococci, *Staphylococcus aureus*	Adherence to valves causing vegetation, bacteremia	Blood culture, 3 within 24 hours	Depends on causative bacterium susceptibility
Rheumatic fever *Streptococcus pyogenes*	Immune-mediated damage following streptococcal pharyngitis	ASO titre, elevated CRP, ESR	Penicillin G for treatment, benzathine penicillin for prophylaxis
Sepsis	Endotoxins mediated, disseminated intravascular coagulation, shock	Blood culture, raised CRP, procalcitonin	Imipenem +/– amikacin
HIV-AIDS	Infects CD4 cells, compromises immunity	Antibody detection by ICT/ ELISA	Antiretroviral therapy
Dengue	Bite of *Aedes*. Primary infection–fever, rash. Secondary infection–antibody dependent enhancement, dengue shock syndrome	NS1 antigen in first week, IgM after 1 week, viral genome by RT-PCR	Supportive therapy
Chikungunya	Bite of *Aedes*, arthritis, arthralgia	IgM or IgG by ELISA, viral genome by RT-PCR	Supportive therapy
Japanese B	Bite of *Culex*, encephalitis	IgM or IgG by ELISA, viral genome by RT-PCR	Killed virus Beijing P3 strain, supportive therapy
Kala-azar *Leishmania donovani*	Bite of sandfly, infects MPS, visceral leishmanaisis	PBS and viscera biopsy amastigotes. NNN medium culture promastigotes.	Sodium stibogluconate
Malaria *Plasmodium* spp.	Bite of female *Anopheles*, infects liver and RBCs	PBS trophozoites, schizonts, gametocytes Antigen HRP2 or pLDH by ICT	Chloroquine, artemisin derivatives
Filaria *Wuchereria bancrofti* *Brugia malayi*	Bite of *Culex*–adult in LN, chronic lymphangitis, lymphatic obstruction	PBS microfilariae LN adult	Diethylcarbamazine
Anemia *Ancylostoma duodenale* *Plasmodium* spp. *Diphylobothrium latum* Parvovirus			

Gastrointestinal Infections and Food Poisoning

The alimentary tract consists of different microbial ecosystems occupying specific sites. The normal stomach being acidic limits the entry of microorganisms, especially bacteria, to the intestines. Bacteria that are pH-sensitive require a larger infective dose to survive the acidic stomach and infect the intestines, e.g., *Vibrio cholerae* requires an alkaline pH.

There exist 400 species of bacteria, some fungi and protozoa that form the resident flora of the normal gastrointestinal tract. The colon has about 10^{12} bacteria/g that have a symbiotic relationship with humans.

> This symbiotic relationship between the intestinal commensal bacteria and humans plays a role in health:
> 1. Convert unabsorbable carbohydrates to absorbable organic acids
> 2. Provide energy for colonic epithelial cells in the form of butyrate
> 3. Provide vitamin K needed for the synthesis of blood clotting factors
> 4. Re-absorb and concentrate estrogens and androgens excreted in the bile
> 5. Prevent colonisation by invading pathogens
> 6. Prevent disease by antimicrobial mucosal soluble factors like IgA and lactoferrin (in breastmilk)

On failure of the barriers to establish gastrointestinal infections, infective diseases of the GIT could either be diarrhea, dysentery, food poisoning or an opportunistic infection. Diarrhea is defined as three or more loose bowel movements in a day. Dysentery is defined as blood in stool, with or without fever. Food poisoning is diarrhea, with or without vomiting, subsequent to the consumption of contaminated food. Opportunistic gastrointestinal pathogens, when infecting, do not cause disease in normal individuals but cause disease in the immunocompromised. With the exception of *Helicobacter pylori* gastritis, the term 'gastroenteritis' is used to describe diarrhea and/or vomiting due to intestinal infections or infestations.

Mechanisms that provide a defence against gastrointestinal infection include an unbroken mucosal epithelium, presence of mucus, motility of the gut, presence of acid in the stomach and bile in the intestine and IgA in the case of infection. The commensal bacteria form colonies on the intestinal mucosa, thereby competing for space and preventing the pathogens from establishing infection.

Bile plays an important role in selecting those bacteria and viruses that are able to colonise the intestine. Enteric viruses like poliovirus and HAV lack the lipid envelope that would make them sensitive to destruction by bile. Similarly, *Salmonella typhi* and enterococci spp. are bile-resistant and can grow in the gall bladder.

At a normal gastric pH (lower than 4), most coliform bacteria are killed, reducing the normal gastric flora to 10 CFUs/mL. Depending on the susceptibility of enteric pathogens to gastric acidity, their infectious dose varies from as low as 10 bacilli in *Shigella* to 10^8 in *V. cholerae*. Gastric acidity also plays a role in parasitic excystation and ova/larvae destruction. Anatomic alterations, changes in stomach acidity (achlorhydria), alterations in the normal flora (due to antibiotic intake) and encounters with pathogens predispose one to gastrointestinal disease.

Once a microbe overcomes the barriers to infection and causes disease, the mechanisms of virulence and pathogenesis are as follows:

1. Exotoxins are secreted by enteric pathogens. They can be neurotoxins, enterotoxins or cytotoxins, with certain toxins having more than one action or bacterium secreting more than one toxin. Neurotoxins are pre-formed toxins of *Clostridium botulinum*, *Staphylococcus aureus* and *Bacillus cereus*. The enterotoxins of *V. cholerae* and enterotoxigenic *Escherichia coli* provoke copious watery diarrhea, leading to profound dehydration, electrolyte loss, depletion of intravascular volume and shock. Cytotoxins are produced by *Shigella* spp. and *Clostridium perfringens*.
2. Attachment and adhesion is the ability to adhere to and colonise the mucosa as seen in *V. cholerae* and pathogenic *E. coli*.
3. Local invasion and inflammation occur due to pathogens like *Shigella* spp. and *Entamoeba histolytica* and cause dysentery. Deep tissue invasion by typhoidal salmonellae results in septicemia and that by *E. histolytica* results in hepatic amoebiasis. Intestinal perforations leading to peritonitis may be sequelae to infection with typhoidal salmonellae.
4. Selective lysis of absorptive villous tip cells leaves the secretory crypt cells intact as seen in rotavirus and Norwalk virus.
5. The disease may result in intestinal obstruction or stricture formation, e.g., ascariasis or tuberculosis, respectively.

Normal microbial flora of the GIT

Oral cavity

The normal defences in the oral cavity are non-pathogenic resident flora, and antimicrobial constituents of the saliva like lysozyme and IgA.

Streptococcus mutans transforms sucrose to sticky polysaccharides causing dental plaque with a dense collection of bacteria that act on sugars to form acids that erode the tooth, resulting in dental caries. Oral α-hemolytic streptococci play a role in subacute bacterial endocarditis and rheumatic fever. Pharyngitis of polymicrobial origins causes Ludwig's angina. Oral thrush caused by *Candida* spp. is a pseudo-membrane formed under predisposing conditions including diabetes, malnutrition, malignancy, immunosuppression and inhaled steroids (in asthmatics).

Esophagus

Candida may also invade the esophagus, an organ that is rarely prone to infection. Mucocutaneous candidiasis, HSV-1 and CMV can affect the esophagus in the immunocompromised host.

Stomach/gastric pouch

The level of normal commensal bacteria in stomach is reduced to 10^3 bacteria/mL due to the presence of acid. These bacteria are predominantly gram-positive *Streptococcus, Staphylococcus, Peptostreptococcus* and *Lactobacillus,* and gram-negative *H. pylori* and anaerobic *Bacteroides* and *Clostridia* spp.

The liver and biliary tree including the gall bladder are free of microbes unless diseased. Cholecystitis or cholangitis can be caused by *E. coli, Klebsiella pneumoniae,* Group D streptococci, *S. aureus, C. perfringens,* typhoidal salmonellae, or *Ascaris.*

Primary bacterial infection of the liver is not common. Intracellular pathogens that survive in macrophages can cause granulomatous infections like salmonellosis, TB and brucellosis. Primary viral hepatitis (HBV, HCV, HAV, HEV), secondary viral hepatitis and parasitic infections like amoebiasis and hydatid disease can affect the liver.

Like the skin, the gastrointestinal tract is heavily colonised with bacteria in variable proportions and variable bacteria at variable sites (Table 26.1).

Bacterial overgrowth in the small intestine results in malabsorption of fat (steatorrhea or fecal fat), deficiencies of vitamin B_{12} as it is utilised by the bacteria, diarrhea and malabsorption of vitamin A and D.

Diarrhea is defined as an increase in the daily amount of stool water, and frequency of motion. It is secretory in nature and caused by *Vibrio* and rotavirus. Dysentery involves the colon, with small-volume stools of blood, pus and mucus. It is caused by invasive, tissue-damaging enteric bacteria, e.g., *Shigella, E. coli* 0157:H7.

Etiologic agents of diarrhea

1. Viruses — Rotavirus, Norovirus, Calicivirus, Astrovirus, Enteroviruses
2. Bacteria — *V. cholerae, E. coli, Salmonella* spp., *Clostridiodes difficile*
3. Protozoans — *Giardia lamblia, Cryptosporidium, Isospora belli, Cyclospora, Microsporidia*
4. Helminths — *Ascaris lumbricoides, hookworm, Strongyloides stercoralis, Enterobius vermicularis, Trichuris trichiura, Fasciolopsis, Taenia, Diphyllobothrium latum*

Etiologic agents of dysentery

1. Bacteria — *Shigella,* entero-invasive *E. coli* (EIEC), *Campylobacter jejuni, Vibrio parahaemolyticus*
2. Protozoan — *Entamoeba histolytica*

Table 26.1 The normal microbiome of the gastrointestinal tract

Bacteria sp. in CFU/mL	Stomach	Jejunum	Ileum	Colon
Facultative anaerobes				
Non-pathogenic enterobacteria (e.g., *E. coli*)	10	10^3	10^7	10^{10}
Streptococci	10^2	10^4	10^6	10^9
Staphylococci	10^2	10^3	10^5	10^{10}
Lactobacilli	10^3	10^4	10^5	10^{10}
Strict anaerobes				
Bacteroides	–	10^3	10^7	10^{12}
Bifidobacterium	–	10^4	10^9	10^{12}
Streptococci	–	10^3	10^6	10^{12}
Clostridia	–	–	10^6	10^{12}
Eubacteria	–	–	–	10^{12}

Note: The numbers increase from the stomach to colon.
Strict anaerobes far outnumber the facultative anaerobes, *Bacteroides* spp. being the most predominant bacteria in the intestines. Amongst the facultative anaerobes, *E. coli* is predominant but less in number than *Bacteroides* spp.

Etiologic agents of food poisoning

1. Bacterial — *S. aureus, B. cereus, C. botulinum, C. perfringens* type A
2. Bacterial zoonotic — *Yersinia enterocolitica,* non-typhoidal salmonellae, *C. jejuni, Aeromonas, Plesiomonas* spp.
3. Mycotoxins

The source of these etiologic agents causing gastrointestinal infection varies (Table 26.2).

Infections of the small intestine

1. Viruses that cause death of intestinal epithelial cells in the villi—rotavirus, Norwalk and related viruses.
2. Bacteria that colonise the upper small intestine and produce exotoxins, e.g., *V. cholerae*, enterotoxigenic *E. coli* (ETEC).
3. *Giardia* and *Cryptosporidium*.
4. Toxigenic bacteria that cause food poisoning (those multiplying in food before being eaten).
5. *C. jejuni* and *Yersinia enterocolitica* that cause dysentery. Reactive arthritis (associated with HLAB27) is a sequela within weeks of diarrhea in *C. jejuni, Y. enterocolitica* and non-typhoidal salmonellae (NTS).

Table 26.2 Source of microbial gastrointestinal infection

Microorganism	Source
Bacteria	
E. coli	Human fecal contamination
Salmonella spp.	Human fecal contamination, poultry products
Shigella spp.	Human fecal contamination
*Campylobacter jejuni**	Farm animals, eggs
*Yersinia enterocolitica**	Animal products, raw milk
Vibrio cholerae	Human fecal contamination, water and shellfish
Vibrio parahaemolyticus	Seafood
Bacillus cereus	Pre-cooked rice
Clostridium perfringens	Canned food
Protozoans	
Entamoeba histolytica	Human fecal contamination, water
Giardia lamblia	Human fecal contamination, water
Cryptosporidium	Human fecal contamination, water
Cyclospora	Human fecal contamination, water
Virus	
Rotavirus	Human fecal contamination, water
Norwalk virus	Human fecal contamination, water

*Zoonotic infections

Infections of the large intestine

Epithelial damage with ulceration, mucosal inflammation and blood in stool are seen in invasive pathogens *Shigella* and *E. histolytica*. Pain worsening with bowel movement is called tenesmus. *E. coli* 0157:H7 is a normal inhabitant of meat and dairy cattle.

C. difficile causes antibiotic-associated diarrhea as antibiotics alter resident flora, forming an adherent pseudo-membrane with mucosal inflammation, without tissue invasion. Shigellosis can result in malnutrition, Kwashiorkor and rectal prolapse in children. *E. coli* can cause hemolytic uremic syndrome. *E. histolytica* can cause intestinal perforation and liver abscess.

Other gastrointestinal infections

These are characterised by invasion of gut-associated lymphoid tissue (GALT) of the small intestine, disseminating to the liver and spleen—*S. Typhi* in bloodstream causing septicemia. With sustained bacteremia, the microbe invades the Peyer's patches, causing inflammation, bleeding and perforation. Early use of antibiotic controls acute illness but predisposes one to relapse by limiting the immune response.

Perforation of intestinal wall leads to peritonitis due to resident intestinal flora, both strict and facultative anaerobes.

General approach to laboratory diagnosis of infectious diarrhea

1. History of fever, blood in stool, tenesmus, travel, type of food consumption, antibiotic use and immunosuppression.
2. Macroscopic examination of stool consistency and for the presence of adult helminths, blood, mucus and bile in fecal matter.
3. On microscopic examination the absence of WBCs in stool implies a viral, or toxin-mediated bacterial etiology while the presence of WBCs implies an invasive bacterial etiology.
   a) By wet mount for ova, larvae, cysts and trophozoites of parasites following stool concentration.
   b) By monochrome staining and the hanging drop method for cholera.
4. Culture of stool for bacterial enteric pathogens — *Salmonella, Shigella, Vibrio, Campylobacter.*
5. Immunoassay for detecting the toxin of rotavirus and *C. difficile* in stool.

Diarrhea in children less than 2 years of age is mostly viral, with rotavirus being the commonest causative virus. Most diarrheas are self-limiting in 2–3 days and need rehydration alone. However, in the case of persistent symptoms and dysentery, specific antibacterial or antiparasitic therapy is indicated.

BACTERIAL FOOD POISONING

Clinical Case

An 80-year-old man complains of sudden onset diarrhea and generalised weakness without fever. He gives a history of having a non-vegetarian meal on a flight he undertook the same morning. On examination, he has mild dehydration and tachycardia but systemic examination is normal.

1. *What is the most likely diagnosis and causative agent?*
2. *Name three other exotoxin-mediated pathogens causing this condition.*
3. *What is the antimicrobial of choice in this case?*
4. *What other treatment would you recommend?*

The various mechanisms by which food poisoning occurs are:

1. *Pre-formed toxins in the food:* Cooking can destroy bacteria but toxins are resistant to heat (cooking). *S. aureus* exotoxins occur in 1–6 hours of contamination of the food by food handlers or carriers. The *C. botulinum* toxin is pre-formed in food cans and *B. cereus* in pre-cooked rice. Generally, toxin-mediated food poisoning has a short incubation period and involves the upper gastrointestinal tract, causing vomiting more than diarrhea.
2. *Toxins produced in the human body following consumption of live bacteria*: Watery diarrhea occurs 12 hours after consumption of pre-cooked meat that is reheated before serving in aircrafts, e.g., *C. perfringens*.
3. *Intestinal invasion by bacteria:* Generally has a longer incubation period and involves the lower gastrointestinal tract, causing diarrhea more than vomiting, e.g., *Salmonella* (non-typhoidal) and *C. jejuni*.
4. *Rare pathogens:* These include *Vibiro parahaemolyticus* (shellfish), *Y. enterocolitica* (milk), and EIEC.

Bacteria that cause food poisoning are listed in Table 26.3 with the virulence mechanism, incubation period, source and features of infection. In most cases, food poisoning is self-limiting and needs rehydration. Only invasive pathogens may need antimicrobial therapy. Bacteria that invade the intestinal mucosa are NTS and *C. jejuni, V. parahaemolyticus* and *Y. enterocolitica*.

Staphylococcus aureus

S. aureus is a gram-positive coccus and facultative anaerobe that resides on human skin. Some of the strains secrete an enterotoxin that stimulates intestinal peristalsis and vomiting by acting on the vagus nerve and the central vomiting centre in the medulla oblongata. Food handlers who are carriers of *S. aureus* contaminate food such as dairy products, processed meat and salads. The bacterium produces enterotoxin in the food, which when consumed causes predominantly vomiting. As the exotoxin is pre-formed prior to consuming the contaminated food, the symptoms occur within 1 to 6 hours of the food poisoning. The enterotoxin can be detected in the food by ELISA, LAT or PCR, but there is no role for antimicrobial therapy, only rehydration.

Bacillus cereus

B. cereus is a strictly aerobic gram-positive bacillus that inhabits the soil and contaminates food. It produces either a pre-formed emetic enterotoxin or a secretory diarrheal

Table 26.3 Bacteria that cause food poisoning

Bacterium	Mechanism	Incubation period (hours)	Source	Features
S. aureus	Exotoxin (enterotoxin)	1–6	Food handler	Vomiting
B. cereus	Exotoxin	1–6	Pre-cooked rice, vegetables, cereals, spices	Vomiting, diarrhea
Non-typhoidal *Salmonella*	Invasive	16–48	Eggs, poultry	Fever, diarrhea
C. jejuni	Invasive	16–48	Poultry, milk, beef	Fever, diarrhea, dysentery
V. parahaemolyticus	Invasive	16–48	Shellfish	Fever, diarrhea
*Y. enterocolitica**	Invasive	16–48	Milk	Diarrhea
E. coli O157:H7	Exotoxin	16- 48	Beef, raw milk, salad	Fever, diarrhea
C. perfringens type A *C. botulinum*	Exotoxin Neurotoxin	8–12 18–36 hours	Re-heated meat Canned food, honey	Diarrhea Diarrhea, flaccid paralysis

*Rarely reported in India

enterotoxin that causes vomiting or diarrhea, respectively. *B. cereus* emetic food poisoning is seen within 1–6 hours of consuming pre-cooked rice, while diarrheal food poisoning occurs in 8–16 hours of consuming vegetables, meat or cereals as the enterotoxin is secreted by the bacterium in the intestine. *B. cereus* can be cultured from the stool sample of the patient and is sensitive to erythromycin.

Non-typhoidal salmonellae (NTS)

Salmonellae are gram-negative bacilli and facultative anaerobes that belong to *Enterobacteriaceae*. Non-typhoidal salmonellae are salmonellae that do not cause enteric fever and are not exclusively found in humans. They cause zoonotic infections, acquired by ingestion of contaminated eggs, poultry, meat and dairy products. They invade the intestinal mucosa and cause gastroenteritis due to food poisoning, rarely followed by bacteremia. *Salmonella enterica* serovar Typhimurium and *Salmonella Enteritidis* are the most commonly encountered amongst many other species. They invade the intestinal epithelium and present with fever and diarrhea 16–48 hours after consumption of the contaminated food.

Campylobacter jejuni

Campylobacter is a genus of microaerophilic, motile, gram-negative bacilli, appearing curved, wavy or spiral. They are zoonotic and found in the intestines of many birds and animals. The infection is transmitted by ingestion of contaminated food, handling of pets or during travel. In India, cases of *C. jejuni* causing bacterial enteritis have been reported, although the incidence is lower than that in developed countries.

C. jejuni secretes an enterotoxin and a cytotoxin, causing diarrhea, dysentery and fever. *Campylobacter fetus* causes extraintestinal disease like meningitis, bacteremia and vascular infections in the immunocompromised. *Campylobacter* infection is also associated with reactive arthritis in persons with HLA-B27 and may trigger immune-mediated Guillain–Barré syndrome. The bacterium can be demonstrated in stool by smear microscopy and culture identification on specialised media. Treatment is by rehydration and antimicrobials erythromycin or azithromycin.

Vibrio parahaemolyticus

V. parahaemolyticus is an aerobic, gram-negative bacillus, motile and capsulated. In India, it has been found in the crustacean shellfish (crabs, shrimps) from the East coast and reported in cases from Kolkata. It is an invasive bacterium causing self-limiting acute gastroenteritis 16–48 hours after consumption of contaminated seafood. *V. parahaemolyticus* can be isolated from stool on culture.

Enterohemorrhagic *E. coli* (EHEC) 0157:H7

This causes food poisoning by the consumption of contaminated meat and vegetables, and has a low infective dose. The bacterium secretes Shiga-like toxins—Stx 1 and Stx 2—that cause hemorrhagic colitis or hemolytic uremic syndrome, characterised by dysentery, renal failure and encephalopathy. The condition is diagnosed by stool culture, and toxin detection by ELISA or PCR for toxin genes.

Clostridium perfringens type A

C. perfringens is a sporulating anaerobic gram-negative bacillus found in the intestine of humans and animals. It causes food poisoning on consumption of non-vegetarian food that has been cooked, kept warm for a long time and re-heated. On reaching the intestine, it secretes an exotoxin causing diarrhea 8–12 hours following infection. The diarrhea is self-limiting and antimicrobials are not indicated. Cooked food needs to be refrigerated if not consumed and re-heated properly before consumption.

Clostridium botulinum

C. botulinum is a sporulating anaerobic gram-negative bacillus found in nature, in soil and vegetables. On ingestion either of the bacterium or pre-formed toxin, especially in canned food or honey (infant botulism), a lethal neurotoxin 'botulinum toxin' (BT) is produced intracellularly that is released only on cell lysis. The BT travels through blood to peripheral (not central) cholinergic nerve terminals at the neuromuscular junctions, ganglions and post-ganglionic neurons. It binds to Ach receptors, blocking the release of Ach in nerve terminals, resulting in flaccid paralysis of voluntary muscles. The patient experiences blurred vision, dysphagia, slurred speech to begin with and may progress to respiratory muscle paralysis and death, if not treated.

The bacterium can be demonstrated by food or stool microscopy or culture. As the disease is exotoxin-mediated, immediate administration of botulinum antitoxin is important to prevent the disease from progressing. Botulism, as the disease is called, can also result from wound infection, in which case there is no diarrhea and the incubation period for flaccid paralysis to set in is from 4–14 days. Botulinum toxin or botox is also used for cosmetic purposes.

Mycotoxicosis

Food poisoning due to fungi can be either due to toxins released by fungi—mycotoxicosis—or due to consumption of poisonous mushrooms—mycetism. Mycotoxins are produced mostly by *Aspergillus* spp. (aflatoxin), *Penicillium* spp. and *Fusarium* spp. (fumonisins). They contaminate grains (wheat, maize), nuts (groundnuts), cereals and bread, causing a variety of disease ranging from hepatitis to nephropathies.

Answers

Clinical Case

1. *Acute food poisoning; Clostridium perfringens*
2. *Staphylococcus aureus, Bacillus cereus, Clostridium botulinum*
3. *Antimicrobial therapy not indicated*
4. *Oral rehydration to correct dehydration*

Questions

1. Name two facultative anaerobic and two strict anaerobic bacteria that are commensals in the intestine. Explain the role of intestinal commensals in health.
2. Discuss the innate defence mechanisms in the gastrointestinal tract against microbial infection.
3. Discuss the mechanisms of virulence and pathogenesis of gastrointestinal infections.
4. Enumerate bacteria causing food poisoning with an emphasis on the source and incubation period as a clue to the causative agent.

Infective Diarrhea

BACTERIAL DIARRHEA

Diarrheal disease is a major public health problem in India, accounting for 1.5 million annual deaths in children under 5 years of age, especially infants. Ninety percent of these deaths could be prevented by early rehydration using an oral rehydration solution. The bacteria causing diarrhea are *Vibrio cholerae*, diarrheagenic *Escherichia coli*, *Salmonella* spp. and *Clostridioides difficile*.

Cholera

Clinical Case 1

There is a sudden increase of cases with acute watery diarrhea with dehydration presenting at a rural public health centre (PHC). On history, it is found that those affected belong to a geographic cluster around a common well providing water.

1. *What is the history suggestive of?*
2. *What immediate laboratory test can confirm the diagnosis?*
3. *What steps are indicated to manage this situation?*
4. *Name four viruses causing diarrhea*

Epidemiology

V. cholerae infection ranges from asymptomatic to life-threatening acute diarrhea, that can be sporadic, endemic, or epidemic, especially as a localised outbreak. In the past, it has caused seven pandemics, the last one between 1961 and 1991 by the El Tor biotype as opposed to the classical biotype in the earlier six.

Presently, cholera is a notifiable disease, seen worldwide, especially in the developing countries. In India, though it can occur in any state, the number of cases and mortality has significantly reduced over the years. In 2020, as of March, one case each in Kerala, Assam and Karnataka and 25 cases in a localised outbreak in Bengaluru were reported. The current serogroup in India is predominantly O1 and occasionally O139, both of biotype El Tor. The O139 or Bengal strain was first documented in Chennai in 1992 as a new serogroup and differed from the other serogroups as it is invasive and can cause bacteremia or extraintestinal disease. The national reference centre for cholera in India is the National Institute of Cholera and Enteric Diseases (NICED), Kolkata.

Human carriers are the only reservoir of *V. cholerae*, who shed bacilli in the stool. Carriers can be healthy individuals who are asymptomatic or patients during convalescence who continue to shed the bacilli after recovery. Rainfall with flooding, poor sanitation, unhygienic practices and mobile mass gatherings are conducive for transmission through contaminated food and water.

Classification of vibrios

Vibrios are classified as halophilic bacteria that require salt for their growth and can tolerate high salt concentrations, e.g., *Vibrio parahaemolyticus* causing seafood-associated gastroenteritis and non-halophilic bacteria that can grow without salt, e.g., *V. cholerae*.

V. cholerae is a comma-shaped gram-negative bacillus with a single polar flagellum, exhibiting darting or extremely active motility (Fig. 27.1). By the Gardner and Venkatraman classification, *V. cholerae* is further classified into serogroups, biotypes and serotypes for epidemiological purposes.

Serogroups

There are 200 serogroups of *V. cholerae* based on the somatic O antigen—O1 to O200. The O1 serogroup is agglutinated by the O1 antiserum and the most prevalent serogroup across epidemics and pandemics. The O139 or Bengal strain was first documented in Chennai in 1992 as a new serogroup also capable of causing epidemics. The serogroups other than O1 and O139 cause sporadic cases or local outbreaks of diarrhea following raw oyster consumption, or ear and wound infection acquired from seawater exposure, but have not been implicated in an epidemic.

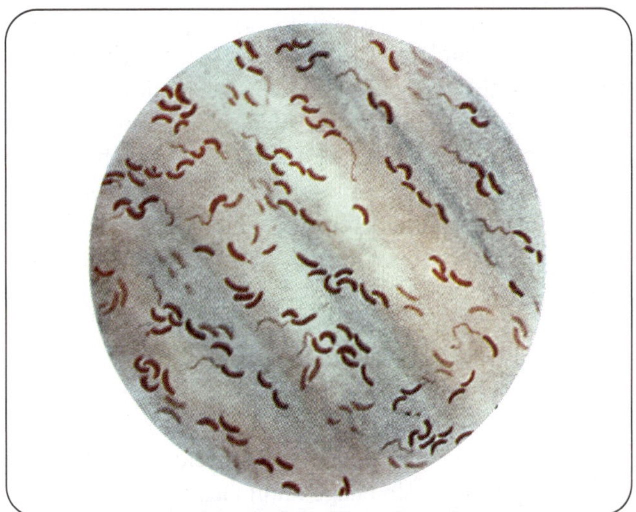

Fig. 27.1 Curved, comma shaped gram-negative bacilli, *Vibrio cholera*, Gram stain, 100X (*Source*: Public Health Image Library, ID 5324 / Centers for Disease Control and Prevention [CDC])

Biotypes

Serogroup O1 has two biotypes—classical and El Tor—the latter being less virulent. The first six pandemics are attributed to the classical biotype and the seventh to the El Tor biotype. The two can be distinguished by a variety of laboratory characteristics and phage typing, but there are variants that do not fit into either and are called El Tor variants.

Serotypes

Serogroup O1 can be further divided into three serotypes, i.e., Ogawa, Inaba and Hikojima, based on differences in the somatic O antigen varying in their frequency of occurrence.

Pathogenicity and clinical features

Being acid-labile, the infective dose of *V. cholerae* is 10^8 bacilli, in order for it to escape the gastric acidity. On ingestion and its survival of the gastric acidity, the bacterium arrives in the small intestine. Due to its motility and secretion of enzymes—mucinase and cholera lectin—it penetrates the mucus to reach the epithelial cells. It then adheres to the epithelium by a fimbria called toxin coregulated pilus (TCP). Once the bacteria adhere to the intestinal epithelium, they produce copious amounts of an exotoxin—an enterotoxin called cholera toxin (CT). A single gene—*ToxR*—regulates the synthesis of TCP and CT. The cholera toxin CTX gene is phage-mediated, the TCP on the bacterial surface being its receptor.

Mechanism of action of cholera toxin

The CT molecule consists of two peptides—A and B. The B fragment binds to the receptors on the intestinal epithelium, following which the A peptide enters the epithelial cell and is cleaved into A1 and A2 peptides. The A1 peptide activates the adenylate cyclase resulting in the accumulation of cyclic adenosine monophosphate (cAMP) intracellularly. The cAMP inhibits the absorption of sodium in villus cells and secretes chloride from crypt cells, causing accumulation of sodium chloride in the intestinal lumen. Subsequently, in order to maintain the osmolarity, a large volume of fluid moves passively into the intestinal lumen, presenting as a watery diarrhea, dehydration and hypovolemic shock.

Other virulence factors that facilitate the pathogenesis are accessory colonisation factors and accessory toxins that disrupt the junctions (zona occludens) between mucosal cells and help phage multiplication.

Most infections are asymptomatic but can result in carriers. Symptomatic infections range from mild to acute life-threatening diarrhea, 24 to 48 hours after infection. The diarrhea is typically voluminous, watery with mucus, has a "rice water" appearance, and without blood or odour. Complications are only due to loss of fluid and dehydration and include hypovolemic shock and renal failure. The complications can be avoided or corrected by immediate administration of ORS or intravenous fluids and electrolytes.

Laboratory diagnosis

1. Demonstration of darting motility or comma-shaped gram-negative bacilli in the watery stool sample.
2. Use of alkaline transport media—Venkatraman-Ramakrishnan (VR), alkaline salt or peptone water (APW), Cary Blair or autoclaved sea water—to transport stool to the laboratory for culture. On isolation, the bacterium is biochemically identified and serogrouped using specific O1and O139 antisera and then serotyped with specific antisera.

Prevention and treatment

Control measures include improved sanitation and hygiene, and vaccination. As vaccination provides short-lived protection it is indicated in travellers, mass religious residential gatherings, and in an outbreak.

The following oral vaccines are currently available for cholera:

1. Killed whole cells (WC) of *V. cholerae* O1 + recombinant CT B subunit — Two doses 10 days apart are given and three doses are given in children aged 2–5 years. It is not for use in those aged < 2 years. It confers 60% protection for 3 years.
2. Live attenuated CVD 103 — A single dose of this HgR strain vaccine confers 80% protection. Oral, live attenuated vaccines with mutant strains are under trial but not available.
3. Killed, whole-cell parenteral vaccine — This may be used 1 month prior to any expected outbreak in large gatherings, and is not recommended in the event of an outbreak.

The mainstay of managing a case of cholera is immediate replacement of fluids and electrolytes, with a limited role for antimicrobial therapy. Antimicrobials may decrease the bacterial load in stool to prevent a person from being a carrier and shorten the duration of infection. Doxycycline or azithromycin can be used in patients and tetracycline for chemoprophylaxis during an outbreak.

Enterobacteriaceae

Enterobacteriaceae are gram-negative bacilli, aerobes or facultative anaerobes, non-fastidious in culture, with specific biochemical characteristics. They are motile with peritrichous flagellae, except *Klebsiella* and *Shigella* that are non-motile. *Enterobacteriaceae* is further classified on the ability of the bacteria to ferment lactose.

They are commensals in the human intestine—coliform bacilli, e.g., *E. coli, Klebsiella, Enterobacter, Serratia, Proteus, Providencia, Morganella,* and *Citrobacter*—but can be pathogens in sites other than the intestine. Enteric pathogens include *Shigella, Salmonella, Yersinia enterocolitica* and some strains of *E. coli. Yersinia pestis* causes the zoonotic disease plague.

Escherichia coli

The presence of strict anaerobes by far outnumbers the presence of aerobes or facultative anaerobes as commensals in the human intestine. The anaerobic *Bacteroides* is the most predominant commensal in the intestine and amongst the aerobes, *E. coli* is the most common.

The virulence factors of *E. coli* are either structural or secretory:

1. Certain *E. coli* serotypes possess fimbriae with increased adherence to colonise the intestinal epithelium (enterotoxigenic *E. coli*) or vaginal and uroepithelial cells to cause urinary tract infection (uropathogenic *E. coli*). Specific adhesins, coded by specific genetic sequences, adhere to specific epithelial receptors to cause disease.
2. The structural cell wall LPS or O antigen is endotoxic. The flagella (H antigen) render it motile and the polysaccharide capsule (K antigen) inhibits phagocytosis.
3. The secreted exotoxins are enterotoxins, hemolysins and cytotoxins (that damage the bladder and kidney).
4. Siderophores are iron-chelating molecules secreted by bacteria and fungi, e.g., enterobactin secreted by *E. coli* can scavenge ferric iron from iron proteins in the immune system and evade the immune response.

E. coli is frequently associated with a variety of diseases, predominantly urinary tract infection, diarrhea, neonatal meningitis, sepsis and secondary infections like peritonitis, surgical site infection, and ventilator-associated pneumonia. Sample smear examination by Gram stain, culture and antimicrobial susceptibility are the mainstay in the laboratory diagnosis of all *E. coli* infections, irrespective of the site/disease.

Diarrhea caused by *E. coli*

Although *E. coli* is a commensal known to colonise the human intestine, some serotypes are distinct from the commensal *E. coli* and cause diarrhea. Table 27.1 shows the characteristics of diarrheagenic *E. coli*.

Diarrheas caused by *E. coli* are self-limiting and need maintenance of hydration as treatment.

Table 27.1 Characteristics of diarrheagenic *E. coli*

E. coli	Disease	Virulence factor	Laboratory diagnosis
Enteropathogenic (EPEC)	Outbreaks in infants Sporadic in adults	Adhesion to epithelium, increased secretion, watery diarrhea	Stool culture
Enterotoxigenic (ETEC)	Traveller's diarrhea	Adhesion and secretion of LT and ST toxins	Detection of toxins
Entero-invasive (EIEC)	Dysentery, intestinal ulceration	Invasion of epithelium mediated by a virulence marker antigen (VMA)	Detection of VMA by ELISA
Enterohemorrhagic (EHEC) O157:H7	Food poisoning with meat, vegetables, low infective dose	Secretes Shiga-like toxins Stx 1 and Stx 2 that cause hemorrhagic colitis or hemolytic uremic syndrome*	Culture, toxin detection by ELISA or PCR for toxin genes
Enteroaggregative (EAEC)	Acute diarrhea	Adhesion, enterotoxin Causes aggregates in Hep-2 cells	Stool culture
Diffusely-adherent (DAEC)	Pediatric diarrhea	Adhesion. Causes diffuse pattern in Hep-2 cells	Stool culture

*Characterised by dysentery, renal failure and encephalopathy.

Enteric fever

Clinical Case 2

A 24-year-old woman presents with a history of persistent fever for 4 days, absence of response to antipyretics, malaise and generalised weakness. On examination, she is febrile, and has relative bradycardia and hepatosplenomegaly.

1. *Name three infections wherein blood culture is indicated.*
2. *Which amongst them is most likely in the above scenario?*
3. *Which cells does it multiply in?*
4. *Name the antimicrobial of choice in treatment.*

Salmonellae are gram-negative bacilli belonging to the *Enterobacteriaceae* group of bacteria but are pathogens and never commensals in the intestine. Clinically, they are classified as typhoidal salmonellae that cause enteric fever and non-typhoidal salmonellae that cause food poisoning. The typhoidal salmonellae are *Salmonella Typhi* and *Salmonella Paratyphi*, exclusive to humans. The non-typhoidal salmonellae that cause food poisoning exist in humans and other animals.

Epidemiology

Humans are the only reservoir of infection, whether a case or a carrier. The source of infection is contaminated food and water, especially as carriers of typhoidal salmonellae exist. Enteric fever is prevalent worldwide, with the highest incidence in Asia due to poor sanitation and hygiene practices, and the consumption of contaminated drinking water. Outbreaks are known to occur and *S. Typhi* is more prevalent than *S. Paratyphi* A or B. The carriers can be convalescent (3 weeks to 3 months after recovery), temporary (3 months to 1 year) or chronic (more than a year). Following recovery, the bacterium multiplies largely in the gall bladder and is shed in the feces or in the kidney and urine. Food handlers as carriers can infect large numbers of people.

In India, typhoid is an endemic disease with epidemic potential, the incidence rate being 100–200 cases per lakh population.

Pathogenesis and clinical features

Salmonellae being antigenically varied in the O (cell wall LPS or somatic) and H (flagellar protein) antigens are classified into 67 distinct serogroups based on the O antigen and further serotypes based on the H antigen. The typhoidal salmonellae *S. Typhi, S. Paratyphi* A, *S. Paratyphi* B and *S. Paratyphi* C belong to serogroups 9, 2, 4 and 7, respectively. Additionally, some species, including *S. Typhi* and *S. Paratyphi* C, possess the Vi (Virulence) antigen of the capsular polysaccharide. Though antibodies to Vi antigen are not used to diagnose enteric fever, the antigen is used for vaccination.

The infective dose of ingested typhoidal salmonellae is a minimum of 10^3 bacilli. On entering the intestines, they attach to the microvilli of the ileal mucosa. They are then phagocytosed by macrophages and resist intracellular killing. They also enter the M cells of the intestinal mucosa by endocytosis.

The macrophages carry the bacterium to the blood-stream via the mesenteric lymph nodes and lymphatics, causing a transient primary bacteremia. This bacteremia seeds the bacterium to the organs of the mononuclear phagocytic system (liver, spleen, lymph nodes and bone marrow) and others organs like the gall bladder, kidneys and lungs.

It multiplies in these organs to release bacteria in large numbers in the blood—secondary bacteremia—when the clinical manifestations appear. The incubation period is 7–14 days following which there is fever, coated tongue, relative bradycardia, abdominal pain, constipation or diarrhea, and hepatosplenomegaly. Peripheral blood smear shows leucopenia with relative lymphocytosis. A roseolar maculopapular rash is seen in some cases at the end of the first week. If not treated, intestinal perforation and hemorrhage can occur in the third week with circulatory collapse. Cholecystitis, arthritis, meningitis and visceral abscesses are rare complications.

Laboratory diagnosis

1. Blood culture that isolates typhoidal salmonellae is the gold standard, presenting direct evidence of enteric fever. If performed in the first week of fever, prior to administering antibiotics, it is likely to be positive in 90% of cases in the first week, 75% in the second week, 60% in the third week and 25% in the fourth week. Blood cultures rapidly become negative with antibiotic therapy.

   Blood culture can be done by manual or automated systems, the time taken for positive results being directly proportional to the bacterial load at the time of inoculation of the blood culture bottle. Therefore, the blood culture bottle is incubated and sub-cultured for 7 days before the test is termed negative.

   Alternatively, lysed blood clot or bone marrow culture can be done. Stool or urine culture is positive only after the third week. It is done to detect carriers. On isolation of the bacterium from any sample, it is important to serotype and establish its species identity for epidemiologic reasons.
2. Demonstration of antibodies in the second or third week is an indirect evidence of infection, subject to endemicity of the infection. Both IgG and IgM antibodies can be detected by slide or tube agglutination, or ELISA. Antibodies to O antigen appear and disappear early, denoting recent infection. Antibodies to H antigen appear late and disappear late, denoting a late infection period or convalescence.

   The Widal is a tube agglutination test that detects IgG and IgM antibodies against O antigen of *S. Typhi* (TO) that is shared and specific H antigens of *S. Typhi* (TH), *S. Paratyphi* A (AH) and *S. Paratyphi* B (BH). The serum is in double dilutions to avoid the prozone phenomenon and detect the antibody titre of each. Either a high titre of antibodies (O>100, H>200) or a four-fold rise in titre in 1 week is suggestive of enteric fever. The Widal test, though widely used, has its limitations. The endemicity of enteric fever, like in India, will determine the significant titre of antibodies. It can be false negative in the first week of fever or later due to antibody excess. Non-specific, anamnestic antibody response may be seen in unrelated fevers like malaria and scrub typhus or following typhoid vaccination. ELISA, Typhidot, Tubex or Dip-stick for IgM antibodies is available to diagnose recent infection in the absence of a positive blood culture.

3. Detection of antigen in serum or urine by ELISA or PCR to detect specific genes is seldom used.
4. To detect carriers, either stool or urine culture or a titre of serum Vi antibodies >10 is done.

Antimicrobial susceptibility

Cotrimoxazole or azithromycin or ceftriaxone is recommended for enteric fever as per the National Treatment Guidelines. Other antimicrobials tested are ampicillin, ciprofloxacin and chloramphenicol. Due to the increased use of ciprofloxacin, fluoroquinolone resistance is common (nalidixic acid resistant [NAR] strains) but resistance to ceftriaxone is rare.

Multi-drug resistant (MDR) *S. Typhi* are defined as those resistant to ampicillin, cotrimoxazole and chloramphenicol. Sensitivity to ampicillin, amoxicillin, cotrimoxazole and chloramphenicol has returned due to these drugs not being in use. This is an example of how antibiotic pressure induces resistance and antibiotic use withdrawal selects sensitivity.

Prevention

Control measures include improved sanitation and hygiene, detection and treatment (ampicillin with probenecid) of food handlers who are carriers, and vaccination. As vaccination provides short-lived protection, it is indicated in travellers, mass religious residential gatherings, and in an outbreak.

Two vaccines are currently available for enteric fever:

1. Parenteral Vi polysaccharide (25 μg) vaccine single dose IM or SC for children more than 2 years old, providing protection for 2 years after administration. If the Vi antigen is conjugated with the recombinant *Pseudomonas aeruginosa* exotoxin A protein, its antigenicity is enhanced for administration to children less than 2 years in age.
2. Typhoral, the oral live attenuated strain of *S. Typhi* Ty21a, that lacks the gal epimerase enzyme is available as a tablet. It is indicated for children above 6 years as a single tablet on days 1, 3, 5 and 7 with a booster every 5 years. The bacterium temporarily multiplies in the intestine to produce an immune response.

The whole cell killed (TAB) vaccines that were in use earlier have been discontinued.

Clostridioides difficile or antibiotic-associated diarrhea or pseudomembranous colitis

The prolonged use of broad-spectrum antibiotics during hospitalisation results in the suppression of normal intestinal flora and the selective growth of *C. difficile* in the patient. It is a sporulating, gram-positive bacillus, an obligate anaerobe, which is found in the hospital environment and as an intestinal commensal. The antimicrobials associated are broad-spectrum cephalosporins, clindamycin and ciprofloxacin, but all antibiotics have the potential to cause pseudomembranous colitis on prolonged use. Pseudomembranous colitis is seen to occur more in the Western countries than in India.

On the selective overgrowth of *C. difficile* in the intestine, the toxigenic strains secrete exotoxins A (enterotoxin) and B (cytotoxin). Both these bind to the specific receptors on the intestinal epithelium and disrupt the epithelial cells, causing diarrhea and formation of a pseudomembrane. The pseudomembrane, consisting of cellular debris, mucus and fibrin, is attached to the mucosa as a white plaque. The disease is influenced by the strain of the bacterium and immune response to the bacterium.

C. difficile is difficult to isolate in culture, hence the name 'difficile'. The disease is diagnosed in the laboratory by detecting the presence of glutamate dehydrogenase (GDH) **and** toxins A or B in stool by rapid ICT or ELISA. GDH is found in toxigenic and non-toxigenic strains of *C. difficile*. Alternatively, specific genes coding for the toxin A and B can be detected by PCR. Diagnosis by histopathologic examination of the colonic biopsy of the pseudomembrane is also possible. Depending on the severity of the symptoms, the condition is treated with metronidazole and/or vancomycin, but can relapse following treatment. It is pertinent to prevent pseudomembranous colitis by preventing unnecessarily prolonged use of antibiotics and by infection control measures in the hospital.

VIRAL DIARRHEA

Diarrhea is a leading cause of childhood mortality in the world. A number of viruses are known to cause diarrhea, especially in children (Table 27.2). In early childhood, the single most important cause of severe dehydrating diarrhea is rotavirus infection.

Picornaviruses, influenza virus and HIV may also be associated with diarrhea. Rotavirus is the most common cause of diarrhea in children followed by enteric adenoviruses.

Pathogenesis

Following feco-oral transmission, the virus infects and destroy villus tip enterocytes in the small intestine. The gastric or colonic mucosa is not affected and extraintestinal infection in the immunocompetent is rare. Infection leads to decreased absorption of salts and water

Table 27.2 Viruses causing diarrhea

Virus	Virology	Serotypes	Disease
Rotavirus (Reoviridae)	Segmented dsRNA	Species specific A–G (A, B, C) human	< 5 years
Astrovirus	ssRNA	8	< 5 years
Caliciviruses Norovirus (Norwalk) Sapovirus	ssRNA	–	Outbreak in older children, adults
Enteric adenoviruses	dsDNA	40, 41	< 2 years

and an imbalance in the ratio of intestinal fluid absorption to secretion and malabsorption of complex carbohydrates, especially lactose.

Rotavirus and astrovirus acute diarrhea have an incubation period less than 48 hours; they present with fever, vomiting and frequent watery stools, and last for 5 to 7 days, while the adenovirus diarrhea lasts for 10 to 14 days. The incubation period of norovirus diarrhea is 12 hours. The disease lasts for 1 to 3 days and resembles bacterial food poisoning due to pre-formed toxins. The viruses are excreted in the feces during the illness.

Laboratory diagnosis is by LAT, ICT or ELISA to detect the antigen of rotavirus or PCR to detect viral RNA from stool samples. Rehydration is most important as mortality is due to dehydration and not the organism. Good hygiene practices and sanitation are required for control.

Immunoprophylaxis

Two rotavirus vaccines are available:

1. A live attenuated, oral pentavalent bovine rotavirus vaccine administered at 6, 10 and 14 weeks of age.
2. A live attenuated, oral monovalent human rotavirus vaccine administered at 6 and 10 weeks of age.

Both provide 80% protection against rotavirus diarrhea and herd immunity has been observed due to the vaccine.

PROTOZOANS CAUSING DIARRHEA

With the exception of *Giardia lamblia*, the protozoans causing diarrhea are opportunistic pathogens.

Giardia lamblia

Epidemiology

Giardiasis is globally prevalent, the prevalence rate varying between 35% and 75% in different geographic regions in India. It is more common in children below age 5 years and is transmitted by ingestion of cysts from contaminated food and water. Humans are the only reservoir of infection.

Morphology: Cyst and trophozoite

Trophozoites are located in the duodenum and upper jejunum but encystment happens in the large intestine, hence, it is commonly the cysts that are found in the stool and rarely trophozoites.

The cyst is oval, quadri-nucleate, measures around 12 µm, and has a diagonally running axostyle across the cell, along with the presence of rudimentary flagellae (Fig. 27.2).

The trophozoite has a broader, rounded anterior end and tapering, sharp posterior end with a convex dorsal

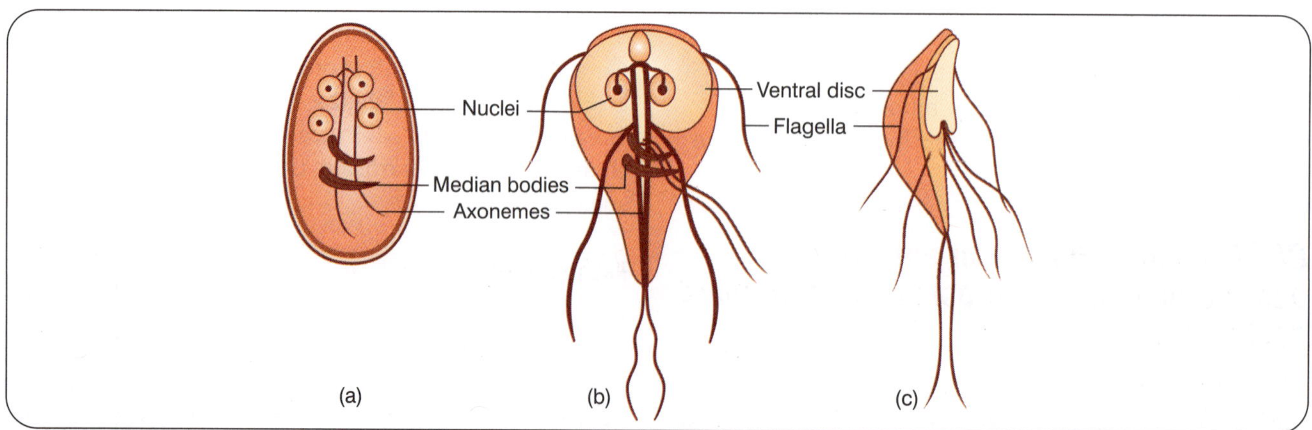

Fig. 27.2 Morphologic forms—trophozoite and cyst of *Giardia lamblia*

surface (Fig. 27.3). The ventral surface is concave with a sucking disc meant for attachment to intestinal mucosa.

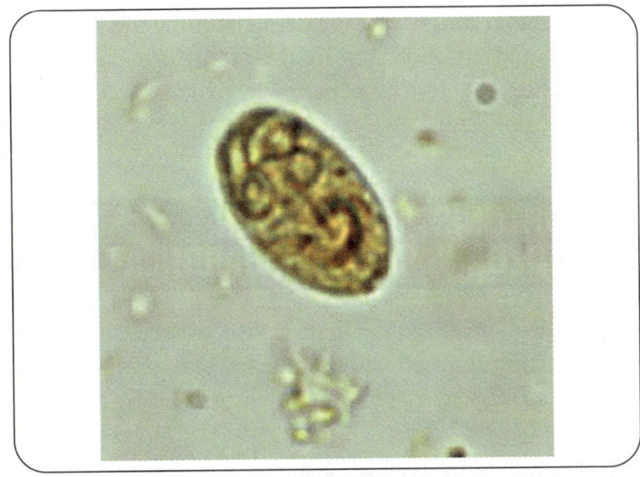

Fig. 27.3 Microscopic appearance of cyst of *Giardia* in stool, Iodine mount, 40X (*Source*: CDC, DPDx)

Life cycle (Fig. 27.4)

Man, the only host, is infected by ingestion of the cyst that reaches the duodenum and excysts into two trophozoites. The trophozoites multiply in numbers by binary fission and colonise the duodenum. Some trophozoites undergo encystation in the large intestine and are excreted in the feces.

Pathogenesis

The parasite does not invade the epithelium but attaches itself by the ventral sucking disc, causing diarrhea, loose stools and malabsorption of fat, leading to steatorrhea.

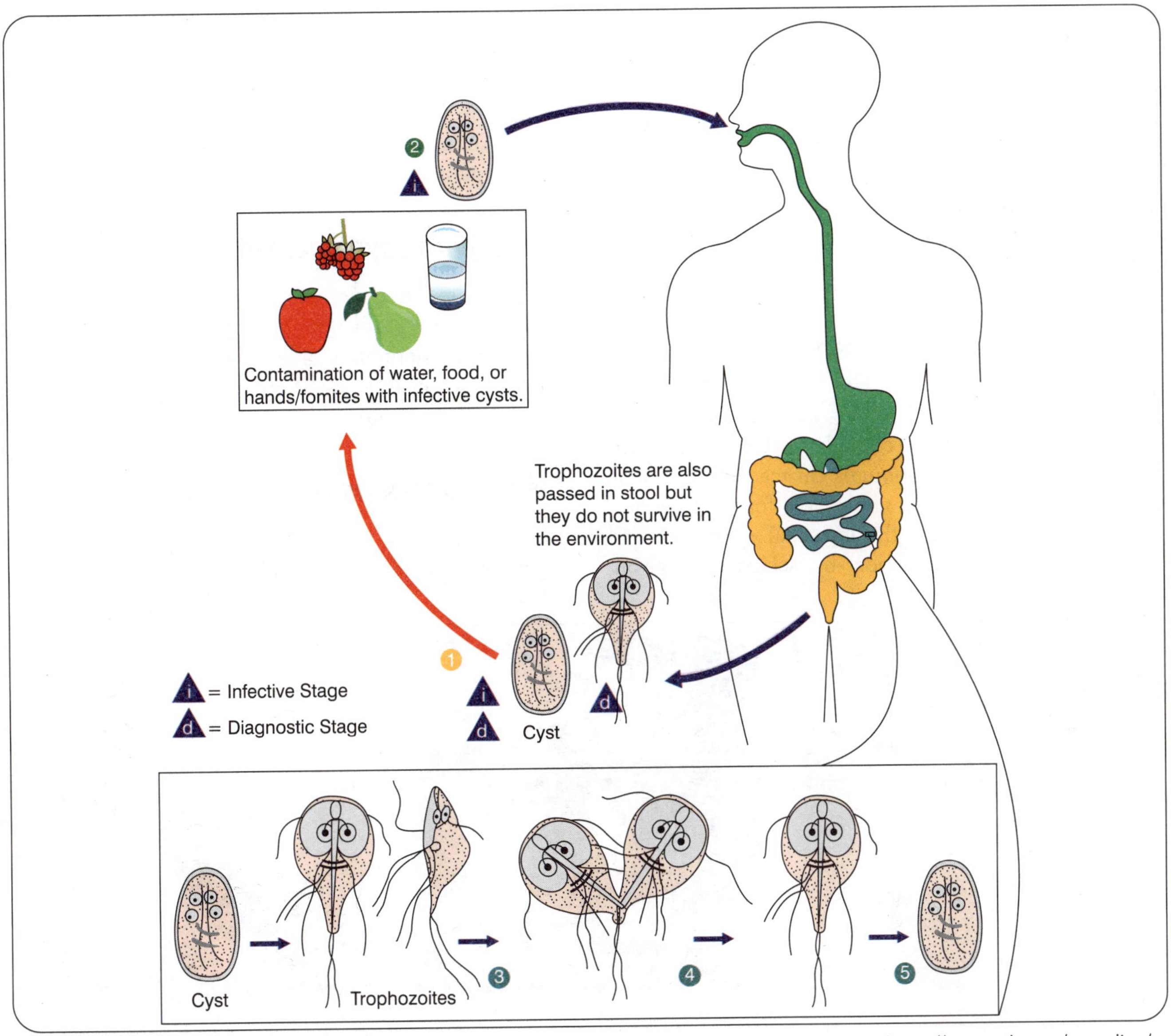

Fig. 27.4 Life cycle of *Giardia lamblia* (*Source:* Adapted from CDC's Parasites and Health page, at https://www.cdc.gov/parasites/giardia/pathogen.html)

Laboratory diagnosis is by demonstrating the presence of cysts and trophozoites in the stool in a saline wet mount. Prevention is by personal hygiene, and treatment with metronidazole.

OPPORTUNISTIC COCCIDIAN PROTOZOANS CAUSING DIARRHEA

These are protozoans that cause diarrhea in the immunocompromised and include *Cryptosporidium* spp., *Cyclospora, Cystoisospora* and *Sarcocystis.*

Cryptosporidium species

Epidemiology

The life cycle requires only one host—either man or animal—and the infection is prevalent worldwide. Man is not only the host but can acquire infection by zoonosis. Infection is largely asymptomatic in the immunocompetent but severe in the immunocompromised.

Morphology: Oocyst (Fig. 27.5a)

The mature, sporulated oocyst is spherical, 4–6 μm in size and acid fast, containing four sporozoites. Each sporozoite is crescent-shaped with a pointed anterior and broad posterior bearing the nucleus.

Life cycle

Man is infected by ingestion of food or water contaminated with oocysts of *Cryptosporidium*. On reaching the small intestine, excystation occurs to release four sporozoites. These sporozoites invade the intestinal epithelium and undergo asexual reproduction by schizogony, giving rise to multiple merozoites. Some merozoites undergo gametogony to form microgametes (M) and macrogametes (F). Sexual reproduction occurs by sporogony wherein the microgamete fertilises the macrogamete, to form a zygote that eventually transforms to a mature sporulated oocyst that is excreted in the stool.

Pathogenesis

In the immunocompromised, especially the HIV-infected with a CD4 count < 100/μL, the infection causes chronic, persistent, watery diarrhea with profuse fluid and electrolyte loss.

Laboratory diagnosis is by demonstration of 1% sulphuric acid-fast oocysts in the stool, specific antigen or 18sRNA in stool by ELISA or PCR, and specific antibody in blood by ELISA. Tests for combined antigen detection of *Cryptosporidium* and *Microsporidia* are possible. Histopathologic examination of intestinal biopsy shows the presence of *Cryptosporidium* within the intestinal epithelium.

Prevention is by good hygienic practices and ensuring consumption of safe drinking water.

Cyclospora cayetanensis

The parasite, life cycle, infection and disease are similar to those in *Cryptosporidium* spp., but man is the only known host. Maturation of oocysts takes place in the soil. The oocysts measure 8–10 μm, and may or may not be acid fast (Fig. 27.5b).

Cystoisospora (Isospora) belli

It has a worldwide prevalence, man is the only host, and the disease is associated with HIV infection. The parasite,

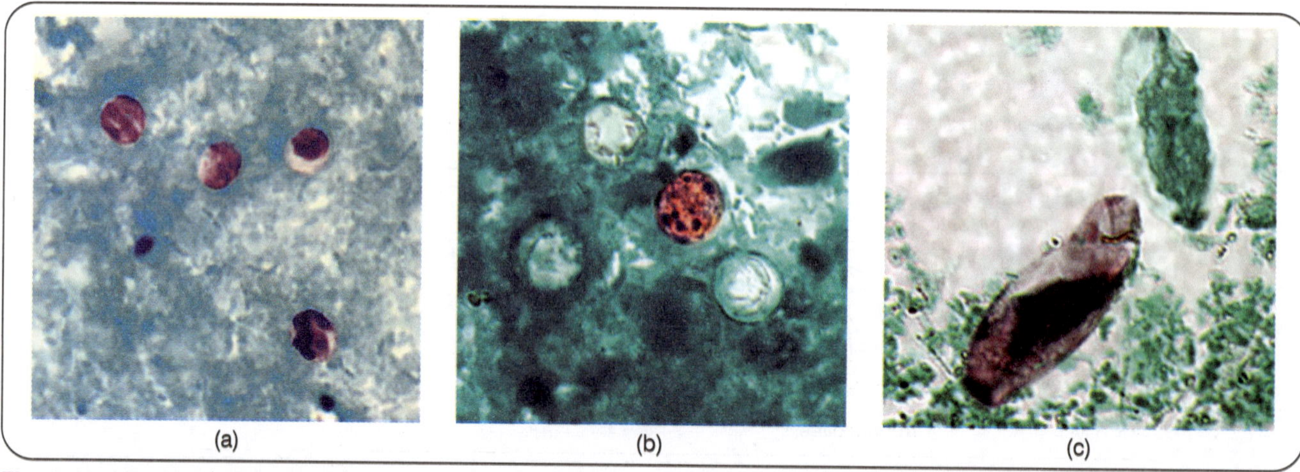

Fig. 27.5 Modified acid fast stain showing oocysts of (a) *Cryptosporidium* spp., (b) *Cyclospora cayetanensis* and (c) *Cystoisospora (Isospora) belli* (*Source*: CDC, DPDx)

life cycle, infection and disease are similar to those in *Cryptosporidium* spp. The oocyst is oval and acid fast, measures 25 μm, and contains two sporocysts with four sporozoites in each (Fig. 27.5c).

The life cycle of ***Sarcocystis* spp.** needs two hosts, including animals (zoonotic). Man serves as a definitive host, with the disease being either intestinal or muscular involving cardiac and skeletal muscle cysts.

Answers

Clinical Case 1

1. *Cholera*
2. *Stool examination to observe darting motility by hanging drop or* comma-shaped *gram-negative bacilli by smear.*
3. *Important to identify the probable source (in this case, the well) and seal it from being used. Manage the dehydration of symptomatic cases with ORS or if required, IV fluids. Prevent cholera in the uninfected by vaccinating the local population.*
4. *Rotavirus, Norovirus, Calicivirus, Astrovirus, Enteroviruses*

Clinical Case 2

1. *Enteric fever, bacterial endocarditis, brucellosis*
2. *Enteric fever*
3. *Macrophages and M cells of intestinal mucosa*
4. *Ceftriaxone*

Questions

1. Classify vibrios stating the epidemiologic significance of the classification in India.
2. Explain the mechanism of pathogenesis of cholera.
3. Outline the laboratory diagnosis, immunoprophylaxis and management of cholera.
4. Explain the role of *E. coli* in human intestinal health and disease.
5. Name two intestinal commensals and pathogens belonging to *Enterobacteriaceae*. Elaborate on the role of *E. coli* in diarrhea and dysentery.
6. Clinically classify salmonellae giving examples.
7. Explain the pathogenesis of enteric fever.
8. Discuss the rationale and interpretation of various laboratory tests in the diagnosis of enteric fever.
9. Write a note on the vaccines, antimicrobial susceptibility and resistance in *S. Typhi*.
10. Explain the pathogenesis of antibiotic-associated diarrhea and its laboratory diagnosis.
11. Enumerate the viruses causing diarrhea, and explain the pathogenesis and immunoprophylaxis of rotavirus.
12. Enumerate the protozoans causing diarrhea and diagrammatically distinguish the morphologic appearance of each in the microscopic examination of stool.
13. Discuss the life cycle, pathogenesis and laboratory diagnosis of giardiasis.
14. Write a brief note on opportunistic protozoans causing diarrhea.

Dysentery and Acid Peptic Disease

Dysentery is an infection of the colon that causes diarrhea containing blood and mucus. Other symptoms can include abdominal cramps, fever and vomiting. Dysentery being highly infectious, it is important for the patient and close contacts to minimise contact, wash hands and ensure proper hygienic practices. The two main causes of dysentery are bacillary, caused by *Shigella* spp., or amoebic, caused by *Entamoeba histolytica*. Rarer causes of dysentery include entero-invasive *E. coli* (EIEC), *Campylobacter jejuni*, *Vibrio parahaemolyticus* and *Neobalantidium coli*.

Shigella

Clinical Case 1

A 24-year-old male tourist in India presents at a PHC with a history of loose stools with blood and mucus, and abdominal cramps for 2 days. On examination, he is febrile, with generalised weakness.

1. *Name three bacteria causing traveller's diarrhea*
2. *What investigation/s will you request for at the PHC?*
3. *If the cause is bacterial, are antimicrobials indicated in treatment?*
4. *What are the precautions in managing this case?*

Shigellae are non-motile, capsulated gram-negative bacilli belonging to *Enterobacteriaceae*. There exist four serogroups based on the somatic O antigenic differences—*Shigella dysenteriae* (Group A), *Shigella flexneri* (Group B), *Shigella boydii* (Group C) and *Shigella sonnei* (Group D). Each serogroup can be further serotyped/biotyped/colicin-typed depending on antigenic or biochemical differences within. *S. flexneri* is the most commonly encountered species, and *S. dysenteriae* type-1 is associated with mortality.

Epidemiology

Man is the only reservoir of infection. As there are no carriers the infection is transmitted exclusively by the ingestion of contaminated food and water from patients. Hand washing and hygienic practices by the patient and close contacts are the ways to prevent spread. Infection can occur as an epidemic in places of mass gatherings but is also seen as an endemic infection causing shigellosis or bacillary dysentery. Worldwide, the microbe mostly infects children < 5 years of age, and can cause mortality.

Pathogenesis and clinical features

The infective dose of Shigella being only 10–100 bacilli, it is highly infectious. On ingestion, the bacilli invade the intestinal mucosa through the M cells (microfold cells are macrophages found in the villi and MALT that are exploited by *Shigella, Salmonella, Yersinia,* poliovirus and reovirus to enter the intestinal epithelium). In the submucosa, the bacterium invades adjacent epithelial cells. The infection triggers a local inflammatory response, release of cytokines resulting in intestinal ulceration and acute colitis. Shigellae also secrete exotoxins—enterotoxins similar to the cholera toxin—and a cytotoxin—Shiga toxin by *S. dysenteriae* type-1, besides the endotoxic LPS.

Following an incubation period of 1 to 4 days, the disease manifests as diarrhea with blood and mucus in stool, fever, vomiting, tenesmus and abdominal cramps. It is mostly self-limiting, with rare cases of complications like intestinal perforation, severe dehydration with metabolic disturbances in blood and metabolic encephalopathy. Rarely, *S. flexneri* infection is followed by autoimmune reactive arthritis, urethritis and conjunctivitis (Reiter's syndrome) in patients expressing HLA-B27.

Microscopic examination of stool wet mount shows the presence of RBCs and pus cells. The stool is cultured without delay or transported in Sach's buffered glycerol saline broth. The bacterium is isolated and identified by its biochemical characteristics and serogrouping with specific antisera.

There is no vaccine for shigellosis, prevention being by hand washing, hygienic practices and improved sanitation. Ciprofloxacin is the drug of choice, alternatively ceftriaxone, with oral or intravenous rehydration being very important.

PROTOZOANS CAUSING DYSENTERY

Entamoeba histolytica

Clinical Case 2

A group of chefs from a restaurant are screened for enteric pathogens as a part of a municipal health camp. In one individual, the microscopic examination of stool reveals the presence of cysts without trophozoites of Entamoeba histolytica.

1. *What is the source of this infection?*
2. *What is the disease status in this case?*
3. *How can you prevent transmission?*
4. *Name three other pathogens that can be transmitted by food handlers.*

Epidemiology

Amoebiasis is endemic in India and in developing countries due to poor sanitation. Its prevalence is 15% in India and it is the second largest cause of parasitic death after malaria. Worldwide, 500 million are infected, with an annual mortality of 70,000 people. The disease is transmitted feco-orally by ingestion of quadrinucleate cysts of the amoebae as humans are the only reservoir of infection.

Morphology: Cyst and trophozoite

The mature morphologic forms of *E. histolytica* are the cyst and trophozoite. The cyst is its vegetative form while the trophozoite is its reproductive form (Fig. 28.1). The mature spherical cyst measures around 10 μ, and possesses nuclei, a chromatoid body and a glycogen mass. It begins as a uninucleate cell and progresses to the binucleate and quadrinucleate forms. The glycogen mass stains brown with iodine, but as the cyst develops from uninucleate to quadrinucleate, both the glycogen mass and the chromatoid bodies disappear. Chlorination of water does not destroy cysts but boiling of water does.

Excystation, the process by which a cyst becomes a trophozoite, occurs only in the intestinal lumen. In excystation, one quadrinucleate cyst forms eight amoebulae, in turn forming eight trophozoites. Encystation, the process by which a trophozoite becomes a cyst, also occurs only in the intestinal lumen. In encystation, one trophozoite forms one cyst. It is not a process of reproduction but a means of protection. However, trophozoites also reproduce by binary fission, thus multiplying the number of trophozoites.

The trophozoite measures around 20 μm, and possesses one nucleus and other ingested material in the inner endoplasm. The outer ectoplasm is capable of projecting pseudopodia as a means of locomotion and motility. Therefore, the trophozoites are motile and not fixed in shape due to the pseudopodia.

Both excystation and encystation can happen only in the intestinal lumen and not in the intestinal wall or in tissues where metastatic invasion by trophozoites occurs. The mature quadrinucleate cysts can infect when transmitted feco-orally.

Life cycle (Fig. 28.2)

The life cycle involves only one host—human with two phases, cyst and trophozoite. The mature quadrinucleate cyst is the infective form that is transmitted feco-orally from man to man by ingestion of contaminated food or water. On ingestion, the cyst resists gastric acidity but is excysted by trypsin in the ileum. Each cyst forms eight actively motile trophozoites that multiply by binary fission. Though cysts transmit infection, acute disease is caused by the trophozoites.

The trophozoite invades the intestinal epithelium, secretes histolysin that causes coagulative cell necrosis and ultimately lodges in the submucosa. It has a tendency to migrate from dead tissue burrowing towards its healthy borders. In the intestine, it results in acute dysentery with ulceration of the colonic mucosa.

The trophozoites can also travel via the portal vein to the liver and multiply in the liver, giving rise to hepatic amoebiasis or amoebic liver abscess. Other organs of extraintestinal amoebiasis are the brain and lungs. Some trophozoites undergo encystation in the intestine and continue the life cycle by excreting the infective cysts in the stool.

Pathogenicity

The incubation period is 4–5 days following consumption of the infective cyst. The disease can be acute amoebic dysentery, chronic intestinal amoebiasis or extraintestinal amoebiasis.

In acute amoebic dysentery, there are generalised ulcers in the colon. A typical amoebic ulcer is transverse, varying in size with ragged undermined edges, and filled with necrotic

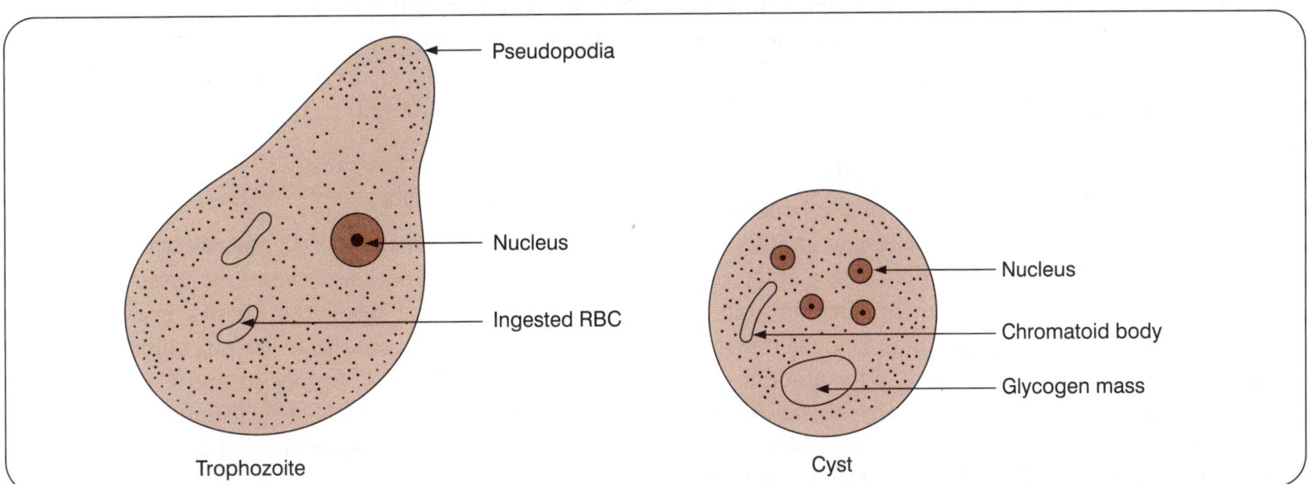

Fig. 28.1 Morphologic forms—trophozoite and cyst of *Entamoeba histolytica*

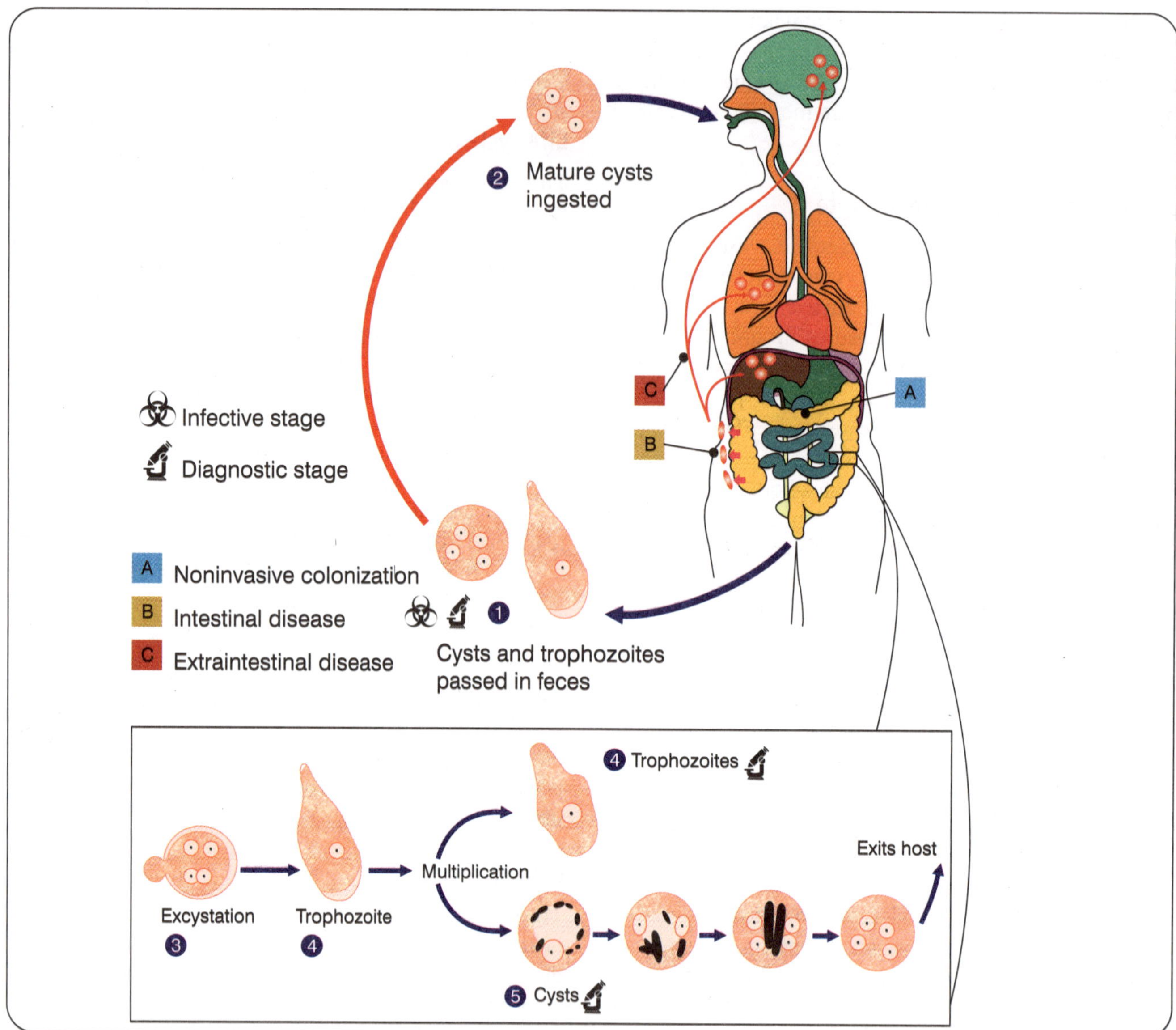

Fig. 28.2 Life cycle of *Entamoeba histolytica* (*Source:* Adapted from CDC's Parasites and Health page, at https://www.cdc.gov/dpdx/amebiasis/index.html)

material. The trophozoites are found in the advancing borders of the ulcers. They can extend to the submucous layer and rupture into the peritoneal cavity. Healing occurs with scarring and strictures. Chronic intestinal amoebiasis varies in its degree of invasion and severity of disease.

Hepatic amoebiasis is not an inflammation of hepatocytes or hepatitis but necrosis of the hepatocytes giving rise to an abscess. It is generally solitary and predominantly in the postero-superior surface of the right lobe. It is the trophozoite that enters the liver through the portal vein and encystation can occur only in the intestinal lumen, therefore cysts are never seen in the abscess.

The abscess fluid is bacteriologically sterile, without pus, and consists of dead hepatocytes and blood cells, giving it a thick texture and brown colour. Trophozoites are generally absent in the aspirate as the trophozoites are present only at the advancing edges of the abscess and not in the abscess fluid.

Some individuals are cyst carriers either convalescent (those who have suffered) or healthy (those who have not suffered).

Laboratory diagnosis

1. **Examination of stool:** The macroscopic appearance of the stool in amoebiasis shows the presence of blood and mucus in semi-solid feces. Microscopic examination with iodine wet mount shows few pus cells, RBCs, cysts, the trophozoites of *E. histolytica* and Charcot–Leyden crystals (Figs. 28.3 and 28.4). Charcot–Leyden crystals are remnants of eosinophils, indicating an immune response to parasitic infection. In asymptomatic carriers, only the cyst of *E. histolytica* will be seen.

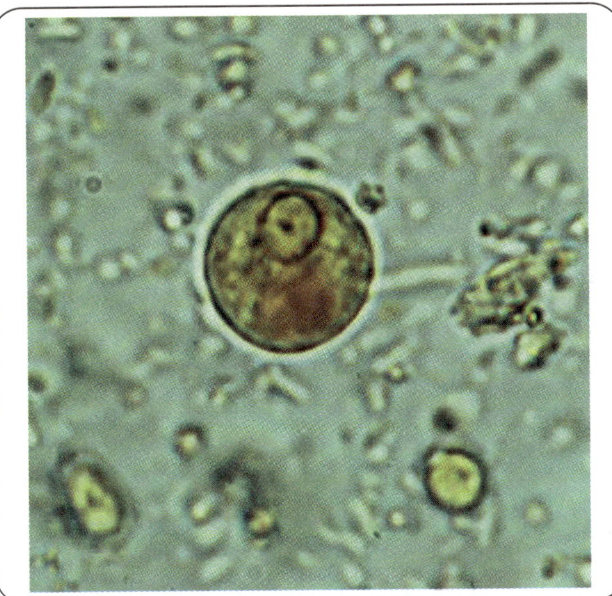

Fig. 28.3 Microscopic appearance of cyst of *Entamoeba histolytica*, Iodine mount, 40X (*Source:* CDC, DPDx)

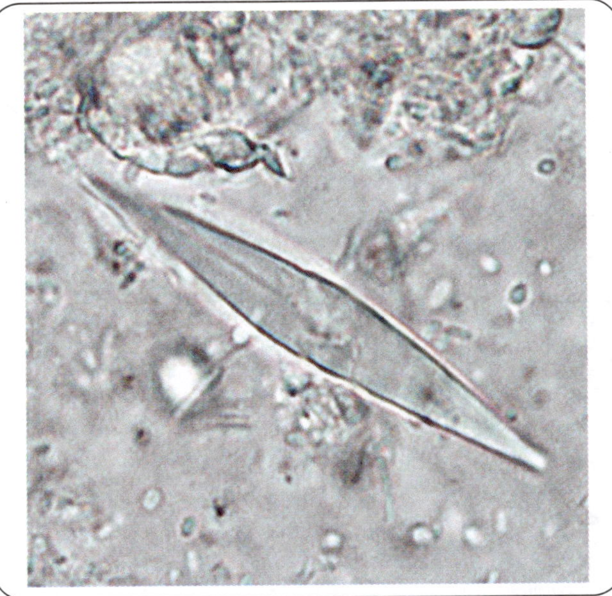

Fig. 28.4 Charcot–Leyden crystals in stool, Wet mount, 40X (*Source:* CDC, DPDx)

2. **Examination of liver aspirate:** The macroscopic appearance of the aspirate is thick and brown due to the presence of necrosed hepatocytes and blood. Microscopic examination with a saline wet mount does not show pus cells or bacteria and rarely shows trophozoites of *E. histolytica*. On culture, the aspirate is bacteriologically sterile.

Prevention and treatment

Amoebiasis is prevented by good hygienic practices and treated with metronidazole. To treat amoebae and cysts in intestinal lumen in carriers, luminal drugs—iodoquinol, paromomycin or diloxanide furoate—are adequate without azoles. It is recommended that food handlers be screened for amoebic cysts and treated to prevent transmission. Other *Entamoeba* spp. like *Entamoeba dispar* are known to exist in intestinal lumen without causing any disease and do not warrant treatment.

Free-living amoebae pathogenic to man—*Naegleria* and *Acanthamoeba*—do not cause gastrointestinal disease. They are known to cause CNS infections.

Neobalantidium coli

Epidemiology

Neobalantidium coli is uncommon in humans and the parasite has low virulence. Close contact between pigs and humans is a major factor in the disease, and cases have been reported from India.

Morphology: Cyst and trophozoite

It is the largest protozoan infecting man, inhabiting the large intestine as a cyst and trophozoite. The cyst measures around 50 μm, contains a macronucleus, a micronucleus, and a refractile body within a thick double-layered cyst wall (Fig. 28.5). The trophozoite measures around 70 μm, contains a macronucleus, a micronucleus, and two contractile vacuoles. The body is covered with fine cilia, with the anterior end leading to a mouth (cytosome) and the posterior end to an anus (cytopyge).

Life cycle (Fig. 28.6)

The life cycle requires only one host, with the pig being the natural host. Man is a rare, accidental host but transmission can occur from pig to man and from man to man. The cyst is the infective form; when ingested, each cyst forms one trophozoite in the large intestine. The trophozoites, the

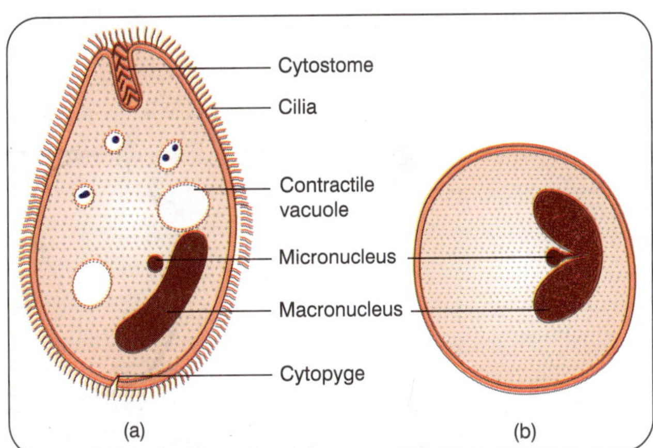

Fig. 28.5 Morphologic forms—trophozoite and cyst of *Neobalantidium coli*

reproductive form, may enter the intestinal submucosa and multiply by binary fission. They also undergo encystation, i.e., each trophozoite forming one cyst.

Pathogenesis

Occupational contact with pigs transmits the infection to man. This happens by ingestion of the cysts. Infection results in ulceration of the colon, similar to amoebiasis, but does not show any extraintestinal infection, invasion or manifestations. Carriers that can transmit infection from man to man and man to pig are known to exist.

Laboratory diagnosis

Macroscopic examination of the stool shows blood and mucus in semi-solid feces. Microscopic examination of stool in a saline wet mount shows cysts and motile trophozoites. Carriers show the presence of cysts.

Prevention and treatment

Neobalantidiasis is prevented by good hygienic practices and treated with diadoquin.

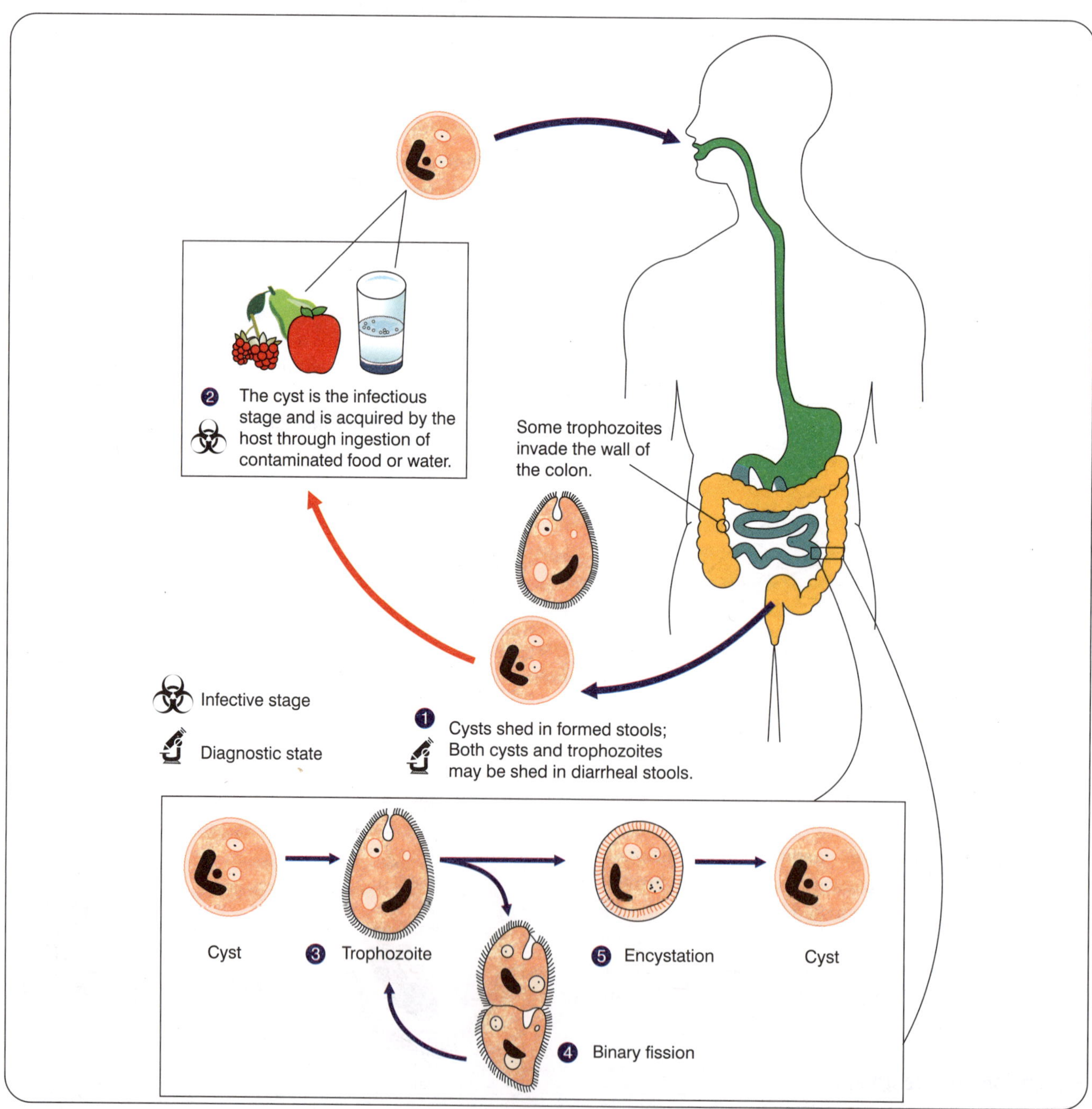

Fig. 28.6 Life cycle of *Neobalantidium coli* (*Source:* Adapted from CDC's Parasites and Health page, at https://www.cdc.gov/parasites/balantidium/biology.html)

ACID PEPTIC DISEASE

Acid peptic disease is a group of disorders related to the upper gastrointestinal tract, largely due to gastric acid and pepsin secretion acting on the mucosa. Colonisation of the stomach by *Helicobacter pylori* has a role to play in acid peptic disease, besides smoking and diet.

Helicobacter pylori

H. pylori is a curved, motile, gram-negative bacillus that can be stained by Warthin–Starry silver staining in a histopathologic specimen (Fig. 28.7). Its virulence factors include motility and resistance to gastric acidity due the presence of enzyme urease that acts on urea to produce ammonia and protective enzymes.

H. pylori causes acute gastritis and peptic ulcer disease—duodenal and gastric ulcers presenting with epigastric pain and a burning sensation following a meal or on an empty stomach, respectively. It is also associated with gastric adenocarcinoma and non-Hodgkin's lymphoma. However, it has a beneficial role in gastroesophageal reflux disease (GERD). The infection is largely diagnosed by the non-invasive urea breath test. The patient consumes radiolabelled urea that is hydrolysed by the urease if *H. pylori* is present and the labelled CO_2 released in the breath is detected by mass spectroscopy. In case an endoscopic biopsy of the gastric mucosa is indicated, *H. pylori* can be demonstrated by histopathological examination, culture or a urease test of the biopsy. Detection of antigen in stool or IgG in serum is done more for epidemiologic screening than for diagnosis. Eradication or treatment of *H. pylori* in asymptomatic carriers or for prophylaxis of acid peptic disease is not recommended. It is indicated for peptic ulcer disease, by multi-drug therapy with omeprazole, clarithromycin and amoxicillin.

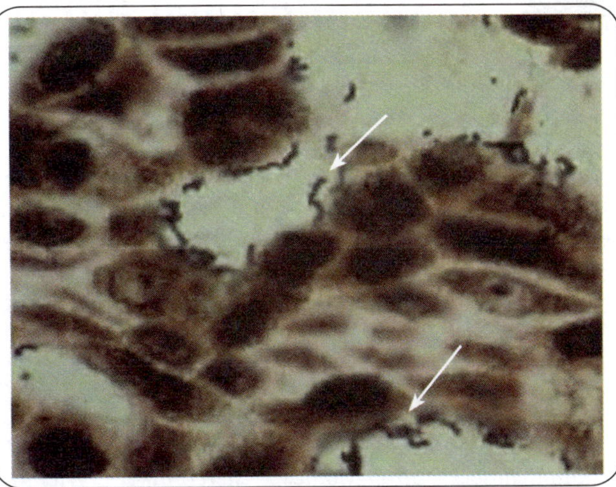

Fig. 28.7 Gastric mucosal biopsy showing black curved rods—*Helicobacter pylori*, Warthin–Starry stain, 40X

ANSWERS

Clinical Case 1

1. *Shigella spp., ETEC, Campylobacter jejuni, Vibrio parahaemolyticus*
2. *Stool microscopy to rule out amoebic dysentery*
3. *Rehydration is more important than antimicrobials as the infection is mostly self-limiting*
4. *Hand washing and hygienic practices by the patient and close contacts to prevent spread*

Clinical Case 2

1. *Man is the only reservoir, feco-oral transmission*
2. *Carrier without disease*
3. *Treatment with luminal drugs—iodoquinol, paromomycin or diloxanide furoate—is adequate without azoles.*
4. *Giardia lamblia, enteric salmonellae, Shigella spp.*

QUESTIONS

1. Enumerate the microorganisms causing dysentery and explain the pathogenesis of bacillary dysentery.
2. Enumerate the microorganisms causing dysentery. Explain the pathogenesis and laboratory diagnosis of amoebic liver abscess.
3. Explain the role of *H. pylori* in acid peptic disease.
4. How can the presence of *H. pylori* be detected in a case of duodenal ulcer?

Gastrointestinal Infestations

Other than protozoans, intestinal helminths too can cause gastrointestinal disease. These are broadly classified as cestodes, trematodes and nematodes. Those that are commonly encountered and seen in India are:

1. Cestodes (Tapeworms) — *Taenia solium, Taenia saginata, Diphyllobothrium latum*
2. Trematodes — *Fasciola* spp., *Schistosome haematobium* (rare)
3. Nematodes — *Ascaris lumbricoides, Ancylostoma duodenale, Strongyloides stercoralis, Trichuris trichiura, Enterobius vermicularis*

Taenia solium and *saginata*

Epidemiology

T. saginata is associated with cattle breeding and beef consumption while *T. solium* is associated with pork consumption. Both cestodes are reported in India, with *T. solium* also presenting as neurocysticercosis.

Morphology: Ovum and adult

The ovum measures around 40 µm, with the embryo or oncosphere containing three pairs of hooklets. It is protected by a thick, brown, radially striated embryophore, and a yolk shell (Fig. 29.1). The ova of *T. saginata* and *T. solium* cannot be distinguished by morphology on stool examination.

The adult parasite has a head (scolex), neck and body (strobila) that consists of numerous segments (proglottids). It is long, flattened dorsoventrally and segmented, with each segment being a hermaphrodite with both sexes present.

The head or scolex is the organ of attachment to the mucosa and bears four muscular suckers. Additionally, in *T. solium,* the scolex has an apical protrusion (rostellum) with hooklets. The morphologic appearance of the adult *T. saginata* and *T. solium* differs (Table 29.1). A narrow neck connects the head to the body. The body consists of immature, mature and gravid segments. The gravid segments are towards the distal end, away from the head.

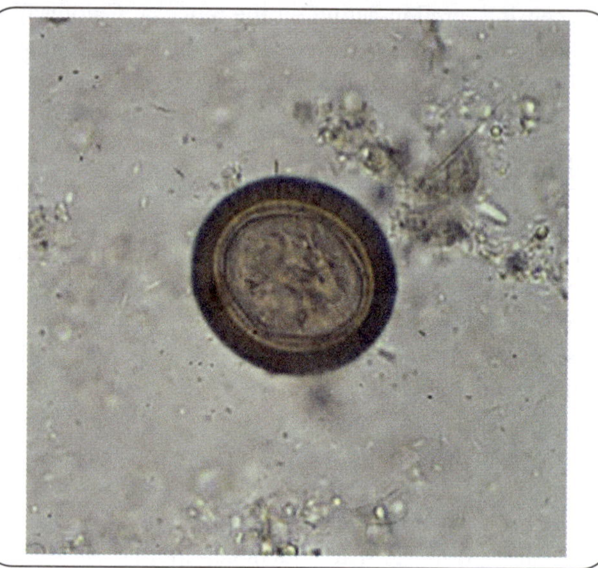

Fig. 29.1 Ovum of *Taenia* spp., Wet mount, 40X (*Source*: Centers for Disease Control and Prevention [CDC], DPDx)

Life cycle

Man is the definitive host, the intermediate host being cattle in *T. saginata* and pig in *T. solium* (Fig. 29.2).

Man acquires infection either by ingestion of ova in contaminated water or food or by consuming undercooked beef (*T. saginata*) or pork (*T. solium*) with the larval form of the parasite. On reaching the small intestine, the ova or larvae grow into adults and reside there. Immature segments become mature, fertilisation occurs within each mature segment to become gravid and the gravid segments release ova in the feces.

The larvae in the intestine may penetrate the intestine and travel via blood to the organs and tissues. Cysticercus is the infection of tissue by the larvae of *Taenia* spp. In *T. saginata*, it occurs only in cattle, and not in human host. In *T. solium* it can be seen in pig and in human host. Cysticercus cellulosae caused by *T. solium* in man is the larval tissue form which makes an abortive end of its life cycle.

Table 29.1 Morphologic differences between *Taenia saginata* and *Taenia solium*

Feature	*Taenia saginata*	*Taenia solium*
Adult length	5 metres	3 metres
Scolex	Large, quadrangular with 4 suckers	Small, globular with 4 suckers
	Rostellum and hooklets absent	Rostellum and hooklets present
Neck	Longer	Shorter
Proglottids	1000–2000	800–1000

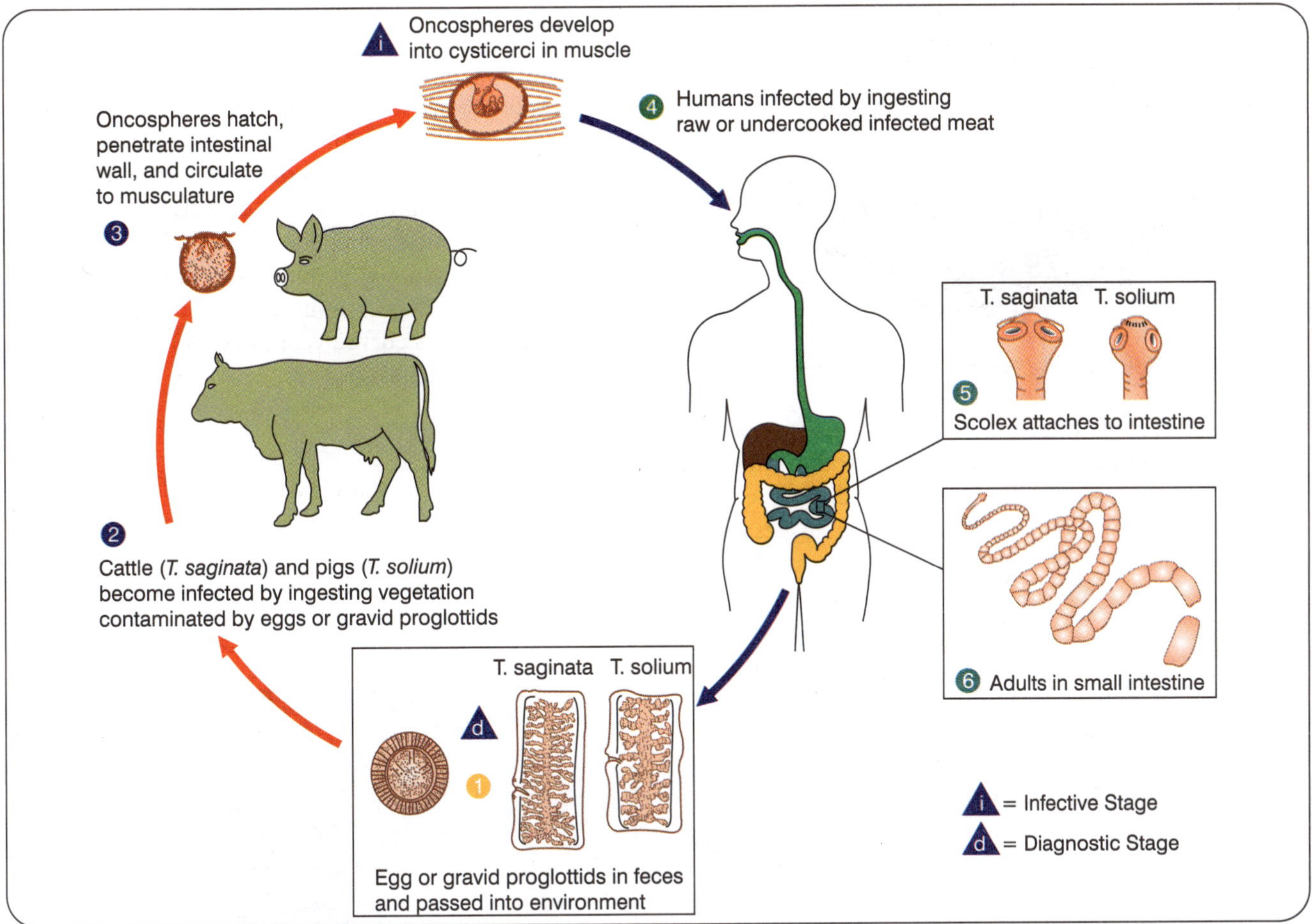

Fig. 29.2 Life cycle of *Taenia saginata* and *Taenia solium* (*Source:* Adapted from CDC's Public Health Image Library, at https://phil.cdc.gov/Details.aspx?pid=3420)

Pathogenesis

Intestinal taeniasis caused by the adult residing in the intestine is characterised by abdominal discomfort and passing of segments in the stool.

Cysticercus cellulosae caused by larvae of *T. solium* forms multiple cysts in the brain (neurocysticercosis), subcutaneous tissue, skeletal muscle or eye. The signs and symptoms will depend on the site of localisation of the cysts. A mature cyst is a 10 mm sac filled with clear fluid containing an invaginated abortive scolex.

Laboratory diagnosis

1. Demonstration of segment of adult or ova in stool.
2. Serology — Cysticercus cellulosae is diagnosed by radiological examination of the affected organ or tissue, and antigen and antibody detection in CSF or serum by ELISA.
3. Histopathological examination of the biopsy of the affected tissue.

Prevention and treatment

Infection is prevented by good hygienic practices and treated with praziquantel.

Diphyllobothrium latum (fish tapeworm)

This parasite occurs worldwide but is rarely reported in India. It is the largest parasite to be found in the intestine and man is the definitive host. The life cycle involves development in cyclops and fish as intermediate hosts. Infection occurs by ingestion of undercooked fish containing the larvae. Disease results in abdominal symptoms and vitamin B_{12} deficiency leading to megaloblastic anemia. Laboratory diagnosis is confirmed by demonstrating bile-stained, operculated ova, measuring around 40 µm, or the presence of segments of the adult, in stool.

Fasciola hepatica (liver fluke)

Rarely seen in India, the adult measures 3 cm, is flat, leaf-like, and resides in the liver (Fig. 29.3). The ova are bile-stained, oval and operculated, measuring around 140 µm. They are excreted in the feces and undergo development in water snails, becoming encysted larvae that are ingested by man with aquatic plants. Excystation occurs in the duodenum and the

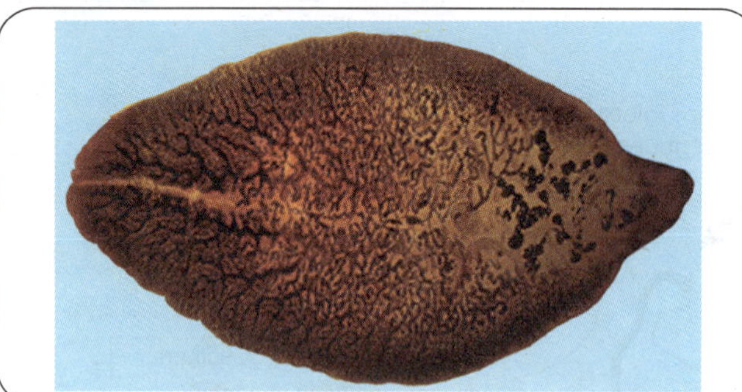

Fig. 29.3 Adult of *Fasciola hepatica* (*Source*: CDC, DPDx)

larva penetrates the intestinal wall to reach the liver and grow into an adult. Adults are hermaphrodites and, on fertilisation, release ova in feces. The disease causes abdominal discomfort, subcapsular liver abscesses or biliary obstruction.

Fasciolopsis buski (intestinal fluke)

Rarely seen in India, the adult measures 7 cm, is fleshy, leaf-like, and resides in the duodenum and jejunum (Fig. 29.4). The ova are bile-stained, oval and operculated, measuring around 140 μm. They are excreted in the feces and undergo development in water snails, becoming encysted larvae that are ingested by man with aquatic plants. Excystation occurs in the duodenum and the larva grows into an adult. Adults are hermaphrodites and, on fertilisation, release ova in feces. The disease causes abdominal discomfort, malabsorption and intestinal obstruction.

Ascaris lumbricoides (roundworm)

Epidemiology

It is the largest intestinal nematode parasitising man. It is seen worldwide and is common in India, especially where open defecation is practised and children (aged 2–10 years) play in the soil.

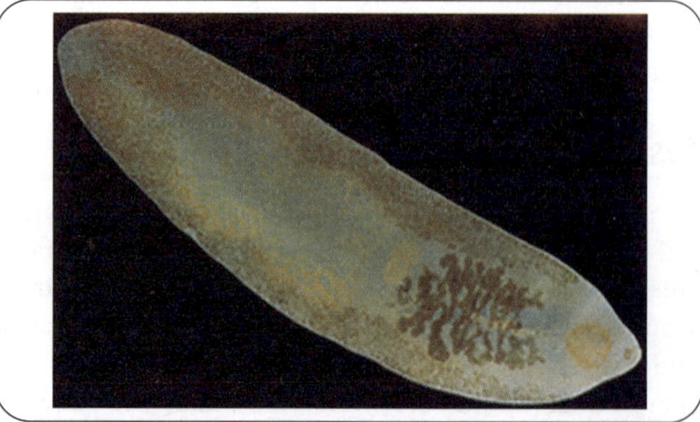

Fig. 29.4 Adult of *Fasciolopsis buski* (*Source*: CDC, DPDx)

Morphology: Ova (Fig. 29.5) and adult (Fig. 29.6)

The ova are excreted in human feces and can be fertilised or unfertilised, the former being decorticated at times. Fertilised ova are bile-stained, oval, measuring 70 μm and contain a single spherical ovum protected by a thick albumin coat that may be lost (decorticate).

The adult is round, white in colour, and tapers at both ends. The male worm is about 15–25 cm in length and the female about 25–40 cm in length. The posterior end of the male worm is curved ventral like a hook and in the female it is conical and straight.

Life cycle (Fig. 29.7)

Man is the definitive and only host, but development of the ovum also needs soil, as the ova excreted by man are not infective unless developed in the soil. Infection starts by ingestion of embryonated ova from contaminated soil (in children), water or food. The ova become larvae in the

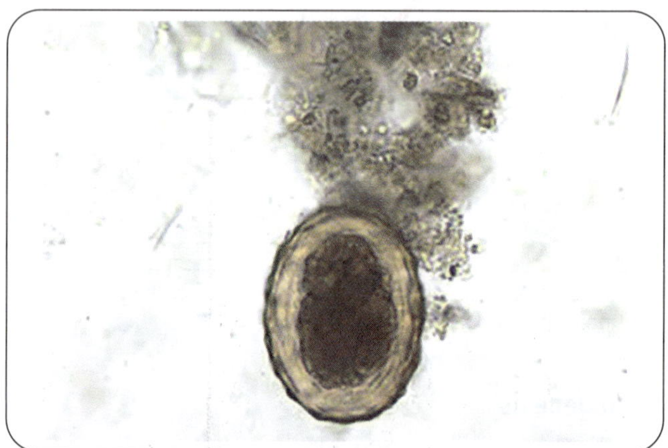

Fig. 29.5 Ovum of *Ascaris lumbricoides*, Wet mount, 40X (*Source*: CDC, DPDx)

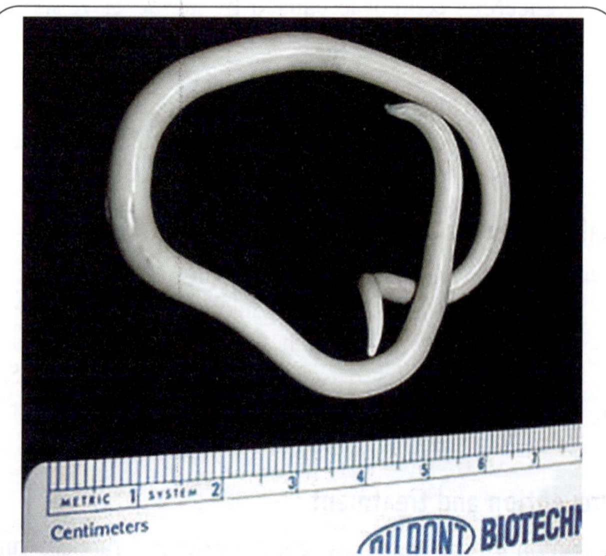

Fig. 29.6 Adult of *Ascaris lumbricoides* (*Source*: CDC, DPDx)

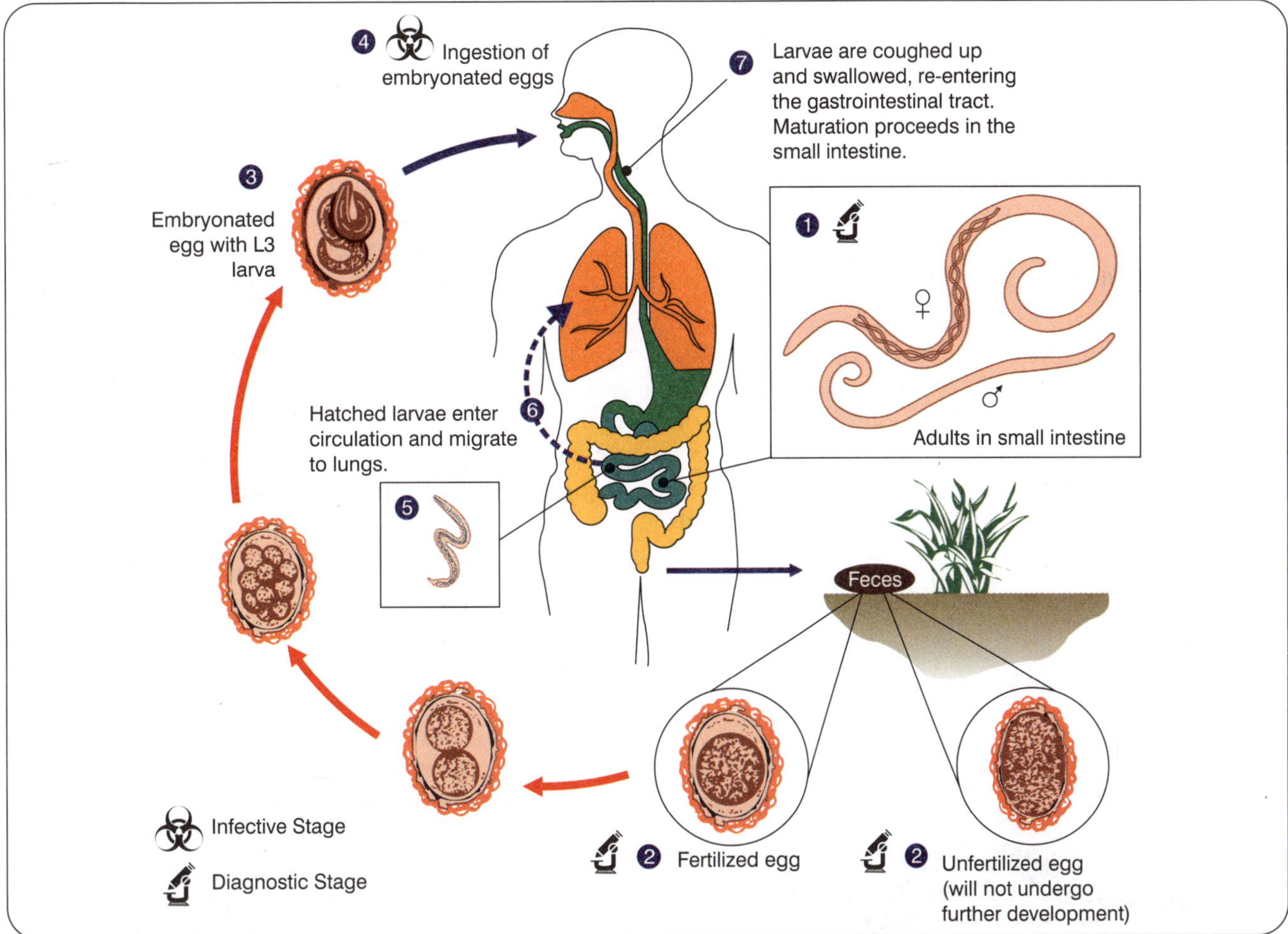

Fig. 29.7 Life cycle of *Ascaris lumbricoides* (*Source:* Adapted from CDC's Parasites and Health page, at https://www.cdc.gov/parasites/ascariasis/biology.html)

duodenum, penetrate the intestine, and reach the lungs via portal circulation. They break the alveoli and travel through the bronchi–trachea–pharynx to re-enter the gastrointestinal tract. Now, the larvae grow into adults, the sexes being separate. Fertilisation occurs and ova are excreted in the feces. The time taken for this life cycle in man is around 8 weeks.

To be infective to man, the fertilised ovum has to undergo embryonation in soil for 2 weeks.

Pathogenesis

The larvae migrating from the intestine to the lungs and back cause discomfort, fever, cough and, in severe cases, breathlessness. A transient pulmonary infiltration with peripheral eosinophilia termed 'eosinophilic pneumonia' or 'Loeffler's syndrome' is self-limiting.

The adult in the intestine causes malnutrition in children and mechanical complications like intestinal, biliary or respiratory obstruction. Allergic manifestations of urticaria and angioedema can occur due to the fluid released by the adult.

Laboratory diagnosis

Diagnosis is by demonstration of ova or adults in the stool. Charcot–Leyden crystals in stool and sputum are seen due to the eosinophilic response to the parasite.

Prevention and treatment

Proper sanitary cover without open defecation, along with the hygienic practice of handwashing before meals can prevent infection. As children are more prone to infection, 10th February and 10th August are observed as National Deworming Day in India. Deworming is advised at least once annually with a single dose of albendazole 400 mg to all children aged 1–19 years.

Ancylostoma duodenale and *Necator americanus* (hookworm)

Epidemiology

Hookworm infestation is worldwide, including in India, with millions affected, and both *Ancylostoma duodenale* and *Necator americanus* are seen. It is an important cause of iron deficiency anemia, especially in children. *N. americanus* is predominant in South India and *A. duodenale* in North India, with more than 20 crore people infected. The incidence is higher in rural than in urban settings, and especially among farmers and those who are in regular contact with soil.

Morphology: Ovum (Fig. 29.8) and adult (Fig. 29.9)

The ovum is oval, not bile-stained but colourless, measuring 60 µm with a thin translucent covering containing segmented (4 to 8) blastomeres. The ova of both hookworm species are indistinguishable on microscopy.

The larva has four stages of development, from rhabditiform larva to filariform larva. The rhabditiform larva, measuring 150 µm, is the first form and not infective to man while the filariform larva measures 700 µm and is the infective form.

The adult is a small greyish-white cylindrical worm measuring about 10 mm in length and has separate sexes. The anterior end is bent, hence the name 'hookworm', and has a buccal capsule bearing six teeth (hooks) that latch onto the intestinal mucosa.

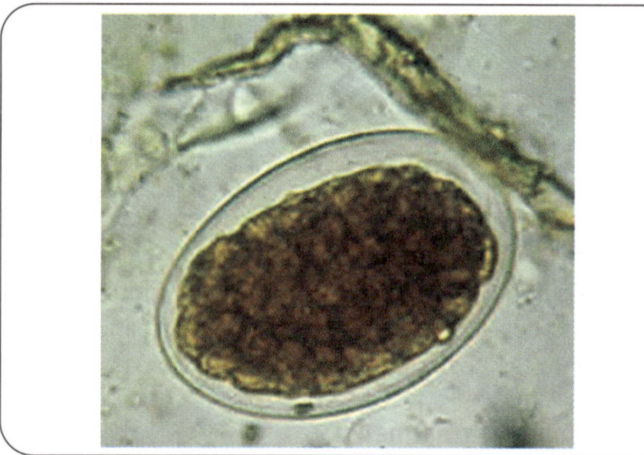

Fig. 29.8 Ovum of hookworm, Iodine mount, 40X (*Source*: CDC, DPDx)

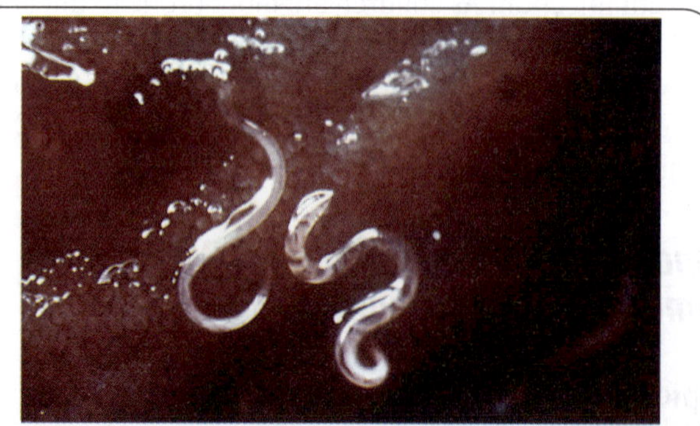

Fig. 29.9 Adult of *Ancylostoma duodenale* (*Source*: Public Health Image Library, ID 5205 / CDC)

Life cycle (Fig. 29.10)

The life cycle takes place only in man, beginning with the penetration of the filariform larva from the soil. From the site of entry, the larvae enter the venous circulation and reach the lungs and from the lungs are swallowed to reach the small intestine. Here, they grow into adults, attach to the mucosa and suck blood from the intestinal vessels. Following fertilisation, the female lays ova that are excreted in stool to reach the soil. Subsequently, the ova become rhabditiform larvae and then filariform larvae that can infect man. Therefore, excreted ova are not infective to man unless they undergo development in the soil.

Pathogenesis and clinical features

1. **Due to the larvae:** At the site of entry of the larvae, there could be pruritic maculopapular eruptions called 'ground itch'. These larvae, instead of entering the venous circulation, may wander along the skin, losing their way—'cutaneous larva migrans'. They may also affect the lungs transiently while migrating through the lungs, causing pneumonitis.
2. **Due to the adult:** The damage is due to the attachment of the adults to the intestinal mucosa, actively sucking blood, leading to anemia. This may be accompanied by abdominal symptoms, diarrhea and blood eosinophilia.

Laboratory diagnosis

1. Stool microscopy: Demonstration of ova or adult in the stool.
2. Other laboratory findings of hypochromic, normocytic anemia and eosinophilia.

Prevention

Improved personal hygiene and sanitary conditions, and walking with covered feet can help in prevention. Infection is treated with albendazole and mebendazole.

Strongyloides stercoralis

Clinical Case

A 40-year-old man presents with a history of persistent, foul-smelling diarrhea for 2 months, and loss of weight since 6 months. On examination, he is afebrile but emaciated. The stool on examination shows the florid presence of motile larvae measuring around 150 µm and no ova.

1. *What is the most likely cause of the diarrhea?*
2. *With the above history, what risk factors would you rule out?*
3. *Name the drug of choice in this infection.*

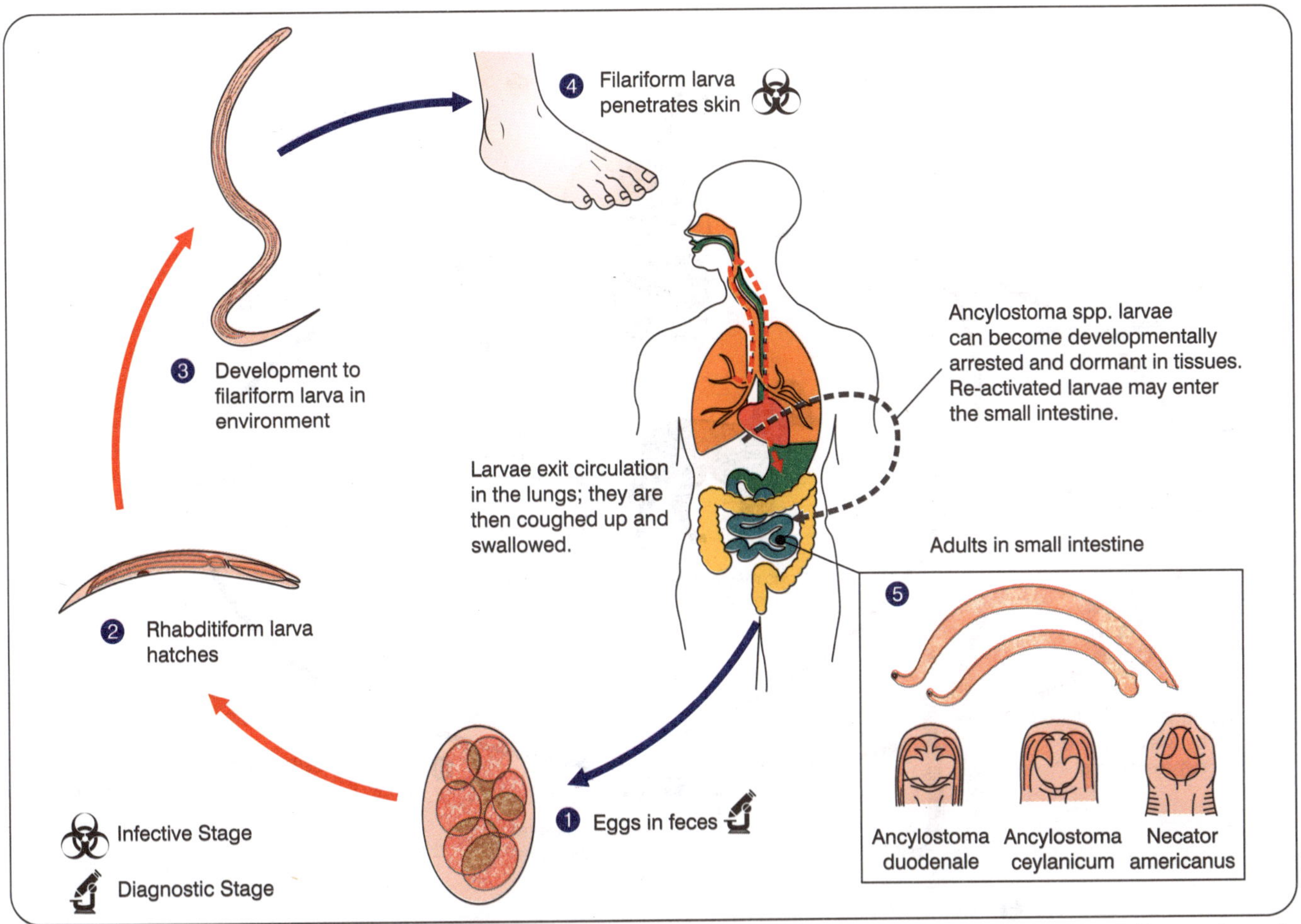

4 Filariform larva penetrates skin

3 Development to filariform larva in environment

Ancylostoma spp. larvae can become developmentally arrested and dormant in tissues. Re-activated larvae may enter the small intestine.

Larvae exit circulation in the lungs; they are then coughed up and swallowed.

Adults in small intestine

2 Rhabditiform larva hatches

5

Ancylostoma duodenale | Ancylostoma ceylanicum | Necator americanus

1 Eggs in feces

Infective Stage

Diagnostic Stage

Fig. 29.10 Life cycle of *Ancylostoma duodenale* (*Source:* Adapted from CDC's Parasites and Health page, at https://www.cdc.gov/parasites/hookworm/biology.html)

Epidemiology

Seen in a hot, humid tropical climate, including in India.

Morphology: Larva and adult (Fig. 29.11)

The ovum is oval, not bile-stained but colourless, measuring 60 μm with a thin translucent covering containing a larva. It is seen in the gravid female adult and not in the stool.

The larvae have four stages of development: L_1 or rhabditiform larvae are 150 μm in size and are first hatched in the intestine and excreted in stool (diagnostic). They undergo moulting in the soil (L_2) to become filariform larvae. Filariform larvae are lengthier and infective as they penetrate the human skin (L_3).

The female adult is a small, greyish-white cylindrical worm measuring about 3 mm in length. The anterior end has a buccal cavity that can latch onto the intestinal mucosa.

Life cycle (Fig. 29.12)

Man is the definitive host, but the larvae shed by humans are not infective and need to undergo development in the soil or environment to become infective to humans. After penetrating the skin, the larvae enter the blood circulation to reach the lung alveoli and moult (L4). From the alveoli, they enter the bronchus, then trachea, to be swallowed to reach the duodenum and upper jejunum. Here, they develop into adults. As only female adults have been

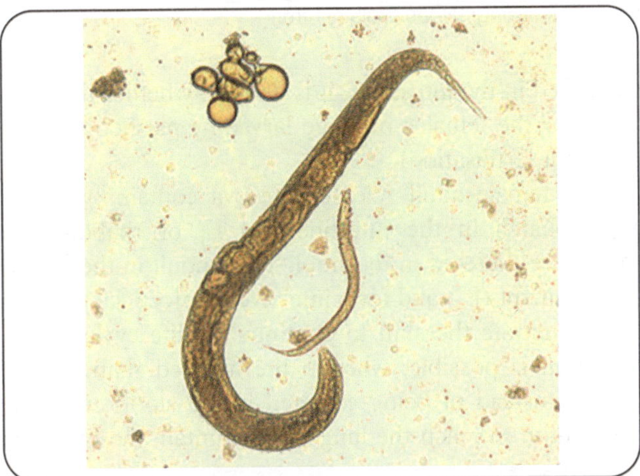

Fig. 29.11 Adult and rhabditiform larva of *Strongyloides stercoralis* (*Source*: CDC, DPDx)

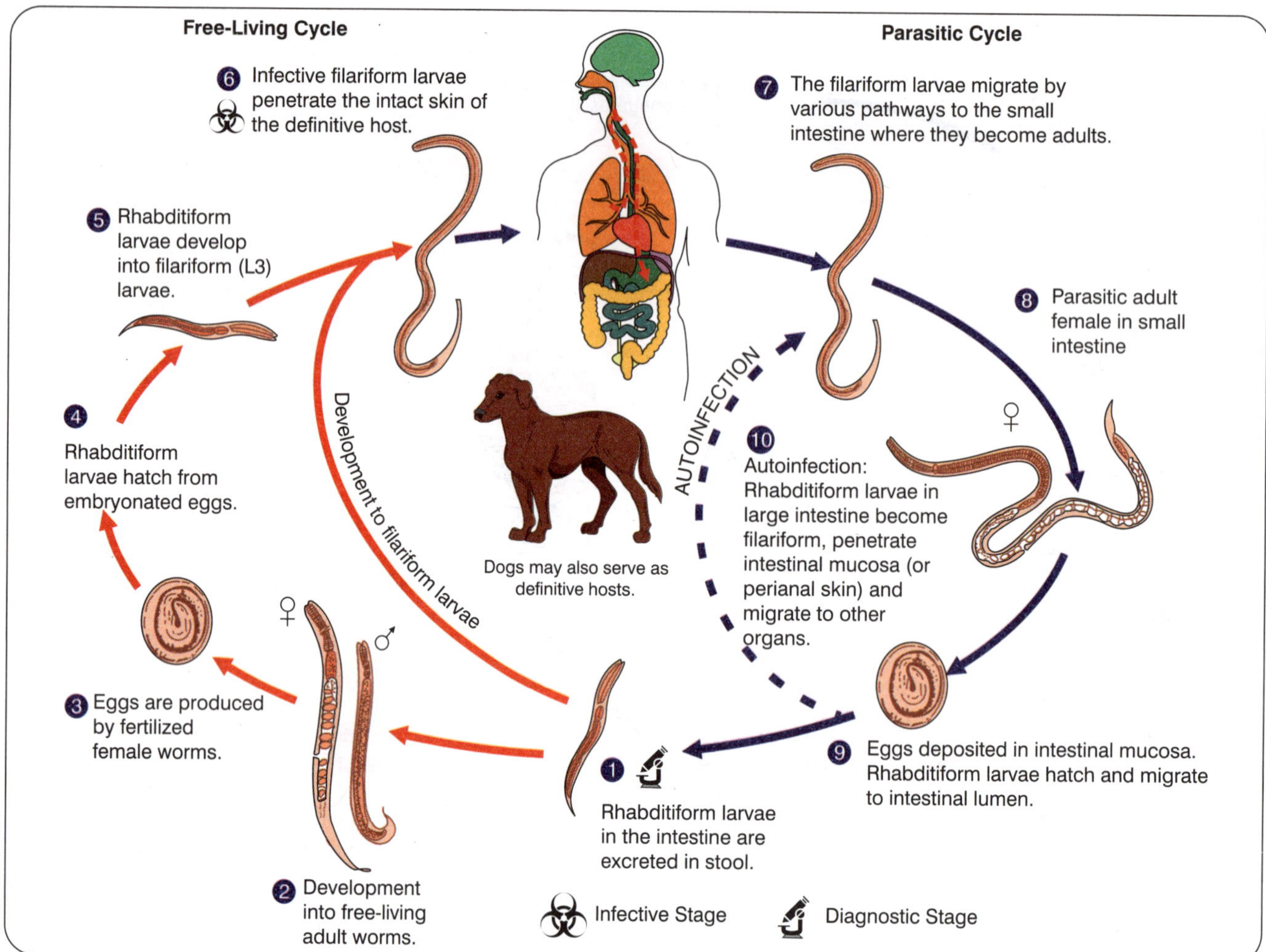

Fig. 29.12 Life cycle of *Strongyloides stercoralis* (*Source:* Adapted from CDC's Parasites and Health page, at https://www.cdc.gov/parasites/strongyloides/biology.html)

identified in the intestines, it is debated whether the male exists or the females produce larvae by parthenogenesis (without fertilisation).

The female adult releases the ova containing larvae which hatch in the intestine and L_1 or rhabditiform larvae are excreted in the stool. They moult in the soil or environment (L_2) and form infective filariform larvae (L_3) that penetrate the skin to continue the life cycle. Auto-infection is possible, wherein the hatched rhabditiform larvae, instead of being excreted in the stool, enter the circulation to reach the lungs and maintain the life cycle for years.

Pathogenesis and clinical features

On penetrating the skin, the migrating larvae may lose path and instead of entering the circulation, wander in the skin. This can cause a serpiginous urticarial rash on the buttocks, perineum, thigh or lower extremities called cutaneous larva migrans or larva currens (if fast-moving).

The adult causes abdominal symptoms—discomfort, pain, diarrhea or constipation. In the event of underlying risk factors due to any cause of immunosuppression, the larval load becomes very high, leading to dissemination of the larvae to other organs—brain, liver and kidneys. The risk factors include immunosuppressive conditions like HIV infection or use of steroids.

Laboratory diagnosis

1. Demonstration of the motile rhabditiform larvae in stool by wet mount either directly or following stool concentration. To distinguish *S. stercoralis* larvae from hookworm larvae, a variety of in vitro culture techniques are available but not routinely employed in diagnostic laboratories. A microscopic examination of stool shows the presence of larvae alone (not ova) in *S. stercoralis* infection. In hookworm infection, the stool predominantly reveals ova with blastomeres and rarely larva.

2. Immuno-diagnosis by detecting antibodies in serum or antigen in stool, or PCR to detect specific gene sequences in stool is useful in the absence of microscopic demonstration of larvae in stool.

Prevention and treatment

Improved personal hygiene and sanitary conditions, such as walking with covered feet can aid in prevention. Ivermectin is the drug of choice.

Trichuris trichiura (whipworm)

Epidemiology

This parasite is seen worldwide, with a higher prevalence in children. It is seen in India and is transmitted feco-orally through contaminated food and water.

Morphology: Ovum (Fig. 29.13) and adult (Fig. 29.14)

The ovum found in stool is unembryonated, bile-stained, barrel-shaped and 50 µm in size. It is covered by a shell with bipolar mucus plugs. In the intestine the embryonated ovum hatches into a larva that grows to be an adult.

The adult resembles a whip, measuring 2–4 cm. The anterior three-fifth is very thin and hair-like and the posterior two-fifth is thick and stout. The anterior end lies embedded in the intestinal mucosa while the posterior end is free in the lumen. The sexes are separate.

Life cycle (Fig. 29.15)

Man is the definitive host and the infection begins by ingestion of embryonated eggs in contaminated food and water. The ovum hatches in the small intestine

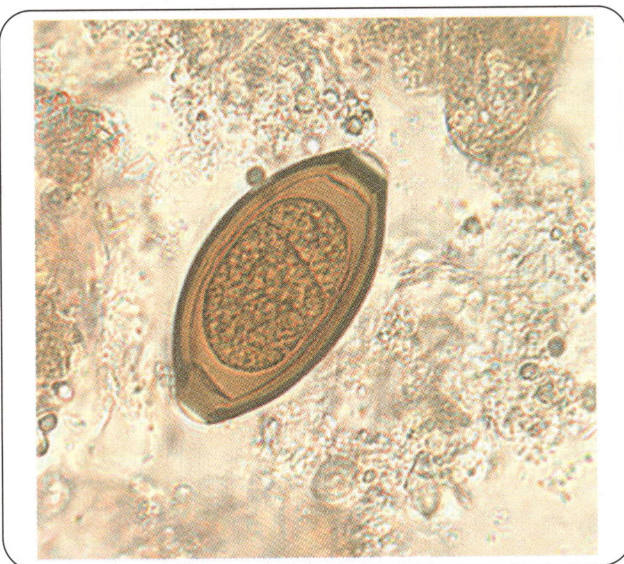

Fig. 29.13 Ovum of *Trichuris trichiura*, Wet mount, 40X (*Source*: CDC, DPDx)

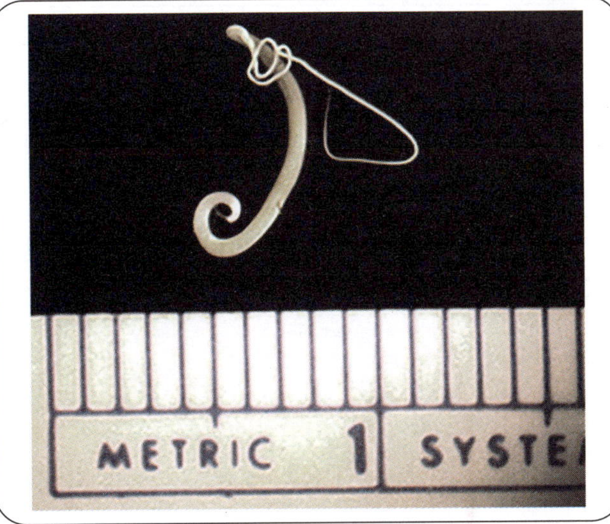

Fig. 29.14 Adult of *Trichuris trichiura* (*Source*: CDC, DPDx)

and the larva travels to the cecum or appendix and becomes an adult. Following fertilisation, the female lays unembryonated ova that are shed in the stool and are not infective. Embryonation occurs in the soil and only embryonated eggs can infect man.

The incubation period is around 3 months, following which abdominal discomfort, pain, diarrhea with mucus and or blood, and anemia may be seen. The infection is diagnosed by examining the stool for the adult worm macroscopically and ovum microscopically. Mebendazole is effective in treatment.

Enterobius vermicularis (pinworm or threadworm)

Epidemiology

This parasite has a global prevalence (including India), and is seen more in children. It is transmitted feco-orally by contaminated fingers.

Morphology: Ovum (Fig. 29.16) and adult (Fig. 29.17)

The ovum is not bile-stained but colourless, plano-convex, 50 µm in size and double-walled, with the presence of larvae.

The adult is spindle-shaped, 2–4 mm in length, with the female lengthier than the male.

Life cycle (Fig. 29.18)

Man is the only host, with children acquiring infection by ingestion of embryonated eggs through contaminated fingers. The ovum hatches in the small intestine to release the larva that migrates to the cecum, appendix or colon and develops into an adult male or female. Following

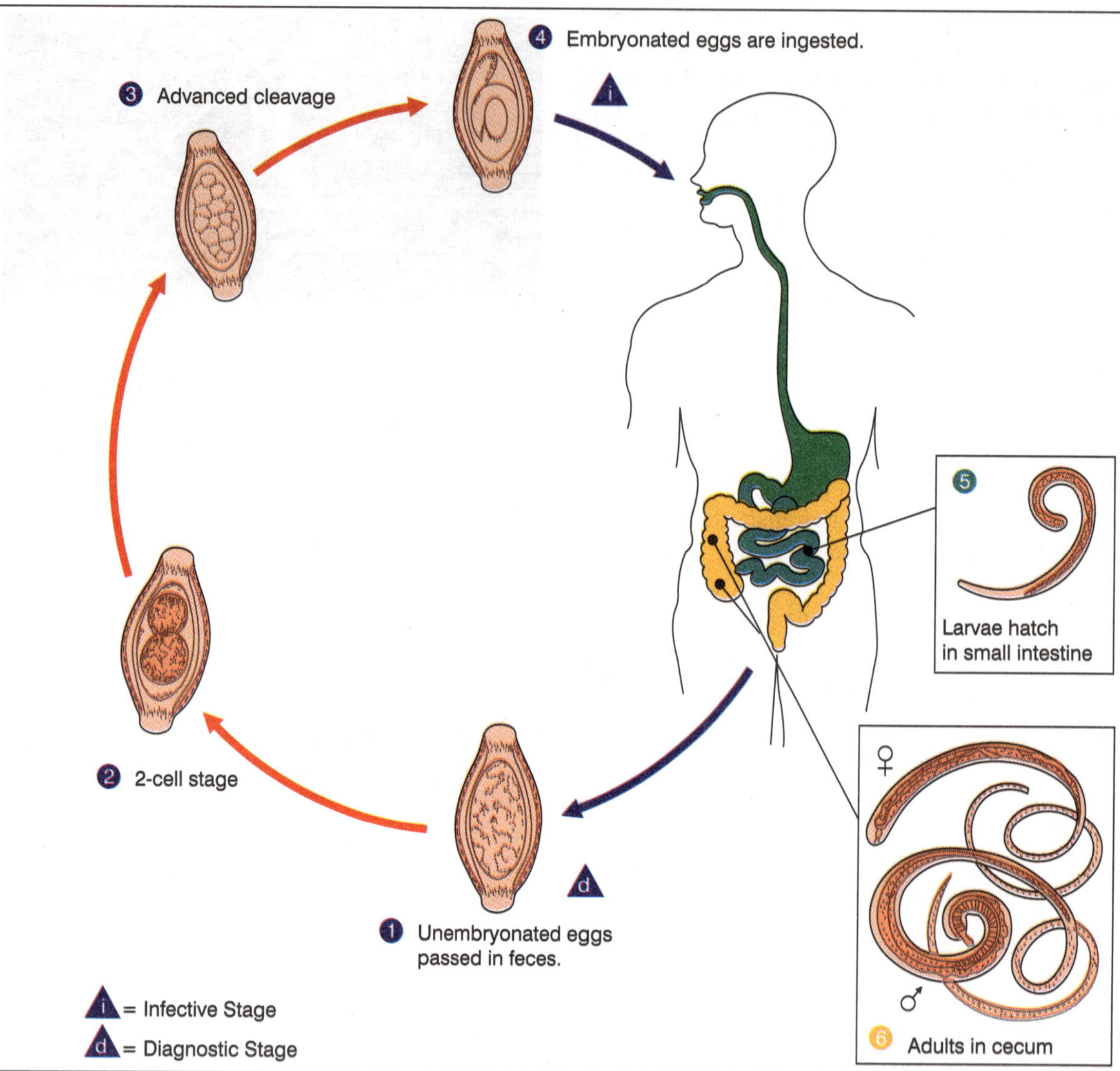

Fig. 29.15 Life cycle of *Trichuris trichiura* (*Source*: Adapted from CDC's Public Health Image Library, at https://phil.cdc.gov/Details.aspx?pid=3424)

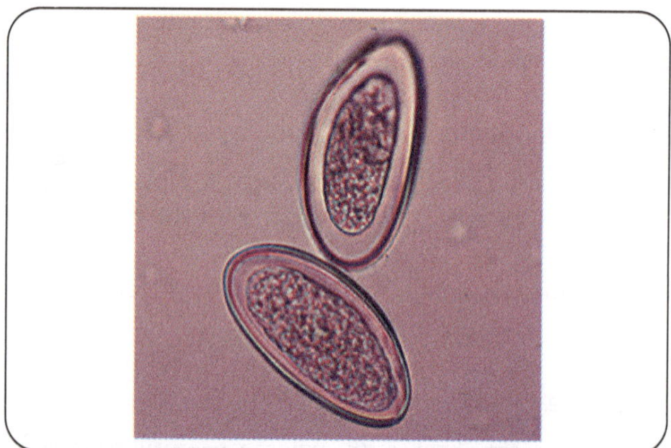

Fig. 29.16 Ovum of *Enterobius vermicularis*, Wet mount, 40X (*Source*: CDC, DPDx)

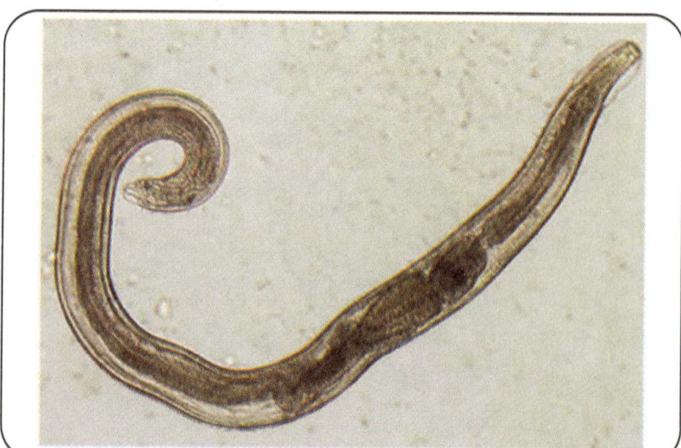

Fig. 29.17 Adult of *Enterobius vermicularis* (*Source*: CDC, DPDx)

fertilisation, the male dies and the gravid female migrates to the perianal region to lay embryonated eggs. The life span of the female is 2 months, but auto-infection can occur to continue the life cycle. Auto-infection is either

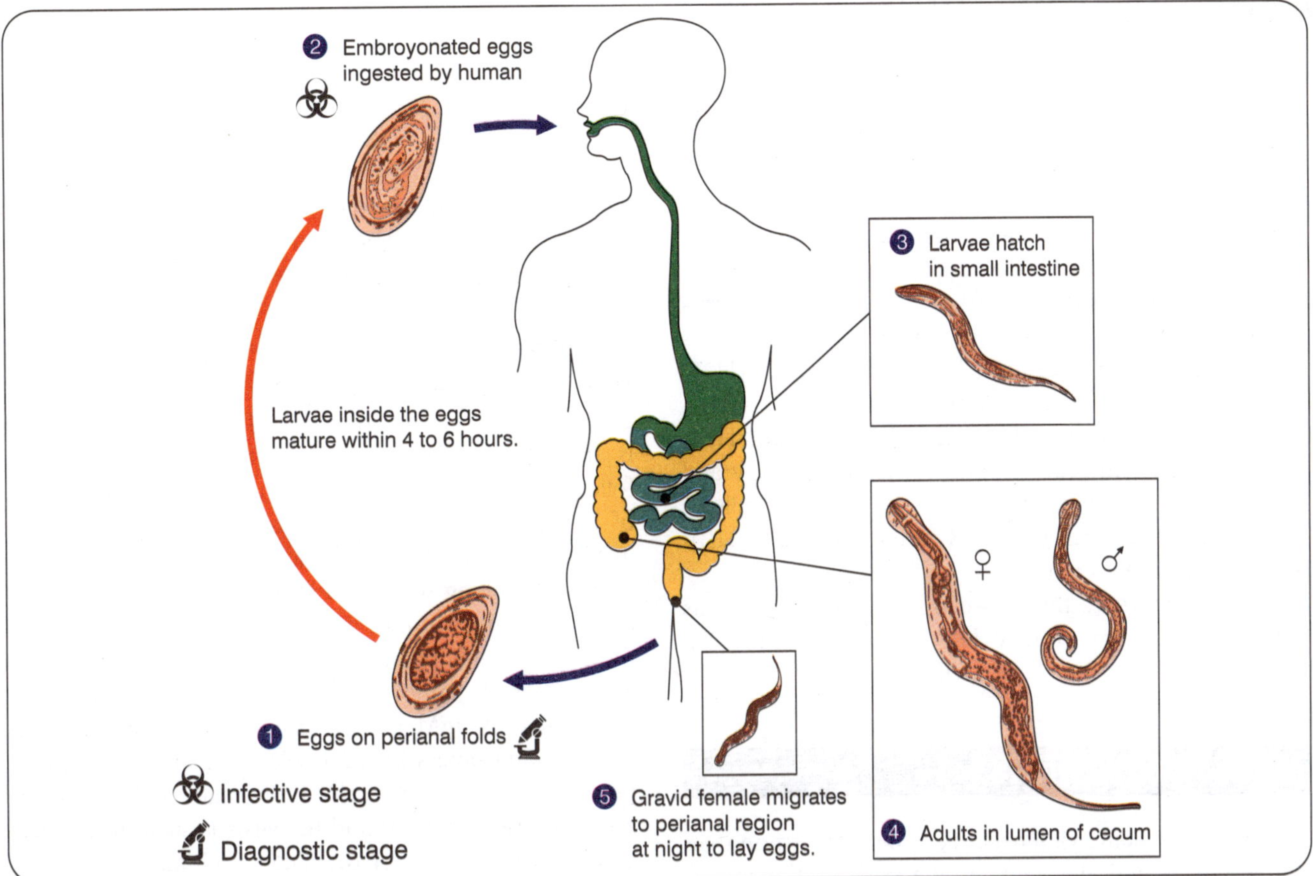

Fig. 29.18 Life cycle of *Enterobius vermicularis* (*Source*: Adapted from CDC's Parasites and Health page, at https://www.cdc.gov/parasites/pinworm/biology.html)

by retrograde migration of the hatched larvae from the perianal region to the intestine or by the contaminated fingers of the child after scratching the perianal region, with the larvae subsequently entering the mouth.

The infection may be asymptomatic and the adult worm discovered on the undergarments of the child, or can cause intense nocturnal perianal pruritus. It is diagnosed by examining the stool for the adult worm macroscopically and ovum microscopically, or by collection of the perianal sample by a cellophane tape or the NIH swab. Mebendazole, albendazole or pyrantel pamoate is effective in treatment.

Answers

Clinical Case

1. *Strongyloides stercoralis*
2. *The risk factors include immunosuppressive conditions like HIV infection or use of steroids.*
3. *Ivermectin*

Questions

1. Compare and contrast infestation with *Taenia solium* and *Taenia saginata*.
2. What is larva migrans? Name the helminths that can cause disease due to migrating larvae.
3. Enumerate parasites causing anemia. Discuss the life cycle and pathogenesis of *Ancylostoma duodenale*.
4. Explain the genesis of complications of *Ascaris* infestation.
5. Enumerate auto-infective helminths and their laboratory diagnosis.

Chapter 30

Section V: Gastrointestinal and Hepatobiliary Infections

Viral Hepatitis

A number of viruses can cause inflammation of hepatocytes, but viral hepatitis refers to those that primarily affect the liver and not viruses that cause disease elsewhere as well as the liver. These five hepatotropic viruses are hepatitis A virus (HAV), hepatitis B virus (HBV), hepatitis C virus (HCV), hepatitis D virus (HDV) and hepatitis E virus (HEV). They are transmitted either enterally (HAV, HEV), causing infective hepatitis, or parenterally (HBV, HCV, HDV), causing serum hepatitis. HAV and HEV are associated with acute hepatitis, while HBV, HCV and HDV with chronic hepatitis, cirrhosis and hepatocellular carcinoma (HCC). All except HBV are RNA viruses, HBV being the only DNA virus.

The other viruses that can cause hepatitis but not as the primary disease are EBV, HSV, VZ, mumps, rubella, rubeola, coxsackie B and adenovirus.

INFECTIVE HEPATITIS

Asymptomatic acute viral hepatitis is more common than symptomatic, and the younger the patient, the milder the disease. The earliest symptoms are non-specific and constitutional, and the illness has phases—incubation, pre-icteric, icteric and convalescence. The symptoms are malaise, anorexia, nausea, vomiting and abdominal discomfort, followed by jaundice and pruritis. Clinical recovery follows biochemical recovery as the weakness and fatigue may persist.

Hepatitis A virus (HAV)

Clinical Case 1

A 6-year-old girl presents with sudden onset malaise, nausea and occasional vomiting since 2 days. On examination, she appears hydrated with generalised weakness, with raised liver enzymes: AST 1100 and ALT 990 units/L.

1. *What is the most likely causative agent?*
2. *Name the marker to confirm your diagnosis.*
3. *What is thc source and route of infection?*
4. *State the vaccines available to prevent this infection.*

Epidemiology

HAV is transmitted feco-orally by ingestion of contaminated food or water and is the commonest cause of acute viral hepatitis in children. In children, most infections are asymptomatic but transmissible. In adults, HAV is more likely to be symptomatic with a higher case fatality rate. Humans are the only reservoir of infection, and outbreaks due to overcrowding can occur. In India, this infection is mostly seen in children and they acquire immunity by the age of 10 years.

Pathogenesis

HAV is an ssRNA virus belonging to the picornaviruses group, with only one serotype. The incubation period is around 1 month, with the acute onset of symptoms, followed by jaundice and complete recovery. The course of HAV infection and the associated markers is shown in Fig. 30.1. Following infection, viremia occurs from 2 to 4 weeks along with fecal shedding of the virus which continues longer, until 7 weeks. A month after infection, the liver enzymes begin to rise, followed by the appearance of symptoms or jaundice and the appearance of antibodies. The anti-HAV IgM is diagnostic of infection, peaks at 6 weeks and tapers by 12 weeks. The anti-HAV IgG is protective, peaks at 8 weeks and remains lifelong. There are no carriers of HAV and the virus is shed in the feces before, during and after the illness and jaundice. Rarely, the infection may relapse or cause cholestatic jaundice. Fulminant hepatitis may be seen in adults.

Laboratory diagnosis

This is ideally confirmed by the presence of anti-HAV IgM in serum, by ELISA, or by HAV antigen in stool by ELISA. The presence of anti-HAV IgG denotes past infection or vaccination.

Prevention

This includes hygienic and sanitation practices, and having safe drinking water. Two vaccines are available, both with long-lasting immunity.

1. Killed vaccine for children more than age 12 months, two doses IM at 0 and 6 to 12 months, confers protection for 15–20 years.
2. Live attenuated H2 strain of HAV vaccine, given as a single dose subcutaneously, confers protection for 15–20 years.

HAV antiserum is indicated for post-exposure prophylaxis within 2 weeks of confirmed exposure.

Hepatitis E virus (HEV)

HEV is a non-enveloped, ssRNA hepevirus, with a specific HEV antigen and one serotype. Besides humans,

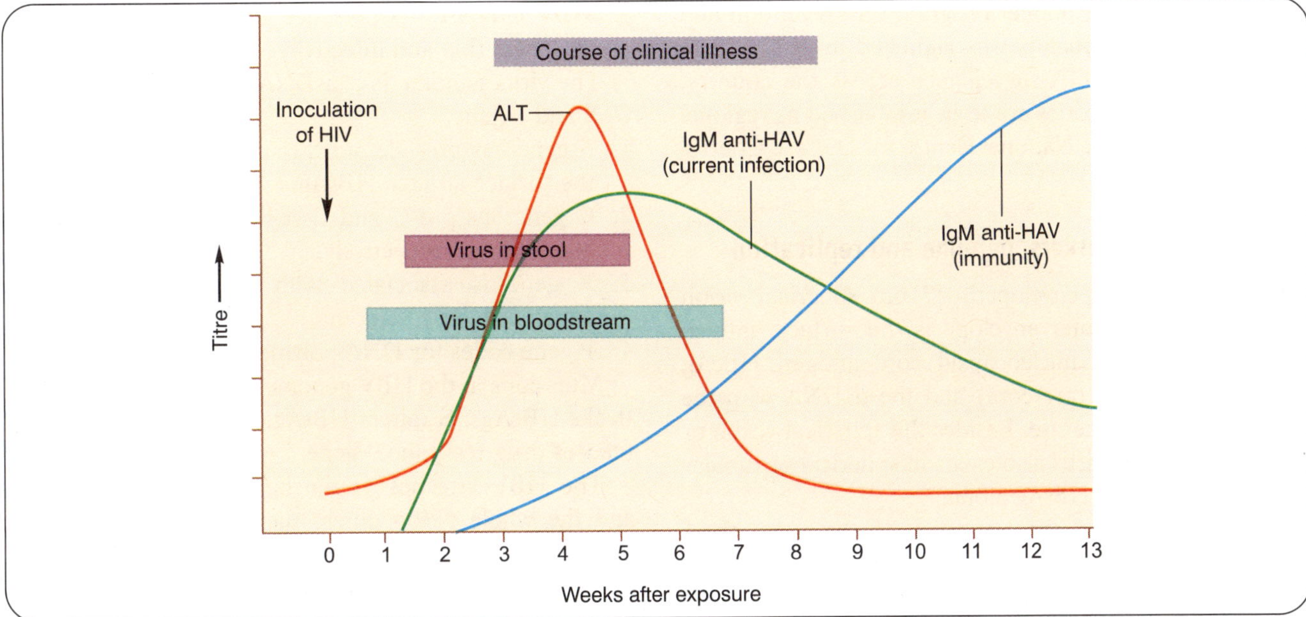

Fig. 30.1 Course of HAV infection

dogs, cats, pigs and monkeys are also reservoirs, but there are no human carriers. It is seen sporadically or as an epidemic, affects young adults more than children, and is transmitted feco-orally by contaminated food and water. Following an average incubation period of 40 days, acute hepatitis presents with jaundice and is self-limiting. Fulminant hepatitis is rare. It is diagnosed in the laboratory by IgM anti-HEV in serum or HEV RNA in stool or serum. Prevention is by general hygiene and sanitation measures as the vaccine is not available in India.

SERUM HEPATITIS

Chronic viral hepatitis describes hepatocellular inflammation and necrosis that last longer than 6 months, caused by HBV, HCV or HBV-associated HDV. The liver injury does not result from the virus but from the host's T cell response to the infected cells.

Hepatitis B virus (HBV)

Clinical Case 2

A 40-year-old surgeon presents with symptoms of nausea and generalised weakness. On investigation, the serum bilirubin and liver enzymes are raised. He recalls attending to an emergency suturing of a wound with profuse blood loss, with bare hands, 2 months back.

1. *What is the most likely cause of infection?*
2. *What viral markers will you test for and why?*
3. *Could this infection have been prevented? How?*
4. *If not treated, what are the likely outcomes of this infection?*

Epidemiology

Humans are the only source of HBV infection and it is transmitted by a case of HBV or a carrier. Patients with chronic inactive HBV infection possess HBsAg alone and are 'simple carriers'. Patients with active hepatitis possess HBsAg, HBeAg, DNA polymerase and HBV DNA, and are termed 'super carriers' with high infectivity.

HBV is transmitted via parenteral contact with blood, blood products and body fluids. Transfusion-associated factors, surgical procedures, healthcare exposure (needle stick injury) and IV drug abuse are possible risks of transmission. HBV is more infectious than HIV or HCV as the number of virions found in blood in HBV is higher. The risk of transmission of HBV, HIV and HCV is 30%, 3% and 0.3%, respectively, following exposure to a contaminated needle stick injury. Therefore, healthcare workers, commercial sex workers, blood and organ recipients and IV drug addicts are considered high risk groups in HBV.

HBV is also transmitted perinatally either in utero, during delivery or at breastfeeding, sexually and by close contact in institutionalised children having breached skin or sharing food.

Globally, 2 billion people are HBV-infected, of whom 350 million have chronic disease, with 1 million deaths annually due to cancer and cirrhosis. HBV prevalence depends on the HBsAg carrier rate in a population: < 2% low, 2–8% intermediate and > 8% high endemicity. India has an intermediate prevalence with a 3.7% carrier rate, as 40 million are infected. The prevalent serotypes are adr and ayw and genotypes A and D. HBV-HIV co-infection is also known, that enhances the risk of HCC and cirrhosis. The National Program on Prevention and Control of Viral Hepatitis was launched in 2012–17 to establish viral hepatitis surveillance across the country. Ten regional laboratories are to be established as regional centres including the National Centre for Disease Control (NCDC), Delhi.

Structure, viral markers, genome and replication

HBV is a dsDNA, enveloped, 42 nm spherical virion (Fig. 30.2). The outer envelope is the surface antigen HBsAg covering a nucleocapsid core antigen HBcAg (core) and HBeAg (pre-core) and the dsDNA with the enzyme DNA polymerase. Besides the spherical virion or Dane particle, the virus also exists in spherical or tubular forms consisting of HBsAg alone.

Corresponding to the structure of the virus there exist viral markers, the presence or absence of which determines the infection, infectivity and the course of the disease. These are useful in the diagnosis and treatment of HBV infection. These markers are:

1. **HBsAg:** The HBV surface antigen that is responsible for gaining entry into the hepatocytes. Its presence is therefore associated with viral multiplication and carriage. There exist four serotypes of HBV determined by its HBsAg—adw, ayw, adr and ayr.
2. **Anti-HBs:** The presence of antibody to the surface antigen also signifies infection with HBV. However, since HBsAg is the portal of entry of the virus in a hepatocyte, anti-HBs presence following vaccination blocks the viral entry and is protective.
3. **HBcAg:** The presence of core antigen is restricted only to hepatocytes and not found circulating in blood as it is associated with the virus core and not secreted.
4. **Anti-HBc:** The antibody to core antigen is seen in blood and signifies infection, present or past.
5. **HBeAg:** The pre-core nucleocapsid antigen is secreted during active multiplication of the virus and is therefore associated with high infectivity if present in blood.
6. **Anti-HBe:** The antibody to HBeAg, its presence reduces infectivity.
7. **HBV dsDNA:** Its presence is associated with active multiplication and infectivity.

The virus genome is a dsDNA that consists of the S, C, X and P gene:

1. S gene has pre-S1, pre-S2 and S regions coding for the surface antigen HBsAg.
2. C gene has pre-C and C regions coding for HBeAg and HBcAg, respectively.
3. X gene is associated with chronic hepatitis and hepatocellular carcinoma.
4. P gene codes for DNA polymerase.

Mutations in the HBV genome are known in the pre-C (lacks HBeAg), S (alters HBsAg, escape mutants) and P (causes drug resistance) gene.

The HBV attaches to the hepatocytes with HBsAg, and the whole virion enters the hepatocyte. On entry, the virion is uncoated of its HBsAg, and only the virion core enters the nucleus of the cell. Transcription occurs within the nucleus to release virion core particles into the cytoplasm. These particles containing the dsDNA and nucleocapsid bud out of the hepatocyte acquiring the envelope HBsAg. The new virion thus formed may either infect other hepatocytes or re-enter the parent hepatocyte to continue replication of the virus.

Pathogenesis and clinical features

The disease caused by HBV infection is due to the host immune response to the viral replication in the hepatocytes. The host immune response responds to the intracellular virus multiplying in the hepatocytes with a cell-mediated attack on the hepatocytes to contain the virus. This consists of NK cells and HBV-specific CD8 T cell cytotoxicity against the hepatocytes, resulting in killing of the hepatocytes. Therefore, the quality of the immune response determines the course of the infection and disease outcome. The outcome of acute HBV hepatitis is: recovery 95%, chronic hepatitis 5%, and rarely HBV carrier (<5%) or fulminant hepatitis (<1%)

A vigorous cytotoxic response to the virus eliminates the virus and contains the course while a lack of immune response leads to a carrier state and chronicity of the disease with its subsequent sequelae. Most primary

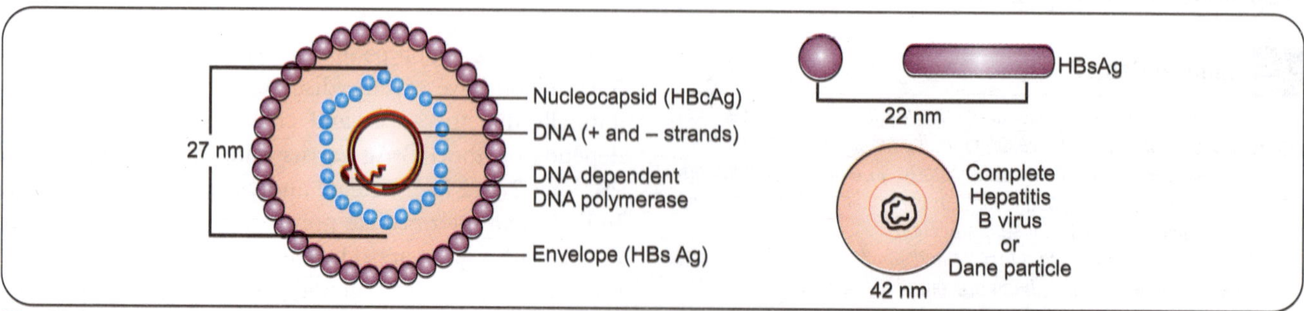

Fig. 30.2 Structure of HBV

HBV infections are self-limiting and resolve within 6 months of onset. Those that fail to resolve persist and may continue for years, and require antiviral therapy. A unique feature of HBV infection is the presence of HBsAg in both, the self-limiting (acute) and persistent (chronic) infection. HBsAg negative acute or chronic HBV infection is rare and attributed to the S mutant strains of the virus.

Following an incubation period of 1 to 6 months, there is acute hepatitis presenting with nausea, lack of appetite, vomiting or abdominal discomfort and jaundice. Most cases (95%) recover spontaneously, with 5% progressing to chronic hepatitis. It is these chronic hepatitis patients that require antiviral therapy, and may progress to liver cirrhosis in 5 years, subsequently leading to liver failure or HCC over another 5 years.

Laboratory diagnosis

The laboratory diagnosis of HBV is by detection of the viral markers (antigens and antibodies, both IgG and IgM in Table 30.1) and the viral DNA and its interpretation (Fig. 30.3).

HBV DNA detected by PCR is present in blood during both acute and chronic hepatitis, and the viral load is monitored during antiviral therapy in chronic hepatitis. Serum bilirubin and liver enzymes are raised during acute and chronic hepatitis.

Prevention

General measures include protected sex, infection control practices, blood and organ donation and antenatal screening.

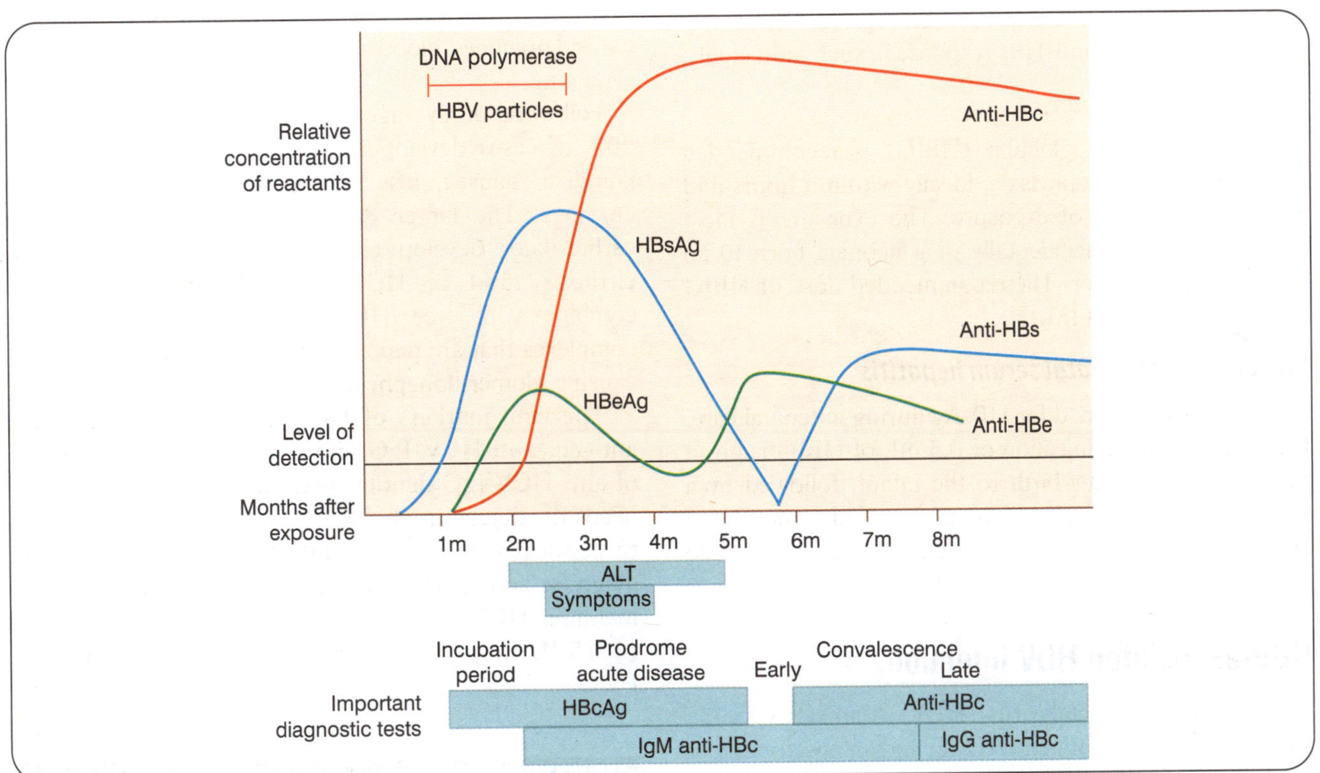

Fig. 30.3 Acute HBV infection

Table 30.1 HBV markers and their interpretation

Virus antigen/Antibody markers					Interpretation
HBsAg	HBeAg	Anti-HBc	Anti-HBs	Anti-HBe	
+	+	IgM	–	–	Acute HBV infection; highly infectious
+	+	IgG	–	–	Late/chronic HBV infection or carrier state; highly infectious
+	–	IgG	–	+/–	Late/chronic HBV infection or carrier state; low infectivity
–	+/–	IgM	–	+/–	Seen rarely in early acute HBV infection; infectious
–	–	IgG	+/–	+/–	Remote HBV infection; infectivity nil or very low
–	–	–	+	–	Immunity following HBV vaccine

Active immunisation

The HBV vaccine consists of 20 μg of recombinant HBsAg given IM in three doses at 0, 1 and 6 months, without a booster. The dose is 10 μg for children below age 10 years. In the national immunisation program, it is given at 6, 10 and 14 weeks. In the event of missing a dose, the schedule can be completed by taking the missed scheduled dose when available. Adult immunisation with HBV is mandated in healthcare workers (HCWs) who are at a high risk.

This results in high titre anti-HBs at 7 months in 95% of vaccinees, with 5% of non-responders (who do not develop anti-HBs antibodies). A minimum anti-HBs titre of 10 mIU/mL is considered protective against HBV infection, and immunity lasts for 5–10 years. Non-response to the vaccine is attributed to the fact that the recombinant vaccine contains only the S region of the virus and not the pre-S1 and pre-S2 regions, which may be necessary for an anti-HBs response in such individuals.

Passive immunisation

Hepatitis B immunoglobulin (HBIG) is recommended as post-exposure prophylaxis, ideally within 6 hours and not after 48 hours of exposure. The exposure is most likely in an HCW accidentally or a neonate born to an HBV-infected mother. The recommended dose of HBIG is 0.06 mL/kg given IM.

Prevention of neonatal serum hepatitis

The mother is screened for HBsAg during antenatal care. If positive, a single injection of 0.5 mL of HBIG is given IM immediately after birth to the infant, followed by a full course of the vaccine given at 0, 1 and 6 months of birth. If negative, HBIG is not required and a full course of vaccine can be given at 0, 1 and 6 months of birth.

HBV-associated HDV infection

The hepatitis D (Delta) virus (HDV) is an ssRNA virus with a protein coat antigen (HDAg) and an envelope from the HBsAg of HBV. Being a defective virus, it can survive only in the presence of HBV, and therefore either causes co-infection or super-infection.

HDV is most commonly transmitted parenterally, followed by sexual and vertical transmission, with 5% of those HBV infected globally being infected with HDV. Though HBV prevalence is intermediate, HBV-HDV infection is not prevalent in India. When a healthy person is simultaneously exposed to HBV and HDV and contracts both together, it is considered co-infection. When a chronic HBV carrier (HBsAg positive) is exposed to HDV and contracts HDV, it is considered super-infection. The risk of developing chronic hepatitis, cirrhosis and HCC is more in super-infection and rare in co-infection. Anti-HDV IgM or HDV RNA are detectable in both co-infection and super-infection, along with HBsAg.

Hepatitis C virus (HCV)

HCV is an ssRNA, enveloped virus with six genotypes and more than 100 subtypes. In India, its prevalence is 1%, with genotypes 1 and 3 being more prevalent. HCV is most commonly transmitted parenterally, followed by vertical and sexual transmission. Prevention is by general measures as there is no vaccine available. General preventive measures include protected sex, infection control practices, blood and organ donation, and antenatal screening.

Following an average incubation period of 2 months, 20% of cases develop acute hepatitis characterised by jaundice, nausea, loss of appetite and raised hepatic enzymes. The larger 80% who remain asymptomatic earlier later develop chronic hepatitis progressing to cirrhosis (5%) or HCC (5%). Anti-HCV antibodies combine with free HCV antigen to form immune complexes that are deposited in the glomerulus or joints, causing glomerulonephritis or arthritis.

The viral markers of HCV infection are HCV core antigen, anti-HCV IgG and HCV RNA. The presence of anti-HCV IgG denotes present or past infection that needs the detection of either HCV antigen or HCV RNA to confirm present active infection. It is important to diagnose active HCV infection as antiviral treatment to maintain HCV RNA in blood to an undetectable level (< 15 IU/mL) prevents the progression to cirrhosis or HCC.

HEPATIC BACTERIAL INFECTIONS

Bacterial infection of the liver results in pyogenic liver abscess, acute hepatitis or chronic granulomatous disease. Acute hepatitis is seen in infection with *Salmonella, Borrelia, Leptospira Brucella, Neisseria* and *Campylobacter*. Chronic hepatitis is seen in infection with *Mycobacterium, Treponema* and *Rickettsia*.

Answers

Clinical Case 1

1. *Hepatitis A virus*
2. *Anti-HAV IgM in serum by ELISA*
3. *Ingestion of contaminated food or water; feco-oral*
4. *Both killed and live attenuated vaccines are available that produce long-lasting immunity.*

Clinical Case 2

1. *Acute viral hepatitis caused by HBV*
2. *HBsAg, HBeAg, HBV DNA viral load*
3. *Vaccination with HBsAg when joining healthcare services. HBIG administration within 6 hours of exposure.*
4. *Outcomes of acute HBV hepatitis: Recovery 95%, chronic hepatitis 5%, rarely HBV carrier (<5%) or fulminant hepatitis (<1%)*

Questions

1. Name the feco-orally transmitted viruses causing acute hepatitis, and outline their laboratory diagnosis and prevention.
2. Enumerate viruses causing serum hepatitis and the predominant routes and risk of transmission of each.
3. Explain the structure of HBV in relation to its viral markers and their interpretive role in laboratory diagnosis.
4. Enumerate viral infections that can progress to chronic hepatitis and their pathogenesis.
5. Explain the clinical and serologic events in acute HBV hepatitis.
6. Discuss pre-exposure and post-exposure prophylaxis of HBV infection.
7. How will you diagnose HCV infection in a case of acute hepatitis?

Hepatic Cyst and Abscess

Hepatitis refers to inflammation of the hepatocytes resulting in derangement of liver function and systemic effects. However, there also exist parasitic infections of the liver, wherein localised cysts or abscesses are formed with localised inflammatory response, without hepatitis of the liver. The parasites affecting the liver are *Echinococcus granulosus* causing hydatid disease, *Entamoeba histolytica* causing hepatic amoebiasis or amoebic liver abscess and the adult liver flukes.

HYDATID DISEASE

Clinical Case

A 24-year-old man presents with difficulty in breathing but without fever. On examination, the respiratory system is clear and the liver is not tender but palpable. Ultrasonography of the liver shows a 10 cm unilocular cyst under the diaphragm in the right lobe.

1. *Name three parasites that can affect the liver.*
2. *What is the most likely cause of the above disease?*
3. *What is the mode of transmission, infective form and reservoir of this disease?*
4. *Mention the laboratory findings on microscopic examination of the cystic fluid.*

Echinococcus granulosus (dog tapeworm)

Epidemiology

Hydatid disease, caused by *Echinococcus* spp., is a zoonotic infection, wherein man is an accidental host. Amongst four species that have been reported to cause human disease worldwide, *E. granulosus* or the dog tapeworm is the commonest, and is also seen in India. Infection occurs by ingestion of ova excreted by the adult residing in the dog's intestine, by consumption of contaminated food or in the handling of the dog. Dogs and other canines serve as the reservoir of infection.

Morphology

Neither the ovum nor the adult is seen in man, as both occur in the dog. Only the abortive larval form is seen in man as scolices of *E. granulosus* within the hydatid cyst.

The ovum in dog feces has an oncosphere with six hooklets surrounded by a radially striated embryophore. The adult residing in the dog is 5 mm in length with a head or scolex, neck and body or strobili of three segments or proglottids.

Life cycle (Fig. 31.1)

In the life cycle in nature, dog is the definitive host and sheep is the intermediate host. Accidental infection in man results in a dead-end cycle with an abortive larval hydatid cyst.

On ingestion, the infective ova excreted by the adult residing in the dog hatch into oncospheres in the human intestine and enter the blood circulation to reach, commonly, the liver or, rarely, other soft tissues and develop into a hydatid cyst.

Pathogenesis and clinical features

The oncosphere, on reaching the liver, forms a fluid-filled unilocular cyst encysted by fibrous tissue. Over months, as it matures and increases in size from 1 cm to 30 cm, it develops into three layers (Fig. 31.2):

1. The outer pericyst of the host-derived fibrous tissue.
2. The intermediate parasite-derived ectocyst that is acellular, hyaline and elastic.
3. The inner parasite-derived endocyst that secretes the hydatid fluid and eventually forms the "brood capsule". The brood capsule is the germinal epithelium of the endocyst that produces 'daughter cysts' (within the primary cyst) that contain multiple scolices. The daughter cysts arise from the same endocyst, hence the name "brood".

The hydatid fluid is clear, colourless, bacteriologically sterile and anaphylactic to man. It consists of sodium salts and deposits of the abortive scolices forming a granular "hydatid sand". The primary hydatid cyst may spontaneously resolve, calcify or rupture causing anaphylaxis.

Besides the liver, especially the right lobe, a hydatid cyst may be rarely seen in lung, kidney, brain, spleen, muscle or bone. Most cases are asymptomatic and symptoms depend on the location and size of the hydatid cyst. Symptoms arise due to the enlarged visceral organ and its pressure effects. It can cause obstruction and if it ruptures, it can cause anaphylaxis. A hydatid cyst in the liver may present as hepatomegaly, abdominal discomfort, difficulty in breathing or portal hypertension.

Laboratory diagnosis

The diagnosis is largely based on ultrasonography or CT scan examination followed by surgical excision of the cyst or puncture, aspiration, injection and re-aspiration

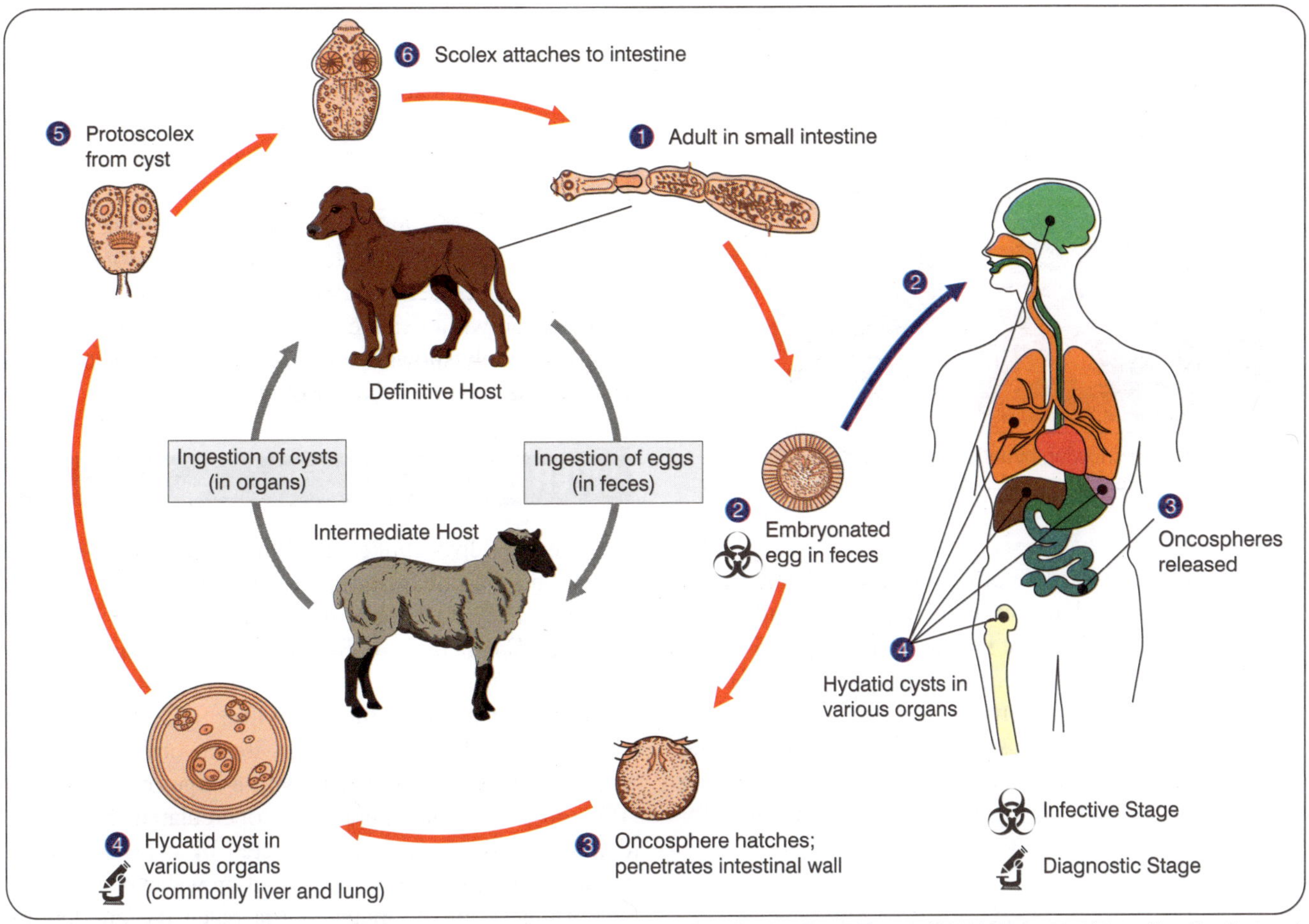

6 Scolex attaches to intestine
1 Adult in small intestine
5 Protoscolex from cyst
Definitive Host
Ingestion of cysts (in organs)
Ingestion of eggs (in feces)
Intermediate Host
2
2 Embryonated egg in feces
3 Oncospheres released
4 Hydatid cysts in various organs
4 Hydatid cyst in various organs (commonly liver and lung)
3 Oncosphere hatches; penetrates intestinal wall
Infective Stage
Diagnostic Stage

Fig. 31.1 Life cycle of *Echinococcus granulosus* (*Source*: Adapted from CDC's Parasites and Health page, at https://www.cdc.gov/parasites/echinococcosis/biology.html)

Host parenchyma
Host collagen
Cyst wall
Germinal layer
Brood capsule
Protoscolices
Hydatid fluid
Daughter cysts

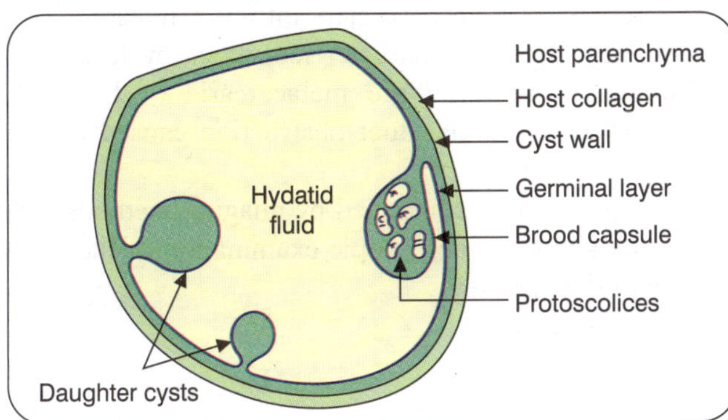

Fig. 31.2 Structure of hydatid cyst

(PAIR) for a single, accessible hepatic cyst. Due to the risk of anaphylaxis, diagnostic aspiration is not recommended.

Following excision or PAIR, the hydatid fluid wet mount on microscopic examination in the laboratory shows the presence of multiple abortive scolices or protoscolices (Fig. 31.3). Alternatively, the histopathologic examination of the cyst will show the three layers with daughter cysts and scolices (Fig. 31.4). Casoni's test is a rarely practised method of diagnosing hydatid disease.

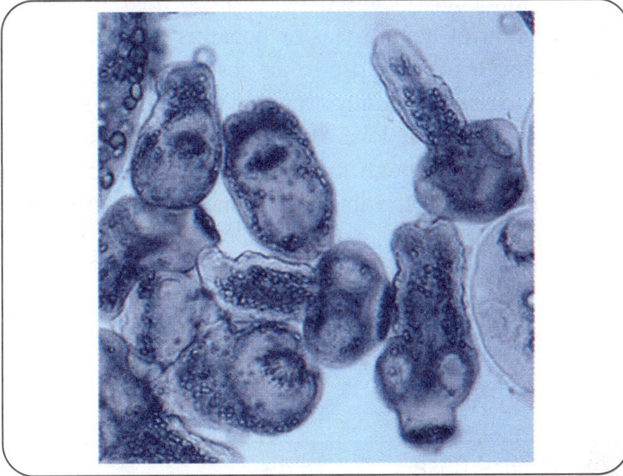

Fig. 31.3 Hydatid fluid showing multiple abortive scolices (*Source*: CDC, DPDx)

Casoni's test is an immediate hypersensitivity skin test. The antigen used is the sterile hydatid fluid from unilocular hydatid cysts. For the test, 0.2 mL of the antigen is injected intradermally on the forearm. For control, an equal volume of sterile water is injected intradermally on the other arm. The control fades immediately while the tested

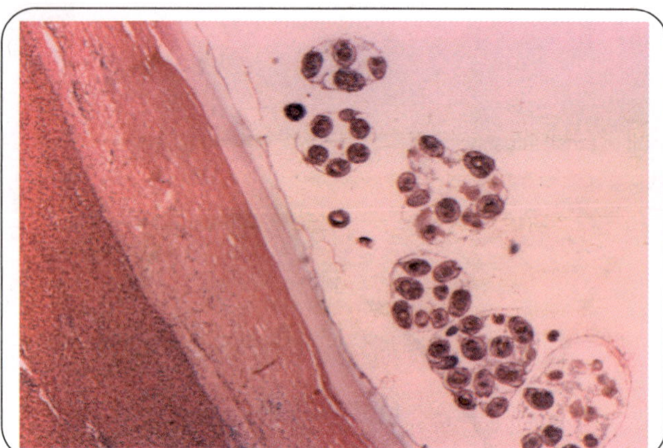

Fig. 31.4 Tissue showing hydatid cyst, H&E stain, 40X (*Source*: Public Health Image Library, ID 910 / CDC / Dr. Peter Schantz).

site in positive cases develops a large wheal measuring 5 cm or more in diameter with multiple pseudopodia within 30 minutes. This test has low sensitivity (55–70%) and gives false-positive reactions in patients suffering from other cestode infections.

Antibody detection assays using hydatid fluid components as the antigen for ELISA or immunoblotting are useful for epidemiologic surveillance but not for diagnosis.

AMOEBIC LIVER ABSCESS

Hepatic amoebiasis caused by *E. histolytica* is not an inflammation of hepatocytes or hepatitis but necrosis of the hepatocytes giving rise to an abscess. It is generally solitary and predominantly occurs in the postero-superior surface of the right lobe. As it is the trophozoite that enters the liver through the portal vein and encystation can occur only in the intestinal lumen, cysts are never seen in the abscess.

The abscess fluid is bacteriologically sterile, without pus, and consists of dead hepatocytes and blood cells, giving it a thick texture and brown colour. Trophozoites are generally absent in the aspirate as the trophozoites are present only at the advancing edges of the abscess and not in the abscess fluid.

Examination of liver aspirate: The macroscopic appearance of the aspirate is thick brown due to the presence of necrosed hepatocytes and blood. Microscopic examination with a saline wet mount does not show pus cells and rarely shows trophozoites of *E. histolytica*. On culture, the aspirate is bacteriologically sterile.

LIVER FLUKES

A variety of liver flukes are known to cause disease depending on the geographic location of the patient. In India, liver flukes are uncommon, with few cases of *Fasciola hepatica* and *Fasciola gigantic* having been reported.

Fasciola hepatica is the sheep liver fluke. It is 3 cm in size, flat, leaf-like and brown, with an oral and ventral sucker. It is a hermaphrodite with a large ovum (around 100 μm), oval, bile-stained and operculated. Sheep are definitive hosts and man is infected by ingestion of the larval form—metacercaria—found on water plants.

On ingestion, the metacercaria reach the duodenum, excyst and penetrate the intestine and liver to reach the hepatic duct and mature into an adult in 2 months. On reproduction, ova are excreted into the intestine and feces. The disease is characterised either by fever and eosinophilia caused by the metacercaria or hepatitis, hepatic abscess, or bile duct obstruction caused by the adult.

Diagnosis can be confirmed by imaging methods that detect the adult or microscopic examination of the stool for ova of the parasite.

ANSWERS

Clinical Case

1. *Echinococcus granulosus, Entamoeba histolytica, Fasciola hepatica*
2. *Echinococcus granulosus—hydatid cyst*
3. *Ingestion, ova, dog*
4. *Abortive scolices or protoscolices seen in the fluid wet mount*

Questions

1. What is hydatid disease? Explain its pathogenesis and laboratory diagnosis
2. How does *Entamoeba histolytica* cause hepatic amoebiasis?
3. Write a note on liver fluke infestation seen in India

SECTION SUMMARY

Section V: Gastrointestinal and Hepatobiliary Infections

Infection/pathogen	Virulence/pathogenesis	Laboratory diagnosis	Vaccine/recommended antimicrobial
Food poisoning *S. aureus*	Exotoxin (enterotoxin)	ELISA or PCR for enterotoxin	No role for antimicrobials
B. cereus	Exotoxin	Stool culture	
Non-typhoidal *Salmonella*	Invasive	Stool culture	
Campylobacter jejuni	Invasive	Stool culture	
V. parahaemolyticus	Invasive	Stool culture	
Y. enterocolitica	Invasive	Stool culture	
E. coli 0157:H7	Exotoxin dysentery, renal failure, encephalopathy	Stool culture, ELISA for exotoxin PCR	
C. perfringens type A	Exotoxin	Stool culture	
C. botulinum	Neurotoxin	Stool culture	Botulinum antitoxin
Infective diarrhoea *Vibrio cholerae*	Infective dose 10^8 exotoxin, cell cAMP increase, sodium chloride in lumen, watery stool dehydration	Comma-shaped on gram stained stool, darting motility, culture	Oral vaccines: killed whole cell + CTB subunit, live attenuated CVD 103- HgR strain No role for antibiotics. Azithromycin in select cases.
E. coli Enteropathogenic (EPEC)	Adhesion to epithelium increased secretion, watery diarrhea	Stool culture	No role for antibiotics
Enterotoxigenic (ETEC)	Adhesion and secretion of LT and ST toxins	Detection of toxins	
Enteroinvasive (EIEC)	Invasion of epithelium mediated by a virulence marker antigen (VMA)	Detection of VMA by ELISA	
Enterohemorrhagic (EHEC) 0157:H7	Secretes Shiga-like toxins Stx 1 and Stx 2 that cause hemorrhagic colitis or hemolytic uremic syndrome	Culture, toxin detection by ELISA or PCR for toxin genes	
Enteroaggregative (EAEC)	Adhesion, enterotoxin Causes aggregates in Hep-2 cells	Stool culture	
Diffusely-adherent (DAE)	Adhesion	Stool culture	
Enteric fever Salmonellae	Infective dose 10^3 Invasive, macrophages, bacteremia	Blood culture, Widal test, ELISA	Parenteral Vi, Oral live attenuated Ty21a strain Ceftriaxone, azithromycin
Clostridiodes difficile	Enterotoxin, cytotoxin	Toxin in stool by ELISA or ICT	
Rotavirus	Villous enterocytes, malabsorption	Stool ELISA to detect antigen	No role for antibiotics

Infection/pathogen	Virulence/pathogenesis	Laboratory diagnosis	Vaccine/recommended antimicrobial
Giardia lamblia	Ingestion of cyst, excystation in duodenum, malabsorption in children	Cysts in stool	Metronidazole
Cryptosporidium *Cyclosora* *Cycloisospora*	Opportunistic in HIV-AIDS	Stool microscopy	Nitazoxanide
Dysentery *Shigella*	Infective dose 10–100 bacilli, ingestion, invasion of M cells, exotoxins, ulceration of mucosa	Stool culture	Ciprofloxacin Ceftriaxone
Entamoeba histolytica	Ingestion of cysts, carriers, trophozoites invade mucosa, amoebic liver abscess	Iodine mount stool, liver aspirate mount for trophozoites	Metronidazole Secnidazole for acute infection, diloxanide furoate for cyst carriers
Neobalantidium coli	Ingestion of cyst, ulceration of colon mucosa	Stool microscopy	Diadoquin
Acid peptic disease *Helicobacter pylori*	Invasion. Acute gastritis and peptic ulcer disease	Urea breath test, Gastric biopsy HPE Warthin—Starry stain	Omeprazole+ clarithromycin + amoxicillin/metronidazole
G.I infestations *Taenia saginata*	Ingestion of ova, adult in intestine	Stool for ova or segment of adult	Praziquantel
Taenia solium	Ingestion of ova, adult in intestine, cysticercus cellulosae in tissue	Stool for ova or segment of adult	Praziquantel
Diphyllobothrium latum	Ingestion of fish with larvae, vitamin B_{12} deficiency	Stool for ova or segment of adult	Praziquantel
Hymenolepsis nana	Ingestion of ova, adult in intestine	Stool for ova/ adult	Praziquantel
Fasciola hepatica	Ingestion of larvae from plants, penetrate duodenum to reach liver-adult	Stool for ova	Praziquantel
Fasciola buski	Ingestion of larvae from plants, adult in intestine	Stool for ova/adult	Praziquantel
Schistosoma haematobium	Venous plexus of urinary bladder, hematuria	Antigen detection in urine	Praziquantel
Ascaris lumbricoides	Ingestion of ova, larva penetrate intestine to reach lungs, re-swallowed to become adult in intestine Eosinophilic pneumonia	Stool for ova/adult	Albendazole Mebendazole
Ancylostoma duodenale/Necator americanus	Larvae penetrate skin to enter blood and reach lungs. Swallowed to be adult in intestine. Anemia, larva migrans.	Stool for ova/adult	Albendazole Mebendazole
Strongyloides stercoralis	Larvae penetrate skin to enter blood and reach lungs. Swallowed to be adult in intestine. Larva migrans/currens	Stool for larvae/adult	Ivermectin

Infection/pathogen	Virulence/pathogenesis	Laboratory diagnosis	Vaccine/recommended antimicrobial
Trichuris trichiura	Ingestion of ovum. Adult in caecum/appendix	Stool for ova/adult	Albendazole Mebendazole
Enterobius vermicularis	Ingestion of ovum Adult in cecum/colon/appendix nocturnal enuresis	Stool for ova/adult	Albendazole Medendazole
Hepatobiliary viral hepatitis HAV	Feco-oral, acute hepatitis-jaundice, no carriers	Anti-HAV IgM ELISA	Killed and live vaccine-H2 strain
HEV	Feco-oral, acute hepatitis-jaundice, no carriers	Anti-HEV IgM ELISA	Self-limiting
HBV	Parenteral, sexual, vertical transmission, HBsAg carriers, acute hepatitis jaundice, Chronicity and HCC	HBsAg, HBeAg ELISA, HBV DNA PCR-viral load quantification	HBsAg active HBIG passive Self limiting Antiviral therapy in chronic hepatitis
HDV	Associated with HBV co-infection or superinfection	Anti-HOV IgM or HOV RNA	
HCV	Parenteral, sexual, vertical transmission, HBsAg carriers, acute hepatitis jaundice, chronicity and HCC	HCV RNA	Antiviral therapy in chronic hepatitis
Hydatid disease *Echincoccus granulosus*	Feco-oral from dog ova, liver cyst with daughter scolices	Radiography, HPE, Casoni's test	Albendazole

Chapter 32

Section VI: Central Nervous System Infections

Meningitis

Like the vascular compartment, the CSF circulation is sterile. Infection is rare but serious and may leave neurological deficits. The blood–brain barrier inhibits passage of microorganisms into the brain and CSF. The same also hinders the passage of antimicrobials and immune response elements. The brain and spinal cord are suspended in the CSF and covered by the layers of the meninges—the piamater, arachnoid and duramater. Infection of the brain parenchyma is encephalitis, meninges, i.e., meningitis, and spinal cord tissue, i.e., myelitis. Besides these three entities, there exist other chronic neurologic infections and cysts and abscess in the brain.

The CSF has poor defence mechanisms, low complement levels and low blood penetration. However, it possesses an intrinsic immunosurveillance mechanism in the microglia (phagocyte) and a lymph-like system of perivascular sheaths surrounding the blood vessels as they enter the brain.

Tissue damage in the CNS can be due to bacterial toxins, lytic cycles of the virus or host inflammatory response. This is characterised by infiltration of microglia and proliferation of astrocytes that cause cerebral edema.

Specific infections cause specific functional impairments of the CNS, e.g., psychosis, impaired memory and seizure suggest HSV infection of the temporal lobe, neck stiffness suggests meningitis, while flaccid paralysis of the lower extremities suggests poliovirus infection.

The route of entry for hematogenous microbes from the blood circulation to the CSF is presumed to be the choroid plexus where most of the CSF is formed. The chance of CNS infection correlates with the microbial load of the blood. By the neural route, rabies travels to the anterior horn of the spinal cord via the peripheral nerves on entering the axon of a peripheral nerve. HSV ascends via the trigeminal neuron root or the olfactory nerve endings. *Naegleria* spp. travel via the cribriform plate following trauma.

Depending on the distribution of receptors on specific host cells, viruses show tissue tropism for certain neural cell types—poliomyelitis and motor neuron of the spinal cord and medulla, mumps and ependyma of the fetus, rabies and hippocampus gyrus, HSV and temporal lobe of the cerebrum.

Route of infection of the nervous system

1. Through colonised respiratory epithelium, e.g., *Neisseria meningitidis*
2. Through blood, following bite of a vector mosquito, e.g., Japanese B virus
3. Zoonotic — bite of a dog, e.g., Rabies
4. Transplacental, e.g., Rubella
5. Feco-oral, e.g., Poliovirus
6. Inhalation, e.g., Tuberculosis
7. Post-traumatic tissue infection or abscess

The causative agents of CNS infections

1. Acute meningitis — Bacteria: *Neisseria meningitidis*, *Haemophilus influenzae* type b, *Streptococcus pneumoniae*, Group B streptococci, *Escherichia coli*, *Listeria monocytogenes*; viruses: mumps virus, enteroviruses; parasites: *Naegleria*, *Acanthamoeba*
2. Chronic meningitis — *Mycobacterium tuberculosis*, *Cryptococcus*
3. Acute encephalitis — HSV, Japanese B virus, rabies virus, enteroviruses
4. Acute brain abscess — *Staphylococcus*, *Streptococcus* A or D, non-sporing anaerobes
5. Chronic brain abscess — *Mycobacterium tuberculosis*, cysticercus cellulosae, *Toxoplasma*

Meningitis is inflammation of the membranes that cover the brain and spinal cord—mostly the piamater and the arachnoid—while encephalitis is inflammation of the brain parenchyma. Meningitis can be caused by bacteria (pyogenic), viruses (aseptic) or fungi. Bacteria can cause meningoencephalitis, but encephalitis is largely caused by viruses and can be primary (directly infecting the brain) or secondary (subsequent to infection elsewhere).

Meningitis can be classified by clinical presentation as acute or chronic, by etiology as bacterial, viral, fungal or parasitic, and by epidemiology as sporadic or epidemic.

Acute meningitis presents with fever, headache, meningismus and altered sensorium within hours or a few days. The microorganisms causing acute meningitis are mostly viruses—enteroviruses, HSV, mumps virus, HIV; bacteria—*S. pneumoniae, N. meningitidis, H. influenzae, L. monocytogenes;* and rarely parasites—*Naegleria, Acanthamoeba, Angiostrogylus*.

Chronic or subacute meningitis presents over weeks, months or a year or in immunocompromised individuals, with low-grade insidious onset fever and other symptoms of meningitis. The most likely microorganisms are bacteria—*M. tuberculosis, Treponema pallidum, Borrelia burgdorferi,* or fungi—*Cryptococcus neoformans*.

Aseptic meningitis implies that the CSF is sterile on routine bacterial culture due to infection with viruses, fungi, *Leptospira* or *Treponema*, while TB infection is granulomatous. **The CSF examination varies according to the infectious cause of meningitis** (Table 32.1).

Table 32.1 The CSF examination varies according to the infectious cause of meningitis

Cause	WBCs/mm³	Primary cell type	Glucose, mg/dL	Protein, mg/dL
Viral	50–1000	Mononuclear	>45	<200
Bacterial	1000–5000	Neutrophilic	<40	100–500
Tuberculous	50–300	Mononuclear	<45	50–300
Cryptococcal	<20	Mononuclear	<40	>45

BACTERIAL OR PYOGENIC MENINGITIS

Bacterial meningitis implies pus in the subarachnoid spaces, causing an increase in the CSF pressure, leading to vomiting, headache, meningismus and altered vision. Meningococcal meningitis may not leave any neurodeficit, but *H. influenzae* may leave behind residual mental retardation and deafness. In 50% of cases, Gram stain of centrifuged CSF will reveal the causative bacteria. Rapid antigen detection in CSF and vaccines are available for *S. pneumoniae, N. meningitidis* and *H. influenzae*.

In children and adults, the bacteria causing pyogenic meningitis are *H. influenzae*, *S. pneumoniae* and *N. meningitidis*. In neonates and infants, they are Group B streptococci, *L. monocytogenes*, *E. coli* and *H. influenzae* (Table 32.2).

Clinical Case 1

A 1-year-old baby presents with acute onset fever, vomiting and altered sensorium for 2 days. On examination, she has neck rigidity and her CSF examination shows pus cells and gram-positive, capsulated, lanceolate diplococci.

1. *What is your diagnosis and the causative agent?*
2. *Name three other bacteria commonly implicated in this infection.*
3. *What is the source of this infection in this case?*
4. *Name the drug of choice in treatment.*

Streptococcus pneumoniae (pneumococcus)

Epidemiology

Most children between 6 months and 5 years of age are carriers of *S. pneumoniae* in the upper respiratory tract or nasopharynx, which is transmitted by inhalation. Risk factors associated with pneumococcal disease are extremes of age (< 2 years, > 60 years), underlying chronic lung disease or viral respiratory infections, splenectomy or diabetes mellitus. Around 90 distinct serotypes of pneumococci have been identified worldwide that vary in geographical distribution. In India, serotypes 1, 6, 19A and 19F are common.

Morphology

Gram-positive lanceolate diplococci, with a clearing around them due to the presence of a capsule (Fig. 32.1). The capsule can be demonstrated by negative staining with India ink.

Pathogenesis and clinical features

The capsular polysaccharide prevents phagocytosis, to escape the immune response. It also possesses another C antigen that precipitates with a protein that is synthesised by the liver in acute inflammations. This protein is therefore called 'C-reactive protein' (CRP), which is a non-specific marker in inflammations and not an antibody to the C antigen of *S. pneumoniae*.

Other virulence factors include the exotoxins pneumolysin and autolysin, surface proteins and adhesion

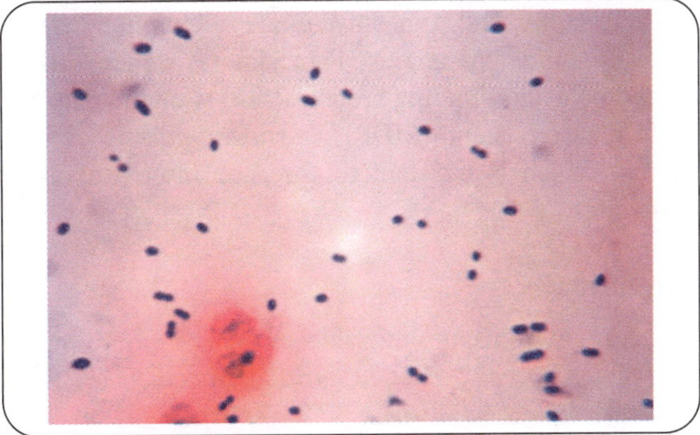

Fig. 32.1 Gram-positive cocci in pairs—*Streptococcus pneumoniae*, Gram stain, 100X (*Source:* Public Health Image Library, ID 18081 / Centers for Disease Control and Prevention [CDC] / Dr. Thomas Sellers, Emory University, Atlanta, Georgia)

Table 32.2 Causative agents of bacterial meningitis

Age	Underlying disease	Causative bacterium	Others
0–2 months	None	Group B streptococci	*E. coli, Listeria*
2 m–5 years	None	*H. influenzae*	*N. meningitidis, S. pneumoniae*
>5 years	None	*S. pneumoniae*	*N. meningitidis**

*epidemic potential

molecules. The IgA protease cleaves the respiratory mucosal IgA to enter the respiratory mucosa.

From the nasopharynx, the bacterium travels locally to cause otitis media or pneumonia and through the bloodstream to cause infections at distant sites. Local non-invasive respiratory infections are lobar pneumonia, empyema, otitis media and sinusitis. The most common cause of lobar (alveolar) pneumonia is *S. pneumoniae*. Distant invasive infections are pyogenic meningitis, endocarditis, bacteremia, osteomyelitis and septic arthritis.

Laboratory diagnosis

1. Direct/primary smear examination of sputum (pneumonia), CSF (meningitis) or other body fluids by Gram stain to demonstrate lanceolate, gram-positive diplococci surrounded by a clear halo due to the capsule.
2. Detection of the polysaccharide capsular antigen in CSF by a latex agglutination test (LAT).
3. Culture and biochemical identification of the growth from any sample. Or demonstration of the presence of capsule by Quellung reaction i.e., when type-specific antiserum is added to colonies of *S. pneumoniae*, the capsule swells.
4. RT-PCR for *S. pneumoniae*-specific genes or multiplex PCR to detect bacteria causing pyogenic meningitis.
5. Inflammatory markers like CRP and procalcitonin are raised, with leucocytosis.

Ceftriaxone and vancomycin are recommended for *S. pneumoniae* meningitis. Penicillin resistance due to altered penicillin-binding protein (PBP) is on the rise, and multi-drug resistant (MDR) *S. pneumoniae* is defined as that which is resistant to three or more classes of antibiotics.

Prevention

There are two vaccines available:

1. *Pneumococcal Polysaccharide Vaccine PPSV23 (Pneumovax 23) with 23 serotypes:* It is less immunogenic in children and, hence, indicated in adults, providing immunity for 3–5 years.

   The PPV23 consists of a purified capsular polysaccharide of 23 serotypes, accounting for 80% of the invasive serotypes. Like other polysaccharide vaccines, it induces a T cell-independent immune response without immunological memory and is not immunogenic in children less than 2 years. It is, therefore, recommended only in the elderly and high-risk populations as a single dose, without a booster. It does not provide long-term protection.
2. *Pneumococcal Conjugate Vaccine PCV13 (Prevnar 13) with 13 serotypes:* It is conjugated with diphtheria toxin for enhanced immunogenicity, and is indicated in children and adults, providing longer immunity.

   In PCV, the individual polysaccharides are conjugated to a carrier protein to make it more immunogenic and induce immunological memory. Three of these, PCV7, PCV10 and PCV13, contain polysaccharides of the 7, 10 and 13 pneumococcal serotypes, respectively. The WHO recommends PCV10 or PCV13 at 6, 10 and 14 weeks, without a booster. It is not included in India's Universal Immunization Program.

   The indications for vaccination include young age (primary immunisation with PCV13 at 2, 4, 6 and 12 months) and age > 65 years (PCV13 followed by PPSV23 after 1 year). Other indications are splenectomy or splenic dysfunction, an immunocompromised state, and an immunocompetent state with cochlear implants and CSF leaks, diabetes mellitus, and chronic lung, liver or heart disease.

Group B *Streptococcus* (*Streptococcus agalactiae*)

It is a β-hemolytic, gram-positive coccus arranged in chains, that is carried in the vagina or rectum in women. From the birth passage, it is transmitted to the newborn, causing neonatal meningitis or sepsis. Antenatal screening at 36 weeks is recommended with prophylactic penicillin or ampicillin in the case of carriers, to prevent neonatal infection.

Neisseria meningitidis

Clinical Case 2

A 4-year-old boy, the youngest of a family with three children presents with sudden onset fever, headache and vomiting. On examination, the neck is stiff and there is a petechial rash on the extremities.

1. *What is the most likely causative agent?*
2. *Enumerate three other bacteria causing pyogenic meningitis.*
3. *How will you confirm your diagnosis in the laboratory?*
4. *What would you recommend to the boy's siblings?*

Epidemiology

Neisserial meningitis is seen in early childhood or adolescence and can occur sporadically, endemically or as an epidemic due to overcrowding. It is more common in sub-Saharan Africa and 5 lakh cases occur worldwide, annually, with a 10% mortality. There are 13 serogroups, of which six cause disease—A, B, C, W, X and Y, distributed geographically. India has reported sporadic cases, especially from the North, and an outbreak was reported in Delhi in 1983. Human carriers are the only source of infection that is transmitted by inhalation.

Morphology

Gram-negative diplococci (adjacent sides flat), capsulated and intracellular (Fig. 32.2).

Pathogenesis and clinical features

The polysaccharide capsule is virulent and protects the bacterium from phagocytosis. As the complement functions as an opsonin to enhance phagocytosis, complement deficiency is a risk factor for *N. meningitidis* infection. The gram-negative cell wall lipopolysaccharide (LPS) is endotoxic and binds to the host cell surface. This activates the endothelial cells and releases cytokines, leading to increased vascular permeability and intravascular thrombosis and disseminated intravascular coagulation. The distinctive petechial, blanching rash on the skin appears from the bleeding under the skin.

It is a commensal in the nasopharynx of children and adolescents, but only capsulated strains of six pathogenic serotypes cause invasive disease. From the nasopharynx, the bacterium reaches the meninges via the bloodstream or along the olfactory nerve through the cribriform plate. Both blood and meninges being sterile compartments, this results in meningococcemia and pyogenic meningitis. The Waterhouse–Friderichsen syndrome is a severe form of fulminant meningococcemia, characterised by large purpuric spots, endotoxic shock, disseminated intravascular coagulation (DIC), bilateral adrenal hemorrhage and multi-organ failure that is fatal.

Laboratory diagnosis

The CSF sample should not be refrigerated as *N. meningitidis* and *H. influenzae* die on refrigeration.

1. Gram stain of CSF showing gram-negative diplococci (adjacent sides flat), capsulated and intracellular.
2. Detection of capsular antigen in CSF by the latex agglutination test.

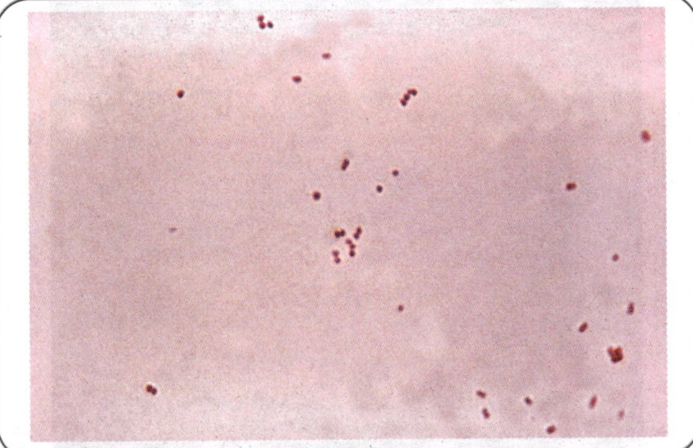

Fig. 32.2 Gram-negative cocci in pairs—*Neisseria meningitidis*, Gram stain, 100X (*Source:* Public Health Image Library, ID 6423 / CDC / Dr. Brodsky)

3. Culture of CSF and/or blood in a blood culture bottle or CSF on solid media.
4. RT-PCR or multiplex PCR on CSF-targeting specific genes.

Prevention

In the event of a local epidemic spread or sporadic cases with close contacts, single-dose ceftriaxone IM is used for chemoprophylaxis. Ceftriaxone and vancomycin are recommended for *N. meningitidis* meningitis.

The vaccines available are:

1. Meningococcal capsular polysaccharide bivalent (serogroup A and C) or quadrivalent (serogroup A, C, Y and W 135) is given in two doses 2 months apart at 3–18 months or a single dose, if older. It provides protection for 3–5 years. Serogroup B is not immunogenic and cross-reacts with neuronal molecules.
2. Conjugate vaccine is one which consists of the capsular antigen with a protein carrier to enhance immunogenicity.
3. Recombinant serogroup B antigen is given to adolescents.

Haemophilus influenzae

The genus *Haemophilus* (meaning blood loving) consists of capsulated, pleomorphic, gram-negative bacilli that require blood factor X (hemin, porphyrins) and V (nicotinamide adenine dinucleotide) for their growth. *H. influenzae* causes pneumonia and meningitis in children and *Haemophilus ducreyi* causes sexually transmitted genital ulcers.

H. influenzae is a commensal in the upper respiratory tract, and is transmitted by inhalation amongst close contacts. *H. influenzae* has six capsular serotypes, of which serotype b (Hib) is common and invasive. The capsular polysaccharide, endotoxic LPS and outer membrane proteins are virulent. The Hib capsule consists of the polyribosyl ribitol phosphate (PRP) antigen that is immunogenic and therefore used in the Hib vaccine. The annual incidence of *H. influenzae* meningitis in India is 50 in 1,00,000 children below 5 years of age.

Hib causes pyogenic meningitis in children less than 2 years of age presenting with fever, headache, vomiting and neck rigidity. It can also cause croup or epiglottitis with acute airway obstruction in older children and pneumonia in infants. Non-typeable, non-capsulated *H. influenzae* are non-invasive and locally spreading, and cause disease in adults. They cause exacerbations of chronic obstructive pulmonary disease (COPD) and pneumonia in adults and otitis media and sinusitis in children.

The sample of CSF, pus or sputum to detect *H. influenzae* should not be refrigerated. Microscopic examination of the CSF shows pleomorphic gram-negative coccobacilli and the bacterium can be cultured on fastidious media. Antigen detection in CSF by the latex agglutination test is available for rapid diagnosis. Additionally, specific capsular genes can be detected in multiplex PCR for pyogenic meningitis.

The bacteria are susceptible to ceftriaxone, quinolones and macrolides. Ceftriaxone and vancomycin are recommended for *H. influenzae* meningitis. The Hib vaccine consists of the PRP antigen and is given at 6, 10 and 14 weeks in combination with the DPT and HBV vaccines. Oral rifampicin is used for chemoprophylaxis of close contacts.

Listeria monocytogenes

This is a gram-positive, motile, intracellular bacillus that is found in the environment, i.e., in birds, animals and human carriers. It can cause disease at extremes of age, in pregnancy and in the immunocompromised. In neonates, it can cause early onset (< 5 days) sepsis acquired from the maternal birth canal or late (> 5 days) neonatal meningitis acquired from the environment. In pregnancy, it can cause maternal meningitis or sepsis, or preterm labour or abortion. It can cause sepsis or meningitis in the immunocompromised, and food poisoning (gastroenteritis) on consumption of dairy products stored under refrigeration. The infection is diagnosed by demonstrating the bacterium in CSF or amniotic fluid by Gram stain or culture or by blood culture. The bacterium is sensitive to ampicillin and cotrimoxazole.

Other bacteria that can cause meningitis are *E. coli*, *M. tuberculosis*, *T. pallidum* (STD) and the zoonotic *Leptospira interrogans*. Rarely do *Brucella*, *Mycoplasma pneumoniae*, *B. burgdorferi*, *Salmonella typhi*, *Nocardia actinomyces* and *Legionella* cause meningitis.

Neurosyphilis

The neurological involvement in early syphilis is meningitis, and neurosyphilis in tertiary syphilis, especially in the HIV-infected. Tertiary syphilis presents as meningovascular syphilis, tabes dorsalis and general paresis. CSF leucocyte count, CSF VDRL, CSF FTA and PCR are diagnostic.

VIRAL OR ASEPTIC MENINGITIS

Viruses usually cause meningoencephalitis, predominantly lymphocytic, which is milder than bacterial meningitis. Acute viral neurological disease can be aseptic meningitis, poliomyelitis, encephalitis or post-infectious encephalomyelitis due to host immune response. Aseptic meningitis shows a predominance of lymphocytes in the CSF, and is caused by enteroviruses (echovirus and coxsackievirus) or is secondary to mumps, measles, chickenpox or EBV infection.

Picornaviruses are grouped as those that are transmitted feco-orally (enteroviruses) or by inhalation (rhinoviruses) causing common cold. Enteroviruses multiply in the intestine but cause disease elsewhere—poliovirus causing poliomyelitis, and coxsackievirus and echovirus causing aseptic meningitis.

Coxsackieviruses group A can also cause hand foot mouth disease, vesicular pharyngitis, and acute hemorrhagic conjunctivitis. Group B is associated with myocarditis, pericarditis, pneumonia, pleurodynia, pancreatitis and hepatitis.

Echoviruses (enteric cytopathogenic human orphan viruses) are associated with aseptic meningitis, encephalitis, conjunctivitis and common colds.

CRYPTOCOCCAL MENINGITIS

C. neoformans, a capsulated true yeast, causes meningitis in HIV-AIDS and other immunocompromised patients. It is found in pigeon and other bird droppings and infects by inhalation of the yeast cells or its reproductive spores.

Morphologically, it a spherical yeast cell measuring around 6 µm, surrounded by a polysaccharide capsule of 15–30 µm thickness (Fig. 32.3).

On inhalation, the infection begins in the lungs and is disseminated through the bloodstream to the CNS where it crosses the blood–brain barrier to cause meningitis.

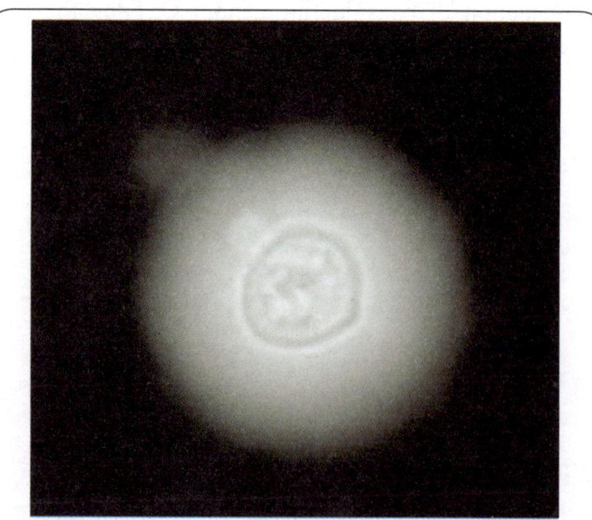

Fig. 32.3 *Cryptococcus*—India ink staining showing clear halo of capsule surrounding spherical budding yeast cells

It can also cause pulmonary, skin or bone lesions. The capsule and other enzymes are its virulence factors.

The disease is diagnosed in the laboratory by Gram staining of the yeast cells or demonstrating the presence of capsule by negative staining (nigrosin or India ink). A rapid LAT detects the capsular antigen in CSF or serum. Histopathologic examination of brain tissue with mucicarmine stains the cell wall. It can be grown on culture on SDA in 7 days.

Invasive candidiasis can involve the meninges, causing meningitis that can be diagnosed by demonstration of the candidal yeasts cells by microscopy or culture.

TUBERCULOUS MENINGITIS

Tuberculous meningitis is one of the extrapulmonary manifestations of *M. tuberculosis* infection. On inhalation of MTB bacilli, localised infection in the lungs is seeded to the brain parenchyma, resulting in metastatic caseous lesions that rupture into the subarachnoid space causing meningitis. The CSF examination shows high protein, low glucose and lymphocytes. MTB can be demonstrated by CBNAAT or MGIT culture with greater sensitivity than microscopic examination of centrifuged CSF stained by ZN stain.

PRIMARY AMOEBIC MENINGOENCEPHALITIS

There exist amoebae that are not strictly parasitic in human tissue and can be found in the environment. They are known rarely to infect humans and cause meningoencephalitis. *Naegleria fowleri* and *Acanthamoeba* spp. are aquatic amoebae that contaminate water resources or are airborne. Though rare, cases of *Naegleria* and *Acanthamoeba* infection have been reported in India.

N. fowleri exists in two morphologic forms—motile trophozoite and non-motile cyst. The trophozoite form, in turn, has two morphologic forms—the amoeboid form, due to the presence of pseudopodia and the flagellate form, due to the presence of two flagellae. The amoeboid form is seen in human tissue and multiplies by binary fission while the flagellate form is found in water. Encystation of the amoeboid form into cysts occurs in human tissue.

The flagellate forms are thermotolerant, destroyed by salination, and found in freshwater lakes contaminated by factory effluents that increase the water temperature. While swimming in such waters, they enter the nasal cavity. Rarely are they airborne. They are transformed to amoeboid forms that invade the brain via the olfactory neuroepithelium and tissue. The disease is rapidly progressive, with a weak immune response and is usually fatal.

The lesions are exudative in nature and restricted to the olfactory bulbs, maxillary and sphenoid sinuses, meninges and brain, rarely affecting the heart. The condition is diagnosed by demonstrating the cysts and motile amoeboid trophozoites in CSF examined as a wet mount or stained by Wright's stain, or by histopathological examination of tissue biopsy.

Acanthamoeba has four pathogenic species and exists in two morphologic forms—the motile amoeboid trophozoite and the non-motile cyst. Unlike *Naegleria*, there does not exist any flagellate form. The infective form is the cyst that is predominantly airborne in humidifiers and in dialysers and gains entry through skin and mucosal surfaces. The excystation occurs in tissue, forming amoeboid trophozoites.

On entering via the nasal mucosa or eye or lungs or genitourinary tract, tissue invasion is slow, with a prominent immune response. It results in granulomatous lesions of secondary meningoencephalitis with a gradual onset and chronic course. Besides the CNS, it can cause secondary chronic granulomatous lesions in the skin, kidney, liver, spleen, uterus, prostate and eye. The condition is diagnosed by demonstrating the cysts and motile amoeboid trophozoites in CSF examined as a wet mount or stained by Wright's stain, or by histopathological examination of diseased tissue biopsy.

EOSINOPHILIC MENINGITIS

Eosinophilic meningitis is defined as the presence of > 10 eosinophils/μL in the CSF or 10% eosinophils in the total CSF leucocyte count. Infective causes of eosinophilic meningitis suggest nematodes *Parastrongylus cantonensis* or *Gnanthostoma* species disease. Both are zoonotic infections wherein man is an accidental host.

Parastrongylus infection is acquired by ingestion of food contaminated with the larval form from rats and is diagnosed by demonstrating the presence of the larva or adult in the CSF or detecting specific antibodies in serum by ELISA. *Gnathostoma* spp. is acquired by ingestion of food contaminated with the larval form from cats or dogs and is diagnosed by demonstrating the presence of the larva in CSF but in man it does not develop into an adult.

POLIOMYELITIS

Virology

Poliovirus is a small, spherical, ssRNA virus with a capsid. It possesses a type-specific D (dense) antigen with a type-specific antibody response. There are three serotypes—types 1, 2 and 3 of poliovirus.

Epidemiology

It is possible to eradicate polio with an effective vaccine because man is the only reservoir and there are no human carriers. Children are usually affected and patients alone are infective from a week before symptoms appear till 3 weeks later. In India, 28 cases were reported in 2004 and the WHO declared us to be polio-free in 2014, the last case being reported in 2011. Surveillance for Acute Flaccid Paralysis is ongoing to ensure detection of cases, if any.

Pathogenesis and clinical features

It is transmitted feco-orally and multiplies in the mucosa-associated lymphoid tissue of the intestine and pharynx. This is followed by a primary hematogenous viremia, seeding the virus into the mononuclear phagocytic system (MPS). After multiplication in the MPS, the secondary hematogenous viremia leads to infection of the anterior horn cells of the spinal cord. The virus-infected neurons degenerate, causing muscle weakness and acute flaccid paralysis (AFP). The incubation period is 10 days and most cases of infection are asymptomatic or non-paralytic. Paralytic poliomyelitis occurs in 1% of cases, beginning with the hip muscles and progressing distally in the lower limb, without any sensory loss.

A rare form of poliomyelitis is that occurring post tonsillectomy, where the virus multiplies in the tonsillar lymphoid tissue and spreads via the glossopharyngeal nerve, causing bulbar poliomyelitis. Bulbar poliomyelitis results from paralysis of muscle groups innervated by the cranial nerves, especially those of the soft palate and pharynx, which may present as dysphagia, nasal speech, and sometimes dyspnea.

Laboratory diagnosis

1. In the case of AFP, two stool samples weighing 10 g each, collected 24 hours apart, within 14 days of the onset of paralysis need to be sent to a WHO-accredited laboratory through district health centres, for the isolation of poliovirus. If the case is clinically suspected but the virus is not isolated, it is considered as a 'compatible' case of polio and reviewed by experts for 60 days and subsequently decided as 'confirmed' or 'discarded' as polio.
2. Demonstrating raised titres of antibodies in paired serum.
3. Multiplex RT-PCR to detect and differentiate disease and vaccine poliovirus strains in stool.

Vaccines

There are two polio vaccines:

1. *Injectable polio vaccine (IPV) or Salk vaccine, named after Jonas Salk.* It is a formalin-killed whole virus given intradermally at 6 and 14 weeks along with OPV. It confers immunity for 5 years. It neither causes vaccine-associated paralytic polio (VAPP) nor does it provide herd immunity.
2. *Oral polio vaccine (OPV) or Sabin vaccine, named after Albert Sabin.* It is a live attenuated vaccine available as a trivalent, bivalent and monovalent preparation. Presently bivalent (1 and 3), two drops are given orally in five doses—at birth and 6, 10 and 14 weeks with IPV at 6 and 14 weeks, and a booster between 16 and 24 months. It is easy to administer, confers lifelong immunity and contributes to herd immunity. Being a live vaccine, it requires stringent storage and transport conditions and can be eliminated by diarrhea in the child or by breastfeeding before or after the vaccination.

Being a live vaccine, OPV can rarely cause VAPP and vaccine-derived poliomyelitis viruses (VDPVs). VAPP strains are genetically similar to OPV strains, while VDVPs are divergent and similar to the wild polioviruses.

Globally, the polio immunisation program is targeted to eradicate the disease. India has been polio-free since March 2014, due to the Pulse Polio Immunisation (PPI) since 1995. PPI included two additional doses of OPV 6 weeks apart to all children under age 5 years.

Answers

Clinical Case 1

1. *Acute pyogenic meningitis, Streptococcus pneumoniae*
2. *Neisseria meningitidis, Haemophilus influenzae, Streptococcus agalactiae*
3. *Nasopharynx of the patient*
4. *Penicillin G*

Clinical Case 2

1. *Neisseria meningitidis*
2. *Streptococcus pneumoniae, Haemophilus influenzae, Streptococcus agalactiae*
3. *Gram stain of CSF showing gram-negative diplococci, capsular antigen detection by latex agglutination test*
4. *Single-dose ceftriaxone and meningococcal vaccine*

QUESTIONS

1. Enumerate four bacteria most likely to cause pyogenic meningitis and outline their pathogenesis.
2. Enumerate four bacteria most likely to cause meningitis and outline the laboratory diagnosis of pyogenic meningitis.
3. Write a brief note on aseptic meningitis and its causative agents.
4. Name the microbes most likely to cause meningitis in the immunocompromised host. In this case how will you confirm the causative agent in the laboratory?
5. Compare and contrast, in brief, the pathogenic features of protozoans causing primary amoebic meningoencephalitis.
6. Define eosinophilic meningitis. State its cause and source of infection.
7. Explain the pathogenesis and immunoprophylaxis of poliomyelitis, in the context of its global elimination.

Encephalitis

Acute encephalitis is rare. It features a fever, a headache, an altered sensorium, and may result in seizures or coma. It is mostly caused by viruses—HSV, arboviral Japanese B encephalitis (JE) virus, rabies virus, Nipah virus, enteroviruses, CMV, VZ, HHV-6 and HIV. Its non-viral causes are *Listeria*, *Rickettsia*, *Mycoplasma* and *Toxoplasma*.

VIRAL ENCEPHALITIS

HSV-1 and HSV-2

Both HSV-1 and HSV-2 can cause acute viral encephalitis and meningitis. The primary infection is exogenous as a result of oropharyngeal secretions in HSV-1 or it could be due to a secondary reactivation of the virus. HSV-2 is more likely to cause encephalitis in neonatal herpes that can be prevented by administering acyclovir to the mother. Laboratory diagnosis is by detecting viral antigen by IF, HSV-DNA by RT-PCR or specific antibodies by ELISA.

Japanese B encephalitis (JE)

Arbovirus JE is a zoonotic infection found in the reservoir of pigs and birds. Japanese B encephalitis is transmitted by the mosquito *Culex tritaeniorhynchus* or *Culex vishnui*. It presents with fever accompanied by disorientation, delirium or coma. Humans are infected by a mosquito bite. After an incubation period of 5–15 days, the patient develops fever and an altered sensorium. This infection is more common in children. The cases may recover with residual neuro-deficit, with the case fatality being 20–40%.

The public health importance of JE is because it has epidemic potential, a high case fatality rate, permanent sequelae and no treatment. Infection is preventable by vector control. In India, the incidence is low—1 in 1,000 mosquito-bitten persons, implying that there is subclinical infection. It is endemic in UP (Gorakhpur district), Assam, Manipur, TN, Tripura, Bihar, Odisha and WB. In 2011, 6,297 cases were reported in India with 861 deaths. It is worth noting that 75% of these cases were from Uttar Pradesh.

JE can be diagnosed in the laboratory by detecting IgM in serum by ELISA or the viral envelope E gene in blood by RT-PCR. A live vaccine is given to children between 1 and 15 years of age in the endemic districts of UP, WB and Assam. It is a single dose given subcutaneously followed by a booster dose given after 1 year. An inactivated vaccine of Beijing P3 strain, prepared at the Central Research Institute, Kasauli, India, is also available.

Rabies

Clinical Case

A 10-year-old boy is admitted to the emergency service with hallucinations, clouding of consciousness, muscle spasms, seizures and delirium. According to the history provided by the patient's family, he had had a dog bite on his right ankle 2 weeks back that had been dressed locally.

1. *What is the most likely cause of his condition?*
2. *What immediate steps will you take?*
3. *Where in the brain is the virus predominantly found?*
4. *Describe its characteristic histopathologic feature.*

Epidemiology

Rabies is a lethal zoonotic encephalitis caused by the bite of a rabid animal. In India, it is largely caused by the bite of dogs and rarely cats or other canines. Other than a dog bite, it is rarely transmitted by a dog licking a wound or through mucosal contact, inhalation of infected bat aerosols or a corneal transplant. It is endemic worldwide, including India, except in the Andaman, Nicobar and Lakshadweep islands. It is a fatal disease accounting for 30,000 annual deaths in India, and 3 million cases receiving anti-rabic treatment.

The virus is present in the saliva of the infected dog for up to 4 days before the onset of symptoms. Dogs that remain healthy for 10 days after biting can be regarded as rabies free. It is about 15% of people bitten by a rabid dog who develop the disease.

Virology

The rabies virus is a large, bullet-shaped, enveloped ssRNA virus. The ssRNA is helical while the envelope is lipid with glycoprotein spikes (Fig. 33.1). It has only one serotype. The virus present in susceptible animals—canines, cats and bats—is referred to as a 'wild' virus. When the rabies virus is freshly isolated in the laboratory, it resembles the wild virus and is referred to as **'street' virus**. On undergoing serial passage in culture, the street virus is rendered non-pathogenic and is referred to as **'fixed' virus** that is used in vaccines.

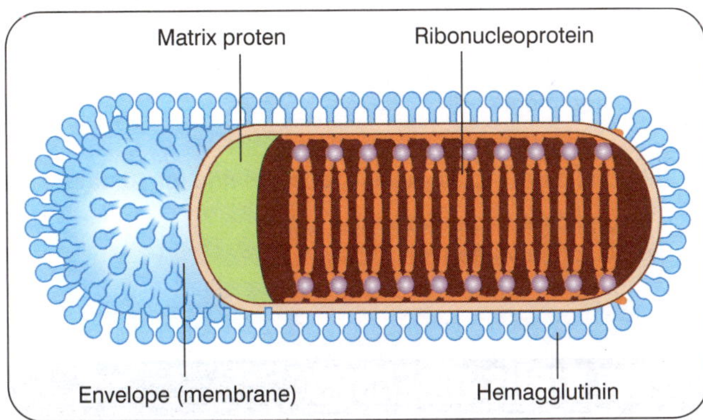

Fig. 33.1 Structure of rabies virus

Pathogenesis and clinical features

On the bite of an infected dog, the virus multiplies locally at the site of the bite. The outer glycoprotein of the virus binds to the acetyl choline receptors at the neuromuscular junctions and enters the peripheral neurons. On entering the neurons, it travels rapidly centripetally via the axons to reach the dorsal root ganglion of the spinal cord and ascends to the brain, the most preferred sites being the hippocampus and cerebellum. From the brain, it then travels centrifugally along the autonomic and sensory nerves to reach the salivary glands, skin and cornea.

The entire course of viral infection and disease being restricted to the nervous system, the virus does not enter or multiply in the blood. As in the dog, the virus is shed in human saliva too but man-to-man transmission has not yet been documented.

The intracellular multiplication of the virus results in the presence of intracytoplasmic, eosinophilic inclusions bodies in the brain, predominantly in the hippocampus and cerebellum, called Negri bodies.

The incubation period of rabies is directly proportional to the site of the bite and the viral load delivered at the bite. The closer the site of the bite to the brain, the lesser the travel distance and shorter the incubation period. With bites on the head, neck, upper limbs and lower limbs, the incubation period varies from 4 to 12 weeks. Severely lacerated bite injuries have a larger viral load than minor bite injuries.

If not vaccinated, the disease presents with non-specific prodromal symptoms of paraesthesia in the wound followed by the classical 'furious' rabies symptoms of hydrophobia or aerophobia, hyperactivity and excessive salivation in most cases. The act of swallowing causes spasm of the respiratory, laryngeal and pharyngeal muscles. In some cases, paralytic or 'dumb' rabies presents as flaccid paralysis of the limbs and face in the absence of other classical symptoms. In both presentations, the patient eventually progresses to coma and death within 2 to 4 weeks of the onset of symptoms.

Laboratory diagnosis

1. Viral antigen can be detected by immunofluorescence in hair-bearing skin (back of neck), or corneal impression smears.
2. Antibody detection in CSF and serum, preferably CSF, as serum antibodies may be seen following immunisation.
3. RT-PCR to detect viral RNA from saliva or skin biopsy.
4. Postmortem demonstration of Negri bodies in histopathologic examination of brain biopsy by H&E or Seller's stain shows intracytoplasmic, eosinophilic inclusion bodies measuring 5 to 10 µm (Fig. 33.2).

Current practice of post-exposure prophylaxis (PEP) in rabies

Rabies being fatal, PEP in suspected dog bites is of utmost importance to save lives. Local wound care is most important followed by vaccination and anti-rabies serum if required.

Local wound care includes washing thoroughly with soap and water and the use of antiseptics with gloved hands to decrease the viral contamination or load in the wound. Suturing is contraindicated. The WHO risk categorisation and PEP are shown in Table 33.1.

Rabies immunoprophylaxis

1. Cell culture vaccines: Human diploid cell strain vaccine (HDCS), purified chick embryo cell culture vaccine (PCEC), and purified Vero cell vaccine (PVC) are used. These vaccines are highly immunogenic and not associated with neurological complications, and are widely used nowadays.
2. Subunit vaccines: These are vaccinia virus recombinants expressing rabies virus glycoproteins.
3. Wash the wound with soap and running tap water immediately as this reduces the chances of the virus entering the tissue.

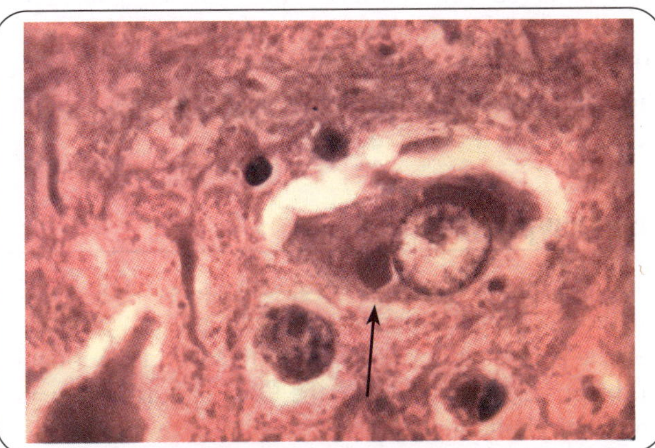

Fig. 33.2 Negri bodies in brain biopsy, Seller's stain, 100X

Table 33.1 The WHO risk categorisation and PEP in rabies

Risk category	Type of exposure	Recommended PEP
Category I (No risk)	Touching/feeding animal Licks on intact skin	No treatment
Category II	Minor scratches or abrasions without bleeding	Wound care, rabies vaccine, observe dog for 10 days*
Category III	Transdermal bites, bleeding Licks on non-intact skin	Wound care, rabies vaccine HRIG, observe dog for 10 days*

*To be discontinued if dog is healthy after 10 days

Active immunisation is with HDCS/PCEC/PVC. The schedule (for any vaccine) is six doses at 0, 3, 7, 14, 30 and 90 (optional) days for 5-year protection. The vaccine is given IM/SC on the deltoid (adults) and anterolateral aspect of the thigh (children).

Passive immunisation is with human rabies immunoglobulin (HRIG) 20 IU/kg. Half is to be injected at the site of the wound and the other half IM. It is to be given simultaneously with or before the vaccine but not after the vaccine.

Nipah virus (NiV)

Three outbreaks of Nipah virus encephalitis have been reported in India—from Siliguri in 2001, Nadia in WB in 2007, and Kozhikode, Kerala in 2018. Bats residing on palm trees are the reservoir of infection. Inhalation of aerosols or ingestion of bat saliva or the dropping of contaminated fruit or palm sap is considered to be the mode of transmission. Laboratory diagnosis is by RT-PCR to detect viral RNA from CSF, throat/nasal swab, urine and blood in the early stages and IgG and IgM antibodies by ELISA in the later stages.

Cerebral malaria

It is the most severe complication of *Plasmodium falciparum* malaria with high mortality. It is associated with long-term neuro-cognitive impairments and presents with impaired consciousness, even coma, that can be rapidly reversed with effective antimalarials. Parasitised RBCs get sequestered in the cerebral microcirculation as they adhere to the endothelium, causing inadequate tissue perfusion and hypoxia to neurons.

Neurosyphilis

Neurosyphilis and cardiovascular syphilis, referred to as quaternary syphilis, manifest as meningitis or vasculitis or general paresis of insane or tabes dorsalis (demyelination of the posterior columns) in the CNS or aortic aneurysm and regurgitation in the CVS.

SLOW VIRUS DISEASES

These neurodegenerative viral diseases have a long incubation period and a protracted slow development of symptoms but are eventually fatal. Slow virus disease can be due to conventional or unconventional agents (Table 33.2).

SSPE is a rare, slow, progressive demyelination of the CNS caused by a defective measles virus, diagnosed by high titre anti-measles antibodies in the CSF. It is a rare complication of measles after 1 week of disease, occurring in 1 in 1,000 cases.

PML is demyelination of the CNS only in the immunocompromised, caused by the reactivation of JC virus infecting the brain. It is usually fatal in 3–4 months. Patients with primary or secondary immunodeficiency can develop PML by activation of a polyoma virus called CJV. CJV is usually acquired early in childhood, when it causes a self-limiting disease. The virus then becomes latent within the CNS. Any immunodeficiency reactivates the virus.

Slow virus disease due to an unconventional agent or prion disease in humans is Creutzfeldt-Jakob disease, showing subacute degeneration of the brain and spinal cord with spongiform encephalopathy. Prions are infectious proteins that lack nucleic acids, and are filterable and resistant to disinfection and sterilisation. Once they infect the neurons, they cause spongiform encephalopathy. Prion diseases exist in animals, e.g., scrapie in sheep, minks, mad cow disease. Diagnosis in humans is by brain MRI or brain biopsy, or detecting the prion protein by immunoassay.

Table 33.2 Slow virus disease due to conventional agents

Virus	Disease
Measles	Subacute sclerosing panencephalitis (SSPE)
Rubella	SSPE following congenital infection
Human polyoma virus (JC virus)	Progressive multi-focal leucoencephalopathy (PML)

Answers

Clinical Case

1. *Rabies*
2. *Wash and decontaminate the wound thoroughly. Give human rabies immunoglobulin (HRIG) 20 IU/kg, half injected at the site of the wound and the other half IM with simultaneous vaccination.*
3. *Hippocampus and cerebellum*
4. *H&E or Seller's stain shows intracytoplasmic, eosinophilic, inclusion bodies measuring 5 to 10 μm.*

Questions

1. Explain the pathogenesis of rabies. How will you prevent rabies following exposure to a dog bite?
2. Enumerate zoonotic viral encephalopathies. Write a note on Japanese B encephalitis.
3. Write a note on slow virus disease.

Brain Abscess and Cyst

Focal CNS infective lesions include the brain abscess and cyst caused by bacteria, fungi *Mucorales* or parasites causing toxoplasmosis or neurocysticercosis.

BRAIN ABSCESS

It is, at times, a sequela to congenital heart disease (the source being infected valves from endocarditis) or chronic pharyngeal infections (middle ear, mastoid, sinus infections). It may be located in the brain parenchyma, subdural or epidural space.

An acute abscess is pyogenic or bacterial, usually caused by mixed bacteria that are facultative and strictly anaerobic. A chronic abscess is usually due to *Mycobacterium tuberculosis*, *Cryptococcus* or other fungi. An abscess following trauma or surgery (post-operative) is caused by gram-negative bacilli and *Staphylococcus aureus*. Microscopic examination and culture of the aspirate confirm the laboratory diagnosis.

RHINOCEREBRAL MUCORMYCOSIS

On inhalation of the spores of mucormycetes from the environment, in patients of diabetic ketoacidosis, a rapidly progressive, fatal disease occurs, affecting the brain. The spores develop into mycelia that are broad, aseptate and thin-walled, showing obtuse branching. They are angio-invasive and rapidly invade the orbit, paranasal sinuses and brain. Besides the brain, mucormycosis can also affect the lung, skin or intestines.

Laboratory diagnosis is by histopathologic examination of tissue by Gomori's methenamine silver (GMS) stain and culture on SDA. Mucormycosis (*Mucor, Rhizopus* and *Absidia*) grows rapidly in 24 hours and the LPCB mount shows broad, aseptate and thin-walled hyphae, with obtuse branching and sporangia (Fig. 34.1)

BRAIN CYSTS

The parasites causing cysts in the brain are *Toxoplasma gondii* and *Taenia solium* (neurocysticercosis), and rarely *Echinococcus granulosus* (hydatid disease).

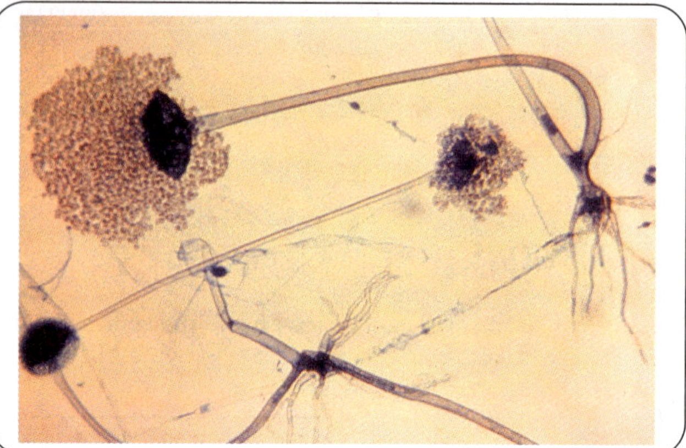

Fig. 34.1 LPCB mount of *Rhizopus*, 40X (*Source*: Public Health Image Library, ID 17589 / Centers for Disease Control and Prevention [CDC] / Dr. Hardin)

Toxoplasmosis

Toxoplasma gondii is an intracellular protozoan that causes asymptomatic infection in the healthy and disease in the immunocompromised. This occurs worldwide.

Cats, contaminated soil or water or meat (chicken, lamb, mutton, pork) are the reservoirs of infection. It is transmitted by ingestion or from mother to child, and rarely by blood transfusion and organ transplantation. Man-to-man transmission does not occur.

Morphologic forms—Tachyzoite, bradyzoite, oocyst

Tachyzoites are oval or crescent-shaped uninucleate cells that measure around 7 µm (Fig. 34.2). They are found in

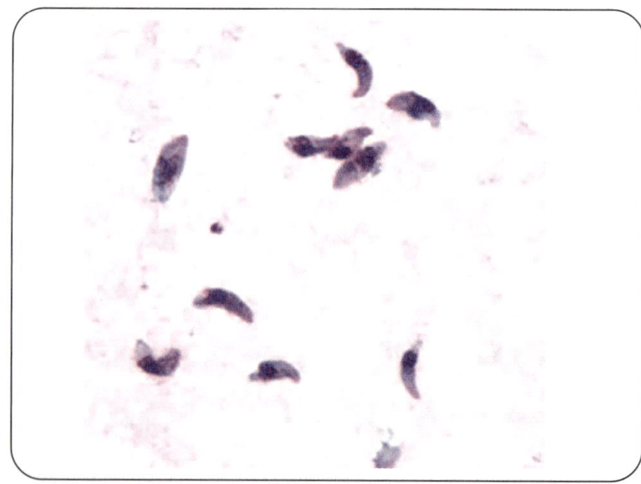

Fig. 34.2 Tachyzoites of *Toxoplasma gondii*, Giemsa stain, 100X (*Source*: CDC, DPDx)

the acute stage of infection. Tachyzoites multiply rapidly and invade all cells except RBCs.

Bradyzoites measure around 5 µm, resemble tachyzoites but multiply slowly. They are found within the true cysts that measure 10–50 µm.

Oocysts are not seen in man. They are seen in the intestinal epithelium of cats and measure around 10 μm. The infectious oocyst has two sporocysts with four sporozoites in each.

Reproduction

T. gondii reproduces asexually by endodyogeny in man, the intermediate host, and by schizogony in cats. Sexual reproduction of gametocytes by sporogony occurs extracellularly only in cats, the definitive host. Endodyogeny is seen in tachyzoites and bradyzoites in man wherein two progenies are formed within the parent cell and released on death of the parent cell. Schizogony is intracellular wherein one cell divides into more than two progenies.

Life cycle

Infection occurs either by consuming an infective oocyst from a cat or by consuming bradyzoite in a tissue cyst from infected meat (Fig. 34.3). Sporozoites liberated from the oocyst penetrate the ileal mucosa and proliferate in the lymphoid macrophages as tachyzoites, spreading via blood and the lymphatics.

In the immunocompetent host, 2 weeks after infection, 'pseudocysts' of tachyzoites are formed in multiple tissues with inflammation. They can be seen in the liver, spleen, lungs, heart, brain, lymph node, eye or skeletal muscle. Subsequently, the infection forms 'true cysts' of bradyzoites in the brain and skeletal muscle, without inflammation. In pregnancy, the tachyzoites can cross the placenta to infect the fetus causing congenital toxoplasmosis.

In the immunocompromised, these bradyzoites are reactivated to cause encephalitis, pneumonitis and myocarditis. The cycle has a dead end in man as there is no man-to-man or man-to-cat transmission.

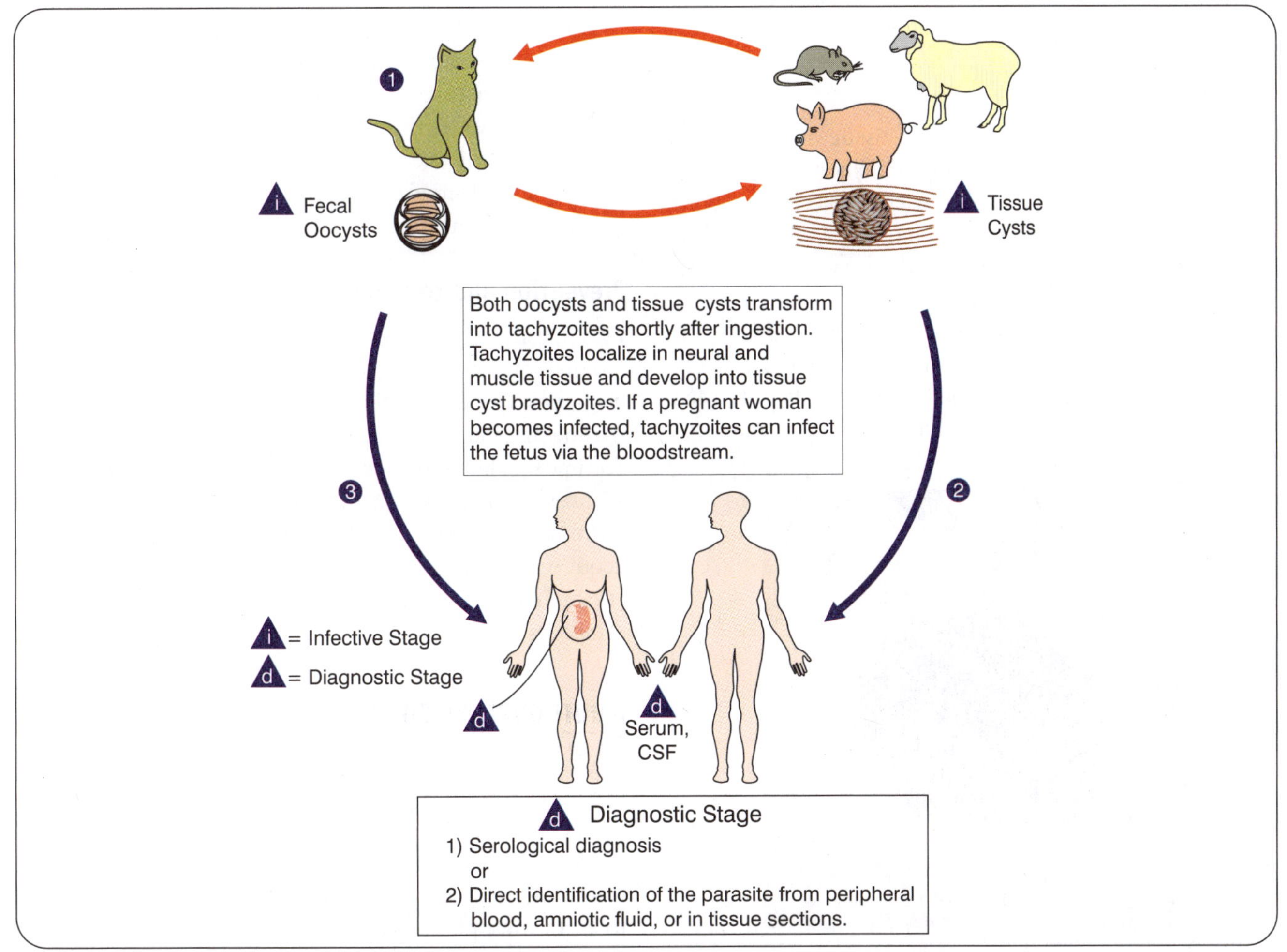

Fig. 34.3 Life cycle of *Toxoplasma gondii* (*Source:* Adapted from CDC's Public Health Image Library, at https://phil.cdc.gov/Details.aspx?pid=3421)

Pathogenesis and clinical features

Acute acquired toxoplasmosis in the immunocompetent is largely asymptomatic but may show transient, self-limiting, cervical, non-tender lymphadenopathy. In the immunocompromised, acute acquired toxoplasmosis may be due to newly acquired infection or reactivation of the latent infection. The spread of tachyzoites causes meningoencephalitis, cerebral cysts, pneumonitis and myocarditis.

Infection during pregnancy may be asymptomatic in 10% of cases, but the fetus is at risk for transmission in utero or during vaginal delivery. Transmission occurs if the mother acquires infection during pregnancy and most often results in an uninfected infant (two-third of cases), but may cause spontaneous abortion or an infected preterm or term infant (one-third of cases).

Congenital toxoplasmosis is the least common (17%) in the first trimester but the disease is more severe. Conversely, it is more common in the third trimester (65%) but is usually asymptomatic. The classical triad of congenital toxoplasmosis is hydrocephalus, cerebral calcification and chorioretinitis. It does not cause any fetal malformation as it is not teratogenic.

Laboratory diagnosis

1. Demonstration of tachyzoites in tissue or body fluids in acute toxoplasmosis. Samples include either biopsy of lymph node, skeletal muscle, spleen, or bone marrow, or body fluids like blood, cord blood, CSF, aqueous humor or amniotic fluid by Romanowsky stains.
2. Demonstration of tissue cysts (Fig. 34.4) by histopathological examination using periodic acid–Schiff (PAS), Gomori's methenamine silver (GMS) or immunofluorescent antibody test (IFAT) staining.

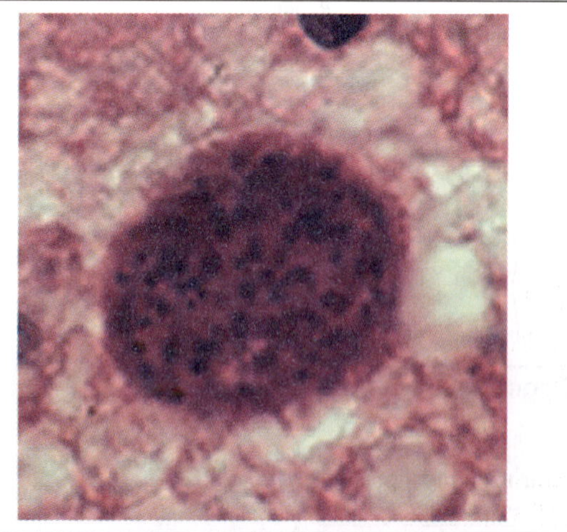

Fig. 34.4 Cyst of *Toxoplasma gondii*, H&E, 40X (*Source*: CDC, DPDx)

3. Detection of IgM or IgA or rise in the titre of IgG. Combined testing of *Toxoplasma*, rubella, CMV and HSV (TORCH) for antenatal screening in cases with a bad obstetric history.
4. Molecular test — PCR to detect the *Toxoplasma* B1 gene is used in CSF and amniotic fluid.

The above tests can be utilised as follows:

1. Acute acquired toxoplasmosis in the immunocompetent — A negative IgG test excludes infection with *Toxoplasma*. One single high titre is suggestive of infection, but not diagnostic. Diagnosis is possible by demonstrating seroconversion from negative to positive or a two-fold rise in titre at 3-week intervals of testing.
2. Acute acquired toxoplasmosis in the immunocompromised is done either by detecting the antigen or by PCR as antibody response is not reliable.
3. During pregnancy, it is important to determine the probable time of infection by routine screening either before pregnancy or at the first ANC visit. Subsequently, a positive IgM test or a two-fold rise in IgG, or the presence of the *Toxoplasma* B1 gene in the amniotic fluid in cases with a gestational age of 18 weeks or more signifies maternal infection.
4. Congenital toxoplasmosis is confirmed if IgM is found in blood or CSF 7 to 10 days after birth or by the presence of the *Toxoplasma* B1 gene. The passive IgG level decreases over 6–12 months after birth, and active IgG appears by 3 months, if untreated.

Prevention and treatment

Prevention is by good hygienic practices, especially among cat pet owners, and by ensuring consumption of well-cooked meat. Antimicrobials used for treatment are pyrimethamine, folinic acid and sulfadiazine/clindamycin/cotrimoxazole/azithromycin. Treatment during pregnancy reduces the risk of transmission, but therapeutic medical termination of pregnancy (MTP) is indicated when women acquire infection in the first trimester. All neonates with congenital toxoplasmosis need to be treated, whether symptomatic or asymptomatic.

Neurocysticercosis

Adult *T. solium* resides in the intestine and may or may not give rise to symptoms. Occasionally, man serves as the host for *T. solium* larvae, by consuming water or vegetables contaminated with *T. solium* ova. These larvae migrate from the intestine to any organ or tissue to form multiple, localised cysts. The common sites are skeletal muscle, subcutaneous tissue, brain and the eye. In neurocysticercosis, the brain shows multiple cysts.

A mature cyst measures 5 X 10 mm. It is filled with an albuminous fluid and has a dense white spot at the point where the scolex remains invaginated. Eventually, the cyst gets calcified over 5–6 years. The symptoms depend on the site of localisation of the cyst especially if it is in the brain or eye.

The diagnosis of cerebral cysticercosis is based on radio-imaging and confirmed by detection of specific antibodies in the CSF or histopathologic examination of a brain biopsy (Fig. 34.5).

Hydatid cyst

This is caused by *E. granulosus* and is usually seen in the liver and rarely in the lung or brain. It is solitary, unilocular, bacteriologically sterile, and consists of hydatid fluid enclosed in a brood capsule. The brood capsule shows the presence of protoscolices of the parasite. Diagnosis is based on radio-imaging.

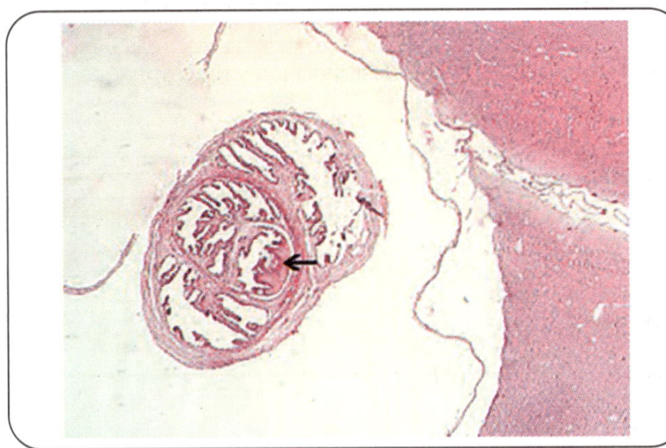

Fig. 34.5 Histopathologic appearance of cysticercus cellulosae, H&E stain, 40X (*Source*: CDC, DPDx)

Questions

1. Explain the pathogenesis and laboratory diagnosis of rhinocerebral mucormycosis.
2. Enumerate microbes causing focal infective lesions in the brain with a brief note on brain abscess.
3. Explain the pathogenesis and laboratory diagnosis of antenatal and congenital toxoplasmosis.
4. Explain the pathogenesis and laboratory diagnosis of neurocysticercosis.

SECTION SUMMARY

Section VI: Central Nervous System Infections

Infection/pathogen	Virulence/pathogenesis	Laboratory diagnosis	Vaccine/recommended antimicrobial
Meningitis Pyogenic			
Streptococcus pneumoniae	Capsule, exotoxins, adhesion, URT	Smear, culture, antigen detection, PCR, CRP, procalcitonin	Pneumococcal PolySaccharide Vaccine PPSV23 Pneumococcal Conjugate Vaccine PCV13, penicillin, amoxicillin, ceftriaxone vancomycin
Neisseria meningitidis	Capsule, endotoxic LPS, URT	Smear, culture, antigen detection, PCR, CRP, procalcitonin	Meningococcal capsular polysaccharide bivalent or quadrivalent Ceftriaxone
Haemophilus	Capsule, endotoxic LPS and OMPs, URT	Smear, culture, antigen detection, PCR, CRP, procalcitonin	Hib vaccine with PRP antigen Ceftriaxone
GB streptococci	Colonises cervix, neonatal meningitis	Smear, culture	Penicillin, ampicillin

Infection/pathogen	Virulence/pathogenesis	Laboratory diagnosis	Vaccine/recommended antimicrobial
Listeria monocytogenes	Early onset neonatal meningitis	Smear, culture	Ampicillin, cotrimoxazole
Tuberculous	Resist killing by macrophages	Smear ZN/AO, MGIT culture, CBNAAT	HRZE (INH, rifampicin, pyrazinamide, ethambutol)
Viral/Aseptic		Routine CSF examination	
Cryptococcal	Capsule, inhalation to meninges	Smear, India ink, culture, antigen detection	Antifungal
Primary amoebic *Naegleria* *Acanthamoeba*		Microscopy	
Eosinophilic *Parastrongylus*		> 10 eosinophils/μL in CSF Microscopy	
Poliomyelitis	Feco-oral, MALT, viremia, MPS, viremia, anterior horn cells, acute flaccid paralysis	Stool RT-PCR Serum antibodies	Injectable Polio Vaccine (IPV) or Salk-killed Oral Polio Vaccine (OPV) or Sabin-live
Encephalitis HSV-1 and HSV- 2	Secondary reactivation	Detect viral antigen by IF or HSV DNA by RT-PCR or specific antibodies by ELISA	
Japanese B	Zoonotic, reservoir pigs and birds, bite of *Culex*	IgM by ELISA or the viral envelope E gene in blood by RT-PCR	Killed Beijing P3 strain Live vaccine
Rabies	Zoonotic, canine bite, peripheral neurons-hippocampus, cerebellum-salivary glands, cornea	Antigen by IMF in cornea, CSF antibodies, post-mortem HPE-Negri bodies	Wash wound, active HDCS/PCEC/PVC killed cell cultures. Passive HRIG
Nipah virus	Zoonotic–bats, aerosols	RT-PCR to detect viral RNA from CSF, throat/nasal swab, urine and blood in early stages and IgG and IgM antibodies by ELISA in later stages.	
Cerebral malaria	*P. falciparum* residing in endothelium of brain capillaries	PBS for *P. falciparum*, HRP2 in blood	Quinine
Neurosyphilis	*T. pallidum* sequelae–meningitis, general paresis of insane or tabes dorsalis	VDRL in CSF	Penicillin G
Slow virus disease	Measles, rubella (SSPE), JC virus (PML) Prions		
Abscess/cyst *Toxoplasma gondii*	Oocyst ingestion from cat, pseudocyst of tachyzoites in MPS, true cyst of bradyzoites in brain, muscle Congenital toxoplasmosis	HPE of tissue cysts, IgM in acute infection, PCR in CSF and amniotic fluid	Sulfadiazine+ pyrimethamine
Mucormycosis	Predisposes diabetic ketoacidosis	HPE, smear culture	Amphotericin B
Neurocysticercosis	*T. solium* larval cyst	HPE	Albendazole

Upper Respiratory Tract Infections

Acute respiratory infections (ARIs) are common in infants, children and the aged. In India, 2 million deaths annually in children under 5 years of age are attributed to acute respiratory infections. ARIs account for 40% of pediatric outpatients and 20% of hospital admissions.

The clinical syndromes associated with respiratory tract infection (Fig. 35.1) and their likely pathogens are listed in Table 35.1. Lower respiratory infections include tracheitis, bronchitis, bronchiolitis, pneumonia and pleurisy. Risk factors in children are low birth weight, failure of breastfeeding and malnutrition.

Upper respiratory tract infections include the common cold, pharyngitis, acute laryngitis, croup, acute epiglottitis, otitis externa, otitis media, mastoiditis and sinusitis. They are common, as the respiratory tract is in direct contact with air and hence prone to aerosol infections. Being in contact with air, the upper respiratory tract is not sterile. The commensal, resident bacterial flora includes oral streptococci, Group A *Streptococcus* (GAS), *Moraxella catarrhalis*, *Corynebacterium*, *Haemophilus influenzae* and non-sporing anaerobes (*Fusobacterium*).

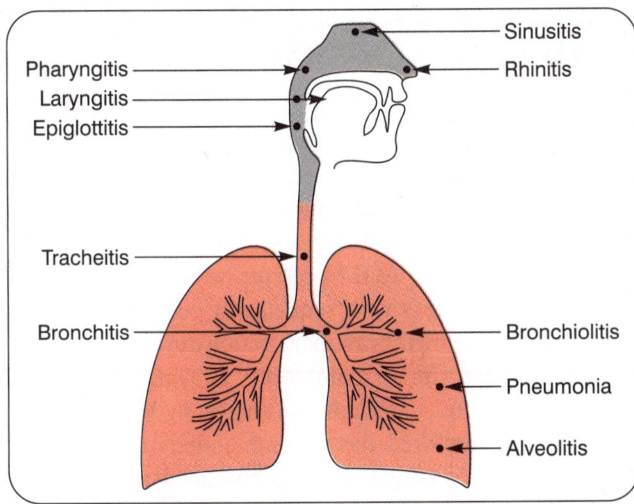

Fig. 35.1 Clinical syndromes associated with respiratory tract infection

THE COMMON COLD

This is a syndrome of upper respiratory symptoms caused by viruses. It occurs with a high frequency in seasons, transmitted from person to person by aerosols or fomites, and is self-limiting. The commonest pathogens are rhinoviruses (50%), followed by coronaviruses (10%) and others (40%)—respiratory syncytial virus (RSV), and rarely metapneumovirus and bocavirus. Influenza virus, parainfluenza virus and adenoviruses too can cause common cold but these may be associated with lower respiratory or systemic symptoms.

Acquired immunity is serotype-specific, therefore, other serotypes remain infective following one episode. Infection begins in the nasal epithelium, followed by an acute inflammatory response, leading to nasal secretions and obstruction. Cough is less common but sore throat is associated with bradykinin release in rhinovirus colds. Sinusitis or otitis media may occur as a complication.

Allergic rhinitis or bacterial causes of upper respiratory infections—streptococcal, pertussis or superimposed

Table 35.1 Respiratory pathogens

Location	Pathogen
Nasopharynx	Rhinovirus, coronavirus, other respiratory viruses *Staphylococcus aureus* (carrier)
Pharynx	*Streptococcus pyogenes* (GAS), *Corynebacterium diphtheriae, Neisseria gonorrheae* EBV, rhinovirus, adenovirus, enterovirus, HSV, influenza, parainfluenza, coxsackievirus
Conjunctiva	*Streptococcus pneumoniae, H. influenzae, N. gonorrheae, Chlamydia trachomatis,* adenoviruses
Middle ear and paranasal sinuses	*S. pneumoniae, H. influenzae, Moraxella catarrhalis, S. pyogenes*
Epiglottis (Croup)	*H. influenzae*
Larynx-trachea	Parainfluenza virus, *S. aureus*
Bronchi	*S. pneumoniae, H. influenzae, Mycoplasma pneumoniae,* influenza virus, measles virus
Bronchitis	RSV
Lungs	Pneumonia (refer Chapter 36)

bacterial infection—need to be ruled out if infection persists for a longer time.

Symptoms are rhinorrhea, nasal obstruction, sore throat, and sneezing or cough with or without fever. These usually begin 3 days after infection, and last for 1 to 2 weeks. Treatment is symptomatic with no role for antibiotics.

ACUTE PHARYNGITIS

Also called sore throat, it is more common in children. The bacterium of concern is Group A *Streptococcus* (GAS) or *Streptococcus pyogenes* because of its role in acute rheumatic fever. Asymptomatic carriage of GAS in the pharynx is also known, with some of the GAS present being rheumatogenic serotypes.

Scarlet fever is caused by pyrogenic exotoxins produced by the pharyngeal strains of GAS. Group C and G too are normal flora in the pharynx and can cause acute pharyngitis but without any sequelae. Penicillin or penicillin derivatives like amoxicillin or erythromycin (penicillin allergy) are effective against streptococcal sore throat.

Complications are suppurative infections, sinusitis, otitis media, abscess around the pharynx, mastoiditis, invasive necrotising fasciitis or toxic shock syndrome, or non-suppurative rheumatic fever or acute glomerulonephritis. Non-suppurative complications have been attributed to the immunological response to the bacterium damaging the host tissue due to molecular mimicry (autoimmune phenomenon).

Other bacteria causing pharyngitis are *S. pneumoniae, H. influenzae, M. catarrhalis* and *C. diphtheriae*, with the presence of a pseudomembrane, and anaerobic *Fusobacterium* spp. (Vincent's angina), and rarely *N. gonorrheae*, *Mycoplasma* spp. and *Chlamydia* spp. Viruses causing common cold are all implicated in causing pharyngitis, besides bacterial causes. Other viruses that can cause pharyngitis are EBV, HIV, enteroviruses, adenoviruses and HSV.

The National Treatment Guidelines do not recommend antibiotics for acute pharyngitis unless there is proof of a bacterial etiology of GAS, *S. pneumoniae, H. influenzae or M. catarrhalis*. If bacterial, amoxicillin is the drug of choice, alternatively, azithromycin is used in case of penicillin allergy. The Centor score (any 3 of the following 4)—exudative pharyngitis, fever, tender cervical lymph nodes and the absence of cough—alludes to a bacterial cause.

ACUTE LARYNGITIS AND LARYNGOTRACHEOBRONCHITIS (CROUP)

Characterised by hoarseness of voice, it is caused by all respiratory viruses, GAS and *H. influenzae*, and predominantly affects children. It is a self-limiting condition, managed by anti-inflammatory therapy with no indication for antibiotics.

ACUTE EPIGLOTTITIS

This presents with respiratory obstruction and is most commonly caused by *H. influenzae*. Maintenance of adequate airway is of primary concern, followed by antimicrobials to cover *H. influenzae*.

OTITIS EXTERNA

Infection of the external auditory canal is infection of the skin and soft tissue and therefore similar in the causative agents, pathogenesis, laboratory diagnosis and antimicrobial therapy.

OTITIS MEDIA

Acute otitis media (AOM) is an acute illness with inflammation of the middle ear mucosa and presence of fluid. Otitis media effusion (OME) is the presence of middle ear fluid without mucosal inflammation. AOM occurs largely in the first 3 years of life following a viral respiratory infection. AOM presents with fever, localised pain and hearing loss with sequelae of persistent hearing loss and tympanic membrane perforation during adulthood.

The bacteria most commonly causing AOM are *S. pneumoniae* and *H. influenzae*, and rarely *M. catarrhalis* and GAS. Therefore, the recommended antimicrobial therapy is amoxicillin, amoxicillin-clavulanic acid or azithromycin in case of penicillin allergy. AOM can be prevented by the use of pneumococcal and Hib vaccines. Fungi *Candida* and *Aspergillus* spp. are associated with otitis externa, and rarely otitis media.

MASTOIDITIS

The mastoid is the portion of the petrous temporal bone that lies superior to the middle ear cavity. Acute mastoiditis is usually accompanied by AOM, and has the same causative microorganisms and antimicrobial recommendations.

SINUSITIS

This is defined as the inflammatory disorder of the paranasal sinuses—maxillary, ethmoid, frontal and sphenoid—that are normally sterile. The obstruction of the normal draining system of the sinuses either by viral infection, allergy, local trauma or mechanical blockage results in mastoiditis. The bacterial etiology is *S. pneumoniae* and *H. influenzae*, followed by *M. catarrhalis*, GAS, *S. aureus* or non-sporing anaerobes. Fungal sinusitis caused by mucormycosis is seen with underlying diabetic ketoacidosis. Due to their proximity to the cranium, complications can be intracranial infection or orbital cellulitis. Amoxicillin or amoxicillin-clavulanic acid is the recommended first choice antimicrobial. Ceftriaxone and doxycycline are alternatives.

Streptococcus pyogenes

Clinical Case 1

A 6-year-old boy presents with exudative pharyngitis, fever and tender cervical lymph nodes. He has difficulty in swallowing but no cough.

1. *What is the most likely diagnosis?*
2. *Name three bacteria likely to cause the condition.*
3. *If proven to be bacterial, what antimicrobial agents would you choose?*
4. *Which of the causative agents has late sequelae to infection? Name the sequelae.*

Streptococci are non-motile, capsulated, gram-positive cocci arranged in chains.

A β-hemolytic GAS is *S. pyogenes*. Streptococci do not colonise the human skin, but *S. pyogenes* colonises the pharynx in some carriers. Respiratory carriers are symptomless but can transmit the bacterium. Infections are common in children of 5–8 years of age, especially in a community setting like a school hostel.

S. pyogenes is associated with suppurative infections of the skin and soft tissue, and sore throat (pharyngitis) in children. Recurrent infections can lead to post-streptococcal non-suppurative complications of rheumatic fever and glomerulonephritis.

Streptococcal virulence factors include a capsule, carbohydrate and protein antigens in the cell wall, and fimbriae. Streptococcal toxins include streptolysin O (hemolysin), streptococcal pyrogenic toxin, streptokinase, streptodornase and hyaluronidase. Streptolysin O is cytotoxic for cardiac tissue and highly antigenic. Antibodies to streptolysin O, termed anti-streptolysin O (ASO), are measurable in blood.

The ASO titres are not raised in streptococcal skin diseases but raised in streptococcal pharyngitis. Post-streptococcal glomerulonephritis may be a complication of streptococcal skin infections while rheumatic fever is not. Rheumatic fever is a complication of streptococcal sore throat with ASO titres > 200 units. As rheumatic fever is an immunological sequela to *S. pyogenes* pharyngitis, the bacterium cannot be isolated in rheumatic fever.

S. pyogenes can also cause skin and soft tissue infection like cellulitis, pharyngitis with local spread, puerperal sepsis, genital infections and abscess in visceral organs.

Laboratory diagnosis of upper respiratory infections caused by streptococci

1. Gram stain examination of the primary/direct smear from the pus or affected lesions. The presence of gram-positive cocci arranged in chains resembling streptococci is of limited value as they could be part of the normal resident flora.
2. Streptococcal antigen can be detected by rapid tests performed on a throat swab.
3. Culture of pus before the antibiotic is administered followed by identification of the bacterium.
4. Antimicrobial susceptibility spectrum — penicillin, cefazolin, erythromycin, clindamycin, ciprofloxacin, cotrimoxazole and tetracycline.
5. ASO titres for streptococcal pharyngitis and rheumatic fever.

All β-hemolytic streptococci are susceptible to penicillin and may be resistant to tetracycline and cotrimoxazole, with variable resistance to clindamycin and macrolides. The National Treatment Guidelines recommend amoxicillin and amoxicillin-clavulanic acid for empiric treatment, with azithromycin or moxifloxacin as alternatives. It is important to treat so as to prevent rheumatic heart disease.

Streptococcus pneumoniae (pneumococcus)

Clinical Case 2

A 3-year-old boy presents with a history of common cold for a week. He complains of intense pain localised in the ear and fever since 1 day.

1. *What is your most likely diagnosis?*
2. *Name the two commonest etiologic agents of this condition?*
3. *What is the course of infection if not treated?*
4. *What antimicrobial agents are recommended for empirical therapy?*

It is a gram-positive lanceolate-shaped diplococcus, with each pair enclosed in a polysaccharide capsule that is highly antigenic (Fig. 35.2). More than 90 serotypes of the capsule are known to exist.

It is a commensal in the upper respiratory tract and transmitted by aerosols. Children are colonised with one or more strains of pneumococci, but nasopharyngeal colonisation does not necessarily progress to disease.

Local spread of pneumococci can cause upper respiratory infections of otitis media, sinusitis, mastoiditis and lower

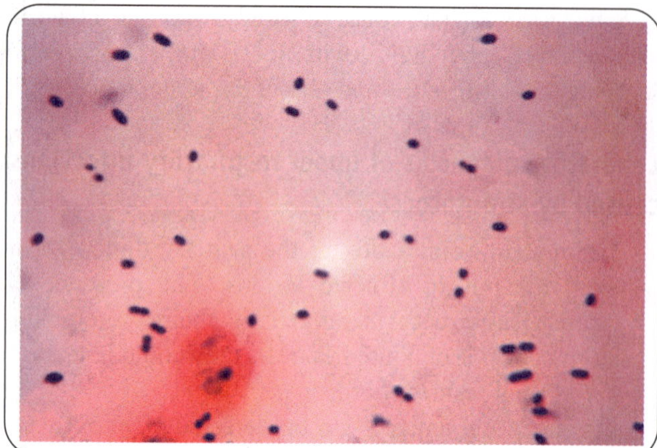

Fig. 35.2 Gram-positive cocci in pairs—*Streptococcus pneumoniae*, Gram stain, 100X (*Source*: Public Health Image Library, ID 18081 / Centers for Disease Control and Prevention [CDC] / Dr. Thomas Sellers, Emory University, Atlanta, Georgia)

respiratory infections of acute pneumonia—lobar and broncho-pneumonia, tracheobronchitis and empyema.

Some strains invade the bloodstream, causing pyogenic infection in other tissue—pericarditis, arthritis, peritonitis—or the eye, but the most important and serious disease is pyogenic meningitis. These are invasive diseases caused by *S. pneumoniae* and are serious and life-threatening.

The serotypes, amongst the known 90, vary in their potential to invade, with only some of the serotypes being invasive . The incidence of pneumococcal disease is highest in children less than 2 years of age and the elderly geriatric population. Risk factors are extremes of age, poor nutrition or immunity, asplenia, chronic lung disease, nephrotic syndrome and diabetes.

The laboratory diagnosis is by Gram or India ink stain showing diplococci or capsule, respectively. The capsular antigen can be detected by a rapid test (especially in the CSF) and the sample can be cultured on blood agar. In bacteremic or septicemic disease, blood culture is indicated. Pro-inflammatory biomarkers like C-reactive protein and procalcitonin are adjuncts to the diagnosis.

C protein is a somatic carbohydrate antigen of *S. pneumoniae*. In bacterial infections and inflammation, an abnormal protein is secreted by the hepatocytes, that has the ability to react with the C protein. It is not produced in response to the C protein as it is also seen in bacterial infections other than *S. pneumoniae*. It disappears when the infection subsides.

Procalcitonin is produced by many cells in response to bacterial infection or tissue injury. The level of procalcitonin in blood can increase significantly in systemic bacterial infections and sepsis. The normal procalcitonin level is 0.50–1.99 ng/mL, and a level > 10.00 ng/mL is due to severe bacterial sepsis or septic shock.

S. pneumoniae is sensitive to penicillin, amoxicillin, ceftriaxone, erythromycin, tetracycline and vancomycin. Resistance to β lactams is due to altered penicillin binding proteins on the bacterium.

Currently, two types of vaccines are available—pneumococcal polysaccharide vaccine (PPV23) and pneumococcal conjugate vaccine (PCV) referred to in Chapter 32.

Haemophilus influenzae

The genus *Haemophilus* means 'blood loving', with the organism so named because of its characteristic growth requirement of Factors X (hemin) and V (nicotinamide adenine dinucleotide phosphate, NADP) found in blood.

H. influenzae is an exclusively human pathogen that is carried as a commensal (non-capsulated strains) in the upper respiratory tract. Transmitted by respiratory aerosols, its capsulated strains cause invasive infections. The capsular polysaccharide is antigenic and therefore can be detected in the CSF in case of pyogenic meningitis caused by *H. influenzae*. Six capsular serotypes, from 'a' to 'f', have been identified, amongst which the *H. influenzae* type b (Hib-CV) vaccine is available.

H. influenzae causes either locally spreading, non-invasive infections like otitis media, sinusitis, acute exacerbations of chronic bronchitis and bronchiectasis, or invasive infections disseminated by blood, such as meningitis (in children less than 3 years of age), croup, endocarditis, pericarditis and pneumonia.

Laboratory diagnosis using a specimen of CSF, blood or sputum is done by microscopy or culture (chocolate agar containing hemolysed blood to release X and V factor). Smear examination will show pleomorphic gram-negative bacilli. It is a fastidious bacterium that is difficult to grow. The antimicrobial susceptibility of *H. influenzae* is to amoxicillin, amoxicillin-clavulanic acid, ceftriaxone, cotrimoxazole and clarithromycin.

H. influenzae, being a leading cause of pneumonia, meningitis and other invasive diseases in children under 5 years of age, the Hib-CV was introduced in the Universal Immunisation Program in the states of Kerala, TN and Haryana to begin with. It contains the Hib polysaccharide with a carrier conjugate protein. The schedule is 6, 10 and 14 weeks with a booster at 16 months.

The HACEK group of bacteria—*H. influenzae, Actinobacillus, Cardiobacterium, Eikenella* and *Kingella* are fastidious, slow-growing bacteria that are commensals in the oral cavity and rarely cause endocarditis.

Haemophilus ducreyi causes chancroid or soft chancre, an STD characterised by localised, tender, non-indurated, irregular ulcers on the genitalia and is discussed in Chapter 41.

Bordatella pertussis

The word 'pertussis' means violent cough, and the disease is called 'whooping cough' because of the characteristic whoop associated with the cough. The whoop is caused by the rush of air in an empty lung during the paroxysmal cough. In India, the incidence of the disease has markedly declined due to effective immunisation.

Jules Bordet received the Nobel prize in 1919 for identifying the bacterium causing whooping cough and other work in immunology. *B. pertussis* is a small (0.5 µm) coccobacillus, and is an exclusive human parasite that is a fastidious, obligate aerobe that grows on selective media.

It has an affinity for the upper respiratory tract mucosa to which it attaches with adhesins. The bacterium does not invade the blood but the 'pertussis toxin' enters the bloodstream and can cause respiratory effects and encephalopathy. Other virulent factors include pertactin, adenyl cyclase, LPS and toxins.

Following infection by inhalation, and the incubation period of 1 week, the disease has three stages that last 1 week each—catarrhal, paroxysmal and convalescent. The catarrhal stage is more infective, with dry cough and fever. However, it is indistinguishable from other upper respiratory infections and therefore escapes early treatment. It advances to the paroxysmal stage, wherein the characteristic whoop appears, with violent spasms of continuous coughing. This is followed by convalescence. Complications are either due to the increased pressure effects during a violent coughing episode or due to the respiratory (pneumonia) or neurological (convulsions, altered sensorium) effects of the PT.

Laboratory diagnosis is by demonstration of *B. pertussis* in throat swab, by Gram stain or culture (Bordet Gengou medium) or by the cough plate method. Antimicrobial therapy with macrolides needs to be initiated within 10 days of appearance of cough.

***B. pertussis* vaccines are of two types:**

1. wP, killed *B. pertussis* whole-cell vaccine
2. aP, acellular purified, with selective components of *B. pertussis*

The components of the aP vary, with multi-component aP vaccines being more effective than the low-efficacy wP but less effective than the high-efficacy wP. The standard primary immunisation schedule is 6, 10 and 14 weeks after birth, and two boosters at 15–18 months and 5 years. The primary immunisa0tion is recommended by wP, avoiding the aP. The protection level drops after 7 years, hence, with the resurgence of the disease, booster doses can be given at 12 and 18 years with aP.

Corynebacterium diphtheriae

Clinical Case 3

A 10-year-old boy presents with fever, cough and difficulty in swallowing for 5 days, 3 days and 1 day, respectively. On examination, the boy is febrile and the pharynx shows a grey membrane over the tonsils with surrounding congestion.

1. *Name three bacteria causing sore throat*
2. *What is the most likely cause of the above infection?*
3. *How will you confirm your diagnosis?*
4. *What is the mechanism of virulence of the bacterium?*

Epidemiology

As man is the only reservoir and an effective vaccine is in use, diphtheria is rare. Children between 1 and 5 years are affected and the infection is transmitted by inhalation of aerosols from convalescent carriers. Adult diphtheria implies failure of immunisation. In India, the annual incidence is less than 1,000 cases.

Morphology

Corynebacterium diphtheriae are gram-positive bacilli arranged in Chinese letter pattern in V or L shape as when the bacterium divides, the cells lie at acute angles to each other (Fig. 35.3). The bacilli store poly-metaphosphate granules at the ends that stain gram-positive and can also be demonstrated by special stains, e.g., Albert stain (Fig. 35.4). On Albert stain, the bacillary body appears green with blue granules.

Pathogenesis and clinical features

Being a gram-positive bacillus, it secretes a powerful, lethal exotoxin that is highly virulent. However, the ability to produce this diphtheria toxin (DT) is conferred only when the bacterium is infected by the β phage. That is,

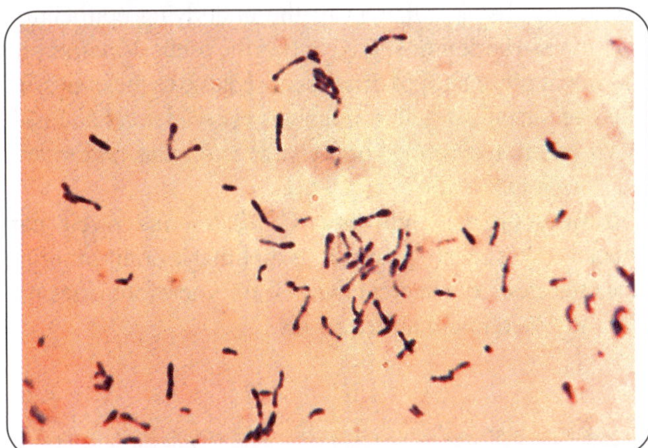

Fig. 35.3 *Corynebacterium diphtheriae*, Gram stain, 100X (*Source*: Public Health Image Library, ID 7323 / CDC / Dr. P.B. Smith)

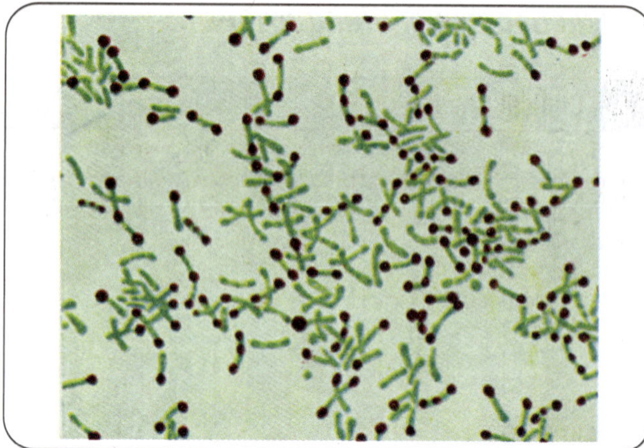

Fig. 35.4 *Corynebacterium diphtheriae,* Albert stain, 100X

the bacteriophage (virus that can infect bacteria) β phage integrates its toxin-producing gene into the bacterial gene (lysogenic conversion) and renders it toxigenic.

The DT is a protein that has two fragments—A (Active) and B (Binding). Once DT is secreted, its B fragment binds to the host cell receptors and the A fragment enters the target cell cytoplasm and inhibits protein synthesis. Death occurs due to necrosis of the heart and liver. Myocarditis due to DT is the most important risk in diphtheria.

The DT is highly antigenic and the diphtheria toxoid used as a vaccine, maintaining its antigenicity but obviating its virulence. Antitoxin antibodies are also utilised to inactivate the toxin in the event of the disease, as passive immunisation.

The incubation period is 1 to 3 days. On inhaling the bacilli, and on lysogenic conversion due to β phage, DT is secreted. The bacilli do not invade other sites or blood and multiply in the pharynx and tonsils, causing faucial diphtheria. It is the exotoxin secreted by the bacteria that causes toxemia in blood and reaches other organs.

Locally, the inflammatory response causes an exudate and necrosis resulting in the formation of a grey pseudo-membrane that is adherent to the underlying epithelium. Bacteria can be demonstrated only in the pseudo-membrane and not at other sites of disease caused due to the exotoxin. Massive tonsillar swelling and pharyngeal edema can result in a swollen appearance of the submandibular region and neck, seen externally as a bull neck.

Besides faucial diphtheria, *C. diphtheriae* rarely can infect the skin causing cutaneous diphtheria or diphtheritic ulcer but does not secrete the DT and therefore does not have any systemic manifestations.

Laboratory diagnosis

1. Microscopic demonstration of *C. diphtheriae* in the smear of a throat swab by Gram or Albert stain. Gram-positive bacilli are seen arranged in Chinese letter pattern in V or L shape, with poly-metaphosphate granules at the ends. On Albert stain, the bacillary body appears green with blue granules.
2. Culture of *C. diphtheriae* followed by confirmation of toxin production by the isolated strain. The methods used to demonstrate toxin production are immunologic (ELISA or ICT) or molecular (Tox gene by PCR). Elek's gel precipitation test is an in vitro toxin–antitoxin precipitation reaction to detect the toxigenic strains.

Prophylaxis

Active immunisation is achieved by diphtheria toxoid given IM at 6, 10 and 14 weeks as a trivalent vaccine—DPT—along with pertussis (killed whole cell) and tetanus (toxoid). Two booster doses are given between 16 and 24 months and at 5 years, with pertussis acting as an adjuvant. The diphtheria toxoid is a formol toxoid adsorbed on aluminium phosphate/hydroxide as an adjuvant. Pentavalent vaccines contain DPT and, in addition, HBsAg (HBV) and Hib.

Passive immunisation with diphtheria antitoxin is advisable only as treatment in a case of diphtheria as it neutralises the circulating DT. It is important to check for hypersensitivity to anti-diphtheritic serum (ADS) prior to and after use. Penicillin or erythromycin is the antimicrobial of choice administered along with the antitoxin.

VINCENT'S ANGINA

It is an acute ulcerative necrotising gingivostomatitis or membranous oropharyngitis caused by two oropharyngeal commensal bacteria causing disease following viral infections or malnutrition. *Borrelia vincentii* and the anaerobic *Leptotrichia buccalis* (formerly called *Fusobacterium fusiforme*) are the two symbiotic causative bacteria that can be demonstrated by Gram stain smear examination of the throat or oral swab. Being an anaerobic infection, penicillin or metronidazole are used in treatment.

Answers

Clinical Case 1

1. *Acute bacterial pharyngitis*
2. *S. pyogenes, S. pneumoniae, H. influenzae*
3. *If bacterial, amoxicillin or amoxicillin-clavulanic acidis the drug of choice; alternatively, azithromycin is used in the case of penicillin allergy*
4. *S. pyogenes. Rheumatic fever and glomerulonephritis*

Clinical Case 2

1. *Acute otitis media*
2. *S. pneumoniae and H. influenzae*
3. *Acute mastoiditis and sinusitis. Persistent hearing loss and tympanic membrane perforation during adulthood. Recommended antimicrobial therapy is amoxicillin, amoxicillin-clavulanic acid or azithromycin in the case of penicillin allergy*

Clinical Case 3

1. *S. pyogenes, C. diphtheriae, B. vincentii*
2. *C. diphtheriae*
3. *Gram or Albert's stain examination and culture of throat swab showing gram-positive bacilli in Chinese letter pattern, with metachromatic granules*
4. *The bacilli in the pharynx secrete a powerful exotoxin that circulates in the blood*

Questions

1. Enumerate microbes causing common cold.
2. What are the causative agents of sore throat in children? Which amongst these is a cause for concern and why? How will you diagnose it?
3. Explain the pathogenesis of acute otitis media and mastoiditis, stating the most likely pathogens.
4. How can infection with *H. influenzae* and *B. pertussis* be prevented?
5. Is blood culture indicated in diphtheria? Explain the rationale of pathogenesis in the laboratory diagnosis of diphtheria.
6. Explain the role of diphtheria toxoid and anti-diphtheria toxin in the immunoprophylaxis of diphtheria.

Lower Respiratory Tract Infections

PLEUROPULMONARY AND BRONCHIAL INFECTIONS

Acute bronchitis

This condition is a self-limiting inflammation of the large and mid-sized airways that is not associated with pneumonia. It is most prevalent in winters and caused by respiratory viruses. The respiratory viruses associated with acute bronchitis are influenza, rhinoviruses, coronaviruses, adenoviruses, respiratory syncytial virus, human metapneumovirus, parainfluenza and measles. Bacterial causes include *Mycoplasma*, *Chlamydia* and *Bordetella pertussis*. The commonest presentation is cough, followed by fever and other common cold symptoms, and the treatment is symptomatic.

Chronic obstructive pulmonary disease (COPD) and acute exacerbations

COPD is defined as a disease characterised by and diagnosed with the spirometric measurement of airflow limitation that is not fully reversible. The risk factors for COPD include smoking, occupational dust exposure and air pollution. Acute exacerbations are acute inflammatory events superimposed on a chronic inflammatory disease, often caused by infectious agents. These are viruses, atypical bacteria, and gram-positive and gram-negative aerobic bacteria.

Unlike the upper respiratory tract, the normal lower respiratory tract is bacteriologically sterile. In the event of an acute exacerbation in underlying COPD, *Streptococcus pneumoniae, Haemophilus influenzae* and *Moraxella catarrhalis* are the predominant microbes, followed by respiratory viruses, and rarely the atypical bacteria—*Mycoplasma pneumoniae* and *Chlamydophila pneumoniae*. The use of inflammatory biomarkers for diagnosis and antibacterial therapy is indicated in these cases.

Bronchiolitis

Bronchiolitis is the most common acute viral lower respiratory tract infection occurring in the first 2 years of life, and is managed with supportive therapy. It presents with wheezing, commonly associated with fever, cough, rhinorrhea and tachypnea. The viruses causing bronchiolitis are RSV, hMPV, influenza virus, rhinovirus, parainfluenza virus, bocavirus, coronavirus and adenovirus.

Acute pneumonia

The pulmonary defence system of ciliated epithelia, mucus-producing cells, colonisation by commensals, IgA, bronchus-associated lymphoid tissue (BALT) and phagocytic cells effectively prevent infection of the lung parenchyma. Particles larger than 10 μm are filtered by nasal hair and the trachea and bronchial tree trap those that are 2–10 μm in size. Most bacteria are 0.5 to 2 μm in size and may reach the alveoli. The alveolar fluid that contains antibacterial molecules and macrophages in the lung-alveolar cells, interstitial cells, dendritic cells, and neutrophils prevents pneumonia. The normal defences of the lung are efficient in maintaining sterility of the lower respiratory tract.

Risk factors that impair these normal defences and thereby contribute to pneumonia are alterations in the level of consciousness, smoking, alcohol, iatrogenic manipulations in the respiratory tract and some prescription medications. Older adults are at an increased risk for the development of pneumonia, especially those with underlying congestive cardiac failure, diabetes and COPD. HIV infection increases the risk of tuberculosis and pneumocystis.

Common causative agents of pneumonia

1. Bacteria — *S. pneumoniae, Staphylococcus aureus, H. influenzae, Klebsiella pneumoniae,* non-sporing anaerobes, *Pseudomonas aeruginosa, Legionella* spp., *Mycoplasma, Chlamydia*
2. Viruses — Influenza virus, parainfluenza virus, RSV, hMPV
3. Fungi — *Histoplasma,* mucormycetes, *Aspergillus, Candida, Pneumocystis*
4. Parasites — *Paragonimus westermanii, Entamoeba histolytica, Echinococcus granulosus*

Laboratory diagnosis of pneumonia

1. Microscopic examination and culture of the expectorated sputum is important for the laboratory diagnosis of pneumonia, though a variety of the organisms may not be observed in a Gram-stained smear.

   The macroscopic examination of the sputum provides clues to the most likely causative bacterium.

The sputum is mucopurulent in infection with pyogenic bacteria, watery and scant in atypical bacteria, rusty in pneumococcus, dark red in *K. pneumoniae*, and foul-smelling in non-sporing anaerobes.

The sputum being examined should not be contaminated by oral secretions. For diagnostic value, when examined under low power (10X), the specimen should have more than 25 pus cells and less than 10 epithelial cells. If the epithelial cells outnumber the pus cells, the sputum smear is considered to be inappropriately collected and cannot be reported on. The bacteria that can be diagnosed by examination of a primary or direct sputum smear are *S. pneumoniae, H. influenzae, S. aureus, K. pneumoniae* and *M. tuberculosis*.

2. Culture of sputum for non-fastidious and fastidious bacteria with standard and special media.
3. Antigen detection can be done for *S. pneumoniae, Legionella pneumophila, Pneumocystis* and a variety of viruses. The detection of nucleic acid in respiratory samples exists for tuberculosis and *Legionella* infection.

Besides sputum examination, a bronchoscopic specimen brush, bronchoalveolar lavage or pleural fluid examination can be done.

Clinical syndromes of pneumonia

1. Acute community-acquired pneumonia
2. Community-acquired pneumonia in the aged
3. Community-acquired pneumonia in HIV
4. Healthcare-associated pneumonia
5. Ventilator-associated pneumonia
6. Atypical pneumonia syndrome
7. Aspiration pneumonia
8. Eosinophilic pneumonia — this has a variety of causes, with the infectious causes being *Ascaris, Strongyloides* and *Filaria*.

Pneumonia is characterised by inflammation of the alveoli and the interstitial tissue of the lungs, caused by inhaled microbes. Clinically, it can be acute or chronic, community or healthcare-associated (VAP). Depending on the causative agent, acute pneumonia can be typical (alveolar) or atypical (interstitial). Typical pneumonia is caused by *S. pneumoniae*, while the atypical causative agents are *M. pneumoniae, C. pneumoniae* and *L. pneumophila*. The agents of chronic pneumonia are *Nocardia* spp., *Actinomyces* spp., *Mycobacterium tuberculosis* and mycobacteria other than tuberculosis (MOTT).

Antimicrobial therapy is a must for pneumonia, amoxicillin-clavulanic acid being the drug of choice for community-acquired pneumonia outpatients. Vaccination against influenza virus and pneumococci helps in preventing pneumonia in older adults.

Pleural effusion, empyema, bacterial lung abscess, infectious chronic pneumonia and cystic fibrosis are other distinct clinical lower respiratory tract infections. Pleural effusion and empyema are largely caused by pyogenic bacteria. Lung abscess results when bacterial infection causes necrosis of the lung parenchyma, producing one or more cavities.

Cystic fibrosis is a recessive genetic disorder identified with the CFTR gene. Clinical manifestations include COPD and bronchiectasis, with *S. aureus* and *P. aeruginosa* infection.

Mycoplasma pneumoniae

Mycoplasma are the smallest bacteria and hence, like viruses, pass through bacterial filters. They lack a rigid cell wall and therefore have no fixed shape, and can be cultured on inanimate media. Pathogenic *Mycoplasma* are *M. pneumoniae* that causes respiratory infections and others that cause sexually transmitted urethritis.

M. pneumoniae can cause upper respiratory infection—pharyngitis and lower respiratory infection—and acute community-acquired atypical pneumonia. It can be diagnosed by culture of the throat swab, sputum, BAL or lung biopsy on specialised culture media, or detecting the antigen in these specimens by immunofluorescence (IMF). Antibodies specific to *Mycoplasma* can be detected in serum by ELISA. Molecular techniques include PCR for *M. pneumoniae* and multiplex PCR for atypical pneumonia causative agents. Erythromycin or azithromycin is the drug of choice.

Legionella pneumophila

Legionella pneumophila is a gram-negative bacillus that inhabits water in cooling towers, fountains, spas and air conditioners. Unlike other pneumonias, *Legionella* pneumonia, also called 'Legionnaire's disease', is not transmitted from person to person but acquired from aerosolised contaminated water in the community and in hospitals.

On inhalation, it either causes a self-limiting, acute upper respiratory infection (Pontiac fever) or is phagocytosed by alveolar macrophages and multiplies intracellularly, causing atypical interstitial pneumonia. For laboratory diagnosis, sputum or BAL can be examined by Gram stain and cultured on specialised media. The specific antigen can be detected in urine by ELISA. Azithromycin is the drug of choice.

Chlamydophila pneumoniae

Chlamydiae are obligate intracellular bacilli that cause a wide spectrum of diseases in man—from the ocular, such as conjunctivitis, trachoma, and urogenital NGU, LGV caused by *Chlamydia trachomatis* to the pulmonary, such as pneumonia caused by *Chlamydophila pneumoniae* and *Chlamydophila psittaci* causing psittacosis.

C. pneumoniae is transmitted from man to man by inhalation, as humans are the only source of infection. It causes atypical or interstitial pneumonia or exacerbations of COPD or asthma. Laboratory diagnosis is by culture, antigen or nucleic acid detection in bronchoalveolar aspirate or IgM detection in serum.

C. psittaci is an avian pathogen associated with pets and poultry. Psittacosis is a rare disease and is transmitted by inhalation of infectious avian-origin aerosols and not from man to man.

Questions

1. What are the pulmonary defence mechanisms and risk factors in acute pneumonia?
2. Enumerate the causative agents and discuss the laboratory diagnosis of pneumonia.

Tuberculosis and Non-Tuberculous Mycobacteria

Tuberculosis (TB) is a chronic granulomatous infection that predominantly affects the lungs. However, it can affect any other tissue or organ as extrapulmonary tuberculosis (EPT).

Clinical Case

A 35-year-old man presents with a history of chronic productive cough for a month. He does not give a history of fever but feels that he has lost weight over the last few months.

1. *What are the two microbiology investigations that you will recommend first?*
2. *If both the tests are positive, what test is mandated next?*
3. *Name two other methodologies employed in the diagnosis of the above disease*
4. *Name the most common extrapulmonary site of infection*

Epidemiology

In India, TB kills more adults than any other infectious disease and accounts for one-fourth of the global TB burden. We have 4 million TB cases, with 2.2 million new cases and 2,20,000 deaths annually. More than 40% of the Indian population is infected with TB, with 6,000 new infections and 600 deaths on a daily basis. India has the highest burden of TB, MDR TB (71,000 annually) and second highest of HIV-associated TB. About 5% of incident TB cases are HIV-associated.

One infectious case of TB can infect 10–15 persons in a year. A person infected only with *Mycobacterium tuberculosis* (MTB) has a 10% lifetime risk of getting TB disease, a person with only HIV infection has a 30% risk, and a person infected with both HIV and TB has a 50% risk. Among diabetics who are infected, the risk of disease is 30%.

In 1993, the National Tuberculosis Program was modified to 'Revised National Tuberculosis Control Program (RNTCP)' to detect at least 70% cases and achieve an 85% cure rate amongst sputum positives across India with Directly Observed Treatment Short-course (DOTS) therapy. Presently, the END TB strategy implies a reduction in the number of TB deaths and in incidence rates by 2035 as compared to 2015. The aim is to achieve a 95% reduction in TB deaths and a 90% reduction in TB cases (< 10/100,000).

Pathogenesis

The bacilli are present in the sputum of an 'open' case of pulmonary TB and the disease is transmitted by inhalation of aerosolised droplets. The bacilli then multiply in the sub-pleural area, mostly in the mid lung zone that has greater airflow. In extrapulmonary tuberculosis (EPTB), the initial focus rarely is skin, intestine or genitalia, all associated with foci in the regional lymph node.

In primary infection, the bacilli are engulfed by alveolar macrophages but resist killing and continue to multiply, destroying the macrophage. Infected blood macrophages carry the bacilli to regional lymph nodes through lymphatics and other organs through blood. This lymphohematogenous spread seeds the bacilli into lymph nodes, bones, meninges and largely to the apical-posterior area of the lungs. This provides the focus for subsequent pulmonary or EPTB either sooner or later after a latency period.

The immune system responds to this infection with the development of cellular immunity and delayed hypersensitivity to the TB bacilli. Thus, the infection is controlled in about 6 weeks and the tuberculin skin test is positive. In few instances, the primary pulmonary focus (Ghon focus) may result in local necrosis and calcification.

Rarely may primary infection progress to primary disease in the very young (known as primary childhood tuberculosis in those less than 5 years) and the immunocompromised, resulting in tuberculous meningitis, pulmonary tuberculosis or miliary tuberculosis. In most cases, the primary infection is controlled, leaving behind tuberculin reactivity and bacilli in the lung apex that may progress to disease later in life. The apical predilection is because MTB is an aerobic bacillus and the oxygen concentration is highest at the apex.

In healthy individuals, primary infection also provides resistance to exogenous reinfection. Subsequent exposure to inhaled bacilli maintains the cellular immunity and hypersensitivity to prevent disease. When the immunity is compromised or when the inhaled inoculum is large, superinfection may happen. General stress, poor health and malnutrition favour the progression of infection to secondary disease.

The immune response that plays a role in the pathogenesis and defence is cell-mediated, with T4 or T helper cells deciding the course. Once activated, the T cells follow either the TH1 or the TH2 pathway. The TH1 pathway activates macrophages and controls the disease while the TH2 pathway induces delayed-type hypersensitivity (DTH) and disease progression.

Pulmonary tuberculosis (PTB)

Post-primary PTB in adults is usually asymmetrical and characterised by granulomas, caseous necrosis, fibrosis, cavitary lesions and calcification. It begins as pneumonitis and progresses to caseation that drains into the bronchi, spreading the infection via bacilli in sputum while coughing. It also leads to local lung cavities with uncontrolled multiplication of bacilli. Coughing aerosolised bacilli can also seed the infection throughout the lungs in the patient. Adult PTB does not resolve only with immunity and requires long-term antimicrobial therapy. Persistent cavities may develop superinfection with *Aspergillus* (aspergilloma) or non-tuberculous mycobacteria (NTM).

Following an incubation period of 2–12 weeks, the disease presents with symptoms of low-grade fever, cough with or without hemoptysis, and loss of weight. For TB screening among PLHIV/DM/smokers, the four symptom complex is applicable. In adults, the symptoms of current cough, fever, weight loss and night sweats are to be screened. In children, the symptoms of current cough, fever, poor weight gain and contact with TB cases are to be considered.

Extrapulmonary tuberculosis (EPTB)

EPTB occurs due to:

1. Lymphohematogenous dissemination either at the time of primary infection or later, e.g., miliary tuberculosis in children and EPTB in adults
2. Local contiguous spread into the pleural space, e.g., tuberculous pleurisy
3. Mucosal infection from the spread of infectious pulmonary secretions, e.g., intestinal tuberculosis

Miliary tuberculosis is so called because of the pathologic resemblance to millet seeds. Acute miliary tuberculosis in children following primary infection is rare in the chemotherapy era. In adults, it is seen with underlying morbidities, and needs prompt diagnosis and treatment.

Lymphadenitis is the most common site of EPTB. Other presentations of EPTB include tuberculous meningitis, brain tuberculomas, skeletal, genitourinary, mediastinal, cutaneous tuberculosis, pericarditis, laryngitis and otitis.

RNTCP case definitions

Presumptive pulmonary TB refers to a person with any of the symptoms and signs suggestive of TB including cough or fever for more than 2 weeks, hemoptysis, significant weight loss or any pulmonary abnormality in chest radiograph.

Presumptive EPTB refers to the presence of organ specific symptoms and signs like swelling of lymph node, pain and swelling in joints, neck stiffness, disorientation etc., and/or constitutional symptoms like significant weight loss, persistent fever for ≥ 2 weeks, night sweats.

Presumptive paediatric TB refers to children with persistent fever and/or cough for more than 2 weeks, loss of weight/no weight gain, and/or history of contact with infectious TB case.

Presumptive DR TB refers to those TB patients who have failed treatment with first line drugs, pediatric TB non responders, TB patients who are contacts of DR-TB, TB patients who are found positive on any follow-up sputum smear examination during treatment with first line drugs, previously treated TB cases, TB patients with HIV co-infection.

Microbiologically confirmed TB case

Presumptive TB patient with biological specimen positive for AFB, or positive for MTB on culture, or positive for TB through quality assured rapid diagnostic molecular test.

Clinically diagnosed TB case

Presumptive TB patient who is not microbiologically confirmed, but diagnosed with active TB by a clinician on the basis of X-ray, histopathology or clinical signs with a decision to treat the patient with a full course of TB treatment.

In children, this is based on the presence of abnormalities consistent with TB on radiography, history of exposure to an infectious case, evidence of TB infection (positive tuberculin skin test) and clinical findings suggestive of TB in event of negative or unavailable microbiological results.

Microbiologically confirmed or clinically diagnosed cases of TB are also classified according to anatomical site of disease, history of previous treatment and whether drug resistance or drug sensitive.

Long-term follow-up

After completion of treatment, the patients should be followed up at the end of 6, 12, 18 and 24 months. In presence of any clinical symptoms and/or cough, sputum microscopy and/or culture should be considered. This is important in detecting recurrence of TB at the earliest.

Contacts

All close contacts, especially household contacts should be screened for TB. In case of paediatric TB patients, **reverse contact tracing** for search of any active TB case in the household of the child must be undertaken. Particular attention should be paid to contacts with the highest susceptibility to TB infection.

Mycobacterium tuberculosis (MTB)

Mycobacteria are a group of gram-positive bacilli that are filamentous and may show branching, thus, morphologically resembling fungi, and hence named 'Myco'bacteria. They are difficult to Gram stain but once stained, resist decolourisation by acid and hence are also called acid-fast bacilli or AFB.

There are three mycobacteria pathogenic to humans—*Mycobacterium tuberculosis* complex, *Mycobacterium leprae* and non-tuberculous (atypical) mycobacteria.

The *Mycobacterium tuberculosis* complex causing tuberculosis in man includes MTB, *Mycobacterium bovis* and *Mycobacterium africanum*. Others in the complex are known to cause disease in other animal hosts but not man. *Mycobacterium leprae* is the cause of leprosy, while the NTM cause a host of other diseases in man. The *Mycobacterium tuberculosis* complex and *M. leprae* are obligate intracellular parasites while the NTM are largely opportunistic saprophytes.

MTB cell structure and characteristics

MTB is a slender AFB measuring 3 X 0.3 µm that resists decolourisation by 20% H_2SO_4 and by alcohol. The difficulty in staining by Gram stain and acid fastness is due to the presence of the fatty acid—mycolic acid—in the cell wall.

The generation time (time taken for one bacillus to become two) of MTB is 15 hours. This slow growth of the bacilli affects their multiplication both in vitro and vivo. Thus, the time taken for visible growth on solid culture is 3–4 weeks and the incubation period for the disease to manifest is 2–12 weeks. MTB is an obligate aerobe, a characteristic that has been utilised in identifying the growth of the microbe in liquid cultures. It is also a contagious respiratory pathogen, necessitating the use of Biological Safety Cabinet (BSC) level 2 for processing clinical samples.

MTB antigens are either molecules that are part of the cell or those that are secreted by the bacilli. The former are associated with both live and dead bacilli while the latter only with live, multiplying bacilli. The cell-associated lipids and peptidoglycan play a role in the formation of granulomas while MTB proteins are antigenic and known to induce delayed hypersensitivity or tuberculin conversion. The secretory antigen MPB64 is unique to MTB and used in the identification of MTB from in vitro cultures.

MTB being a facultative, intracellular bacterium that multiplies in macrophages, antimicrobial therapy for MTB needs to address both intracellular and extracellular populations of the microbe. It is also known to mutate once in 10^8 cell divisions in vivo and 10^5 cell divisions in vitro. This necessitates multi-drug therapy in patients, as the use of a single drug leads to selection of resistant mutants.

Multi-drug resistant (MDR) TB is defined as that which shows **in vitro** resistance to rifampicin and isonicotinic acid hydrazide (INH), with or without resistance to other first-line anti-TB drugs. Extremely drug-resistant (XDR) TB is defined as that which shows **in vitro** resistance to a fluoroquinolone (levofloxacin/moxifloxacin) and to bedaquiline or linezolid. The diagnosis of MDR or XDR TB is a laboratory diagnosis and not a clinical diagnosis.

Classification of drug-resistant (DR) TB

Mono-resistance TB (MR): A TB patient whose biological specimen is resistant to one first-line anti-TB drug only.

Poly-drug resistance TB (PDR): A TB patient whose biological specimen is resistant to more than one first-line anti-TB drug other than both H and R.

Multi-drug resistance (MDR): A TB patient whose biological specimen is resistant to both H and R with or without resistance to other first line drugs.

Pre-extensive drug resistance (Pre-XDR-TB): TB caused by *M. tuberculosis* strains that fulfill the definition of MDR/RR-TB and that are also resistant to any fluoroquinolone.

Extensive drug resistance (XDR-TB): TB caused by *M. tuberculosis* strains that fulfill the definition of MDR/RR-TB and that are also resistant to a fluoroquinolone (levofloxacin/moxifloxacin) and to bedaquiline or linezolid.

In India, the screening, diagnosis and treatment of TB, including drug resistance, is governed by the RNTCP that has laid algorithms and protocols followed across the country. The RNTCP has established District Microscopy Centres (DMCs) to implement the Directly Observed Treatment, Short-course (DOTS) policy with Fixed Drug Combinations (FDCs).

Laboratory diagnosis of tuberculosis

1. Collection of sample

In PTB, the sample is expectorated sputum or swallowed sputum (obtained by gastric lavage) in the case of children. In EPTB, the sample is any body fluid or tissue obtained from the suspected site of disease—CSF, pleural fluid, ascitic fluid, lymph node aspirate, tissue biopsy, etc. EPTB biopsy samples are not to be collected in normal saline and not in formalin.

Sample collection

Obtaining **good quality** specimens of an **adequate volume** is critical to ensure correct diagnosis. Sputum and not saliva is to be collected as one spot and one early-morning sample or two spot specimens collected with a gap of at least 1 hour (if the patient is travelling a long distance or is unlikely to return to give the second specimen).

The RNTCP recommends one or two samples for EPTB. The present policy for screening includes any type of cough that has exceeded a 2-week duration and is not responding to antibiotics and/or if the candidate for screening has had any contact with a known case of PTB. Additionally, vulnerable populations defined as tribals, slum dwellers, the homeless, refugees, diabetics and pregnant women are to be screened if symptomatic. All suspected TB cases are screened for HIV.

2. Demonstration of MTB in clinical samples by staining

The time-tested technique of Ziehl–Neelsen is still being practised in some laboratories. However, the RNTCP has moved to staining by the fluorescent dyes (auramine-O, rhodamine, auramine-rhodamine, acridine orange, etc.), visualised by a light emitting diode (LED) attachment to the light microscope. The results are similar to those in ZN, but the technique is time-saving and low-cost, obviating the need and cost of fluorescent microscopy.

The drawback of demonstrating MTB by staining is its low sensitivity of 10,000 bacilli/mL of sputum. Besides diagnosis, sputum smear microscopy and/or culture is to be done on completion of treatment for every patient.

Principle of ZN staining: Acid fastness is due to the high content of lipids (mycolic acid), fatty acids and higher alcohols found in the cell wall of mycobacterium thereby resisting decolourisation with acids and alcohol.

Procedure of ZN staining

a. Pour strong carbol fuchsin stain on the slide so as to cover the fixed smear and apply gentle heat to the underside of the slide. Leave the carbol fuchsin on the slide for 10 minutes with intermittent heating during this period; take care to ensure that the stain does not dry, but the stain remains steaming hot.
b. Wash with tap water.
c. Decolourise with 25% sulphuric acid till smear becomes colourless and wash with water.
d. Counterstain with methylene blue till the smear becomes blue.
e. Wash with water and dry.
f. Examine under oil immersion (100X) objective.

The MTB load in the ZN-stained smear (Fig. 37.1) **is graded as follows:**

AFB not seen on screening 100 OI (X 100) fields report negative,

1–9 AFB/100 fields (scanty, report actual number),
10–99 APB/100 fields (1+),
1–10 AFB/field (2+),
> 10/field (3+)

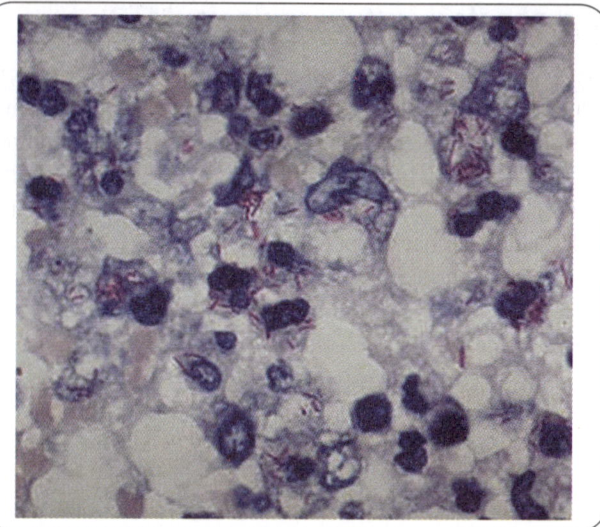

Fig. 37.1 AFB in sputum—*Mycobacterium tuberculosis*, RNTCP grade 3+, ZN stain, 100X

The MTB load in the acridine orange stained smear (Fig. 37.2) is graded as follows:

AFB not seen on screening 40 HP (40X) fields report negative,

1–19 AFB/40 fields (scanty, report actual number),
20–199 APB/40 fields (1+),
5–50 AFB/field (2+),
> 10/field (3+)

3. Culture of MTB

The sensitivity of detecting MTB by culture is superior to staining as it is 10–100 bacilli per mL of sputum. The conventional, manual method is to culture on a solid medium that is both enriched and selective for MTB—Lowenstein–Jensen (LJ). The drawback is that the time taken to visualise growth is 2–4 weeks depending upon

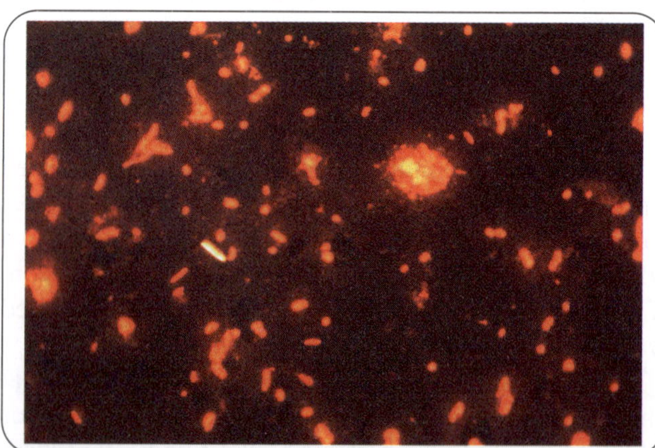

Fig. 37.2 Acridine orange staining of sputum smear showing *Mycobacterium tuberculosis* bacteria glow yellow under ultraviolet light microscopy (*Source*: Public Health Image Library, ID 2189 / Centers for Disease Control and Prevention [CDC] / Ronald W. Smithwick)

the number of bacilli in the clinical sample and the need for cumbersome decontamination of sputum samples prior to culture. Culture on LJ medium can also be used to detect drug resistance.

This has led to the acceptance of automated liquid culture methods that reduce the 'turnaround time' (TAT) by using alternative systems for detecting mycobacterial growth, for example BACTEC 460, BacT/ALERT MB, VersaTREK and BACTEC MGIT 960. Of these, the WHO and RNTCP recommend the BACTEC MGIT 960 and it has been adopted at Intermediate Reference Laboratories (IRLs).

The BACTEC Mycobacterial Growth Indicator Tube (MGIT) 960 works on the principle that MTB, being a strict aerobe, consumes a large amount of oxygen during growth. The liquid culture medium contains oxygen dissolved in the broth and a fluorescent compound at the bottom of the tube that 'indicates' the utilisation of this oxygen. Thus, it does not need to wait for the MTB bacilli, if present, to multiply but detects the presence of live bacilli by its consumption of the provided oxygen. This reduces the TAT from weeks to 4 days for detection of MTB in a clinical sample. Another advantage of the system is its ability to perform a simultaneous drug susceptibility test (DST) for MDR and XDR TB, as it is available for first- and second-line drugs.

All pulmonary and extrapulmonary samples other than blood can be processed by MGIT 960 culture. Currently, the RNTCP recommends smear-negative PT cases and all EPTB cases that are cartridge-based nucleic acid amplification test (CBNAAT)-positive be tested by MGIT 960. The DST for drugs that cannot be tested by LPA can be tested by MGIT, viz. moxifloxacin, linezolid, clofazimine, pyrazinamide, bedaquiline and delamanid.

The rapid identification of the MTB in liquid culture systems, including the BACTEC MGIT 960, is done by the use of an immunochromatography test (ICT) that detects the MPB64 antigen secreted specifically by the MTB complex and not by NTM. This antigen detection test is intended for the identification of MTB invitro only in culture and not in clinical samples.

4. Molecular methods for MTB

At the IRLs under the RNTCP, the genotypic detection of MTB and MDR TB can be done simultaneously. The methods employed are CBNAAT and/or LPA.

Cartridge-based nucleic acid amplification test (CBNAAT): This test detects the rpoB gene that codes for rifampicin resistance. CBNAAT does not detect resistance to drugs other than rifampicin. It is a simple, fully automated, closed system, with no contamination risk and a TAT of 2 hours. All pulmonary and extrapulmonary samples other than blood, stool and urine can be processed by CBNAAT. Currently, the RNTCP recommends smear-negative presumptive PT, all EPTB, and pediatric and HIV-infected cases for CBNAAT.

Line probe assay (LPA): This test is an rpoB/inhA gene amplification and hybridisation technique to detect MTB and drug resistance simultaneously with a TAT of 5 hours. LPA can detect resistance to both first- and second-line drugs in smear-positive cases. All sputum smear-positive pulmonary cases will be tested for first-line drugs. If rifampicin resistance is detected by LPA or CBNAAT (sputum-smear negatives), second-line drugs are tested by LPA, viz. all fluoroquinolones, kanamycin, capreomycin and amikacin. CBNAAT tests that do not detect rifampicin resistance are tested by LPA for first-line drugs.

5. Drug susceptibility testing (DST) in MTB

First-line drugs are rifampicin, INH, pyrazinamide, ethambutol and streptomycin. Second-line drugs are fluoroquinolones, amikacin, kanamycin and capreomycin for MDR TB and bedaquiline and delamanid for XDR TB. MTB DST can be done by both phenotypic (BACTEC MGIT 960) and genotypic (CBNAAT, LPA) methods, the latter being faster.

6. Detection of antigen and/or antibodies in TB

These tests are banned in India by the WHO and the RNTCP. The only WHO-approved test is the detection of lipo-arabino-mannan antigen in the urine of PLHIV cases with a CD4 count < 100 cells/ μL, by ICT or ELISA.

7. Other methods

Methods that are not yet recommended by the WHO or RNTCP include TB loop-mediated isothermal amplification (TB LAMP), microscopic observation of drug susceptibility (MODS), nitrate reductase assay (NRA), colorimetric redox indicator (CRI), thin layer agar (TLA), phage plaque assay (PPA) and IFN gamma release assay (IGRA).

8. Tuberculin skin test (TST)

It is NOT a diagnostic test and has a limited role to play, especially in settings like India where the exposure to TB is high. In the test, a purified protein derivative (PPD) of MTB is injected intradermally in the forearm of the individual and the response is measured after 48–72 hours. The response will show erythema and induration at the site of injection that can be measured in mm. It is a delayed type of hypersensitivity reaction.

Tuberculin-positive (induration size > 10 mm) implies sensitisation to the tubercle bacilli in the individual, while no response or an induration size < 5 mm is considered tuberculin-negative and implies tuberculin anergy. The BCG vaccination results in tuberculin conversion.

The RNTCP utilises a combination of laboratory techniques to diagnose TB and drug resistance (Fig 37.3).

Drug-sensitive tuberculosis not responding to treatment is retested by CBNAAT for rifampicin/INH resistance. If found to be resistant, the isolate is subjected to first-line LPA. Drug-sensitive tuberculosis not responding to treatment implies no change in the grade of the sputum-positive smear, chest X-ray or clinical response after 2 months of intensive treatment.

TB prevention and prophylaxis

General measures include good nutrition and housing (not over-crowded), cough etiquette practice and chemoprophylaxis in specified conditions.

Bacillus Calmette-Guerin (BCG) is a live attenuated vaccine developed in 1921! It does not prevent TB but prevents infants and children from virulent manifestations of the disease like meningitis and miliary tuberculosis. In India, it is administered immediately at birth or at the earliest and before 12 months to the immunocompetent. It sensitises the child to tubercle bacilli, thereby also achieving tuberculin conversion.

Non-tuberculous mycobacteria

Non-tuberculous mycobacteria (NTM), earlier called atypical mycobacteria or mycobacteria other than tuberculosis (MOTT), are saprophytic, opportunistic mycobacteria that differ from MTB.

Based on their culture characteristics, the Runyon's classification of NTM is as shown in Table 37.1.

M. avium and *M. intracellulare* are the most common NTM causing pulmonary disease amongst the HIV-infected. *M. kansasii* is the second most common NTM in lung disease, and is sourced from tap water. *M. fortuitum* and *M. chelonae* are saprophytes contaminating medical devices like endoscopes and injections. Endoscopic site infection and injection abscess have been attributed to them. The laboratory diagnosis of NTM causing pulmonary disease includes staining the clinical sample by ZN for AFB and culture on LJ. Antimicrobials for treating NTM pulmonary disease are azithromycin, amikacin, clofazimine and rifampicin.

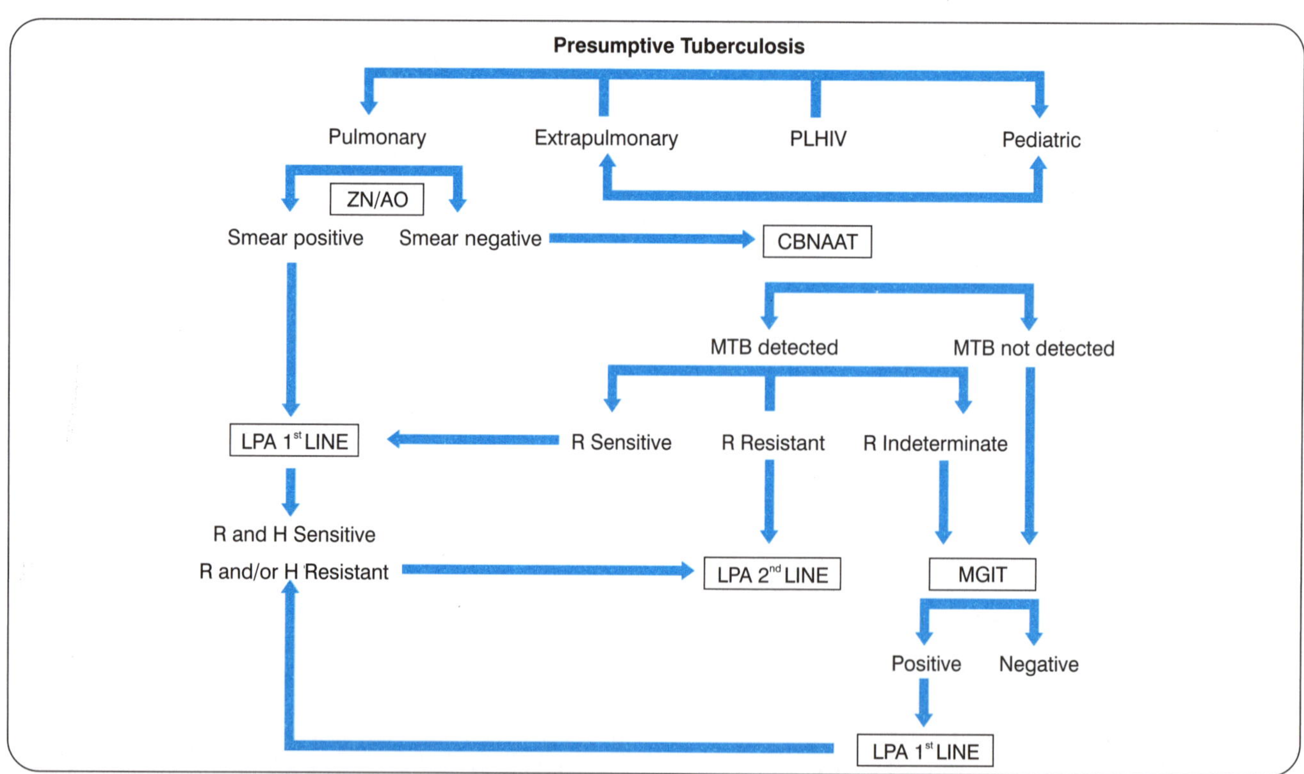

Fig. 37.3 Algorithm for the laboratory diagnosis of TB and drug resistance (as per the RNTCP); *Abbreviations:* PLHIV, people living with HIV; R, rifampicin; H, INH; MGIT, mycobacterial growth indicator tube; LPA, line probe assay; CBNAAT, cartridge-based nucleic acid amplification test

Table 37.1 Runyon's classification of NTM

Group	Name	Culture	Example	Disease
Group I	Photochromogens	Pigmented only in light	*M. kansasii*	Resembling TB
			M. marinum swimming pool granuloma	
Group II	Scotochromogens	Pigmented in the dark	*M. scrofulaceum* cervical lymphadenitis *M. gordonae*	Resembling TB
Group III	Non-photochromogens	No pigment	*M. avium* *M. intracellulare* *M. ulcerans*	Pulmonary disease Pulmonary disease Skin ulcers
Group IV	Rapid growers	Grow in 7 days	*M. fortuitum* *M. chelonae*	HAI-abscess HAI-abscess

Answers

Clinical Case

1. *Sputum microscopy for AFB by Ziehl–Neelsen or acridine orange, and test for HIV antibodies.*
2. *CBNAAT for MDR TB.*
3. *Culture on solid medium (Lowenstein–Jensen) manually or automated liquid MGIT.*
4. *Tuberculous lymphadenitis.*

Questions

1. Elaborate on the pathogenesis of pulmonary and extrapulmonary tuberculosis.
2. Outline the algorithm for laboratory diagnosis of tuberculosis as per RNTCP recommendations.
3. What are the advantages of genotypic methods of diagnosis of tuberculosis over phenotypic methods?
4. What is the role of non-tuberculous mycobacteria in causing human disease?

Respiratory Viral Infections

Viruses that affect the respiratory tract infect both the upper and lower respiratory tracts, often simultaneously. Respiratory infections are by far the most common infections because the very simple act of inhalation/exhalation helps spread them rapidly within a population. The common cold is perhaps the most common infection suffered by humans several times over in a life time. The commonest causes are the rhinoviruses (50%), followed by the coronaviruses (10%) and others (40%)—respiratory syncytial virus (RSV), metapneumovirus and bocavirus. The reason for it being that there exist a number of viruses with many serotypes causing the common cold. As common colds are caused by viruses and not bacteria, antibiotics have no role to play in the treatment which is largely symptomatic. Influenza, parainfluenza and adenoviruses too can cause common cold but these may be associated with lower respiratory or systemic symptoms.

Viruses that affect the respiratory tract include RNA viruses—influenza virus, parainfluenza virus, RSV, rhinoviruses, adenoviruses, coronaviruses, coxsackieviruses, echoviruses and the Epstein–Barr virus (EBV).

The reproduction number R_0 (pronounced "R naught") indicates the transmissibility of the disease in a non-immune population. If R_0 is less than 1, each existing infection can cause less than one new infection, and the disease will die out. If R_0 equals 1, each existing infection can cause one new infection, and the disease will be stable, without an outbreak. If R_0 is more than 1, each existing infection can cause one new infection, and the disease will be stable, without an outbreak. Amongst respiratory viral infections the R_0 number is: MERS 0.8, influenza 1.5, SARS 3.5, mumps 4.5, COVID-19 5, rubella 6, and measles 16.

INFLUENZA

Myxoviruses are viruses that can bind to mucin or sialic acid receptors on RBCs resulting in hemagglutination. They have two families—*Orthomyxoviridae* (influenza virus) and *Paramyxoviridae* that are differentiated by segmented and unsegmented (single) RNA, respectively. The parainfluenza virus, mumps virus, measles virus, RSV and metapneumovirus belong to *Paramyxoviridae*. Influenza is an acute sporadic respiratory disease capable of causing epidemics with a pandemic record due to the ability of the virus to mutate antigenically into new strains. It also exists in non-human hosts like birds and animals, but is species-specific. However, a potential danger of genetic recombination from avian and swine influenza to create new human strains always exists.

Clinical Case 1

An 80-year-old man presents with fever and sudden onset of breathlessness. On examination, he is found to be hypoxic with SpO2 < 90% and there is radiographic evidence of pneumonia. He gives a history of being vaccinated for Streptococcus pneumoniae.

1. *What is your most likely diagnosis?*
2. *Categorise the above case as per national guidelines.*
3. *What test would you recommend?*
4. *How can this disease be prevented in the elderly?*

Epidemiology

Influenza can be human (seasonal), avian or pandemic. Human (seasonal) influenza is endemic, resulting in a seasonal epidemic cluster due to antigenic drift. Avian influenza (bird flu) is a disease in poultry/birds that rarely jumps to humans and needs to be watched. Pandemic influenza occurs due to antigenic shift resulting from a re-assortment of the avian, pig (swine flu) and human influenza virus genome, e.g., the 2009 pandemic of influenza A H1N1 (Table 38.1).

Virology

Influenza is an orthomyxovirus (affinity for mucin) that has eight separate strands of RNA. Spherical in shape, it contains an envelope of radially projecting spikes of hemagglutinin (H) and neuraminidase (N) (Fig. 38.1).

Table 38.1 Influenza pandemics

Pandemic	Year	Virus subtype	Infected	Deaths
Spanish flu	1918	A (H1N1)	>50 million	>20 million (6 million in India)
Asian flu	1957	A (H2N2) 1,000 million	1,000 million	2 million (767 in India)
Hong Kong flu	1968	A (H3N2)	—	1 million
Swine flu	2009	A (H1N1)	1,62,380 (959 in India)	1,154 (6 in India)

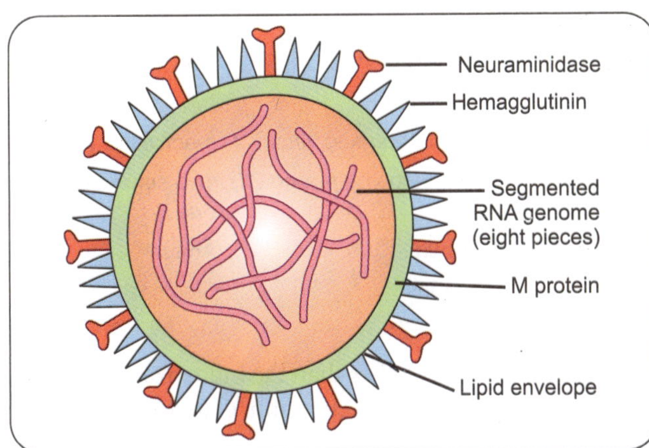

Fig. 38.1 Structure of influenza virus

Hemagglutinins have the ability to hemagglutinate (HA) RBCs of various animal species due to adsorption to specific RBC receptors composed of neuraminic acid. Fifteen HA subtypes have been identified in nature, of which four exist in human infection. Neuraminidase is a glycoprotein enzyme that has the ability to destroy neuraminic receptors on RBCs, thereby causing the virus to elute from the RBC by hemagglutination inhibition (HAI). Nine variants of neuraminidase are found to exist in nature. As hemagglutinin and neuraminidase are antigenic, both these phenomena—hemagglutination and hemagglutination inhibition—are strain specific.

There are three serotypes of influenza viruses—A, B and C. A is the predominant cause of epidemic influenza, B is associated with milder disease and C may not be pathogenic in man.

Antigenic structure

Influenza viruses have three antigens:

1. 'S' or soluble antigen is the protein component of the ribonucleoprotein core of the virus and is shared only within each type, i.e., type A, B and C have a different S antigen that is common only within each type.
2. Hemagglutinin of the envelope is a strain-specific neutralising antigen that is responsible for immunity to the virus.
3. Neuraminidase of the envelope is strain specific and has a minor role in immunity to re-infection.

Antigenic variations

These viruses are unique in their ability for antigenic variations due to re-assortment of the RNA segments from time to time. Epidemics and pandemics of influenza type A are due to emergence of new strains with novel hemagglutinin and neuraminidase that have no antibodies/herd immunity in a population. There exist two kinds of antigenic variations—major antigenic shift and minor antigenic drift.

Antigenic shift

This occurs as a result of RNA re-assortment when two different influenza viruses (possibly of human and non-human origin) infect and replicate in the same host cell. It reflects in the replacement of the neutralising antigen H and at times in N. The antigenic shifts that have occurred in the past pandemics are: 1957 H2N2, 1968 H3N2 and 1977 H1N1. At present, the prevalent strain is the type A subtype H1N1, with the last pandemic having occurred in 2009–10. Once a new antigenic strain occurs, it completely replaces the old strain.

Antigenic drift

This occurs as a result of minor changes in the amino acid sequence of the hemagglutinin protein due to spontaneous mutation in the virus. Influenza type B also shows antigenic variations but the changes are not so marked as in type A.

Pathogenesis and clinical features

Transmission is by inhalation of respiratory secretions from an infected patient, and the incubation period is 1–4 days. The virus multiplies in the epithelium of the upper and predominantly lower respiratory tract, and causes damage to the cilia and desquamation of the epithelium.

The patient presents with fever, malaise, cough, sore throat and cold. Most cases are self-limiting, and a small proportion progresses to pneumonia, either primary or secondary. Primary pneumonia is congestion of the lungs caused by the virus. It is associated with severe respiratory distress, hypoxia, circulatory failure and mortality. Secondary pneumonia develops later in the course of infection and is due to secondary bacterial infection, the most common being that caused by *Streptococcus pneumoniae, Staphylococcus aureus* and *Haemophilus influenzae.* Reye's syndrome, in which there is cerebral edema and fatty degeneration of the viscera, especially the liver, is a rare complication of influenza in children. The high-risk groups for influenza complications include children less than 5 years of age, adults above the age of 65, pregnant and postpartum women (within 2 weeks of delivery), immunosuppression, and underlying chronic medical disorders like obesity, asthma, diabetes, malignancy, or blood, kidney, liver or neurological disease.

Based on the clinical presentation, the patient is categorised as per Ministry of Health and Family Welfare (MoHFW) guidelines as:

Category A (uncomplicated/mild) is a patient with a mild fever and cough/sore throat, in whom laboratory testing and antiviral oseltamivir is not recommended. The patient is to be isolated at home and disease progress monitored.

Category B (uncomplicated but severe/high-risk group) is a patient with fever > 102°F and cough/sore throat OR a Category A patient in the high-risk group. Here, laboratory testing is not recommended, but oseltamivir is, as is isolation at home.

Category C (complicated) is a patient of category B with additional symptoms of breathlessness, hemoptysis, altered sensorium, seizures, poor urine output with signs of tachypnea, SpO_2<90%, cyanosis and hypotension. In these cases, oseltamivir is the empirical therapy, along with a laboratory test for influenza and hospitalisation.

Laboratory diagnosis

Laboratory confirmation of influenza is recommended only for Category C cases.

1. The sample is a nasopharyngeal swab collected in a viral transport medium (VTM). Multiplex PCR assays are used to detect virus RNA subtypes in the specimen.
2. Serological demonstration of a four-fold rise in specific neutralising antibodies can be done at designated reference laboratories.

Prophylaxis

Prevention of the spread of influenza by practising cough/sneeze etiquette, handwashing and no touching of fomites is recommended.

The strains included in the vaccine are annually updated for anticipated strains by the WHO. Trivalent inactivated influenza vaccines (TIVs) given IM are most commonly used. Live attenuated influenza vaccines (LAIV) given intranasally are also available (Table 38.2).

If a double dose is recommended for children between 6 months and 8 years, a second dose is to be given 4 weeks after the first. LAIV is recommended for healthy individuals and healthy contacts between 2 and 49 years, and TIV from 6 months to any age. In persons with chronic medical disorders, only TIV is recommended and not LAIV.

Table 38.2 Comparison between the LAIV and TIV

Property	LAIV	TIV
Route	Intranasal	IM
Content	Live attenuated virus	Inactivated virus
No. of strains	Two serotype A, one B	Two serotype A, one B
Frequency	Annual	Annual
Dose	Single	Single

Respiratory viruses other than the influenza virus cause a range of diseases from common cold to pneumonia (Table 38.3).

The close interaction between humans and birds/animals has over the years seen the emergence of epidemics of respiratory infections other than influenza but presenting as influenza-like illness (ILI). In recent times, they have been identified as coronaviruses, namely SARS-CoV, MERS-CoV and SARS-CoV-2 (COVID-19, 2020)

CORONAVIRUSES (COV)

Virology

These are a zoonotic group of spherical, enveloped RNA viruses with surface projections resembling solar corona. Most common colds are caused by rhinoviruses and coronaviruses. They are classified into three types: mammalian type 1 and 2, and avian type 3. However, in this century there have been epidemics caused by novel coronaviruses—SARS-CoV in 2002, MERS-CoV in 2012 and SARS-CoV-19 in 2019.

SARS-CoV

The 'Severe Acute Respiratory Syndrome' pandemic that began from China affected over 30 countries, with over 800 deaths. The new SARS virus was distinct and named coronavirus 4, which could be a recombinant from some

Table 38.3 Viruses that affect the respiratory tract

Virus	Class	Genome	Serotypes	Disease
Influenza virus	Orthomyxovirus	8 ssRNA	Variety of H and N	Influenza
Parainfluenza virus	Paramyxovirus	ssRNA	4 types	Colds, croup, LRTI in children
RSV	Paramyxovirus	ssRNA	1 type	Colds, bronchiolitis, pneumonia
Rhinoviruses	Picornavirus	ssRNA	100+	Colds
Coxsackievirus	Picornavirus	ssRNA	type A21, B3	Colds
Echoviruses	Picornavirus	ssRNA	type 11, 20	Colds
Adenoviruses	—	DNA	41 types	Pharyngitis, conjunctivitis
Coronaviruses	—	ssRNA	3 types	Colds, pneumonia

civet cats and human viruses. SARS was an ILI spread by inhalation, followed by an incubation period of 7 days, and then by fever and respiratory symptoms. Mortality was due to pneumonia leading to respiratory failure. Cases were confirmed by viral RNA-specific RT-PCR, and a rise in viral-specific antibodies. Neither vaccine nor antiviral therapy is available, and prevention is by hygienic measures in the healthy and patient quarantine.

MERS-CoV

The Middle East Respiratory Syndrome (MERS) began in Saudi Arabia, probably from camels. It spread by inhalation from patients having respiratory symptoms. Neither vaccine nor antiviral therapy is available. Prevention is possible by following hygienic measures and by quarantining the patient.

India escaped SARS and MERS by control measures, including quarantine. Several coronaviruses are circulating in a variety of animals that have not yet infected humans.

SARS-CoV-2 (COVID-19)

Clinical Case 2

A 40-year-old healthy virology laboratory technician engaged in COVID-19 testing presents with fever and myalgia for 1 day. Her SpO2 is 98% and RR 16/min. She lives in a family of four. None of her family members have symptoms.

1. *What test will you advise her?*
2. *When will you test her family? Provide the rationale for your decision.*
3. *If the family tests negative, what advise will you give her?*
4. *What are the chances of her developing severe disease? Enumerate its risk factors.*

Epidemiology

This novel coronavirus probably mutated from a bat. Human infection began from Wuhan, China in December 2019 and was declared a global pandemic by WHO in March 2020. Since then a number of variant strains of the virus have occurred across the world. As of July 2021 it has affected most countries, with 19.4 crore cases and 41.5 lakh deaths globally. In India, there were totally 3.14 crore cases and 4.21 lakh deaths till July 2021. India's first case was imported into Kerala in January 2020 followed by a pan-India outbreak of 4,000 cases from a single cluster in Delhi that lead to several clusters across India. Maharashtra, TN, AP and Karnataka were the worst hit with Delhi, Mumbai, Pune, Chennai and Bengaluru having large numbers.

The infection is transmitted by inhaling droplets from an infected person when in close proximity (less than 1 metre) or by touching inanimate fomites. It is also airborne, especially in rooms with poor ventilation. The transmissibility of the infection is markedly reduced by wearing a mask, maintaining a social distance and handwashing. Aerosol-generating procedures in a healthcare setting—centrifuging, surgical drilling and all anesthesiologic respiratory maintenance procedures—mandate the use of an N95 respirator.

Virology

The SARS-CoV-2 is an ssRNA enveloped virus in helical symmetry. The lipoprotein envelope consists of three structural proteins—spike (S), membrane (M) and envelope (E)—and other non-structural proteins. The spike protein binds the virus to the host cell receptor.

Pathogenesis and clinical features

On the virus envelope S protein binding the angiotensin converting enzyme-2 (ACE-2) receptor, the virus enters the oropharyngeal epithelium and multiplies. The ACE-2 receptors are found on the oropharyngeal epithelium, alveoli, vascular endothelium, heart, kidney, liver and gastrointestinal tract. The spike protein has two sub-units: S1 that recognises ACE-2 and S2 that once attached helps the virus to fuse with the host cell membrane. The viral genome then enters the host cell to hijack its machinery and generate new viruses. This invasion necessitates the S1/S2 to be cut apart at the 'furin cleavage' site.

The incubation period is 5 to 6 days before the onset of an ILI, if symptomatic. The virus is shed during the incubation period, the infection being transmissible even if asymptomatic. The patient remains infectious for 3 weeks after infection and is therefore quarantined during this time. Re-infection with the virus is rare.

Of the COVID-19-infected, 80% are asymptomatic or mildly symptomatic, 15% develop severe infection, requiring oxygen, and 5% are critical, requiring ventilation. The severity of the disease depends on the risk factors for age (> 60 years) and underlying comorbidities like hypertension, diabetes, chronic obstructive pulmonary disease (COPD), chronic kidney disease and the host immune response to the virus.

The ILI presents with fever, myalgia, sore throat, cough, anosmia, loss of taste or diarrhea. Most cases recover in 7 to 10 days. In others, the disease progresses to COVID-19 pneumonia, as the virus causes the alveoli to collapse due to a decrease in the pulmonary surfactants and subsequent edema in the interstitial lung space. This may result in an acute respiratory distress syndrome (ARDS). The host immune response to the virus also results in a proinflammatory 'cytokine tsunami or storm' and microemboli that have deleterious systemic consequences causing multi-organ failure.

Clinically, the disease is categorised as mild (ILI), moderate (pneumonia with $SpO_2 < 94\%$, RR > 24/min) and severe acute respiratory illness (SARI with $SpO_2 < 90\%$, RR > 30/min).

Laboratory diagnosis

1. The rapid antigen test (RAT) on an oropharyngeal swab by immunochromatography is a point-of-care test with a 50% sensitivity and 100% specificity. RAT-positive cases are considered as confirmed COVID-19 cases but RAT-negative cases need to be tested by a molecular method to rule out false-negatives.
2. Molecular methods to detect viral RNA—RT-PCR, CBNAAT and TrueNAT kits—are commercially available that detect specific COVID-19 genes. The RT-PCR has the advantage of testing 90 samples at a time with a turnaround time of 5 to 6 hours and requires a BSL 2 laboratory. The 'Cycle Threshold (CT)' value generated in the RT-PCR has no bearing on the course or prognosis of the disease. The CBNAAT and TrueNAT take 1 hour but can test only one to four samples at a time.
3. IgG detection in serum by ELISA is used only for sero-surveillance and not for diagnosis.

The antigen or viral RNA can be detected from a nasopharyngeal or throat swab collected in a VTM. It is positive from 5 days to 20 days after infection, while the IgM antibodies are detectable 10 days after infection and IgG appears 14 days after infection. The period following appearance of IgG and until its disappearance is variable and needs further study. Herd immunity data reported has varied from 20% to 40% in various geographic populations studied across the timeline in India.

As per Indian Council of Medical Research (ICMR) guidelines, testing is recommended for all symptomatic ILIs and SARIs, asymptomatic high-risk contacts of COVID-19-positive cases (once between 5 and 10 days after contact) and ANC cases at delivery. Besides specific viral markers, elevated IL6, D-dimer, CRP and ferritin are prognostic laboratory markers. In moderate cases, CT lung showing COVID pneumonia may necessitate administration of oxygen, steroids and anticoagulants.

Prevention

Masking, handwashing and social distancing are recommended. If infected, the patient is quarantined (home or facility) for 2 weeks to prevent transmission to the uninfected. A number of vaccine strategies have been employed in the COVID-19 2020 pandemic:

1. **Nucleic acid vaccines:** mRNA vaccines teach our cells to make a microbe protein that triggers an immune response, e.g., COVID-19 vaccine.
2. **Viral vector vaccines:** They use a modified version of a different harmless virus as a vector to deliver instructions to our cells to produce a virus protein, e.g., COVID-19 vaccine.
3. **Whole killed virus:** The whole cell of the killed virus is injected intramuscularly.

The vaccine strategies for COVID-19 will continue to evolve with respect to dose timing, frequency and efficacy of response. Presently the efficacy of response varies from 70 to 90% on completion of the course of immunisation. In general, the available vaccines have a role in largely preventing the severity and mortality but do not prevent infection or transmission of infection.

MUMPS

Mumps, measles, rubella, chickenpox and parvovirus infection are viral childhood fevers but unlike the others mumps does not involve the skin. The mumps virus, a paramyxovirus with a single strand of RNA, has an affinity for the salivary glands and CNS. It is spread by inhalation of droplets from infected cases, and is common between age 5 and 15 years. After an incubation period of 20 days, the disease manifests as a febrile illness with swelling and painful inflammation of the parotid and submaxillary salivary glands. Mumps is self-limiting, but may cause complications like aseptic meningitis, meningoencephalitis, orchitis, pancreatitis, oophoritis or thyroiditis. Orchitis is seen to occur in adult males who contract mumps. On recovery, the immunity is lifelong. The diagnosis is clinical, without laboratory support. There is an effective live attenuated vaccine for mumps combined with measles and rubella (MMR) given in two doses at 9 months and 16–24 months.

EPSTEIN–BARR VIRUS

Infection with EBV is very common across the world. It is acquired during childhood and adolescence by contact with oral secretions, most adults being seropositive. EBV is a dsDNA, enveloped virus belonging to *Herpesviridae* with one serotype and oncogenic potential.

On contact with oropharyngeal secretions, it binds to, enters and replicates in the epithelial cells and B lymphocytes in the pharynx and salivary glands. In children, this primary infection is usually asymptomatic and self-limiting but transmissible. The virus is seeded to other sites through the B cells but remains quiescent and may be reactivated later in life.

In adults, the primary infection presents with fever, pharyngitis and cervical lymphadenopathy. The blood shows atypical lymphocytosis and auto-antibodies cross-reacting with sheep RBC antigens. The disease is called infectious mononucleosis or glandular fever or kissing disease. EBV infection can induce malignant transformation in the infected cells, especially in the immunocompromised host. The malignancies associated with persistent EBV infection are nasopharyngeal carcinoma, Hodgkin's and non-Hodgkin's lymphoma, and Burkitt's lymphoma, In the immunocompromised, it is associated with lympho-proliferative disorders and oral hairy leucoplakia.

The laboratory diagnosis of EBV infection is indicated in infectious mononucleosis and EBV-associated malignancies. Detection of EBV-specific IgM and IgG antibodies by ELISA denotes present and past infection. Detection of EBV-specific antigens by immunofluorescence, or DNA or RNA by PCR denotes present infection. A non-specific heterophile antigen agglutination test that detects auto-antibodies cross-reacting with sheep RBC antigens, such as the Paul-Bunnell test, is no longer in use. There is no vaccine for EBV infection.

ANSWERS

Clinical Case 1

1. *Influenza*
2. *Category C*
3. *Nasopharyngeal swab collected in a viral transport medium (VTM) for PCR*
4. *Trivalent inactivated influenza vaccine to be given IM, single dose annually.*

Clinical Case 2

1. *RT-PCR for COVID-19*
2. *Family to be tested immediately as they are high-risk close contacts from the day she was infected, i.e., 5 days before the onset of her symptoms.*
3. *To quarantine herself for 2 weeks from her family.*
4. *Unlikely, as she is young and healthy with mild symptoms. Risk factors are age (> 60 years) and underlying comorbidities like hypertension, diabetes, COPD, and chronic kidney disease.*

QUESTIONS

1. Explain the rationale in not prescribing antibiotic in the treatment of common cold.
2. Categorise influenza with respect to the indication for testing and antiviral treatment.
3. Provide reasons why influenza has the potential to become an epidemic or a pandemic.
4. Can influenza be prevented?
5. Name the three coronaviruses that have caused an epidemic. Outline the pathogenesis and laboratory diagnosis of the coronavirus epidemic seen in India.
6. Elaborate on the COVID-19 vaccine strategies and the role of vaccines in the pandemic.
7. Name the childhood viral fevers. Why should mumps be prevented with vaccination?
8. Enumerate oncogenic viruses and write a brief note on EBV infection.

CHAPTER 39

Fungal and Parasitic Pulmonary Disease

The fungi infecting the lungs include *Aspergillus* spp., *Candida* spp., *Cryptococcus* spp. and *Pneumocystis jirovecii*. Aspergillosis and pneumocystis are predominantly pulmonary diseases, while candidiasis can affect the skin, endocardium, urinary bladder and meninges. *Cryptococcus* spp. enter by inhalation but predominantly cause meningitis in the immunocompromised. Mucormycosis infection begins in the nasal and paranasal sinuses but largely affects the brain. Parasitic infections include rhinosporidiosis in the nasal cavity and paragonomiasis in the lung.

ASPERGILLOSIS

Of the many species of the fungus *Aspergillus*, *Aspergillus fumigatus* is the commonest pathogen. Other species associated with infection include *Aspergillus flavus* and *Aspergillus niger*. Aspergillus is a saprophyte in the environment, causing opportunistic infection in man.

In humans, *Aspergillus* spp. are associated with pulmonary disease and superficial infections. The manifestation of pulmonary disease caused by *Aspergillus* spp. depends upon the pathologic interaction between man and the fungus.

Pulmonary aspergillosis

In allergic aspergillosis, inhaled spores of the fungus cause an atopic type I hypersensitivity like asthma. IgE antibodies and eosinophilia can be detected in blood but the fungus does not colonise the respiratory tract.

Allergic bronchopulmonary aspergillosis is an exaggeration of allergic asthma, where the *Aspergillus* may grow on the surface of the respiratory epithelium but does not invade it. The hypersensitivity is both type I and type III, resulting in pulmonary eosinophilia. The fungus can be grown on culture of the sputum.

An 'aspergilloma' is a fungal ball that colonises and grows within a pre-existing pulmonary cavity, following tuberculosis or bronchiectasis. The space in the cavity permits the growth of sporulating aerial hyphae of the *Aspergillus*. Culture is generally negative as the cavity may not be in continuity with the bronchus. Radiographic evidence combined with IgG antibodies in blood are diagnostic.

Invasiveaspergillosisisuncommon,andseenpredominantly in the lungs but can involve the brain, kidney or heart. It is a pneumonia that occurs only in immunocompromised patients on immunosuppressive therapy or cytotoxic drugs for malignancies. Histopathological examination shows septate, thin-walled hyphae with dichotomous branching, radiating from a central focus.

Superficial aspergillosis

Cutaneous aspergillosis is rare, but otomycosis, mycotic keratitis and infection of the paranasal sinuses can occur.

Laboratory diagnosis

As *Aspergillus* is a saprophytic contaminant, a single sample may not be adequate to signify infection. Repeated isolation of *Aspergillus* from more than one sample is important to identify it as a causative agent.

1. Demonstrate *Aspergillus* in a 10% KOH mount of a clinical sample, showing narrow, septate hyphae with dichotomous branching
2. Culture sample on SDA; the fungus generally grows in 3 days; examine and identify on an LPCB mount (Fig. 39.1).
3. Histopathologic evidence of aspergilloma or invasive pulmonary aspergillosis on H&E or GMS examination of lung tissue.

Other fungi causing pulmonary infection include *Candida* spp. and *Cryptococcus* spp.

PNEUMOCYSTIS PNEUMONIA

Pneumocystis pneumonia, caused by *Pneumocystis jirovecii*, is an opportunistic infection in HIV/AIDS.

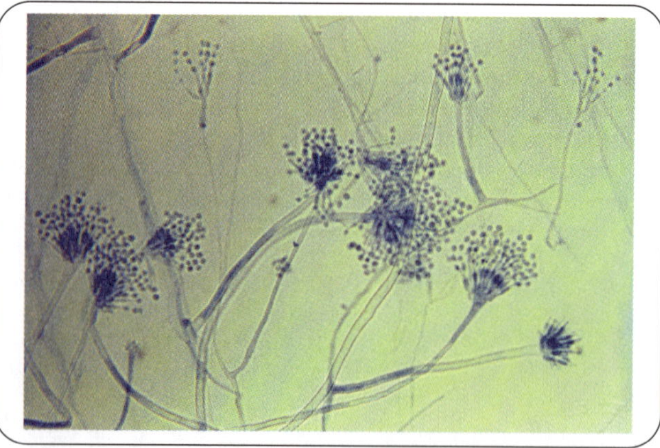

Fig. 39.1 LPCB mount showing mycelium of *Aspergillus flavus*, 40X (*Source*: Public Health Image Library, ID 19407 / Centers for Disease Control and Prevention [CDC] / Dr. Lucille K. Georg)

Though classified as a fungus, *Pneumocystis jirovecii* bears resemblance to a protozoan and occurs in two morphologic forms—cyst and trophozoite. Cysts inhaled from the environment develop into trophozoites in the lung in the immunocompromised. Both cysts and trophozoites are seen in the lung, causing inflammatory exudate in the alveoli, resulting in pneumonia.

Toluidine blue staining of the expectorated sputum or bronchoalveolar lavage shows lavender-coloured, cup-shaped cysts, 5 μm in size with a distinct outline. Histopathologic examination of lung tissue stained by H&E shows the cysts to confirm the diagnosis (Fig. 39.2). The fungus cannot be cultured or tested by serological tests. Cotrimoxazole is the drug of choice.

MUCORMYCOSIS

Mucormycetes are a group of saprophytic fungi that are capable of acute, rapidly fulminant infection in debilitated patients. They are ubiquitous but can cause disease only in the presence of predisposing factors, the commonest being rhinocerebral mucormycosis in diabetic ketoacidosis. Other predisposing factors include leukemias and chemotherapy.

In diabetic ketoacidosis, the fungus enters through the nasal cavity. It has a predilection for blood vessels, and rapidly proceeds towards the paranasal sinuses and eye, and onwards to the brain. Mucormycosis of the paranasal sinuses, ocular cellulitis and rhinocerebral infection have poor prognoses and are generally fatal.

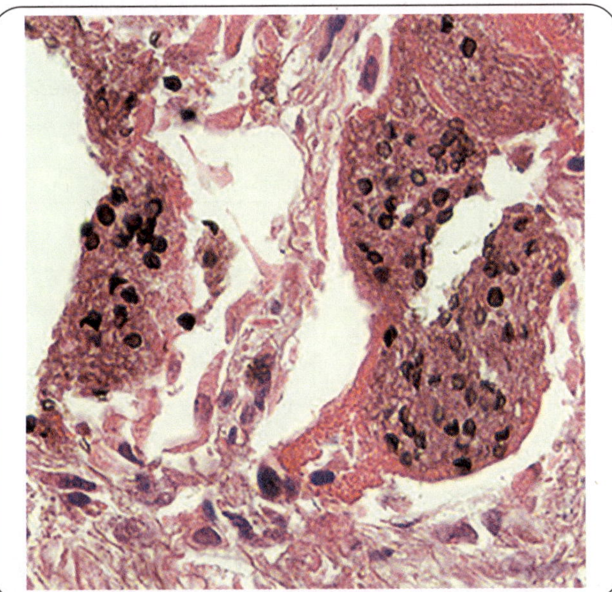

Fig. 39.2 Histopathologic examination showing cysts of *Pneumocystis jirovecii*, H&E, 40X (*Source*: CDC/DPDx)

Laboratory diagnosis

1. Demonstrate mucor in a 10% KOH mount of a clinical sample, showing broad, non-septate hyphae with obtuse branching.
2. Culture sample on SDA; the fungus generally grows in 24 hours; examine and identify on an LPCB mount.
3. Histopathologic evidence on H&E and GMS.

RHINOSPORIDIOSIS

Rhinosporidiosis is a chronic granulomatous infection of mucocutaneous tissue characterised by large polyps that are hyperplastic, vascularised, friable and sessile or pedunculated. Earlier considered a fungus, *Rhinosporidium seeberi*, the causative agent is now classified as a coccidian protozoan.

The most predominant site is the nose, followed by the conjunctiva, and rarely the anus, penis, vagina, ears, pharynx or larynx. The infecting spore penetrates the mucosal epithelium and matures in the subepithelial tissue to form a large spherule which in turn ruptures to release more spores.

Diagnosis is by histopathological examination of the affected tissue as it cannot be cultured invitro, and treatment is by surgical excision.

PARAGONIMIASIS (LUNG FLUKE)

Paragonimiasis is endemic in Manipur and Northeast India and is acquired by ingestion of improperly cooked crab carrying the infective larva (metacercaria). The larva travels from the intestinal lumen to the lung parenchyma penetrating the tissues. It matures into an adult near a bronchus and excretes eggs that are swallowed and excreted in the stool.

The adult is an 8 × 16 mm, thick, fleshy, oval, hermaphrodite with one oral anterior and one ventral sucker (Fig. 39.3). The average life span of the adult is 6 years. The ovum excreted in stool measures 100 μm; it is oval in shape and operculated. The ova undergo development in two intermediate hosts—first snail and second crab to form the larva infective to man.

The adult residing in the lung causes eosinophilic granulomatous inflammation and cysts, resulting in productive cough and hemoptysis. The adult can migrate to the brain, liver or spleen. The adult can be located on imaging or histopathological studies. In the microbiology laboratory, the ova can be demonstrated in sputum or stool, or antigen or antibody demonstrated in serum by ELISA. Praziquantel is the drug of choice.

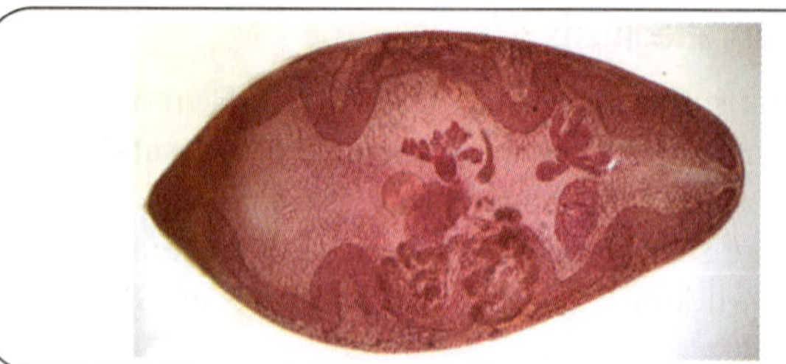

Fig. 39.3 Adult of *Paragonimus westermani* (*Source*: CDC/DPDx)

EOSINOPHILIC PNEUMONIA

The migrating larvae of *Ascaris lumbricoides*, hookworm and *Strongyloides stercoralis* (visceral larva migrans) can affect the lungs. They cause an immune-mediated hypersensitivity leading to an eosinophilic pneumonia. It presents with respiratory symptoms of cough and dyspnea, lung infiltrates being detected on chest X-ray with peripheral eosinophilia.

QUESTIONS

1. Enumerate fungi affecting the respiratory system. Explain the pathologic forms of pulmonary aspergillosis.
2. Enumerate fungi affecting the respiratory system and write a note on pneumocystis pneumonia.
3. Enumerate parasites affecting the respiratory system and write a note on rhinosporidiosis.

SECTION SUMMARY

Section VII: Respiratory Tract Infections

Infection/pathogen	Virulence/pathogenesis	Laboratory diagnosis	Vaccine/recommended antimicrobial
Upper RTI			
Viral common cold	Rhinovirus, coronavirus, metapneumovirus, bocavirus, RSV	Not indicated	Not indicated
Pharyngitis			
Streptococcus pyogenes	Carriers in pharynx, capsule, toxins	Smear culture throat swab	Amoxicillin, azithromycin
Streptococcus pneumoniae	Capsule	Smear culture throat swab	Pneumococcal polysaccharide, Pneumococcal conjugate, penicillin, amoxicillin
Haemophilus influenzae	Capsule	Smear culture throat swab	Hib-CV, amoxicillin, amoxicillin-clavulanic acid
Bordetella pertussis	Adhesion, toxemia-catarrhal, paroxysmal, convalescent	Smear culture throat swab	wP killed, aP acellular, erythromycin, azithromycin
Corynebacterium diphtheriae	β phage, pseudo-membrane, exotoxins-toxemia	Smear culture throat swab, Albert stain	Toxoid, ADS, penicillin, erythromycin
Borrelia vincentii with *Leptotrichia buccalis*	Vincent's angina	Gram stain of throat swab	Penicillin, metronidazole
Lower RTI Pneumonia			
Streptococcus pneumoniae	Community-acquired lobar pneumonia	Sputum smear and culture	Pneumococcal polysaccharide, pneumococcal conjugate, amoxicillin-clavulanic acid
S. aureus			
Haemophilus influenzae	Hospital-acquired pneumonia	Sputum smear and culture	Hib-CV

Infection/pathogen	Virulence/pathogenesis	Laboratory diagnosis	Vaccine/recommended antimicrobial
Klebsiella pneumoniae *Pseudomonas aeruginosa*			
Mycoplasma pneumoniae *Chlamydophila pneumoniae* *Legionella* *Pneumophila*	Atypical pneumonia	Culture, antigen detection, PCR	Azithromycin
Tuberculosis	Inhalation, resists intracellular killing by macrophages, pulmonary and EPTB	Sputum AO/ZN. MGIT culture, CBNAAT, and LPA diagnose and detect MDR/XDR TB	BCG at birth for tuberculin conversion First line: rifampicin, INH, PAS, ETB
NTMs	*M. avium-intracellulare* *M. kansasii*	Sputum smear ZN, LJ culture	Azithromycin, amikacin, clofazimine, rifampicin
Influenza	URTI may lead to pneumonia Cat A, B and C Hemagglutinin neuramindase receptors	Nasopharyngeal swab in VTM for PCR Rise in antibody titre by ELISA	Trivalent inactivated influenza vaccines (TIVs) IM, live attenuated influenza vaccines (LAIV) intranasal, oseltamivir
Coronaviruses SARS-CoV-1 MERS-CoV **SARS-CoV-2 (COVID-19)**	URTI, asymptomatic, highly transmissible, cytokine-mediated damage and microemboli, fatal with co-morbidities	Rapid antigen test, RT-PCR from nasopharynx IL-6, D-dimer, CRP and ferritin as prognostic markers	Whole killed virus or subunit (mRNA) or modified viral vector, remdesivir
Mumps	Inflammation of the parotid and submaxillary salivary glands, rare complications	Clinical diagnosis	Live attenuated (MMR)
EBV	Oropharyngeal, infectious mononucleosis, carcinogenic potential	Antibody detection, PCR	
Aspergillosis	Allergic Type I Allergic bronchopulmonary Aspergilloma Invasive	IgE, eosinophilia Pulmonary eosinophilia IgG HPE	Amphotericin B
Pneumocystis jirovecii	Immunocompromised, HIV	Cysts-toluidine blue smear, HPE	Cotrimoxazole
Mucormycosis	Diabetic ketoacidosis	Smear culture HPE	Amphotericin B
Rhinosporidiosis	Chronic granulomatous nasal polyp	HPE only Cannot be cultured	
Paragonimus westermanii	Ingestion of larvae in crab-intestine-lung-adult, eosinophilic granulomatous	Ova in stool	Praziquantel

Chapter 40

Urinary Tract Infection

The urinary tract is lined with an epithelium that is continuous with the skin. The periurethral area has fecal bacteria like *E. coli* that give rise to ascending urinary tract infection. Hematogenous infection of the kidney is rare.

In terms of the rate of incidence, urinary tract infection (UTI) is second to upper respiratory tract infection (URTI). Since the female urethra is shorter than the male urethra, women are more prone to UTI. Bacteriuria (presence of bacteria in urine), whether symptomatic or asymptomatic, is common in women of all ages, including in those with recurrent episodes of UTI. UTI can be infection of the bladder (cystitis) or the kidney (pyelonephritis) or renal abscess. Urethritis and vaginitis are sexually transmitted diseases and prostatic infection is separate from UTI. UTI can be uncomplicated or complicated. Uncomplicated UTI is infection in the absence of functional or anatomical impairment that would promote UTI. Complicated UTI is infection associated with factors that increase colonisation and decrease the efficacy of therapy. Bacteria causing UTI are listed in Table 40.1.

Table 40.1 Bacteria causing UTI

Uncomplicated	Complicated or Nosocomial
E. coli	*E. coli*
Other *Enterobacteriaceae* (*Klebsiella, Proteus*)	*Klebsiella*
Staphylococcus saprophyticus	*Pseudomonas*
Enterococcus spp.	*Acinetobacter*
Group B streptococci	*Serratia*
Chlamydiae	

Pyelonephritis is an invasive infection with fever, flank pain, tenderness and peripheral leucocytosis. The microscopic examination of urine in pyelonephritis shows WBC casts—elongated structures composed of cells packed in tubules and excreted in a proteinaceous matrix. Pyelonephritis may lead to bacteremia, or renal abscess or pyogenic perinephric abscess. It is usually treated in a hospital with 2–6 week of antibiotic therapy.

Recurrence of UTI (cystitis) can be a relapse (of the original infection by the same strain) or re-infection. Relapse occurs shortly after cessation of antibiotics and is due to the persistence of the bacteria in the urinary tract or a structural abnormality. Re-infection can be prevented by breaking the cycle of colonisation of the periurethral area.

Clinical Case

A 45-year-old married lady presents with sudden onset hematuria, urgency to micturate and low-grade fever.

1. *What is the most likely diagnosis and the most likely microbe source of infection?*
2. *How will you confirm the diagnosis?*
3. *Name the drug of choice.*
4. *What is the follow-up investigation recommended?*

Epidemiology

Commensals in the human intestine, referred to as coliform bacilli, e.g., *E. coli, Klebsiella, Proteus, Providencia, Morganella, Enterobacter* and *Citrobacter* are all capable of causing endogenous ascending UTI. Other bacterial causes include *Pseudomonas* spp., gram-positive cocci—*Staphylococcus aureus* or *Staphylococcus saprophyticus* or *Staphylococcus epidermidis*—and *Enterococcus* spp. Amongst all these bacteria, *E. coli* accounts for 70% of UTIs, known as uropathogenic *E. coli* (UPEC) with numerous serotypes.

Bacteriuria means 'bacteria in urine'. Urine in the bladder is sterile but the urethra and periurethral areas are not and may contaminate the urine during collection. Hence, the term 'significant bacteriuria' ($> 10^5$ bacteria/mL) describes the number of bacteria in voided urine, taking into account this contamination.

Urinary tract infection (UTI) may refer to infection of the lower urinary tract or both the upper and lower urinary tracts. Cystitis refers to bladder infection and pyelonephritis of the kidney, while urethritis or vaginitis are sexually transmitted infections. Uncomplicated UTI is infection in a structurally and neurologically normal urinary tract. Complicated UTI refers to infection in the presence of calculi, obstruction, in-dwelling catheters, or pregnancy. Recurrences of UTI may be relapses or re-infections. Relapse is the persistence of the bacterium in the urinary tract and is therefore seen with the same bacterium as the primary infection. Re-infection is a new infection, generally from a different bacterium from the primary infection.

Uncomplicated cystitis in females is the most common UTI in a community setting while catheter-associated UTI is the most common complicated UTI.

Pathogenesis

Risk factors are age, as the elderly are inclined towards incomplete emptying and urine stasis, females due to the presence of a short urethra, sexual intercourse without voiding of urine after, use of spermicidal contraceptives, estrogen deficiency in menopause, perineal contamination with intestinal flora, pregnancy, urinary obstruction, and diabetes.

UTI can be ascending or descending (hematogenous). Ascending UTI is when the route of entry of bacteria is from the urethral meatus upwards towards the bladder and is more common. The urethra is colonised by uropathogenic *E. coli*. Once within the bladder, bacteria can multiply and pass up the ureters. Descending UTI is the seeding of bacteria via the bloodstream to the kidneys causing pyelonephritis. Catheter-associated UTI (CAUTI) is a hospital-acquired infection and is considered as complicated UTI.

Most UTIs are caused by distinct urovirulent serotypes or uropathogenic *E. coli* present amongst the commensals in the intestine. They are selected from the commensal by virtue of the presence of virulence factors that enhance colonisation and invasion of the urinary tract. *E. coli* causing cystitis and pyelonephritis are genetically distinct in the O (cell wall), K (capsular), and H (flagellar) antigens.

Virulence factors for uropathogenicity are a variety of adhesins corresponding with vaginal and uroepithelial receptors for adherence, resistance to serum bactericidal activity, increased K antigen, presence of hemolysin, cytotoxic necrotising factor and also aerobactin (an iron-chelating agent), and the siderophore (an iron-chelating agent) receptor. Structural abnormalities in the form of obstruction to the flow of urine due to calculi, anatomic defects, strictures, and prostate enlargement contribute to the occurrence of UTI.

Host factors also play a role in preventing UTI—flushing of urine, pH of vagina, and antibacterial mucosal secretions and immune response.

Lower UTI manifests with increased frequency and painful or burning micturition, the urine being turbid or blood-tinged. Upper UTI manifests with fever, flank pain and lower UTI symptoms.

Laboratory diagnosis

Urine sample collection: Clean catch, early morning, mid-stream urine to be collected in a sterile container. Avoid contamination of the sample by cleaning the genital parts prior to collection and voiding the initial stream of urine. In catheterised patients, urine should be collected from the catheter and not from the urobag.

Urine sample needs to be transported immediately to the laboratory to avoid multiplication of bacteria after collection.

1. *Microscopic examination* — Examination of a wet mount is done to screen for pus cells and bacteria, with pyuria defined as the presence of 10 WBCs/mm^3 of urine. Rapid dipstick tests detect leukocyte esterase from WBCs or nitrite from bacteria in urine. Examination of Gram staining for pus cells and bacteria is done, and the presence of at least one bacterium/OI field in an uncentrifuged urine specimen correlates with significant bacteriuria. The absence of bacteria in the sediment of centrifuged urine rules out UTI.
2. Urine in the bladder is sterile but can be contaminated with urethral and periurethral bacteria during voiding; hence, the concept of 'significant bacteria' by quantifying the number of bacteria in an uncentrifuged urine specimen.

Urine culture on CLED (cysteine lactose electrolyte deficient), or MacConkey and blood agar for qualitative and semi-quantitative analysis is done using a calibrated loop. The isolate is identified using standard biochemical tests following overnight incubation. Interpretation of a single bacterium growth > 10^5 CFU/mL is considered as significant. Isolation of more than three types of bacteria and or less than the significant number implies contamination.

The interpretation of pyuria without significant bacteriuria is possible if the patient is already on antibacterial therapy. The presence of pus cells >10^4/mL of uncentrifuged urine corresponds to one pus cell in seven HP fields, and this is considered significant even in the absence of bacteriuria.

Treatment

All cases of symptomatic UTI need antimicrobial therapy. Antimicrobials effective in the empirical treatment of uncomplicated UTI are nitrofurantoin and cotrimoxazole. Treatment or prophylaxis of asymptomatic bacteriuria is not recommended in the healthy and the elderly but requires specialised concerns in case of pregnancy and diabetes.

Antimicrobial therapy of uncomplicated UTI results either in cure, persistence, relapse or re-infection. 'Cure' is defined when the bacterial count decreases within 48 hours of initiation of the antimicrobial, and culture remains negative for 2 weeks even after cessation of the antimicrobial. 'Persistence' is defined when the bacterial count remains significant after 48 hours and later of the antimicrobial agent and implies resistance to that agent. 'Relapse' occurs within 1 to 2 weeks of cessation of therapy, and infection with the same bacterium. 'Re-infection' occurs after cure, and is with another bacterial species or another serotype of *E. coli*.

Enterobacteriaceae are gram-negative bacilli that share certain common biochemical properties. They include:

1. Commensals in the human intestine, referred to as coliform bacilli, e.g., *E. coli, Klebsiella, Proteus, Providencia, Morganella, Enterobacter, Citrobacter*.

2. Enteric pathogens, *Salmonella* and *Shigella*.
3. Genus *Yersinia* causing zoonotic plague and enteric infection.

Klebsiella

This is a gram-negative, non-motile, capsulated bacillus that is a commensal in the human intestine and saprophyte in the soil.

Klebsiella pneumoniae is pathogenic and has three sub-species.

Klebsiella pneumoniae subspecies *pneumoniae* causes lobar pneumonia, UTI, neonatal meningitis septicemia, abscess surgical site infection and ventilator-associated pneumonia. It is a nosocomial pathogen and a multi-drug resistant organism (MDRO). *K. pneumoniae* subspecies *ozanae* causes atrophic rhinitis while *K. pneumoniae* subspecies *rhinoscleromatis* causes rhinoscleroma, a chronic granulomatous hypertrophy of the nasal mucosa and rarely the upper respiratory mucosa.

Sample smear examination by Gram stain, culture and antimicrobial susceptibility testing are the mainstays in the laboratory diagnosis of all *K. pneumoniae* infections irrespective of the site/disease.

Other rarely encountered gram-negative bacilli are *Enterobacter, Proteus, Morganella, Providencia, Serratia, Hafnia* and *Pantoea,* especially in urinary, respiratory and wound nosocomial infections, or septicemia. They can all be identified by culture with biochemical characterisation and antimicrobial susceptibility testing.

Bacteria and viruses causing urethritis are sexually transmitted. Parasites demonstrable in urine are *Trichomonas vaginalis* causing urethritis or vaginitis, both of which are transmitted sexually. One of them is *Schistosome haematobium*, rarely seen in India (but exceptionally reported in Ratnagiri, Maharashtra), and the other is *Wuchereria bancrofti*, seen when presenting with chyluria in filariasis. Amongst fungi, *Candida albicans* can cause UTI.

Answers

Clinical Case

1. *Uncomplicated urinary tract infection, most likely cystitis*
2. *E. coli from intestinal flora contaminating the periurethral area*
3. *Nitrofurantoin or cotrimoxazole*
4. *Abdominal and pelvic sonography to rule out the presence of calculi*

Questions

1. Enumerate the risk factors for urinary tract infection.
2. Explain the pathogenesis of UTI with regard to *E. coli* virulence and host interactions.
3. Name bacteria causing urinary tract infection. How will you interpret a quantitative and qualitative urine culture report?
4. Name antimicrobials indicated in treating uncomplicated urinary tract infection. Define what is meant by *cure, persistence, relapse* and *re-infection* of UTI.

Sexually Transmitted Infections

Sexually transmitted infections (STIs) are infections transmitted through exchange of body fluids during sexual intercourse, and can either cause genital ulcer, mucosal infection or systemic infection in blood or other organs. In India, the annual incidence of STI is estimated to be 5% of the population, i.e., 50 million cases. High-risk groups include those who travel for long periods, commercial sex workers, and homosexuals.

The microorganisms that can be transmitted sexually are listed in Table 41.1. With the exception of *T. pallidum*, HIV, HBV and HCV, most STD-causing microorganisms enter through the local mucosal site (vagina, cervix, urethra, rectum or oral pharynx) during sexual contact with an infected partner. HIV, HBV and HCV can also be transmitted through blood transfusion or contact with contaminated needles, or from mother to child.

All STD-causing microorganisms are sensitive to chemical disinfection and physical factors and are not found free in the environment. There are no animal reservoirs and transmission is exclusively man to man. Therefore, they are ideal candidates for eradication in the event of an effective vaccine. There are no vaccines for STDs other than hepatitis B. As all are sensitive to chemical barriers, the development of a safe and effective female-controlled topical microbicidal shall prove effective by interrupting transmission of the STD.

Acute manifestations of an STD

1. Mucopurulent cervicitis and urethritis, e.g., gonorrhoea, chlamydia
2. Genital ulcer disease, e.g., syphilis, chancroid, HSV
3. Dysplasia and cancer, e.g., genital warts, cervical cancer by HPV

With the exception of HIV, HBV and HCV, all sexually transmitted microorganisms cause primary lesions at or near the site of entry. HIV infects the T4 cells while HBV and HCV infect hepatocytes, irrespective of their route of entry, sexual or otherwise.

The serious consequences of sexually transmitted diseases due to chronicity are:

1. Pelvic inflammatory disease (PID), an ascending infection of the uterus and fallopian tubes caused by gonococci and chlamydia that may result in chronic pelvic pain, infertility and ectopic pregnancy. Risk factors for PID include multiple sexual partners, use of an IUD, young age at first sexual intercourse, high frequency of sexual intercourse and vaginal douching.
2. Anogenital and cervical cancer: HPV
3. Secondary and tertiary syphilis
4. Recurrent herpes infection
5. Congenital disease of the newborn: syphilis, herpes, chlamydia and HPV
6. Increased risk of HIV infection
7. Adverse outcomes of pregnancy: bad obstetric history, low birth weight, premature rupture of membrane

Table 41.1 Microbes that are transmitted sexually

Microbe	Disease	Microorganism	Laboratory diagnosis
Bacterium	Gonorrhea	*Neisseria gonorrhoeae*	Gram stain
	Lymphogranuloma venereum (LGV)	*Chlamydia trachomatis*	Inclusion body Serology
	Chancroid	*Haemophilus ducreyi*	Culture
	Syphilis	*Treponema pallidum*	VDRL, TPHA
	Bacterial vaginosis	*Gardnerella vaginalis*	Microscopy-clue cells
	Granuloma inguinale	*Donovania granulomatis*	
	Non-gonococcal urethritis	*Mycoplasma, Ureaplasma*	Serology
Viral	AIDS	HIV	Serology, PCR
	Serum hepatitis	HBV, HCV	Serology, PCR
	Genital herpes	HSV Type I and II	Serology, PCR
	Warts	HPV	Serology, PCR
Fungi	Vaginitis, moniliasis	*Candida albicans*	Microscopy, Culture
Protozoa	Vaginitis	*Trichomonas vaginalis*	Microscopy

The current practice in the diagnosis of STIs is the syndromic approach (Table 41.2) wherein the history and clinical examination takes precedence over etiologic microbial diagnosis. It includes treatment as per the syndrome, treating both partners, prevention by use of a condom and screening for blood-borne sexually transmitted viruses. The presence of STI increases the risk of HIV infection.

URETHRITIS

Gonococcal urethritis

 **Clinical Case 1**

A 25-year-old man presents with a complaint of yellow urethral discharge. He gives a history of being sexually active with commercial sex workers.

1. *What is your provisional diagnosis?*
2. *How will you proceed to investigate and subsequently treat this case?*
3. *If untreated, what complications can occur?*

The bacterium most commonly associated with urethritis is *Neisseria gonorrhoeae* and the disease is called gonorrhea. Urethritis caused by microorganisms other than *N. gonorrhoeae* is referred to as non-gonococcal urethritis (NGU).

The genus *Neisseria* comprises gram-negative aerobic diplococci that consists of two pathogens: *N. meningitidis* causing pyogenic meningitis and *N. gonorrhoeae* causing urethritis, cervicitis and neonatal conjunctivitis.

N. gonorrhoeae is a non-capsulated, gram-negative, kidney-shaped diplococcus measuring 0.6 μ that can be seen intracellularly and extracellularly (Fig. 41.1). The source of infection is an infected woman who is either an asymptomatic carrier or has the disease. Gonorrhea is transmitted from female to male and vice versa through sexual contact and conjunctivitis from mother to neonate during normal delivery.

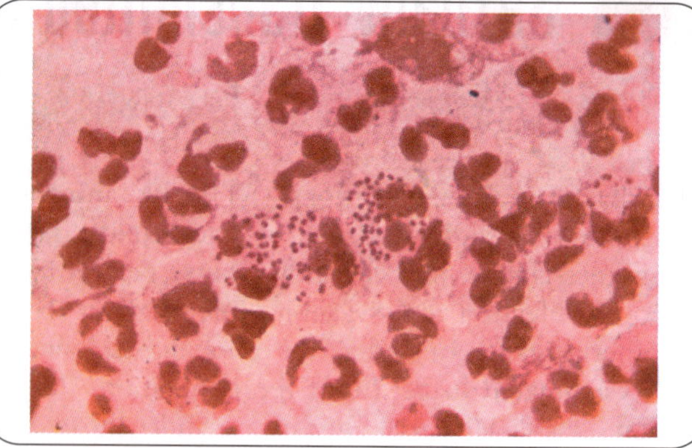

Fig. 41.1 Gram-negative cocci in pairs—*Neisseria gonorrhoea*, Gram stain, 100X (*Source:* Public Health Image Library, ID 15018 / Centers for Disease Control and Prevention [CDC] / Bill Schwartz)

The bacterium possesses a pilus or fimbria which is its organ of adhesion and with which it adheres to the genital mucosa. Besides pili, its virulence is due to the presence of proteins and protease, and a modified gram-negative LPS—lipo-oligosaccharide—that is endotoxic.

In males, acute urethritis manifests as a purulent urethral discharge a week after exposure. If untreated, it can result in prostatitis, epididymitis, balanitis or periurethral abscess with sinus tracts. In females, the acidic pH of the vagina prevents vaginitis but can cause cervicitis, salpingitis and pelvic inflammatory disease leading to sterility. Anorectal and pharyngeal gonorrhea are manifestations following anorectal or oro-genital sex.

Conjunctivitis in the neonate acquired from the birth passage of a female carrier mother is called **ophthalmia neonatorum**, presenting as a purulent eye discharge 4 days after birth. In the mother, it can cause chorioamnionitis or premature delivery.

Disseminated gonococcal infection is extremely rare and associated with specific serotypes resulting in polyarthritis, dermatitis or endocarditis. The concomitant existence of gonorrhoeae is known to enhance the transmission of HIV during sexual intercourse.

Table 41.2 Syndromic approach in sexually transmitted infections

Clinical feature	Common cause	Treatment recommended
Genital ulcer syphilis, chancroid	HSV, LGV, donovanosis	Penicillin/ceftriaxone for bacterial, acyclovir for HSV
Urethral discharge (males)	Gonorrhea, non-gonococcal urethritis (NGU)	Azithromycin/ceftriaxone for bacterial, metronidazole/ tinidazole for trichomoniasis, fluconazole for *Candida*
Vaginal discharge	Same as above	Same as above
Pelvic pain in females	Gonorrhea, chlamydia, anaerobic infection	Cefixime, doxycycline, metronidazole, respectively
Inguinal lymphadenopathy	LGV	Doxycycline
Scrotal swelling	Gonorrhea, NGU	Azithromycin
Ophthalmia neonatorum	Gonorrhea, chlamydia	Baby—ceftriaxone/erythromycin Both parents—azithromycin

Laboratory diagnosis

1. The purulent discharge from the urethra/cervix/conjunctiva collected by a sterile swab is transported to the laboratory immediately or alternatively kept in transport media or devices.
2. Gram stain of the pus shows a non-capsulated, gram-negative, kidney-shaped diplococci, measuring 0.6 μ, seen intracellularly and extracellularly.
3. The pus can be cultured, and bacterial growth identified biochemically and tested for antimicrobial susceptibility.
4. The PCR assay detects specific rRNA genes of gonococci.

Over years of use, gonococci have shown resistance to penicillin, sulphonamides and quinolones, attributed to the production of enzymes like penicillinase. Presently, ceftriaxone or cefixime is recommended to treat, and it is essential to treat both partners.

Non-gonococcal urethritis (NGU)

NGU is a chronic urethritis, wherein the causative agent can be varied, but other than gonococci. The microbes associated with NGU are either bacteria—*Chlamydia trachomatis, Mycoplasma hominis, Ureaplasma urealyticum*, viruses—HSV or CMV, a fungus—*Candida albicans,* or a protozoan—*Trichomonas vaginalis.* The laboratory confirmation of the causative agent is mandated before treatment with the drug of choice (Table 41.3).

Reactive arthritis, formerly referred to as **Reiter's syndrome**, is arthritis along with conjunctivitis and urethritis. The disease affects different organs, and symptoms may or may not appear at the same time. It is mostly seen in HIV-infected individuals following concomitant NGU or gastroenteritis, and linked to HLA B27.

VAGINITIS

Bacterial vaginosis

Bacterial vaginosis is vaginal inflammation due to disruption of the normal vaginal flora by suppression of lactobacilli and proliferation of others, like anaerobes (*Mobiluncus*), and *Gardnerella vaginalis,* a gram-negative bacillus, without inflammation. It is generally seen in women of reproductive age. Risk factors are unprotected sex and vaginal douching.

Symptoms include foul-smelling vaginal discharge, itching and burning micturition. *G. vaginalis* is seen as typical 'clue cells' in the vaginal discharge. Clue cells are vaginal epithelial cells with adherent bacilli (Fig. 41.2). The presence of bacterial vaginosis increases the risk of transmission of other STDs and infections following gynecologic surgery, pelvic inflammatory disease, and preterm labour if during pregnancy.

Bacterial vaginosis is diagnosed by Amsel's criteria if any three of the following four are seen: profuse, thin, white vaginal discharge, pH > 4.5 on a litmus paper test, fishy odour when vaginal discharge is mixed with 10% KOH (whiff test), and presence of clue cells on wet mount microscopy. The Nugent score is the number of bacteria present in the Gram-stained smear of the vaginal discharge. A Nugent score more than 7 is consistent with bacterial vaginosis even in the absence of culture.

Vaginal candidiasis

Also known as vulvovaginal candidiasis or candidal vaginitis or moniliasis, this disease is caused by the yeast *Candida albicans*. The risk factors are pregnancy, diabetes, hormonal contraceptives, and the intake of steroids or antibiotics. It presents with intense itching around the vulva, burning micturition or curdy white

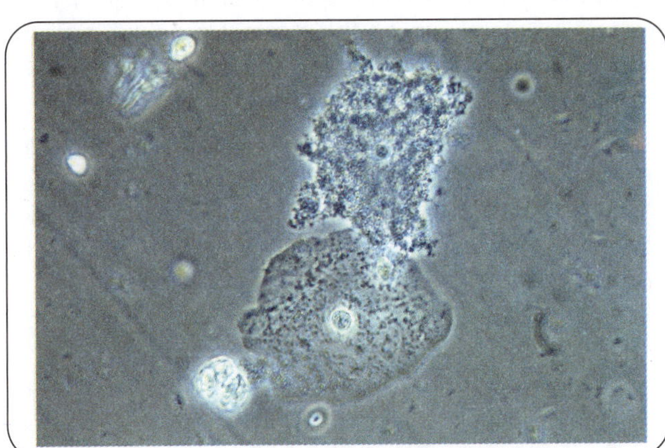

Fig. 41.2 Clue cells, Gram stain, 100X

Table 41.3 Causative agent of NGU—laboratory diagnosis and drug of choice

Microorganism	Diagnostic test	Drug of choice
Chlamydia	Antigen detection	Doxycycline
HSV or CMV	PCR	Antiviral agent
Candida albicans	Culture	Clotrimazole
Trichomonas vaginalis	Trophozoites in urine wet mount	Metronidazole

vaginal discharge. Laboratory examination of the vaginal discharge by smear and culture can confirm candidiasis. Treatment includes a single oral dose of fluconazole with an antifungal vaginal pessary.

Trichomonas vaginitis

T. vaginalis, the causative agent, is a flagellated protozoan that parasitises the vagina and prostate gland. The acidic pH of the vagina is conducive to its growth. It feeds on the mucosal surface of the vagina, causing vaginitis, resulting in a white discharge and intense pruritis. It can also cause urethritis in males and females.

Transmitted at sexual intercourse, it exists only as a motile trophozoite and not as a cyst. The trophozoite measures around 10 µm. It is pear-shaped, with a single nucleus, has an axostyle from the anterior end to the posterior end in the middle of the body, and ends in a pointed tail. The motility is due to the presence of three to five free flagellae arising from the anterior broader end, and one of the flagella being thicker than the others forms an undulating membrane along the body (Fig. 41.3).

The infection is diagnosed by examining a wet mount of vaginal discharge or urine sediment for the motile trophozoites. Treatment is with azole compounds but both partners need to be treated.

GENITAL ULCER DISEASE

Any of the several infections that are characterised by the presence of ulceration on the external genitalia increase the risk of transmission of HIV. Microbes that can cause genital ulceration are bacteria—*T. pallidum* (syphilis), *C. trachomatis* (lymphogranuloma venereum), *Klebsiella granulomatis* (donovanosis), *Haemophilus ducreyi* (chancroid) and viruses—HSV-1 and HSV-2.

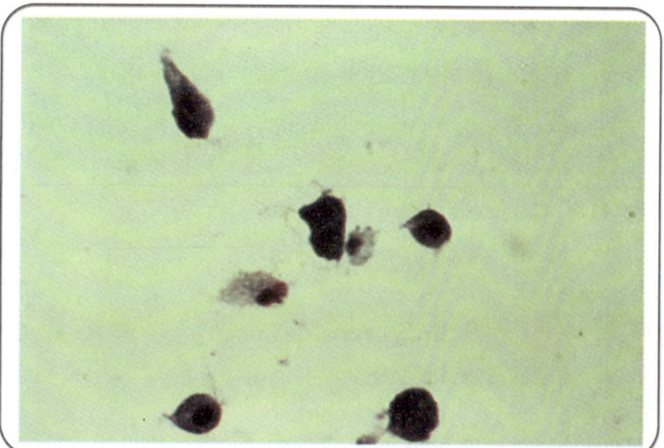

Fig. 41.3 Trophozoite of *Trichomonas vaginalis* (*Source*: Public Health Image Library, ID 5237 / CDC)

Syphilis

Clinical Case 2

A 30-year-old male presents with a painless, hard penile ulceration covered by an exudate (Fig. 41.4). He gives a history of being sexually active and has had the ulcer for 2 weeks.

1. *What steps will you take to examine the patient?*
2. *What microbiologic investigations would you request for?*
3. *What is the drug of choice for this disease?*
4. *If untreated, what is the course of the disease?*

Spirochetes are fine, long, spirally coiled bacilli that exhibit corkscrew or flexion-extension motility due the presence of endoflagella. Being thin, they are difficult to stain by Gram stain and require special silver deposition staining techniques to be visible. The three human pathogenic spirochaetes are *Treponema, Borrelia* and *Leptospira*, the morphology of each being distinct. Treponemes are known to cause syphilis that is transmitted sexually, and yaws and pinta by direct skin contact. *Borrelia* spp. are known to cause relapsing fever, Lyme's disease that is arthropod-borne and Vincent's angina, while *Leptospira* causes Weil's disease that is zoonotic.

T. pallidum is a highly invasive bacterium that can, on contact, penetrate the human skin and mucosa and within hours invade the blood and lymphatics. Venereal syphilis is acquired by any of these routes: (i) sexual contact, (ii) the non-venereal route, i.e., by direct touch, (iii) blood transfusion or (iv) the transplacental route. The incubation period is 9–90 days and varies with the route of infection and the number of infecting bacteria.

Venereal syphilis: Following infection, on invading the blood and lymphatics, during the incubation period the blood is infectious and the bacterium metastasises, including in the CNS, before the appearance of any symptoms. Clinically, if untreated, venereal syphilis has four stages: primary, secondary, latent and tertiary (or late).

Primary syphilis begins with a single, painless papule on the external genitalia that ulcerates, forming a hard indurated chancre covered with an exudate (Fig. 41.4). In homosexuals, the primary chancre may be seen in the rectum or oral cavity. As the bacterium is invasive, care should be taken to examine the ulcer with naked hands and to wear gloves for the examination and collection of samples. In non-venereal occupational syphilis, the ulcer can be seen on the fingers or hand due to direct contact, while in blood transfusion there is no ulcer. One week after the ulcer, the inguinal lymph nodes are palpable but painless and non-tender. Irrespective of the mode of infection, the ulcer heals spontaneously within a month but the lymphadenopathy persists. The presence

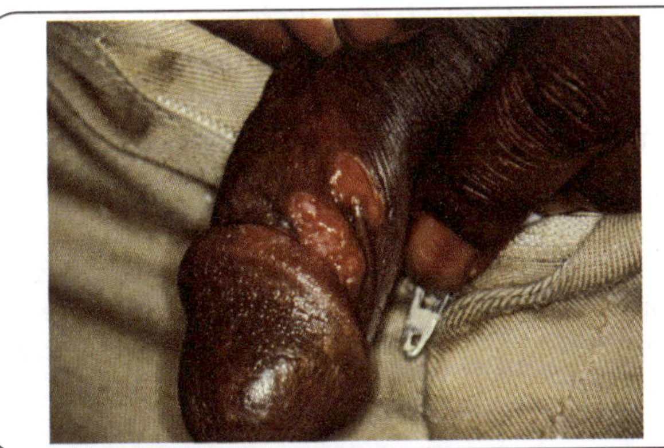

Fig. 41.4 Primary syphilis hard chancre

of syphilitic ulcer facilitates the transmission of HIV and HIV hastens the progress of syphilis.

A month after the ulcer has healed spontaneously, secondary syphilis sets in affecting the skin and mucous membranes, along with generalised lymphadenopathy. Early secondary skin lesions are macular, small papular, follicular or pseudo-vesicular, or lichenoid and can appear anywhere on the body. Late secondary lesions are large papular, annular, pustular or condylomata lata at mucocutaneous junctions at the vulva or anus.

Latent syphilis shows no clinical manifestation but the patient can transmit infection by blood transfusion or transplacentally. It can last lifelong, remit spontaneously or develop into tertiary or late syphilis.

Tertiary or late syphilis is rare and takes several decades to develop if primary syphilis is left untreated. It is characterised either by gumma—granulomatous lesions in skin and bone, or neurosyphilis due to the early invasion of the CNS, or cardiovascular involvement. Neurosyphilis and cardiovascular syphilis, referred to as quaternary syphilis, manifest as meningitis, or vasculitis, or general paresis of the insane, or tabes dorsalis (demyelination of the posterior columns) in the CNS, or aortic aneurysm and regurgitation in the CVS.

Congenital syphilis: Syphilis can be transmitted from mother to fetus transplacentally at any month of gestation. Antenatal screening for syphilis and its treatment, if diagnosed, can prevent congenital syphilis. Early congenital syphilis manifests within 2 years of age while late disease occurs after 2 years. Early congenital syphilis is infectious and characterised by mucocutaneous lesions, hepatosplenomegaly and lymphadenopathy, affecting the bones. Late congenital syphilis is characterised by keratitis, eighth nerve deafness, bilateral knee effusion, saddle nose and malformed teeth.

Laboratory diagnosis of syphilis

The samples to be collected are genital swab, discharge from the ulcer for microscopy and blood for serology. Demonstration of treponemes by microscopy, test for antibodies and molecular detection can be done depending on the stage of disease. *T. pallidum* cannot be cultured in vitro and therefore culture is not an option in the laboratory diagnosis.

1. Microscopy

Demonstration of *T. pallidum* in chancre exudates/blood (primary syphilis) has low sensitivity and therefore is not routinely done. Demonstration of corkscrew motility on dark ground illumination (DGI) microscopy too is rare. Fontana or Steiner silver impregnation stain of smear or Levaditi in tissue sections shows spiral, slender, silver (black) stained bacilli measuring 0.5 × 10 microns (Fig 41.5). Levaditi's silver impregnation staining is employed in tissue sections. Direct smear stained by fluorescent-labelled treponemal-specific antibody is more sensitive and appears as fluorescent green bacilli.

2. Serological tests to detect antibodies

Standard/non-specific tests: These tests are termed 'non-specific' as they do not use a treponeme-specific antigen and instead use the non-specific cardiolipin antigen, e.g., the venereal disease research laboratory (VDRL) test, rapid plasma reagin (RPR) test, unheated serum reagin (USR) test, and toluidine red unheated serum test (TRUST).

The VDRL test is a slide flocculation (precipitation) test. It is the most widely used simple and rapid serological test to detect antibodies in CSF and serum. The test is done with serial dilutions of serum to overcome the prozone phenomenon and determine the reactive titre. Being very sensitive, it is used as a screening test that needs further confirmation by a specific test. In neurosyphilis, reagin antibodies are demonstrable in the CSF. It is reported as reactive/non-reactive, and if reactive has a titre of 1:2, 1:4, 1:8 and so on. Qualitative reactive or low-titre reactive is

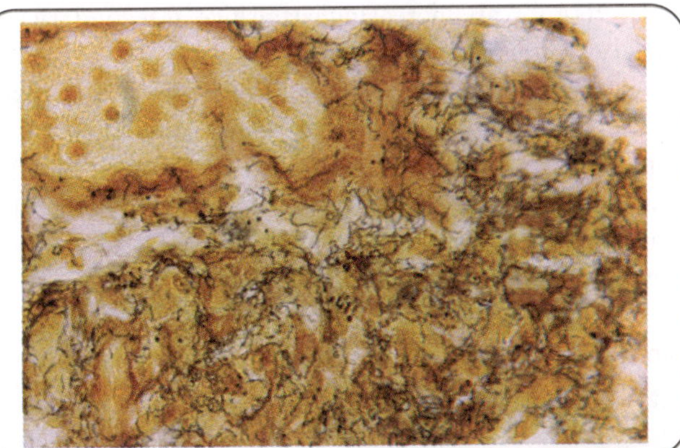

Fig. 41.5 *Treponema pallidum* in Steiner silver stain, 100X (*Source*: Public Health Image Library, ID 21621 / CDC / Skip Van Orden)

considered significant when clinically suspecting syphilis as there is no baseline 'significant titre'. RPR, USR and TRUST are precipitation tests similar to the VDRL.

The IgG antibody that reacts with cardiolipin is known as reagin. Cardiolipin, being non-specific, cross-reacts with other non-syphilis antibodies giving rise to 'biological false positive' (BFP) reactions. BFP reaction is associated with tissue damage and can be acute (disappears in weeks or months) or chronic (persists for more than 6 months). Acute BFP is seen with acute infections and chronic BFP with SLE and collagen disorders. BFP can be ruled out by testing paired sera demonstrating rise in titre or by confirmatory tests. On successful treatment, standard tests become non-reactive and are therefore used to determine response to therapy. Leprosy, malaria, viral hepatitis, HIV and eosinophilia are other conditions associated with a BFP VDRL. Rarely, normal sera can give a BFP due to the presence of reagin antibodies.

Treponemal/specific tests: *Treponema pallidum* hemagglutination (TPHA), *Treponema pallidum* particle agglutination (TPPA), and ELISA all use the extract of *T. pallidum* maintained in animal culture. The fluorescent treponemal antibody (FTA) test uses whole killed *T. pallidum*.

These can be used to detect antibodies in serum and CSF as standard confirmatory tests. They are specific, with high sensitivity, but done only on samples that are positive on any of the non-specific screening tests. Specific tests may not become negative following treatment and remain positive lifelong. As a result they are not useful in assessing response to treatment. Congenital syphilis is diagnosed by demonstration of treponemes in cord blood, a reactive VDRL or IgM in the newborn or by PCR.

The reagin antibodies appear a week to 10 days after the appearance of the primary chancre and can be detected in 70% of cases. The sensitivity of the VDRL increases to 100% in secondary syphilis and then declines. The FTA and TPHA too are increasingly sensitive in secondary and tertiary syphilis (Table 41.4).

As IgG antibodies are passively transferred from mother to child, congenital syphilis is confirmed by detecting IgM and IgG antibodies by Western blotting or *T. pallidum*-specific DNA by PCR. In congenital neurosyphilis, non-specific tests like VDRL on CSF sample are adequate.

Table 41.4 Sensitivity (%) of reactive tests in untreated syphilis

Stage of disease	VDRL	FTA	TPHA
Primary	70	85	55
Secondary	100	100	100
Tertiary	70	98	98

3. Molecular diagnosis of syphilis by PCR is useful in congenital syphilis and neurosyphilis

Treatment

Penicillin is the drug of choice for all stages of syphilis, with tetracycline used in the event of penicillin allergy. Non-specific tests, especially VDRL tests, are used to monitor response to therapy. Syphilis can be prevented by protected sex, contact tracing and treatment of sexual partners.

Non-venereal treponemes

Non-venereal treponemes are *T. pertenue* causing yaws, *T. endemicum* causing endemic syphilis and *T. carateum* causing pinta. They are transmitted by direct skin-to-skin contact in children. Of these, yaws was seen in India and was eradicated in 2016. Non-venereal treponematoses are diagnosed by the same tests as for syphilis and are sensitive to penicillin.

Chancroid or soft chancre

Chancroid is an STD caused by the bacterium *H. ducreyi*, presenting with painful genital ulceration that bleeds easily and enlarged, tender inguinal lymph nodes (Fig. 41.6). As with other genital ulcerations, it facilitates the transmission of HIV. The laboratory diagnosis includes examination of a smear of the exudates showing pleomorphic gram-negative coccobacilli that can be grown on in vitro culture or detection by PCR.

Lymphogranuloma venereum (LGV)

Chlamydiae are obligate intracellular bacilli that cause a wide spectrum of diseases in man that range from the

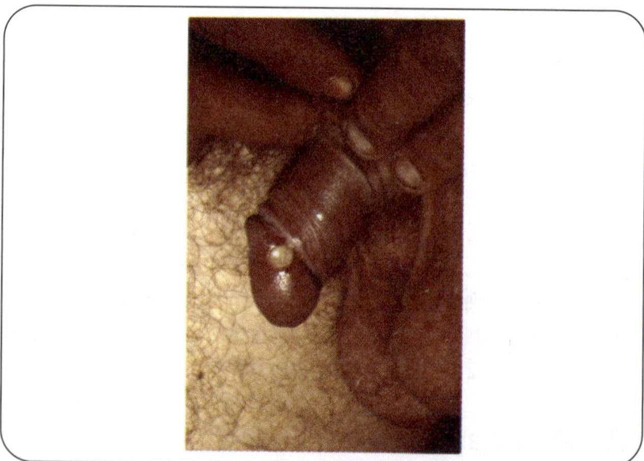

Fig. 41.6 Chancroid or soft chancre (*Source*: Public Health Image Library, ID 3728 / CDC / Joe Miller)

ocular, i.e., conjunctivitis and trachoma to the urogenital i.e., NGU and LGV caused by *C. trachomatis* and to the pulmonary, i.e., pneumonia caused by *Chlamydophila pneumoniae* and *Chlamydophila psittaci* causing psittacosis. *C. trachomatis* has 19 serovars.

Serovars A, B and C cause trachoma or chronic keratoconjunctivitis. They are common in children, transmitted through direct contact, and still a leading cause of blindness.

Serovars D-K cause NGU, PGU and epididymitis in males and cervicitis, endometritis and PID in females, and reactive arthritis. In neonates, the infection can be transmitted from mother to child during birth to cause ophthalmia neonatorum.

Serovars L1, L2 and L3 cause LGV, an STD characterised by a painless papule or vesicle on the genitalia, followed by enlarged, tender, soft inguinal lymph nodes with fever and if untreated leading to fistulae, hypertrophy and elephantiasis of the vulva or scrotum due to blocked lymphatics (Fig. 41.7).

LGV is diagnosed in the laboratory by demonstration of inclusion bodies, antigen or antibody detection using ELISA or PCR to detect specific genes. Macrolides are the drug of choice for *C. trachomatis* infections.

Donovanosis/granuloma inguinale

It is an STD caused by *K. granulomatis*, characterised by a painless papule on the genitalia that ulcerates and bleeds to touch but without lymph node involvement (Fig. 41.8).

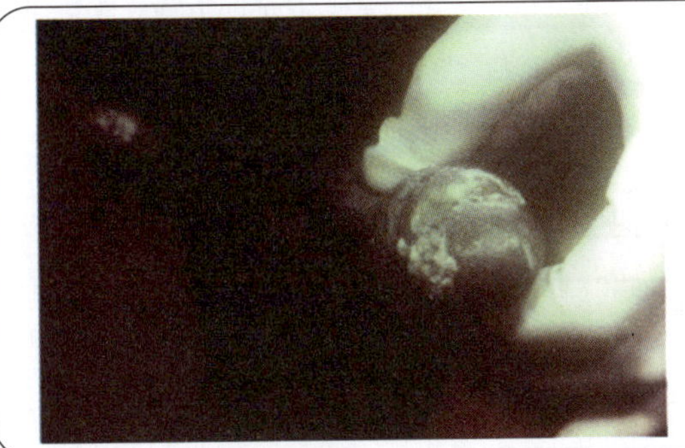

Fig. 41.7 Lymphogranuloma venereum (*Source*: Public Health Image Library, ID 18016 / CDC / Susan Lindsley)

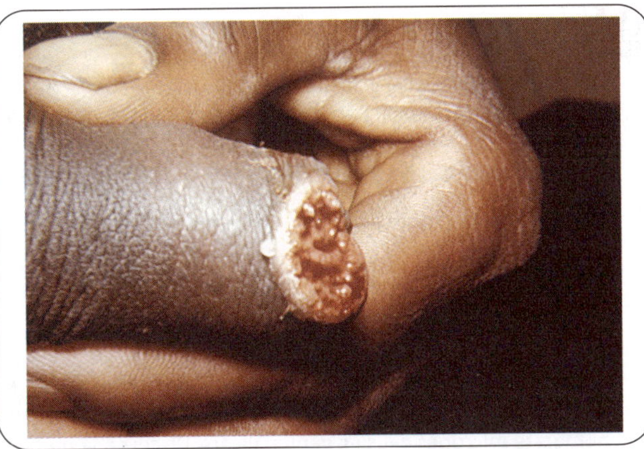

Fig. 41.8 Donovanosis (*Source*: Public Health Image Library, ID 6438 / CDC / Joyce Ayers)

Laboratory diagnosis is done by direct smear examination using Giemsa stain showing bipolar appearance of capsulated gram-negative bacilli or specific PCR (Fig. 41.9).

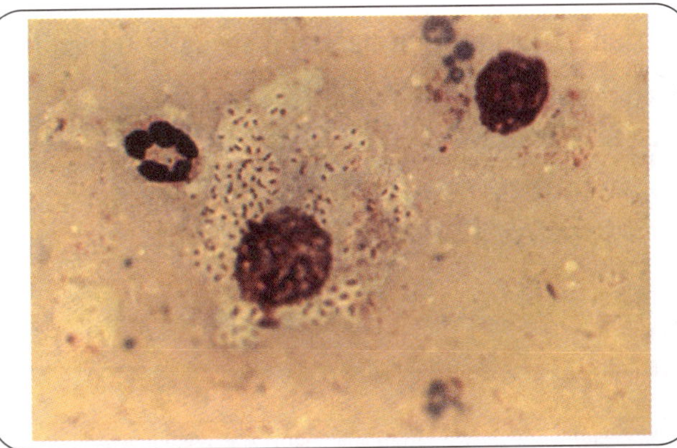

Fig. 41.9 Donovan bodies

SEXUALLY TRANSMITTED VIRAL INFECTIONS

Sexually transmitted viral infections are HBV (hepatitis), HIV (blood), HCV (hepatitis), HSV-1 and HSV-2 (genital lesions), and HPV (warts). A genital ulcer disease multiplex PCR test can simultaneously detect syphilis, chancroid, LGV, donovanosis and HSV.

Answers

Clinical Case 1

1. *Gonorrhea*
2. *Collect urethral swab and examine by Gram stain for gram-negative intracellular diplococci resembling N. gonorrhoeae. Treat with either ceftriaxone or cefixime.*
3. *If untreated, it can result in prostatitis, epididymitis, balanitis or periurethral abscess with sinus tracts.*

Clinical Case 2

1. *Wear gloves and examine the lesions.*
2. *Microscopic examination of the exudates by DGI, Fontana stain or FTA.*
3. *Penicillin.*
4. *Primary syphilis can progress to secondary and tertiary syphilis with late sequelae.*

QUESTIONS

1. Enumerate bacteria causing urethritis and explain the pathogenesis and laboratory diagnosis of urethritis in gonorrhea.
2. List microbes causing vaginitis and outline the laboratory diagnosis of each.
3. Enumerate microbes causing genital ulcers. Discuss the pathogenesis and stages of syphilis.
4. Enumerate microbes causing genital ulcers. Discuss the laboratory diagnosis of syphilis.
5. Enumerate microbes causing genital ulcers. Write a brief note on lymphogranuloma venereum.

SECTION SUMMARY

Section VIII: Genitourinary and Sexually Transmitted Infections

Infection/pathogen	Virulence/pathogenesis	Laboratory diagnosis	Vaccine/recommended antimicrobial
UTI *E. coli* *K. pneumoniae* *S. aureus* Enterococci	Endogenous intestinal uropathogenic *E. coli*, adhesion, ascending infection	Smear, quantitative urine culture Significant > 10^5 bacteria/mL	Nitrofurantoin Cotrimoxazole Norfloxacin
STIs **Urethritis** *N. gonorrheae*	Adhesion to urethral epithelium	Smear, culture	Cefexime or ceftriaxone, both partners
Mycoplasma *Ureaplasma* *Chlamydiae*	Non-gonococcal urethritis	Culture Antigen detection	Doxycycline
Candida albicans		Smear, culture	Clotrimazole
Trichomonas vaginalis		Wet mount trophozoites	Metronidazole
HSV or CMV		PCR	Antiviral
Vaginitis *Gardnerella vaginalis*, aneaerobes	Vaginal discharge thin, profuse	Clue cells	Metronidazole
Candida albicans	Vulval itching, curdy discharge	Smear, culture	Fluconazole
Trichomonas vaginalis	Vulval itching, white discharge	Wet mount trophozoites	Metronidazole
Genital ulcer disease **Syphilis** *T. pallidum*	Invades skin, mucosa, sexual, transplacental or blood transfusion. Multiplies in blood, lymphatics: primary (painless ulcer) secondary, tertiary	Fontana smear Antibody detection Non-specific: VDRL, RPR Specific: TPHA, ELISA	Penicillin G
Chancroid *H. ducreyi*	Painful ulcer, tender inguinal lymph nodes	Smear, culture, PCR	Azithromycin
LGV *C. trachomatis*	Painless papule, tender inguinal lymph nodes, fever	Inclusion bodies, Ag/Ab detection, PCR	Azithromycin
Viral HBV, HCV, HIV, HSV, HPV		ELISA/PCR	Antivirals

Zoonoses and Vector-Borne Infections

The WHO defines zoonoses as diseases that are naturally transmitted between vertebrate animals and humans. There are more than 150 zoonoses worldwide.

Zoonoses classified according to the etiologic agents are:
1. Bacterial zoonoses: Plague, anthrax, brucellosis, leptospirosis, rickettsial fevers, relapsing fever, Lyme's disease, cat scratch disease, rat-bite fever, listeriosis, salmonellosis and tularemia (not seen in India)
2. Viral zoonoses: Rabies, arboviral infections, bird flu, yellow fever, Ebola virus disease and Zika virus disease
3. Fungal: Cryptococcosis, histoplasmosis
4. Protozoal zoonoses: Toxoplasmosis, cryptosporidiosis, leishmaniasis
5. Helminthic zoonoses: Hydatid disease, taeniasis

The modes of transmission can be direct contact with body fluid of the animal, indirect contact with contaminated material, vector-borne, or consumption of infected milk, meat, eggs or raw vegetables.

PLAGUE

Epidemiology

In the year 1994, a plague epidemic originated in Surat, Gujarat with 600 cases and 56 deaths, involving five states but largely Maharashtra. It began with reports of a large number of rats dying in Surat. Wild rodents, bandicoots, field mice and domestic rats are reservoirs of *Yersinia pestis,* the bacterium causing plague with rat flea as the vector. Plague exists in nature as the domestic cycle that involves humans, rat fleas and rodents, and the wild or sylvatic cycle that does not involve humans. It is transmitted by the bite of a rat flea in the case of the bubonic plague and by person-to-person droplet transmission by inhalation in the case of pneumonic plague.

Three pandemics of plague have been reported, the third one having involved India in the early 20th century. The subsequent outbreaks were in 1994 in Surat (bubonic-pneumonic), 2002 in Shimla (pneumonic) and 2004 in Uttarkashi (bubonic). There exist four endemic foci in India—Kolar, Karnataka, Beed-Latur, Maharashtra, Rohru, HP and Dangud, Uttaranchal. Presently, most of the cases reported worldwide are from Africa.

Morphology

Y. pestis is a capsulated gram-negative bacillus with a bipolar or safety pin appearance with central clearing (Fig. 42.1).

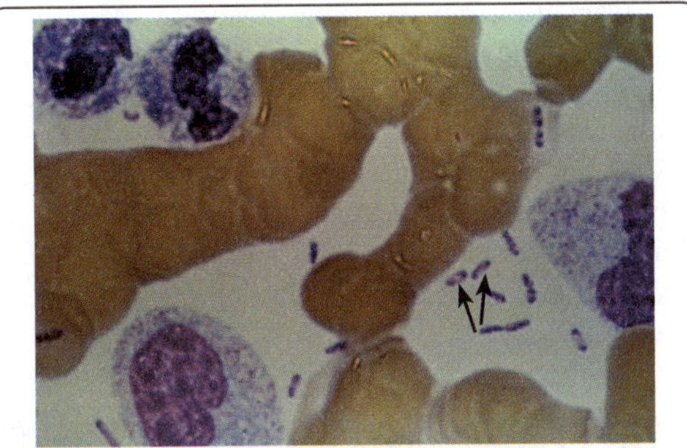

Fig. 42.1 Wayson stain demonstrates bipolar appearance of *Yersinia pestis*

Pathogenesis and clinical features

The virulence factor of *Y. pestis* is the Fraction 1 or F1 antigen which is a heat-labile capsular protein that is immunogenic, inhibits phagocytosis and is a diagnostic marker. Others include the endotoxic lipopolysaccharide, phospholipase, protease and siderophore.

Clinically, plague exhibits three forms—the bubonic, pneumonic and septicemic forms. Bubonic plague, the commonest form, is transmitted by the bite of rat fleas, and not from person to person. The bacilli multiply in the lymphatics, and after an incubation period of 3–7 days result in fever and painful lymphadenitis. Buboes are regional enlarged, tender lymph nodes, the most common site being inguinal. However, they can also be seen in the axillary, cervical or submaxillary lymph nodes, depending on the site of the bite of the rat flea. Secondary spread to pneumonia, meningitis or septicemia can occur.

Pneumonic plague is transmitted from person to person by inhalation of droplets, with a short incubation period resulting in fever, cough, hemoptysis, dyspnea and chest pain. Pneumonic plague is rare but fatal and can be used as a source of bioterrorism. Septicemic plague is rare and secondary to pneumonic or bubonic plague and can lead to gangrene.

Laboratory diagnosis

Plague is easy to diagnose in the laboratory using:
1. Gram or Wayson stain showing gram-negative bacilli with a bipolar or safety pin appearance with central clearing. The sample can be bubo aspirate or sputum.
2. Culture on ordinary media as it is not fastidious.
3. Demonstration of F1 antigen by ELISA or ICT.
4. PCR to detect the F1 antigen gene.

Prevention and treatment

Plague is prevented by rodent or rat flea control. In the event of an outbreak or case, doxycycline or tetracycline is given for chemoprophylaxis and a formol killed vaccine from Haffkine Institute, Mumbai is given subcutaneously at 0, 1 and 6 months. Immunity develops 1 week after vaccination and lasts for 6 months with 50% efficacy.

Two other zoonotic infections caused by the same genus causing gastroenteritis in children following ingestion of animal products contaminated by *Yersinia enterocolitica* or *Yersinia pseudotuberculosis* have been rarely reported from India.

ANTHRAX

Clinical Case 1

A 30-year-old man presents with a painless, black eschar of 1 cm on the left hand (Fig. 42.2). He gives a history of having butchered and de-skinned a dead livestock that had died in his village. He also confides that four others in his village who helped him de-skin also have similar lesions.

1. *What infection does the above history suggest? Name the causative agent.*
2. *What is eschar?*
3. *How will you confirm your diagnosis?*
4. *What is the case fatality rate in this disease?*
5. *Name the antimicrobials of choice for treatment and for post-exposure prophylaxis*

Epidemiology

Anthrax is a disease in animals acquired by ingestion of spores, causing fatal septicemia. The bacterial spores survive in soil. Human contact with an infected or dead animal either by inoculation, inhalation or ingestion causes human anthrax that is cutaneous, pulmonary or intestinal, respectively. It is endemic, but can occur as a local cluster in animals and subsequently in humans. Small local outbreaks of cutaneous anthrax have been reported in India in recent times. It is seen mostly as an occupational disease in persons dealing with agricultural livestock, abattoirs, or those dealing with animal products—leather, wool or hair.

Anthrax is also of historical importance as *Bacillus anthracis* is the first bacterium to be isolated in pure culture by Robert Koch in 1876, proposing the Koch's postulates. The first live attenuated bacterial vaccine available was the anthrax vaccine made by Louis Pasteur in 1881.

Morphology

Bacillus anthracis is a capsulated, rectangular, gram-positive bacillus with spherical, central, non-bulging spores (Fig. 42.3).

Pathogenesis and clinical features

The virulence of the bacterium is due to plasmid-coded exotoxin and plasmid-coded polypeptide capsule. The most common form of anthrax is cutaneous, which is self-limiting and rarely fatal. Pulmonary and intestinal forms are rare but fatal and therefore associated with bioterrorism.

Cutaneous anthrax (hide porter's disease or malignant pustule) begins by the inoculation of spores in abraded skin and the lesions progress from a painless papule to vesicle to a black eschar, with surrounding edema (Fig. 42.2). Eschar is necrotic tissue on the surface of a wound, unlike a scab which is composed of dried exudates and blood. Pulmonary anthrax develops by inhalation of spores (wool sorter's disease) and can lead to fatal meningitis or bacteremia. Intestinal anthrax due to consumption of contaminated meat presents as dysentery.

Laboratory diagnosis

Pus, sputum, CSF, blood or stool can be examined in cutaneous, pulmonary or intestinal anthrax, respectively.

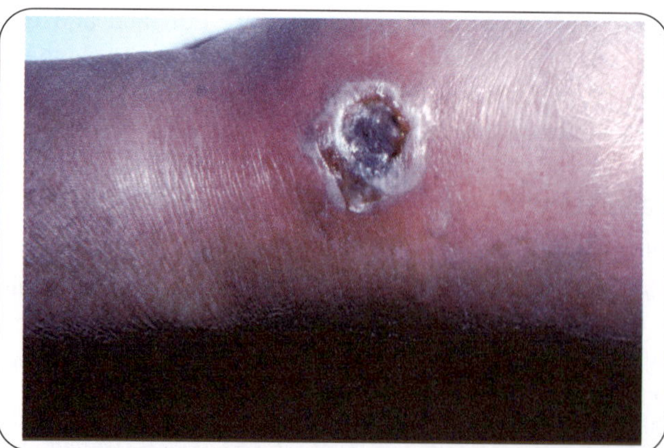

Fig. 42.2 Malignant pustule

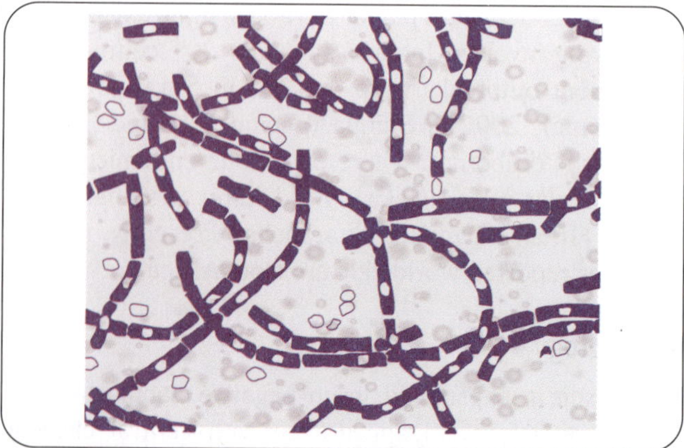

Fig. 42.3 Gram-positive bacilli with spores, Gram stain, 100X

Gram stain shows capsulated, rectangular, gram-positive bacilli with or without non-bulging spores. A purple capsule can be demonstrated by polychrome methylene blue stain (McFadyean's reaction). The capsule can also be demonstrated by direct immunofluorescence or by Ascoli's thermo-precipitation test. The bacterium is easy to grow in culture. Antibody detection in serum by ELISA and specific genomic sequence detection by PCR are possible.

In bioterrorism, bacterial spores can be identified by Gram stain or PCR. Processing of samples needs to be done with requisite biosafety measures to prevent infection in laboratory personnel.

Prevention and treatment

General measures to prevent include deep burial of dead animals, use of protective wear in handling animals and animal products, and vaccination of livestock.

Pre-exposure and post-exposure use of adsorbed anthrax antigen vaccine is available. Antimicrobial therapy is by ciprofloxacin or doxycycline with clindamycin for a minimum of 60 days. Ciprofloxacin or doxycycline is given for 60 days as post-exposure prophylaxis to close family contacts.

Anthracoid bacilli and *Bacillus* species other than *B. anthracis* are mostly non-pathogenic, with the exception of *Bacillus cereus,* which causes food poisoning.

In the World Wars, plague and anthrax bacilli were used as bioweapons. The intentional use of biological agents to cause mass morbidity or mortality is referred to as bioterrorism. Bioweapons can either be naturally occurring microbes or genetically modified microbes. Microbes with a low infective dose, high transmissibility (aerosols) and lack of immunity are likely candidates for bioterrorism, e.g., anthrax virus, plague virus, hemorrhagic viruses, smallpox virus, Nipah virus, Hanta virus, SARS-CoV and MERS-CoV.

BRUCELLOSIS

Clinical Case 2

A 42-year-old man engaged in rural dairy farming presents with complaints of intermittent, low-grade, insidious fever over a month and subsequent knee joint pain. On examination, the patient is febrile and has hepatosplenomegaly.

1. *What additional positive history will you ask for?*
2. *Name four zoonotic bacterial infections.*
3. *What microbiologic investigations would you request for in this case?*
4. *Name the antimicrobials indicated in treating this infection.*

Epidemiology

Brucellosis is an endemic zoonotic disease that is largely occupational involving dairy farmers, shepherds and butchers. It is transmitted to man by contact with infected animal tissue or excretory secretions through abraded skin or mucosa, ingestion of raw milk or unpasteurised dairy products from infected animals or by inhalation of aerosols generated during slaughter. Man-to-man transmission does not occur.

Morphology

Brucellae are non-capsulated, non-motile, gram-negative, small coccobacilli.

Pathogenesis and clinical features

There are six species of *Brucella* pathogenic to animals, of which three are pathogenic to man—*B. melitensis* (sheep, goats), *B. abortus* (cattle) and *B. suis* (pig). On gaining entry, the bacterium enters the lymphatic system followed by the bloodstream and disseminates to organs of the mononuclear phagocytic system (liver, spleen, bone marrow). The bacterial cell wall LPS is a virulence factor and the bacterium resists killing by phagocytes and multiplies intracellularly. Following a variable incubation period of 1 week to 1 month, an insidious 'undulating' fever appears and disappears to reappear—intermittent febrile and afebrile periods. Hepatosplenomegaly and arthritis of the major joints are the typical features. Rarely, meningitis, endocarditis or genitourinary system may be involved.

Laboratory diagnosis

1. Multiple blood cultures (conventional or automated) over 3 consecutive days or bone marrow culture are done to grow the aerobic, capnophilic brucellae, which are very fastidious and incubated for 2–4 weeks.
2. Detection of antibodies is done by a standard agglutination test (SAT) or in serum by ELISA. SAT is a tube agglutination test wherein a titre of 1:160 or greater is considered significant or demonstration of a rise in titre 1–2 weeks apart.
3. PCR to detect specific gene sequences.
4. In animals, brucellae can be isolated or antibodies detected in milk and dairy products.

Prevention and treatment

Human brucellosis can be prevented by control of animal brucellosis by animal vaccination. For humans, a live attenuated *B. abortus* 19-BA is available with a short time cover. Antimicrobials effective in brucellosis are streptomycin or rifampicin with doxycycline.

RICKETTSIAL FEVERS

Clinical Case 3

A 35-year-old woman presents to the hospital with a history of continuous high-grade fever, headache and myalgia for 10 days. She gives a history of visiting a physician and being treated for her symptoms with no relief. She does not show any significant finding on clinical examination and a request is made for blood culture, Widal's test, and peripheral smear for malarial parasite and dengue antibodies. All reports are negative. On thorough questioning, she admits to have had an insect bite on the abdomen while gardening. On examination, the bite lesion has an eschar of 5 mm, surrounded by erythema.

1. *What is the most likely diagnosis? Name the causative organism.*
2. *What is the classic clinical presentation of this disease?*
3. *What investigation will you request for to confirm your diagnosis?*
4. *Name the drug of choice in treating this infection.*

Rickettsiae are obligate intracellular, gram-negative coccobacilli that cannot be grown on inanimate media and are zoonotic bacteria transmitted by arthropod vectors. They are broadly classified in the typhus, spotted fever and scrub typhus groups. Commonly reported emerging and re-emerging rickettsial diseases in India are scrub typhus, murine flea-borne typhus, Indian tick typhus and Q fever (Table 42.1).

The family *Rickettsiaceae* has two pathogenic genera—*Rickettsia* and *Orientia*; *Coxiella* and *Bartonella* are excluded. *Coxiella* is transmitted by inhalation and *Bartonella* grows on inanimate media.

Following the insect bite, the rickettsiae travel through the lymphatics, multiply in the lymph nodes and enter the bloodstream to multiply and target the endothelial cells.

Scrub typhus

Epidemiology

Rickettsioses, of which scrub typhus is the commonest, have been reported from several states in India including Jammu and Kashmir, Himachal Pradesh, Uttarakhand, Bihar, WB, Meghalaya, Rajasthan, Maharashtra, Karnataka, TN and Kerala.

The infection is transmitted through the larval mites that take blood meal. Small rodents are natural hosts for scrub typhus. The field rodent and vector mites act as reservoirs and between the two the infection perpetuates in nature.

Scrub typhus can occur in areas where scrub vegetation, which consists of low-lying trees and bushes, is encountered and also in habitats as diverse as the banks of rivers, rice fields, water pools and gardens. The mites feed usually on rodents and accidentally on humans, and transmit the infection during the prolonged feeding which can last for 1–3 days. The incidence of scrub typhus is higher among the rural population. Cases are more likely to have exposure to rodents at home or at work, and to occupational (farming) or recreational activities.

Pathogenesis and clinical features

Following the insect bite, the rickettsiae travel through the lymphatics, multiply in the lymph nodes and enter the bloodstream to multiply and target the endothelial cells. Acute fever is the most common presenting symptom often associated with breathlessness, cough, nausea, vomiting, myalgia and headache.

Eschar is an early clinical manifestation representing localised necrosis at the site of the bite. However, it may or may not be present. Eschars are painless, punched out ulcers upto 1 cm in width, with a black necrotic centre surrounded by an erythematous margin. Eschar is a pathognomonic sign of scrub typhus. Usually, a single eschar is found on the neck, axillae, chest, abdomen and groin accompanied by regional lymphadenopathy.

Rickettsial infections are generally incapacitating and difficult to diagnose; untreated cases have case fatality rates as high as 30–45% with multiple organ dysfunction, if not promptly diagnosed and appropriately treated. The vast variability and non-specific presentation of this infection have often made it difficult to diagnose clinically.

Rash is considered as the hallmark of rickettsial disease, though it is neither seen at presentation nor in all patients. The presence of rash is common in spotted fever and is extremely rare in scrub typhus. The rash usually becomes apparent after 3–5 days of the onset of symptoms. Initially, the rash is in the form of pink, blanching, discrete maculae which subsequently become maculopapular, petechial or hemorrhagic.

None of these clinical symptoms and signs including eschar are diagnostic of rickettsial disease. Therefore, epidemiological factors pertaining to the geographical

Table 42.1 Rickettsial infections in India

Disease	Bacterium	Vector	Reservoir
Scrub typhus	*Orientia tsutsugamushi*	Mite	Rodents
Murine/Endemic typhus	*R. typhi*	Flea	Rodents
Indian tick typhus (ITT)	*R. conori*	Tick	Rodents, dogs
Q fever	*Coxiella burnetti*	Nil	Cattle, sheep, goats

area, habitat, occupation, and exposure to insect bite could assist in reaching a diagnosis of rickettsial disease with certainty and initiating treatment in time. Jaundice, renal failure, pneumonitis, acute respiratory distress syndrome (ARDS), septic shock, myocarditis and meningoencephalitis are known complications.

Laboratory diagnosis

1. *Weil-Felix tube agglutination test:* The sharing of the antigens between *Rickettsia* and *Proteus* is the basis of this heterophile antibody test. It demonstrates agglutinins to *Proteus vulgaris* strain OX19 and OX2 and *Proteus mirabilis* OXK. Though this test lacks high sensitivity and specificity, it still serves as a useful and inexpensive diagnostic tool for the laboratory diagnosis of rickettsial disease. This test should be carried out only after 5–7 days of the onset of fever. A titre of 1:80 is to be considered indicative of possible infection. However, baseline titres need to be standardised for each region.
2. *IgM and IgG ELISA*: ELISA techniques, particularly IgM capture assays, are probably the most sensitive tests available for rickettsial fever diagnosis and the presence of IgM antibodies indicates recent infection with *Rickettsia*. In cases of infection with *O. tsutsugamushi*, a significant IgM antibody titre is observed at the end of the first week, whereas IgG antibodies appear at the end of the second week. Baseline titres need to be established keeping in view the regional variations.
3. *Polymerase chain reaction (PCR)*: It is a rapid and specific test for diagnosis. It can be used to detect rickettsial DNA in whole blood, buffy coat fraction or tissue specimen. RT-PCR is best done within the first week for blood samples because of the presence of rickettsemia in the first 7–10 days.
4. *Immunofluorescence assay (IFA)*: This is a reference serological method for the diagnosis of rickettsial diseases and is considered the serological 'gold standard'; however, the cost and requirement of technical expertise limit its wide use. Therefore, it is recommended only for research and in areas where the seroprevalence of rickettsial diseases has been established and a reference facility is already available which has the necessary expertise required to conduct these tests.
5. *Indirect immunoperoxidase assay (IPA)*: It gives results comparable to those in IFA but requires special instruments and experienced personnel for interpretation of the test.

Rapid tests for the diagnosis of scrub typhus are not recommended. Doxycycline is the drug of choice, with azithromycin or chloramphenicol as other options. So far, an effective vaccine is not available.

Murine/endemic typhus

Transmitted by bite of the rat flea, it has been reported from Mumbai, Pune, Lucknow, Jabalpur, Shimla and Kashmir. It clinically presents with fever, myalgia, headache and a maculopapular, erythematous rash on the trunk. Laboratory diagnosis is by the Weil-Felix test.

Indian tick typhus (ITT)

Transmitted by a tick bite, it has been reported from Nagpur, Pune, Lucknow, Jabalpur, Sagar, Bengaluru and Secunderabad. It clinically presents with fever, maculopapular, erythematous rash on the extremities and trunk, eschar at the site of bite, and regional lymphadenopathy. Laboratory diagnosis is by the Weil-Felix test.

OTHER ZOONOTIC DISEASES

Q fever

Infection by *C. burnetti* is acquired by inhalation of infected dust or ingestion of contaminated milk of cattle, sheep or goats. It is seen in India in Rajasthan, Punjab, Haryana and Delhi, in animals and humans. Acute Q fever presents with hepatitis, pericarditis, myocarditis, pneumonia or meningitis. The bacterium can remain latent and cause chronic Q fever. Laboratory diagnosis is by detection of IgM by indirect immunofluorescence assay IFA or PCR.

Cat scratch disease

Bartonella henselae is transmitted from cat to man by a cat scratch or bite. It is characterised by painless, erythematous papules or pustules at the site of scratch or bite and regional lymphadenopathy. The bacterium can be demonstrated by lymph node aspirate smear examination with silver stain, culture on inanimate media, detection of antibodies and PCR. Azithromycin is the drug of choice.

Rat-bite fever

It is transmitted to humans by a rat bite or ingestion of water or milk contaminated with rat secretions. *Streptobacillus moniliformis* and *Spirillum minus* are gram-negative bacilli, the two causative agents that are commensals in rats. The disease presents with fever, petechial rash and polyarthralgia, and is treated with penicillin.

Pasteurella spp.

Pasteurella spp. are gram-negative bacilli that are respiratory tract commensals in humans, cats and dogs. Following the bite of an animal, the disease presents as local wound infection with regional lymphadenopathy and fever. It is managed with penicillin.

LEPTOSPIROSIS

Clinical Case 4

A week following rain flooding in Mumbai, a child presents to the OPD with a history of fever, headache and vomiting. On examination, he has jaundice and raised liver enzymes. He does not give any history of food poisoning but confirms having waded through the clogged rain waters during the flood.

1. *What is the history suggestive of?*
2. *What could be the source of infection in this case?*
3. *What microbiological test will you request for?*
4. *What is the reference test for this infection? Where is it available in India?*

Epidemiology

Leptospirosis is a zoonotic infection caused by the spirochete bacterium genus *Leptospira*. It is transmitted from rats, dogs, cattle and pigs. The bacterium excreted in animal urine contaminates water or soil. Man is infected on contact through cuts and abrasions on the skin or intact mucous membrane with contaminated water or soil following rainfall, floods, or occupational exposure. The genus *Leptospira* has two species *L. interrogans* and *L. biflexa*. Within each species exist serogroups and serotypes based on surface antigens. There exist 25 serovars of *Leptospira*, *L.* Andamana and *L.* Pomona being predominant in India.

In India, coastal areas are affected—Andaman and Nicobar, Maharashtra, Gujarat, Kerala, TN, Goa, Karnataka, AP, Puducherry and Odisha. The diagnosis of leptospirosis is based on clinical, epidemiological and laboratory findings—fever following contact with contaminated water/soil with raised liver and renal function test values. The ICMR National Leptospirosis Reference Centre is located in Port Blair, Andaman. Leptospirosis is the most common zoonotic disease, with seasonal outbreaks reported in Mumbai (2005) and Gujarat following heavy rainfall.

Morphology

Leptospira are long (10 μ), fine (0.1 μ), coiled bacilli with hooked ends (Fig. 42.4). Being fine, they cannot be demonstrated by Gram stain. They can be seen either by silver impregnation or immunofluorescence staining techniques or their motility observed by dark ground or phase contrast microscopy.

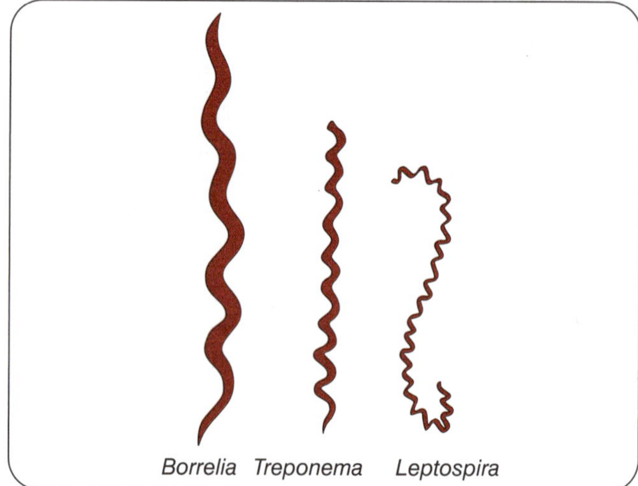

Fig. 42.4 Comparative morphology of different spirochetes

Pathogenesis and clinical features

The bacterium is invasive and enters through abraded skin or conjunctiva or oral mucosa into the bloodstream. In 90% of infection cases, the disease follows a mild, anicteric febrile illness.

The incubation period is 7–10 days, followed by fever, headache, myalgia, conjunctival suffusion, vomiting or abdominal discomfort. The microbe multiples in blood and spreads to invade the brain, liver, heart and kidneys. In this septicemic or first stage, it causes damage to vascular capillaries.

After 10 days of the first stage in the second or immune stage, antibody response to *Leptospira* is seen with disappearance of the bacterium from blood but excretion in urine. Antigen–antibody complexes are formed and deposited in various organs. The disease is characterised by fever, meningitis, rash or ocular inflammations.

In rare cases (10%), the infection takes a severe, icteric form of Weil's disease or hepato-renal-hemorrhagic syndrome. Weil's disease is characterised by fever, jaundice and petechiae, with hemorrhagic manifestations in the lung, conjunctiva, and gastrointestinal tract. Laboratory investigations reveal raised liver and kidney function test values.

Laboratory diagnosis

1. Demonstration of bacilli in CSF, blood or urine is NOT sensitive and generally not done. The pathogens can be seen either by silver impregnation or immunofluorescence staining techniques, or their motility observed by dark ground or phase contrast microscopy.

2. Culture is slow and requires fastidious media, and hence is not routinely done.
3. Demonstration of specific antibodies is most commonly used. IgM appears early by 7 days and can be detected till 4 weeks. IgG appears later but can persist for years. Genus-specific IgM or IgG antibody can be separately detected by ELISA, ICT, dipstick or agglutination assays.

   Serovar-specific antibody detection by the microscopic agglutination test (MAT) is the reference test available at the Regional Medical Research Centre, Port Blair, Andaman.
4. Genus-specific genome detection by PCR can be done in the septicemic stage before antibodies appear.

Prevention and treatment

Avoid physical contact with contaminated waters. A whole-cell serovar-specific vaccine is available, with good efficacy. For treatment, doxycycline, penicillin or ceftriaxone is indicated.

RELAPSING FEVER

Borrelia are spirochete bacilli that are spiral, slender and long. Some are commensals on the oral and genital mucosa. Pathogenic *Borrelia* are *B. recurrentis* causing relapsing fever, *B. burgdorferi* causing Lyme's disease and *B. vincentii* causing Vincent's angina with the anaerobic *Leptotrichia buccalis*.

Relapsing fever comes in recurrent episodes with intermittent afebrile periods caused by arthropod-borne *Borrelia* species. Epidemic relapsing fever is caused by crushing of lice on human skin, transmitting *B. recurrentis*. Endemic relapsing fever is caused by the bite of a rodent tick transmitting *Borrelia* species other than *B. recurrentis*. Rarely seen in India, a few cases of *B. recurrentis* infection have been reported. The disease manifests after an incubation period of a week, during which time the bacterium multiplies in the blood. It multiplies with antigenic variations in order to avoid the immune response. The fever is accompanied by non-specific symptoms, petechial rash and central neurologic involvement.

The disease can be diagnosed by examining a peripheral blood smear by Giemsa stain, detection of specific antibodies by ELISA, or detecting the *GlpQ* gene by PCR. The infection is treated with doxycycline or erythromycin.

Lyme disease

Largely seen in the USA, it is transmitted by tick bite from rodents or deer, usually following outdoor activities like camping and trekking. After a variable incubation period, the disease manifests with a maculopapular rash at the site of the tick bite and spreads hematogenously to other sites on the skin. Fatigue, myalgia, and arthralgia followed by arthritis of the large joints are the predominant symptoms. Rarely, cardiac and neurologic abnormalities may occur. Laboratory diagnosis is by culture of blood in a special medium, DNA detection in joint fluid by PCR, and IgM and IgG detection by ELISA. The infection responds to doxycycline.

Yellow fever

Primarily a zoonotic disease of monkeys, it is transmitted by bite of the *Aedes* mosquito infected with the RNA virus—*Flavivirus fibricus*. The virus enters the bloodstream and affects the liver, heart and kidneys. The liver is the most affected resulting in increased bilirubin and hence termed "yellow fever". Yellow fever has not been reported in India because the causative virus is not present but the virus can gain entry as the monkeys and vectors are present. The Govt. of India mandates the submission of an international certificate of vaccination against yellow fever for travellers arriving from yellow fever endemic countries. The certificate is valid from 10 days after vaccination for 10 years. The vaccine is a single dose of the live attenuated strain 17-D of the virus.

VECTOR-BORNE INFECTIONS

Biological vectors are arthropods or insects that can transmit pathogens from animals to humans or humans to humans. The pathogen undergoes development in the vector and the infection is mostly transmitted by a bite of the vector. The common vector-borne infections seen in India are listed in Table 42.2.

The National Vector Borne Disease Control Program (NVBDCP) was launched in 2003–2004 to encompass malaria, filaria, kala-azar, dengue and Japanese B encephalitis control efforts across India.

Table 42.2 Vector-borne infections in India

Arthropod	Infection
Mosquito	Malaria, filaria, dengue, chikungunya, Japanese B encephalitis
Sandfly	Kala-azar, Oriental sore
Rat flea	Plague
Hard tick	Kyasanur forest disease, typhus fever
Trombiculid mite	Scrub typhus

Answers

Clinical Case 1

1. *Cutaneous anthrax, Bacillus anthracis.*
2. *Eschar is necrotic tissue on the surface of a wound, unlike a scab which is composed of dried exudates and blood.*
3. *Gram stain smear of the wound will show rectangular, gram-positive bacilli, with or without non-bulging spores.*
4. *Not fatal if treated with antimicrobials.*
5. *Ciprofloxacin or doxycycline with clindamycin for a minimum of 60 days for treatment. Ciprofloxacin or doxycycline for 60 days as post-exposure prophylaxis to close family contacts.*

Clinical Case 2

1. *Was raw milk ingested in the course of dairy farming?*
2. *Plague, anthrax, brucellosis, leptospirosis.*
3. *Multiple blood cultures and detection of specific antibodies by the standard agglutination test or ELISA, to rule out brucellosis.*
4. *Streptomycin or rifampicin with doxycycline.*

Clinical Case 3

1. *Scrub typhus, Orientia tsutsugamushi.*
2. *Eschar at the site of insect bite, regional lymphadenopathy and maculopapular rash.*
3. *Detection of specific IgM antibodies by ELISA.*
4. *Doxycycline.*

Clinical Case 4

1. *Leptospirosis*
2. *Contamination of rain water with rat or dog urine.*
3. *Leptospira IgM antibodies by ELISA or ICT.*
4. *Microscopic agglutination test (MAT) to detect serovar-specific L. interrogans.*
5. *Regional Medical Research Centre, Port Blair, Andaman.*

Questions

1. Define zoonoses and classify zoonotic diseases according to causative agents, stating the modes of transmission with examples.
2. Name the three clinical presentations of plague. How do they differ? In the event of an outbreak, what preventive measures can be taken?
3. Name the three clinical presentations of anthrax. How do they differ? In the event of an outbreak, what preventive measures can be taken?
4. What is eschar? Name two bacterial diseases presenting with a skin eschar.
5. Outline the laboratory diagnosis of brucellosis.
6. Enumerate the rickettsial infections seen in India, with their causative agent, vector and reservoir.
7. Outline the laboratory diagnosis of scrub typhus.
8. Draw the morphology of different spirochetes. Discuss the pathogenesis and laboratory diagnosis of leptospirosis.
9. Enumerate vector-borne infections seen in India, specifying the vector. Write a brief note on yellow fever.

Opportunistic, Oncogenic, Perinatal, Teratogenic, Emerging and Ocular Infections

OPPORTUNISTIC INFECTIONS

Infections that do not occur in immunocompetent individuals but are seen in an immunocompromised host are termed opportunistic infections irrespective of the underlying cause of the immune defect. The underlying immune defect can be malnutrition, leucopenia, immunosuppressive therapy, chemotherapy or HIV. However, with the advent of HIV, there has been an increase in the incidence of opportunistic infections.

Microbes causing opportunistic infections and their laboratory diagnosis

1. Bacteria: *Mycobacterium tuberculosis* and MOTT, *Salmonella, Burkholderia* diagnosed by culture
2. Viruses: HSV, CMV, JC polyoma virus diagnosed by ELISA or PCR
3. Fungi: *Candida, Cryptococcus, Penicillium, Fusarium, Histoplasma, Aspergillus, Mucorales* diagnosed by microscopy and culture
4. Parasites: *Strongyloides stercoralis, Cryptosporidium, Isospora, Toxoplasma, Pneumocystis* diagnosed by stool microscopy or histopathologic examination

Burkholderia pseudomallei is a saprophytic, non-fermenting gram-negative bacillus that causes localised skin or visceral abscess, pulmonary or bloodstream infection in patients with HIV infection, diabetes or renal failure, or following post-traumatic inoculation.

Talaromyces (*Penicillium*) *marneffei* is endemic in Southeast Asia as well as in India. It causes a disseminated infection on inhalation of conidia in the immunocompromised. *Fusarium* is known to cause keratitis in the immunocompetent and lung disease in the immunocompromised. Both fungi can be cultured and identified morphologically by microscopic examination.

Histoplasmosis is an opportunistic, granulomatous infection caused by the dimorphic fungus *Histoplasma capsulatum*. In India, cases have been reported from the North and the East in the HIV-infected. Microconidia from the soil are inhaled and phagocytosed by alveolar macrophages. In tissue, they transform into yeast cells and are transported via the bloodstream to lymph nodes and other organs. Clinically, it presents as pulmonary, mucocutaneous or disseminated histoplasmosis. The yeast cells can be demonstrated on histopathologic examination of infected tissue and cultured for identification of its mycelia and conidia.

HIV infects the CD4 T cells resulting in certain opportunistic infections. The diseases arising due to the immunodeficiency in HIV are: mucosal infections (*Candida*), tuberculosis, diarrhea (*Strongyloides, Cryptosporidium, Isospora*), pneumonia (*Pneumocystis*), meningitis (*Cryptococcus*) or toxoplasmosis.

The patient will present with the clinical features of a specific opportunistic pathogen and the laboratory diagnosis will depend on the causative agent.

ONCOGENIC VIRUSES

Viruses that produce tumours in their normal hosts or in experimental animals, or induce malignant transformation of cells on culture are known as oncogenic viruses. The properties of cells that are transformed by viruses are altered cell morphology, growth characteristics, cell metabolism and antigens.

The mechanism of viral oncogenesis is by modification of the host cell genes regulating cell growth. In DNA viruses, the viral DNA gets integrated with the host cell genome to cause malignant transformation. In RNA viruses or retroviruses the RNA introduces into the cellular genome a new transforming genome (oncogene or V-onc), inducing or altering the expression of a pre-existing cellular oncogene (C-onc).

Examples of oncogenic viruses and their role in malignancies:

1. DNA viruses
   HPV — Carcinoma of the cervix, vulva or penis, genital mucosal carcinoma, oropharyngeal carcinoma, laryngeal carcinoma and esophageal carcinoma.
   HBV — Hepatocellular carcinoma
   HSV-2 — B cell lymphoma, carcinoma of the cervix
   EBV — Nasopharyngeal carcinoma, Burkitt's lymphoma, Hodgkin's disease and B cell lymphoma
2. RNA viruses
   HCV — Hepatocellular carcinoma
   HTLV-1 — Adult T cell leukemia
   HHV-8 — Kaposi's sarcoma

Other than viruses, infection with the bacterium *H. pylori* and parasite *S. haematobium* can cause carcinoma of the stomach and bladder, respectively.

PERINATAL AND TERATOGENIC INFECTIONS

Perinatal infection refers to the vertical transmission of infections from mother to child. Teratogenic infections are those that affect organogenesis in the fetus giving rise to congenital anomalies. All teratogenic infections are perinatal but all perinatal infections are not teratogenic. Perinatal transmission of microbes can occur through the placenta (antenatal), during labour or delivery (antepartum) or by breastfeeding (postpartum).

Fetal immunology

Fetal immunity begins to develop by 9 to 15 weeks. The passive transfer of IgG to the fetus begins by 16 weeks of gestation and by 26 weeks is equivalent to the maternal IgG. Breastfeeding provides immunoglobulins that are protective but their levels begin to decline after 2 months of birth. The risk of neonatal infection is higher with prolonged labour, premature rupture of membranes and obstetric manipulations.

Microorganisms causing perinatal infection

1. Viral — HBV, HCV, HIV, HSV, CMV, VZ virus, rubella virus, parvovirus B19, mumps virus, measles virus, Zika virus.
2. Bacterial — Group B *Streptococcus, Neisseria gonorrhoeae, E. coli, Treponema pallidum, Listeria monocytogenes, Chlamydia trachomatis, Mycoplasma* and *Ureaplasma*.
3. Protozoan — *Toxoplasma gondii*

Of the microorganisms causing perinatal infection, HSV, CMV, VZ virus and rubella virus are teratogenic, while others are not.

Perinatal viral infections

Congenital HSV infection of the fetus can occur during the first 20 weeks of pregnancy and present with abortion, stillbirth and congenital anomalies. Neonatal HSV infection antepartum is localised to the eye or brain, or may be disseminated.

Intrauterine CMV infection is the most common infection associated with congenital malformation. Intrapartum infection with CMV in 5% of cases results in microcephaly, intracranial calcification, chorioretinitis, thrombocytopenia and anemia. It is generally asymptomatic at birth with late sequelae of hearing loss, neuro-deficit, psychomotor or learning disability.

Varicella zoster during pregnancy poses a high risk to the fetus at 13–20 weeks, causing congenital varicella syndrome (chorioretinitis, microphthalmos, hydronephrosis, cerebral atrophy). Infection of the mother after 20 weeks has shown no evidence of congenital infection in the fetus. However, peripartum VZIG can be administered to neonates.

Rubella infection in the first trimester leads to abortion or congenital rubella syndrome, and in the second trimester 25% of cases may be affected. Features include cataract, glaucoma, patent ductus arteriosus, pulmonary artery stenosis, microcephaly, encephalitis, retinopathy, purpura or hepatosplenomegaly. A neonate born with congenital rubella is contagious though one in three cases is asymptomatic at birth. Extended rubella syndrome implies encephalitis with type 1 diabetes at 20–30 years of age. Rubella can be prevented by vaccination of women of reproductive age prior to pregnancy. It is contraindicated 1 month before or during pregnancy.

Parvovirus B19 maternal infection is transmitted by one in three cases, at gestational age 13–16 weeks, resulting in abortion, stillbirth and non-immune hydrops. Mumps in the first trimester results in spontaneous abortion, but later has no effect on the pregnancy or the fetus. Measles is not teratogenic, but can cause abortion, or preterm or low-birth-weight babies. If the mother gets measles at delivery, it can cause neonatal measles.

Zika virus infection is rare in India and was reported first in 2017 (three cases in Gujarat and one in TN) followed by an outbreak of 157 RT-PCR-confirmed cases in Rajasthan in 2018. To date, no case of congenital infection has been reported in infected pregnant women. The Govt. of India maintains a laboratory-based Zika surveillance involving viral research and diagnostic laboratories (VRDLs). Zika is an RNA virus transmitted by the bite of the mosquito *Aedes aegypti*, with potential to cause Guillain—Barre syndrome and congenital Zika in case of infection during pregnancy.

The blood-borne viruses HBV, HCV and HIV can be vertically transmitted from mother to child antepartum, during delivery and postpartum.

Antigen p24 and HIV RNA detection is useful in the diagnosis of neonatal HIV infection, as maternal IgG is passively transferred transplacentally to the child, and antibody detection cannot be a diagnostic tool in neonates. Dried blood sample (DBS) is collected by a heel prick on a filter paper and transported to the reference centre for PCR. The first sample of a neonate is to be collected at 45 days of age for PCR and if negative, repeated at 6, 12 and 18 months by the rapid antibody test to be considered HIV-negative. Antenatal screening for HIV in pregnancy is mandated to prevent transmission to the child. Post-exposure prophylaxis with single-dose nevirapine is given at birth and daily for 6 weeks.

Perinatal bacterial infections

Group B streptococci (GBS), *N. gonorrhoeae* and *Chlamydia trachomatis* are carried in the cervix and transmitted to the neonate during delivery causing ophthalmia neonatorum (conjunctivitis). Both GBS and *N. gonorrhoeae* can be detected at 35 weeks of pregnancy by culture and *Chlamydiae* by antigen testing. Neonatal sepsis can be caused by GBS, *Listeria* (food poisoning), *E. coli* and other gram-negative bacilli. Congenital syphilis is acquired intrapartum with early symptoms of secondary syphilis (weeks after birth) or late symptoms (after 5 years).

Congenital toxoplasmosis

It is the least common (17%) in the first trimester but the disease is more severe. Conversely, it is more common in the third trimester (65%) but is usually asymptomatic. The classical triad of congenital toxoplasmosis is hydrocephalus, cerebral calcification and chorioretinitis. It does not cause any fetal malformation as it is not teratogenic.

Congenital toxoplasmosis is confirmed if IgM is found in the blood or CSF 7 to 10 days after birth or by the presence of the toxoplasma B1 gene. The level of passive IgG decreases over 6–12 months after birth, and active IgG appears by 3 months, if untreated.

Laboratory diagnosis of perinatal infections

1. Antenatal screening for HBV, HCV, HIV and syphilis (VDRL)
2. In mothers with a bad obstetric history, serological screen for *Toxoplasma*, others (VZ virus, parvovirus), rubella virus, CMV and HSV (TORCH).
3. Detection of microbe-specific IgM or PCR after birth.

CYTOMEGALOVIRUS

CMV is a dsDNA, enveloped *Herpesviridae* virus that causes enlargement ("cytomegalo") of infected cells. It is transmitted largely by oropharyngeal secretions by inhalation, causing a mononucleosis-like syndrome in immunocompetent adults, followed by latency in the salivary gland and kidney. The CMV infection prevalence is high, with most adults being seropositive.

It can also be transmitted transplacentally, especially in early pregnancy, causing cytomegalic inclusion disease in 5% of cases. In these cases, it presents with microcephaly, intracranial calcification, chorioretinitis, thrombocytopenia and anemia. It is generally asymptomatic at birth with late sequelae of hearing loss, neuro-deficit, and psychomotor or learning disability.

Being a latent virus, it can be reactivated in immunocompromised individuals or in transplant recipients causing pneumonia, or kidney or brain infection. CMV disease is diagnosed by detecting CMV-specific IgM by ELISA or DNA by PCR. Histopathologic examination of infected tissue shows the presence of intranuclear inclusion bodies with a characteristic "owl's eye" appearance (Fig. 43.1).

EMERGING INFECTIONS

In this century, a number of viral infections have made an appearance—Nipah, SARS coronavirus 1, MERS-CoV, COVID-19 and Zika virus. The existing influenza virus appeared as Influenza A (H5N1) and Influenza A (H1N1) in 2003 and 2009, respectively. Dengue, chikungunya and scrub typhus cases are also on the rise. India witnessed Nipah in 2001 (Siliguri) and 2018 (Calicut), Influenza A (H1N1) in 2009 and COVID-19 in 2020–21.

There has been an increase in infections caused by *Acinetobacter*, *Burkholderia* and *CoNS* as pathogens as well as a rise in multi-drug resistant bacterial infections. In addition, localised outbreaks of plague were seen in Surat, Shimla and Uttarakhand.

OCULAR INFECTIONS

Infections of the eye can affect the eyelid, conjunctiva, sclera, cornea, lacrimal system, uvea, orbit and optic nerve. They can either be manifestations of systemic infection or infections of the eye alone. The common microbes affecting the eye are listed in Table 43.1.

The laboratory diagnosis of each depends upon the site and cause of infection. Bacterial and fungal infections are diagnosed by culture, parasitic infections by microscopy and viral infections by serologic or molecular diagnosis.

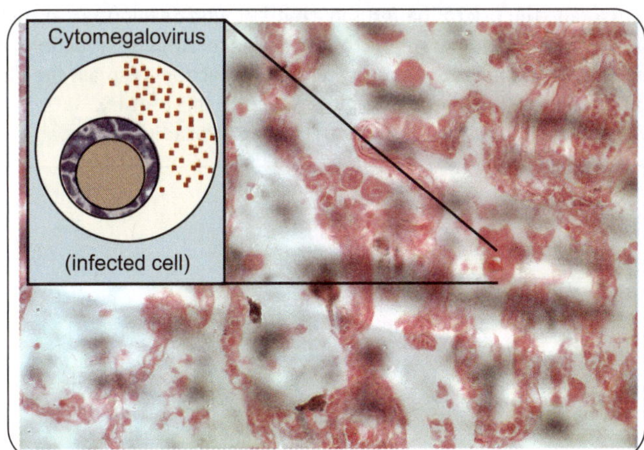

Fig. 43.1 CMV showing intranuclear inclusion bodies, H&E stain, 40X

Table 43.1 Ocular infections

Condition	Microbe
Eyelid	
External and internal hordeolum	*Staphylococcus aureus*
Blepharitis	*S. aureus*
Conjunctivitis	
Viral	Adenovirus (especially types 8,19, 37), herpes simplex or zoster virus
Bacterial	*Haemophilus influenzae, Streptococcus pneumoniae, S. aureus*
Ophthalmia neonatorum	*Neisseria gonorrhoeae, Chlamydia trachomatis*
Scleritis	*Mycobacterium tuberculosis*
Keratitis	
Bacterial	*S. aureus, S. pneumoniae, Streptococcus pyogenes, Enterococcus, Pseudomonas, E. coli, Klebsiella*
Fungal	*Fusarium, Aspergillus, Penicillium, Curvularia, Candida, Cryptococcus*
Viral	Herpes simplex 1 virus, herpes zoster virus, adenoviruses
Protozoa	*Acanthamoeba*
Acute dacryocystitis	*Streptococcus, Pneumococcus, Staphylococcus*
Anterior uveitis	*M. tuberculosis, T. pallidum,* herpes simplex and zoster viruses, *Toxoplasma*
Focal retinitis	*Toxoplasma, T. pallidum,* herpes simplex virus, cytomegalovirus
Choroiditis	*M. tuberculosis, Nocardia, Histoplasma*
Orbital cellulitis	*S. pnuemoniae, S. pyogenes, S. aureus, H. influenzae,* mucormycetes
Congenital rubella	Cataract, glaucoma, retinopathy
Infectious mononucleosis	Keratoconjunctivitis, optic neuritis
Others: HIV, *M. tuberculosis*, cysticercus cellulosae, *Parastrongylus cantonensis*	

Questions

1. Enumerate microbes causing opportunistic infections and outline their laboratory diagnosis.
2. How do viruses cause malignant transformation? Name viruses with their oncogenic potential.
3. Outline the risk to the fetus in a mother who develops an exanthematous fever in the first trimester.
4. Define and enumerate microbes causing teratogenic infections. Explain the diagnosis and prevention of congenital rubella.
5. Enumerate microbes causing perinatal infections. Explain the diagnosis and prevention of HIV infection in a neonate.
6. Enumerate infections emerging in India in this century.
7. Name microbes causing the spectrum of ocular infections.

SECTION SUMMARY

Section IX: Zoonoses and Other Infections

Infection/pathogen	Virulence/pathogenesis	Laboratory diagnosis	Vaccine/recommended antimicrobial
Bacterial **Plague** *Yersinia pestis*	Rat flea bite, F1 antigen, LPS, bubonic, pneumonic, septicemic	Wayson smear, culture, F1 ELISA/PCR	Killed whole cell (only in epidemic), doxycycline, tetracycline
Anthrax *Bacillus anthracis*	Inoculation (cutaneous eschar), inhalation (pulmonary) or ingestion (intestinal) of spores, exotoxin, polypeptide capsule	Gram and McFadyean's smear, culture, ELISA/PCR	Adsorbed anthrax antigen, ciprofloxacin or doxycycline with clindamycin
Brucellosis *Brucella melitensis*	Inoculation/inhalation/ ingestion (dairy) intraphagocytic	Blood culture, antibody SAT/ ELISA, PCR	A live attenuated B abortus 19-BA, streptomycin or rifampicin with doxycycline
Scrub typhus *Orientia tsusugamushi*	Mite bite from rodents, eschar at bite site, multiply in lymphatic, blood, endothelium, fever, rash	IgM/IgG ELISA, IFA, IPA, Weil-Felix test, PCR	Doxycycline, azithromycin
Leptospirosis *Leptospira interrogans*	Invades skin abrasions, bacteremia, Weil's disease	IgM/ IgG ELISA, ICT	Doxycycline/penicillin/ ceftriaxone
Borrelia recurrentis	Relapsing fever	PBS Giemsa, ELISA, PCR	
Borrelia burgdorferi	Lyme's disease	Culture, ELISA PCR	Doxycycline or erythromycin Doxycycline
Viral Rabies	Dog bite, virus in peripheral neuron, ascends to hippocampus and cerebellum neurons (intracytoplasmic, eosinophilic Negri bodies), to salivary glands, skin and cornea via nerves	Antigen by IMF skin, cornea, antibody CSF, serum. RT-PCR saliva.	Wound care. Post exposure cell culture vaccines HDCS/ PCEC/PVC, passive HRIG
Yellow fever	*Aedes* bite, liver (jaundice), heart, kidney Not seen in India		Live attenuated strain 17-D for travellers from endemic countries
Parasitic **Toxoplasmosis** *Toxoplasma gondii*	Oocyst ingestion from cat, pseudocyst of tachyzoites in MPS, true cyst of bradyzoites in brain, muscle Congenital toxoplasmosis	HPE of tissue cysts, IgM in acute infection, PCR in CSF and amniotic fluid	Sulfadiazine+ pyrimethamine
Hydatid disease *Echinococcus granulosus*	Feco-oral from dog ova, liver cyst with daughter scolices	Radiography, HPE, Casoni's test	Albendazole

Hospital-Acquired Infections and Safety Precautions

Definition

Hospital-acquired infection (HAI) is any infection not present or incubating at the time of hospital admission but acquired due to procedures done in the hospital for treating some other condition. To qualify as a hospital-acquired infection, the symptoms should appear at least 48 hours after admission to the hospital, or after discharge from the hospital but linked to the procedures done when in the hospital.

Factors affecting HAIs

1. The incidence of HAIs is associated with an increased length of stay in the hospital and therefore the thrust towards quality care is to minimise unnecessary hospital stays with day care procedures wherever feasible.
2. Extremes of age, immune status and comorbid disease conditions of the patient.
3. Need for therapeutic procedures like intravenous or urinary catheterisation, assisted ventilation, or surgery.
4. The increased use of antibiotics in the hospital.

Source of HAIs

1. Endogenous microbes that are commensals or colonise the patient's skin, or the respiratory or gastrointestinal tract, e.g., CoNS causing bloodstream infection and *E. coli* causing UTI following intravenous and urinary catheterisation, respectively.
2. Exogenous microbes from the hospital environment or carriers amongst healthcare workers (HCWs), e.g., atypical mycobacteria causing endoscopic site infection or *Staphylococcus aureus* causing surgical site infection due to a carrier HCW.

Types of HAIs

1. Catheter-related bloodstream infection (CRBSI)
2. Catheter-associated urinary tract infection (CAUTI)
3. Ventilator-associated pneumonia (VAP)
4. Surgical site infection (SSI)

CATHETER-ASSOCIATED URINARY TRACT INFECTION (CAUTI)

CAUTI is defined as significant bacteriuria in a catheterised patient with or without signs and symptoms of UTI. The presence of a urinary catheter provides a high risk for UTI, the length of duration of catheterisation being important. CAUTI can occur in a single episode, or on short-term (< 30 days) or long-term (> 30 days) catheterisation, and depends upon the asepsis practised during insertion and care thereafter.

Clinical Case 1

A 70-year-old man in the SICU following a head injury has a Foley's catheter placed on day 1 of admission. On day 15, he develops fever with suprapubic tenderness and urine culture is reported as E. coli > 100,000 CFU/mL.

1. *What is the most likely diagnosis?*
2. *What is the source of the bacterium?*
3. *How can it be prevented?*
4. *Name other bacteria that can be implicated in this condition*

The bacteria commonly associated with CAUTI are *E. coli, Klebsiella, Pseudomonas, Acinetobacter*, CoNS and *Enterococcus*. Their source may be endogenous from the patient or exogenous from the hospital environment or HCW.

The laboratory diagnosis of CAUTI needs culture of the urine sample, but the sample should be collected aseptically from the catheter port or in the absence of a port by aspiration from the catheter tube using a needle and syringe. Collection of urine from the urosac or urine bag is not acceptable. The significant count for bacteriuria is 10^3 and 10^5 CFU/mL in symptomatic and asymptomatic cases, respectively. Routine screening of catheterised urine or treatment of asymptomatic catheterised bacteriuria is not recommended. Candiduria may be seen in catheterisation but is usually considered insignificant unless occurring with certain underlying pathologic conditions.

Catheterisation should be done only when absolutely necessary and removed at the earliest feasible. To prevent CAUTI, strict asepsis must be adopted during the procedure, with sterile gloves, periurethral cleaning, untouched catheter insertion and a fixed catheter attached to a closed drainage bag maintained below the level of the bladder. Urine sampling if required should be done from the sampling port or with a sterile needle and syringe. The use of antibiotics or bladder irrigation is not recommended to prevent CAUTI.

CATHETER-RELATED BLOODSTREAM INFECTION (CRBSI)

CRBSI is defined as an intravascular infection wherein the source of infection is laboratory-proven to be the central venous catheter (CVC). The source of bacteria for CVC colonisation can be the skin of the patient or medical personnel or contaminated device/intravenous fluid, or another focus of infection in the patient. The occurrence of CRBSI is related to the duration (> 72 h) and site of catheterisation, aseptic precautions taken and the underlying medical conditions of the patient.

Clinical Case 2

A 20-year-old man with acute lymphoblastic leukemia underwent a bone marrow transplant and was hospitalised with graft versus host disease. A central venous catheter was inserted with weekly surveillance blood cultures. Two months after admission, the patient developed fever with signs of sepsis, and the blood culture grew Staphylococcus aureus.

1. *What is the most likely diagnosis?*
2. *What is the source of the bacterium?*
3. *How can it be prevented?*
4. *What is the significance of methicillin resistance in Staphylococcus aureus?*

The microorganisms causing CRBSI are commonly CoNS, especially *Staphylococcus epidermidis*, *S. aureus* and *Acinetobacter baumanii* from the skin, and rarely, *Pseudomonas aeruginosa*, *Burkholderia cepacia* and *Candida* spp. from intravenous fluid contamination. Staphylococci and *Candida* have the ability to form a biofilm on the device.

In the event of fever in the presence of catheter site inflammation, the laboratory diagnosis of CRBSI requires qualitative and quantitative blood cultures from the CVC, without removal of the catheter. If proven, the treatment varies with the identity of the microbe and the severity of the infection. It may involve removal of the catheter, and systemic antimicrobials based on the susceptibility of the microorganism.

Catheterisation should be done only when absolutely necessary and removed at the earliest feasible. CRBSI can be prevented by hand hygiene, aseptic practices, and insertion site skin antisepsis with regular change of site dressing. Systemic or local use of antimicrobial prophylaxis is not recommended to prevent CRBSI.

VENTILATOR-ASSOCIATED PNEUMONIA (VAP)

Clinical Case 3

An 80-year-old man admitted in the ICU was intubated as he had cardiopulmonary arrest. On the third day after intubation, he developed fever, tachycardia and dyspnea with purulent tracheal secretions. On investigation, his leucocyte and platelet counts were elevated with a raised CRP. The chest X-ray showed infiltration of the right lower lobe.

1. *What is the most likely diagnosis?*
2. *Name three bacteria most likely to cause this condition*
3. *How will you confirm your diagnosis in the microbiology laboratory?*
4. *What are the preventive measures?*

VAP is defined as a pneumonia that occurs 48–72 hours following endotracheal intubation for mechanical ventilation (MV). The source of infection is either aspiration of the endogenous oral or upper respiratory tract microflora of the patient or exogenous contamination of the equipment used for MV. The risk factors being the duration of MV (> 5 days), aseptic precautions taken and the underlying medical conditions of the patient. Early onset VAP (< 5 days) is more likely caused by commensals like *Streptococcus pneumoniae*, *Haemophilus influenzae* and *S. aureus*, while late onset VAP by exogenous hospital-acquired, drug-resistant pathogens like *Klebsiella pneumoniae*, *P. aeruginosa* and *A. baumanii*.

The diagnosis of VAP relies on the clinical, radiological and laboratory findings. The clinical presentation includes fever, dyspnea and tachycardia with the chest X-ray showing infiltration in the lungs. Microbiologic sampling of endotracheal tube aspirate (non-invasive) or bronchoalveolar lavage (invasive) is recommended to be processed without delay.

As per Infectious Disease Society of America (IDSA) guidelines, only a sample with less than 10 epithelial cells/LPF in the presence or absence of pus cells is acceptable. On Gram stain, the absence of bacteria in such a sample implies that VAP is unlikely, while their presence indicates that the diagnosis is open to interpretation by qualitative and semi-quantitative or quantitative culture of the sample.

Endotracheal tube aspirate showing the same bacterium in smear and culture, with a semiquantitative culture (> 10^5 CFU/mL) is adequate for the microbiologic diagnosis of VAP. The inflammatory biomarkers—CRP and procalcitonin, and leucocyte and platelet count are also used in the laboratory diagnosis of VAP, especially in monitoring empirical antimicrobial therapy.

General preventive measures include hand hygiene, use of PPE and other aseptic precautions besides specific measures in the optimal use of the MV apparatus.

SURGICAL SITE INFECTIONS

Clinical Case 4

A 30-year-old woman develops incisional wound infection following thyroid surgery. Pus collected from the wound grows Staphylococcus aureus sensitive to cloxacillin.

1. *Classify this surgical incision*
2. *What pre-operative antimicrobial prophylaxis is indicated?*
3. *Name two pre-operative skin antiseptics*
4. *What is the permissible rate of infection in such surgeries?*

Surgical site infection (SSI) is defined as infection of the localised site of an incision within 30 days of the surgery or within 90 days if surgery had an implant in place. It can be superficial—involving skin and subcutaneous tissue, deep—involving fascia and muscle, or organ space—involving viscera.

The source of bacteria is mostly a patient's endogenous microflora—*S. aureus* and CoNS from the skin or commensals at the mucosa when the mucosa is involved in surgery. The exogenous source can be the operation theatre (OT) environment or instruments or personnel. The risk factors are inadequate asepsis of the patient's (incision site) or surgeon's skin (hands), prolonged operative time, surgical technique, presence of a post-surgical incisional drain, duration of hospital stay, improper disinfection or sterilisation of instruments, and diabetes with poor peri-operative control. Depending upon the degree of bacterial contamination expected during surgery, the surgical incision is classified as clean, clean-contaminated, contaminated, and dirty or infected.

A clean surgical wound is one wherein the skin incision site is uninfected and does not involve the mucosa. Most surgeries (without underlying trauma)—cardiac, neurosurgery, thyroid and breast—are clean and should have an SSI rate of < 2%.

A clean-contaminated surgical wound is one wherein a mucosal tract is breached under controlled operative conditions. Surgery on the vagina in obstetrics and gynecology, appendicectomy, and biliary or oropharyngeal surgeries are clean-contaminated with a permissible SSI rate of 3–11%.

Surgeries following trauma and accidents, gangrene, and colonic surgeries are considered contaminated with high SSI rates while dirty procedures like intra-abdominal perforation or abscess have even greater chances of SSI.

Surgical site infection can be prevented by certain pre-operative, intra-operative and post-operative measures. These are:

1. **Pre-operative** bath with soap and water, screening and treatment of personnel for nasal carriage of *S. aureus*; incision site hair removal, if indicated, with a clipper; antimicrobial prophylaxis for clean surgery with 1 gm intravenous cefazolin, given a minimum of 2 hours prior to incision. The microbial load in the operating room air is also important in the prevention of SSI, which is achieved by regular cleaning and high-level disinfection of the operating room.
2. **Intra-operative** surgical site preparation with alcohol-based chlorhexidine or aqueous povidone iodine. Surgical hand asepsis is achieved with alcohol-based chlorhexidine and gloves.
3. **Post-operative** wound dressing with local antiseptic ointment following wound closure and a reduced length of stay in the hospital.

Pseudomonas, Burkholderia, Acinetobacter

Unlike the non-fastidious, enteric gram-negative bacilli belonging to the family *Enterobacteriaceae* that ferment sugars, non-fermenting gram-negative bacilli (NFGNB) are non-fastidious saprophytes in the hospital environment that do not ferment sugars. The commonly encountered NFGNB are *Pseudomonas, Acinetobacter* and *Burkholderia* that cause hospital-acquired infections.

Pseudomonas aeruginosa is an NFGNB that is extremely motile and adherent due to the presence of flagella and pili. The virulence is attributed to a large number of exotoxins and enzymes secreted by the bacterium, including diffusible pigments pyocyanin (blue), pyoverdin (green), pyorubin (red) and pyomelanin (black). It also possesses a capsule, biofilm, endotoxic cell wall and genes that code for intrinsic resistance to antimicrobials.

It can cause CRBSI, CAUTI, VAP and SSI that may lead to septicemia, and pyogenic infection of skin in burns and practically any other site (ear, eye, blood, bone, meninges). It is a pathogen in hospitalised patients and in cystic fibrosis. A Gram stain examination of the pus sample from the site of infection demonstrates gram-negative bacilli that can be cultured and tested for antimicrobial susceptibility. As the bacterium is intrinsically resistant to antimicrobials, those that can be used in treating *Pseudomonas* infections are: broad-spectrum β-lactams piperacillin, ticarcillin, mezlocillin, ceftazidime, cefoperazone, cefepime, β-lactamase combinations piperacillin-tazobactam and cefoperazone-sulbactam, carbapenems, aztreonam, aminoglycosides, fluoroquinolones and polymyxins.

The *Burkholderia cepacia* complex are NFGNB similar to *Pseudomonas*, that are pathogenic in hospitalised patients and in cystic fibrosis. *Acinetobacter baumanii* is a saprophyte in the hospital environment and commensal in human skin, oral cavity and intestine, and is associated with hospital-acquired infection and outbreaks. Both are intrinsically resistant to antimicrobials, therefore, therapy needs to be based on the sensitivity test results.

STANDARD PRECAUTIONS

Standard precautions are certain protective measures to be practised at all times while providing medical services. They are aimed at preventing transmission of infection from and to a patient and applicable to all HCWs.

Components of standard precaution

1. Hand hygiene
2. Personal protective equipment (PPE)
3. Sterilisation and disinfection of hospital environment and medical devices
4. Safe disposal of biomedical waste
5. Post-exposure prophylaxis

Hand hygiene

In 1846, Ignaz Semmelweiss demonstrated a decrease in deaths due to puerperal sepsis by washing hands with chlorine, while William Halsted introduced rubber gloves in the operating room in 1889. Today, the WHO has defined the steps for hand washing and the 'Five moments of hand hygiene' (Figs. 44.1 and 44.2).

Certifiable Practical Competency of Handwashing (based on WHO guidelines)

WHO Checklist for Hand Wash	
Step 0: Wet hands with water	YES/NO
Step 1: Apply enough soap to cover all hand surfaces	YES/NO
Step 2: Rub hands palm to palm	YES/NO
Step 3: Right palm over left dorsum with interlaced fingers and vice versa	YES/NO
Step 4: Palm to palm with fingers interlaced	YES/NO
Step 5: Backs of fingers to opposing palms with fingers interlocked	YES/NO
Step 6: Rotational rubbing of left thumb clasped in right palm and vice versa	YES/NO
Step 7: Rotational rubbing, backwards and forwards with clasped fingers of right hand in left palm and vice versa	YES/NO
Step 8: Rinse hands with water	YES/NO
Step 9: Dry hands thoroughly with a single use towel	YES/NO
Step 10: Use towel to turn off faucet	YES/NO

The skin carries normal resident bacteria and transient bacteria acquired from contact with the animate and inanimate environment. The resident flora comprises *Staphylococcus epidermidis*, *Cutibacterium acnes* and diphtheroids, while the transient flora comprises *S. aureus*, enterococci, *E. coli*, *Klebsiella*, *Acinetobacter* and other gram-negative bacilli. The transient microflora can be removed by handwashing while the resident flora cannot be removed. The human skin can never be sterilised but the microbial load of the resident flora can be reduced by a surgical hand scrub followed by the use of antiseptics before any surgical or aseptic procedure (Refer skin antisepsis in Chapter 45).

The five moments of hygiene as recommended by the WHO are: before touching a patient, before aseptic procedures, after body fluid/risk exposure, after touching a patient, and after touching a patient's surroundings.

Personal protective equipment (PPE)

PPE refers to specific clothing or articles worn by a healthcare worker (HCW) to prevent transmission of infection from patient to HCW or vice versa. PPE consists of gloves, respiratory masks, face shields, goggles, gowns/aprons or suits with caps and footwear. The quantum of PPE needed depends on the specific healthcare situation.

The gloves most commonly used are latex gloves in varying sizes, available as sterile and unsterile. Hand hygiene is necessary before gloving and after de-gloving. Gloves should not be worn on wet hands and should be discarded in a red bag after use.

Unsterile gloves are recommended for procedures involving contact with a patient's skin or mucous membrane, e.g., blood collection, dental examination. Sterile gloves are mandated in all invasive procedures where sterility is to be maintained, e.g., surgery and surgical dressing, lumbar puncture, central line or urinary catheter insertion. During surgery, depending upon the surgical procedure, double gloving or change of gloves during prolonged surgery is recommended.

Reusable rubber gloves and not surgical gloves can be used for all housekeeping and cleaning work.

Surgical face masks are fluid-resistant, disposable, made of three-pleat polypropylene, and shield the HCW and the patient from any exchange of respiratory droplets or blood or body fluid splashes during the procedure. They are worn by the HCW during surgical or endoscopic procedures or any procedure wherein blood or body fluid splash may be anticipated. They are also to be worn by the patient who has either a respiratory infection or is immunosuppressed.

The **N95 mask** is a filtering face piece (FFP) that filters 95% of airborne particles. It is indicated for use by HCWs only in the event of close contact with patients with respiratory infections with high transmissibility and infectivity, especially viruses like influenza and COVID-19. It is not to be worn by the patient or non-HCWs.

Respiratory droplets are large (> 5 μm) droplets generated during coughing, sneezing or talking. They do not remain suspended in the air and generally travel short distances. Transmission of infection requires close (< 3 feet) contact with the patient. Respiratory infections transmitted by droplets include bacterial upper respiratory diphtheria, pertussis, lower respiratory pneumonias caused by *Haemophilus* and *Mycoplasma*, plague and childhood pharyngitis, mumps and rubella.

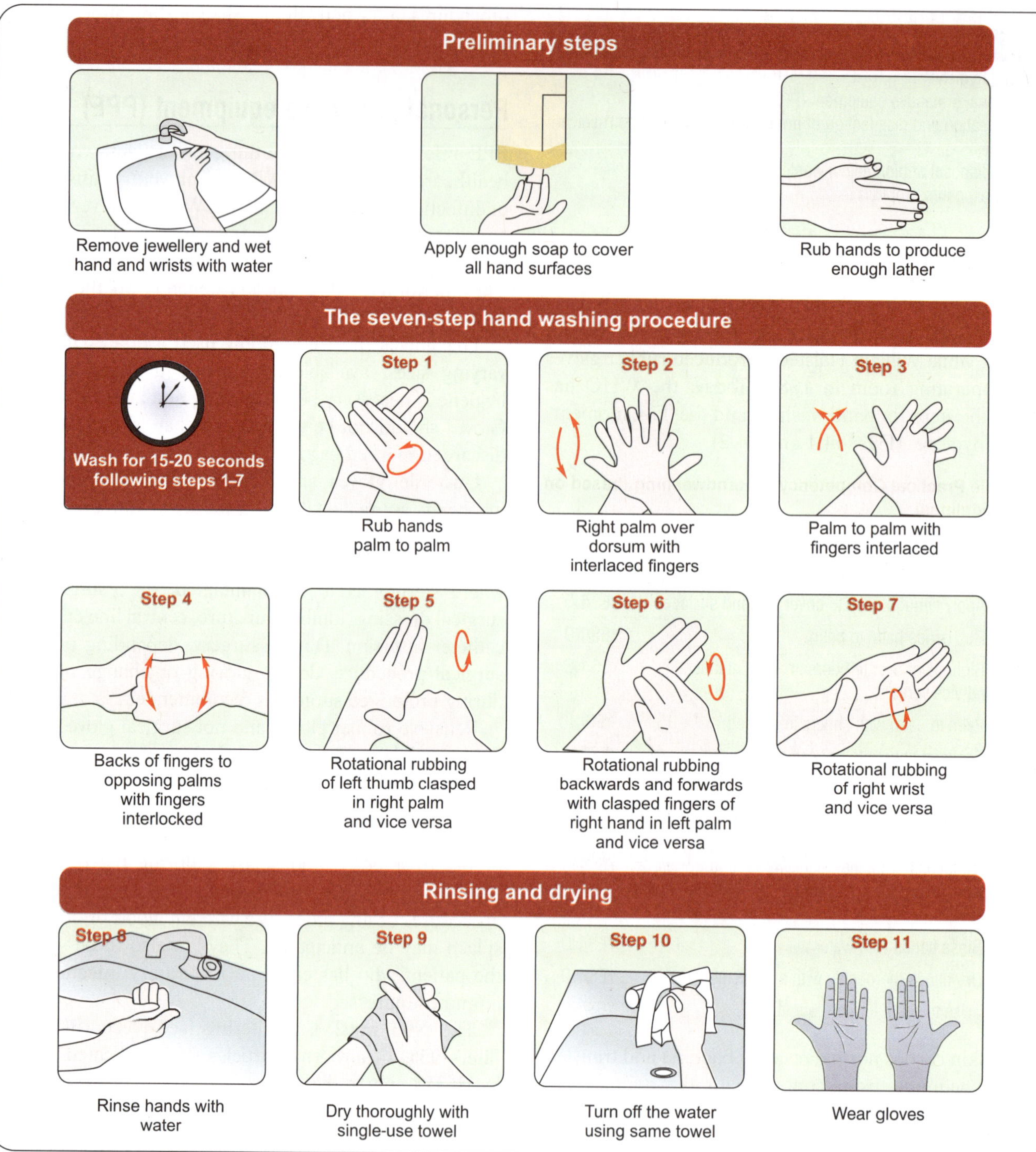

Fig. 44.1 Steps for hand hygiene

Respiratory aerosols are small (< 5 µm) particles generated during coughing, sneezing or talking. Aerosols can also be generated in a hospital care setting during certain procedures, e.g., while centrifuging samples in the laboratory, or intubating patients in OTs and ICUs, or while using a drill in surgery. Unlike respiratory droplets, respiratory aerosols can remain suspended in the air and travel longer distances. Respiratory infections transmitted by aerosols include tuberculosis, influenza, coronavirus, measles and chickenpox.

Other PPE like plastic aprons, surgical gowns, full-body gowns, face shields, goggles, caps and footwear protect the skin and mucous membranes of HCWs from being exposed to blood and body fluids of the patient. The sequence of donning PPE is in the order of hand hygiene, gown, mask, goggles and gloves. The sequence of doffing is in the order of gloves, goggles, gown, mask and hand hygiene.

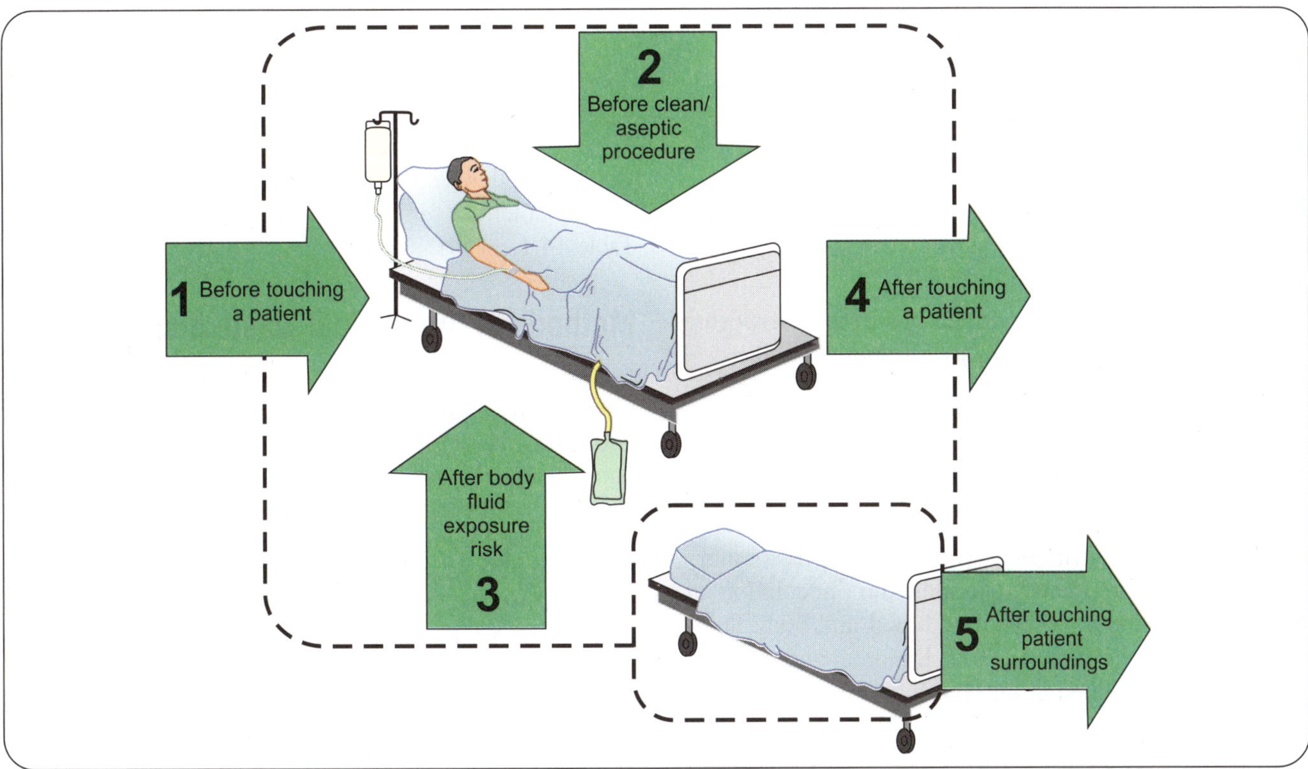

Fig. 44.2 Five moments of hand hygiene

BIOSAFETY LEVELS

The biosafety level at which a laboratory functions is defined and designated keeping in view the variability of the risk of transmission of pathogens while handling patient samples. To reduce the risk of transmission to laboratory workers who process samples, there are four designated biosafety levels (BSLs)—1–4 to be followed.

Certifiable Practical Competency for use of PPE (based on WHO guidelines)

WHO Checklist for Donning of Personal Protective Equipment (Coverall)	
Step 1: Remove all personal items.	YES/NO
Step 2: Put on scrub suit and rubber boots in the changing room.	YES/NO
Step 3: Perform hand hygiene.	YES/NO
Step 4: Put on gloves.	YES/NO
Step 5: Put on coverall.	YES/NO
Step 6: Put on face mask.	YES/NO
Step 7: Put on face shield or goggles.	YES/NO
Step 8: Put on head and neck covering surgical bonnet covering neck and sides of the head (preferable with face shield) OR hood.	YES/NO
Step 9: Put on disposable waterproof apron.	YES/NO
Step 10: Put on second pair of gloves over the cuff.	YES/NO

WHO Checklist for Doffing of Personal Protective Equipment (Coverall)	
Step 1: Ensure that infectious waste containers are available in the doffing area for safe disposal of PPE.	YES/NO
Step 2: Perform hand hygiene on gloved hands.	YES/NO
Step 3: Remove apron.	YES/NO
Step 4: Perform hand hygiene on gloved hands.	YES/NO
Step 5: Remove head and neck covering.	YES/NO
Step 6: Perform hand hygiene on gloved hands.	YES/NO
Step 7: Remove coverall and outer pair of gloves.	YES/NO
Step 8: Perform hand hygiene on gloved hands.	YES/NO
Step 9: Remove eye protection.	YES/NO
Step 10: Perform hand hygiene on gloved hands.	YES/NO
Step 11: Remove the mask.	YES/NO
Step 12: Perform hand hygiene on gloved hands.	YES/NO
Step 13: Remove rubber boots.	YES/NO
Step 14: Perform hand hygiene on gloved hands.	YES/NO
Step 15: Remove gloves carefully and dispose of them safely.	YES/NO
Step 16: Perform hand hygiene.	YES/NO

BSL-1 applies when the worker has no risk in handling the microbe, and can work on an open laboratory bench, e.g., commensal bacteria

BSL-2 applies when there is moderate risk in handling the pathogens, but that the worker can still work on an open laboratory bench with the requisite PPE and decontamination protocols, e.g., pathogenic enteric bacteria

BSL-3 applies when there is a risk of handling microbes with potential aerosol transmission, causing diseases with considerable morbidity or mortality, e.g., *M. tuberculosis*, influenza virus, coronaviruses. The work must be done in a biosafety cabinet (BSC) level I or II in a specifically designed laboratory setting.

BSL-4 applies when the worker has a risk of exposure to life-threatening microbes with potential aerosol transmission and no treatment, e.g., Ebola virus, Marburg virus. The work must be done in a BSC level III in a specifically designed laboratory setting, with full PPE.

Occupational exposure to laboratory-acquired infections can occur by inhaling aerosols generated by procedures like centrifuging while handling respiratory pathogens, and accidental percutaneous inoculation or mucosal exposure while handling blood and body fluids with blood-borne viruses (BBVs). Drinking and eating is to be strictly prohibited in the laboratory.

SAFE DISPOSAL OF BIOMEDICAL WASTE

Biomedical waste is waste generated during diagnostic or research procedures or treatment of patients. It is important to ensure safe segregation and disposal of biomedical waste not only for the safety of HCWs but also to prevent pilferage for reuse. Depending on the kind of biomedical waste generated, it may or may not be treated on-site with chemical disinfectants prior to final disposal. The biomedical waste policy includes specific steps to be followed with segregation at the site of generation, storage at the hospital site, transport to the disposal facility and the final disposal method.

Biomedical waste segregation—Hospital waste generated can be categorised as infectious, non-infectious and sharps (both infectious and non-infectious). Solid waste is not to be disposed in drains, and liquid wastes can be disposed in drains but only after sterilisation or disinfection. Non-infectious glass, plastic and paper articles can be recycled instead of being disposed. The 2016 Ministry of Environment, Forests and Climate Change guidelines for on-site segregation of biomedical waste, amended in 2018 and 2019, are shown in Table 44.1.

Methods of biomedical waste disposal

Incineration is electrical combustion at a high temperature (800°C) that is used for human anatomical and non-plastic infectious waste but it comes with considerable environmental effects. Autoclaving is high temperature (120°C) steam sterilisation to render infectious material sterile, for microbiological and infectious plastic waste. Hydroclaving is low temperature steam decontamination with shredding and drying of the treated products. Microwaves with shredding are also used for treating infectious waste as in autoclaves. Infectious waste when rendered non-infectious can be disposed by deep burial in a landfill along with other non-infectious waste.

OCCUPATIONAL EXPOSURE AND POST-EXPOSURE PROPHYLAXIS (PEP)

Healthcare workers (HCWs) are at an increased risk of acquiring infections due to occupational exposure. The risk of acquiring infections includes respiratory infections, e.g., tuberculosis, influenza, COVID-19 and other respiratory pathogens and blood-borne viruses (BBVs), e.g., HBV, HCV and HIV, and rarely syphilis.

The higher the compliance with safety precautions, the lesser the risk of infection. However, in spite of safety precautions, accidental infection can occur and the concerned HCW needs to be aware of the action to be taken, in the event of an exposure.

Table 44.1 Recommended container and disposal method for medical waste categories

Waste category	Container	Disposal
Anatomical-human and animal waste, non-plastic infectious waste	Yellow plastic bag	Incineration
Infectious plastic waste	Red plastic bag	Autoclaving and shredder/Microwave
Non-infectious	Plastic bag/bin Kitchen/biodegradable waste Bin/basket	Landfill/compost Recycle
Sharps	White puncture proof container	Autoclaving and shredder/Microwave
Glassware	Blue puncture proof container Broken glass-vials, ampoules	Autoclaving/Microwave, Recycle

In case of respiratory pathogens, the use of masks by the patient and appropriate PPE by the HCW, and vaccine in case of influenza, minimises the risk of transmission. In the event of an exposure/contact without PPE, it is advisable to test any symptomatic HCWs for the disease.

In case of blood-borne viruses, the exposure can be percutaneous or mucosal, or through skin (intact or non-intact). Percutaneous exposure occurs while handling a needle—needlestick injury (NSI), or during any surgical procedure requiring handling of sharps. Reuse of syringes to administer injections or vaccines to more than one person is to be strictly avoided. Recapping of needles is the single most common high-risk practice that leads to occupational BBV infection, followed by unsafe segregation, transport or disposal of sharps. Mucosal and skin contact results from the splash of blood or body fluids during endoscopic or surgical procedures. The risk of transmission is higher in case the underlying skin is non-intact due to cuts or abrasions, or disease.

The risk of occupational infection of BBVs can be reduced by covering any existing cut or wound on the hands, handwashing, safe infectious waste disposal, avoiding the reuse or recapping of needles, use of PPE, and HBV vaccination.

The magnitude of risk of acquiring a blood-borne infection is 0.3%, 3% and 30% in the case of percutaneous exposure to HIV, HCV and HBV, respectively. This difference in the transmissibility of the viruses is attributed to the number of virions in blood that determine the infectivity of the virus.

The risk of transmission of BBVs is not absolute, and depends on several factors

1. The nature of exposure—contact with blood or body fluid, the needle being hollow bore or solid, contact being with intact skin, mucosal or percutaneous, or through skin. Exposure to blood and genital fluids carries a higher risk than to other body fluids.

   CSF, pleural, pericardial, peritoneal, synovial and amniotic fluid are potentially infectious. Feces, urine, vomitus, saliva, sputum, sweat and tears are potentially non-infectious. Any body fluid visibly contaminated with blood is considered infectious.
2. The viral load of the source at the time of exposure. The source is the patient—known or unknown.
3. The exposure being with PPE or without, and whether the HCW wore gloves, a mask, goggles, or a face shield during contact.
4. The prevalence of the particular infection in the population, geographical and epidemiological, e.g., the average prevalence of HIV in India is 0.2% but that of HBV is 4%, and the prevalence is higher in commercial sex workers (CSWs) than the general population.

All patients are to be considered potentially infectious with safety precautions and post-exposure prophylaxis strictly followed at all times.

The action to be taken by a HCW in the event of an NSI or any other occupational exposure to BBVs can be divided into **immediate first aid action** and **follow-up action.**

Immediate first aid action is to wash the exposed site or injury with soap and water, or flush the exposed mucosa with water. The reflex action of sucking the pricked finger or squeezing the injury is to be avoided. There is no scientific evidence for either of these actions or the use of antiseptics on the injury reducing the transmission of infection.

Follow-up action is to inform the source patient of the accident, and report it to the designated hospital authority for evaluation of the exposure and initiation of the post-exposure prophylaxis (PEP) if required. Every healthcare setup must have a designated location (example casualty) that is available 24X7 to record and provide PEP for HBV and HIV. The authority will decide on the PEP for HIV after evaluating the type of exposure (Exposure Code) and the HIV status of the source patient (Source Code).

The Exposure Code depends on whether the exposure was to blood or other body fluids, the volume of blood/body fluid and whether the exposure was percutaneous/mucosal/on intact skin. In the case of percutaneous exposure, whether the exposure was through a solid or hollow bore needle, as the volume of blood/body fluid exposed to a solid needle is smaller than that to a hollow bore needle.

The Source Code is generated depending on the serologic status of HIV being unknown/known and, if known, the stage of HIV infection, and, if HIV-positive, whether on ART or not. If the source patient status is unknown, the patient is tested for HBV, HIV and HCV after pre-test and post-test counselling.

The risk of transmission of HIV to the exposed HCW is negligible if **post-exposure anti-retroviral therapy is initiated within 2 hours of the exposure**, and increases thereafter. PEP provided 72 hours after the exposure is unlikely to prevent transmission. A single oral dose of tenofovir 300 mg + lamivudine 300 mg + dolutegravir 50 mg is recommended by NACO in case of any exposure, unless the source patient's status was tested earlier and known to be HIV-negative. The subsequent duration of PEP depends on the expert evaluation of the risk of transmission based on the exposure and source code. Post-exposure prophylaxis for HBV includes HBIG or vaccination, depending upon the previous vaccination status of the HCW.

The source patient and the exposed HCW need to be tested for BBV serology at the time of exposure (to rule out infection prior to this exposure). Follow-up testing is to be done at 6 weeks, 3 months and 6 months for HIV, and at 6 months for HBV and HCV. The tests indicated are anti-HIV antibody, HBsAg, anti-HCV antibody and anti-HBs antibody if the HCW has been previously vaccinated. During this period, the HCW is advised to avoid blood donation, unprotected sex and breastfeeding, and seek medical advice for any febrile illness within 12 weeks of exposure.

The risk of transmission of BBVs from HCWs to the patient is low and rare. HCWs are not restrained from working but monitored for infectivity, and safety precautions are mandated.

Answers

Clinical Case 1

1. *Catheter-associated urinary tract infection.*
2. *The source may be endogenous from the patient or exogenous from the hospital environment or HCW.*
3. *To prevent CAUTI, strict asepsis must be adopted during the procedure—sterile gloves, periurethral cleaning, untouched catheter insertion, fixed catheter attached to a closed drainage bag maintained below the level of the bladder.*
4. *The bacteria commonly associated with CAUTI are E. coli, Klebsiella, Pseudomonas, Acinetobacter, CoNS and Enterococcus.*

Clinical Case 2

1. *Catheter-related bloodstream infection.*
2. *Skin of the patient or carrier HCW.*
3. *Prevented by hand hygiene, aseptic practices, insertion site skin antisepsis with regular change of site dressing.*
4. *MRSA is resistant to all β-lactams including cephalosporins and carbapenems. Antimicrobials effective against MRSA are trimethoprim-sulfamethoxazole, doxycycline, linezolid, vancomycin, teicoplanin, and daptomycin.*

Clinical Case 3

1. *Early-onset ventilator-associated pneumonia.*
2. *Streptococcus pneumoniae, Haemophilus influenzae and Staphylococcus aureus.*
3. *Endotracheal tube aspirate, showing the same bacterium in smear and culture, with a semiquantitative culture (> 105 CFU/mL).*
4. *General preventive measures include hand hygiene and the use of PPE, and other aseptic precautions besides specific measures in the optimal use of the MV apparatus.*

Clinical Case 4

1. *Clean surgery.*
2. *Antimicrobial prophylaxis for clean surgery with 1 gm intravenous cefazolin given a minimum of 2 hours prior to incision.*
3. *Chlorhexidine and povidone iodine.*
4. *Less than 2% in clean surgeries.*

Questions

1. Define 'hospital-acquired infection'. What are the factors affecting HAIs and the sources of HAIs?
2. Mention the four types of HAIs. Write a note on catheter-related bloodstream infection (CRBSI).
3. Mention the four types of HAIs. Write a note on catheter-associated urinary tract infection (CAUTI).
4. Mention the four types of HAIs. Write a note on ventilator-associated pneumonia (VAP).
5. Mention the four types of HAIs. Write a note on surgical site infection (SSI).
6. Enumerate the components of Standard Precautions. Write a note on hand hygiene.
7. Define PPE. Elaborate on the role of masks in a hospital.
8. Differentiate between respiratory droplets and aerosols.

9. Write a note on Biosafety Levels and their implications in laboratory work.
10. Define biomedical waste and provide current guidelines for its segregation and disposal.
11. Write a note on the methods of biomedical waste disposal, citing the uses of each.
12. What is the risk of occupational exposure to HCWs? How can it be minimised?
13. What is the risk of occupational exposure of BBVs to HCWs? How can it be minimised?
14. Elaborate on the steps to minimise risk of BBV infection following an NSI.

Principles and Practice of Sterilisation and Disinfection

Patients visit healthcare facilities for treatment of their disease. It is therefore of paramount importance that in the process of treatment the facility itself does not cause more harm to the patient, i.e., with the patient ending up with a 'hospital-acquired infection' (HAI). HAI or nosocomial infection is an infection that did not exist at the time of access or admission to the hospital and appeared 48 hours after contact with the hospital. It can range from an injection abscess following an injection in the OPD or pneumonia due to ventilator support in the ICU. One of the arms in the prevention of HAI is the use of physical and chemical agents in healthcare facilities, i.e., certain practices that use heat or chemicals. In addition, many medical devices and surgical instruments are reused on patients that need, as the situation demands, to be either sterilised or disinfected.

An object is considered to be sterile if it is completely devoid of the presence of microorganisms in all forms, including bacterial spores. Therefore, such an object is incapable of causing an infection when used on a patient. Sterilisation is the process by which an object is freed of all microorganisms in their vegetative and spore form.

Disinfection means the destruction of all pathogens or organisms capable of causing infection but not necessarily spores. All organisms may not be killed but the number is reduced to a level that is no longer harmful to health. Chemical disinfection is applicable either to inanimate surfaces and objects or to human skin and wounds. The agents used for the former are referred to as 'disinfectants' and those used for the latter as 'antiseptics'.

Antisepsis indicates the prevention of infection usually by inhibiting the growth of bacteria in skin, wounds and tissues. 'Standard precautions' include appropriate use of sterilisation, disinfection and antisepsis procedures. In general, sterilisation utilises the physical process of heat (dry or moist steam), or gas or radiation or filtration as opposed to disinfection and antisepsis that use chemical agents.

Methods of sterilisation in practice with specific use

Heat sterilisation — sterilising glassware and metal surgical instruments
Gaseous sterilisation — to process heat-sensitive devices
Radiation sterilisation — sutures, prosthesis, disposable syringes
Filtration sterilisation — ophthalmic solutions, heat-sensitive injectable solutions

Sterilisation using heat (incineration or autoclave) and gas (ethylene oxide or plasma) is available within a healthcare facility. However, sterilisation by ionising radiation and filtration, though used for medical devices, is not done in-house in a healthcare facility. The material thus processed is provided to the HCW, as pre-sterilised, by the manufacturer. Dry heat is employed in an incinerator to burn anatomical waste and in a hot air oven in the laboratory for glassware.

Spaulding's classification for medical devices and levels of disinfection

The three categories described by Earle H. Spaulding on the basis of the degree of risk of infection involved in the reuse of the medical devices are **critical, semi-critical** and **non-critical**.

Critical items: Objects that breach the skin or mucous membrane and enter a sterile tissue, e.g., surgical instruments, urinary and cardiac catheters, implants, etc. These are essentially available pre-sterilised (syringe and needle) or have to be sterilised in-house by autoclaving (surgical instruments) or by gas (interventional cardiac catheters) prior to use.

Semi-critical items: Objects that come in contact with mucous membranes, e.g., respiratory therapy and anesthesia equipment, endoscopes, laryngoscope blades, etc. These are subjected to high-level disinfection (HLD) by chemical disinfectants like aldehydes.

Non-critical items: Objects that come in contact with intact skin, e.g., bedpans, blood pressure cuffs, bed rails, furniture and floors. These are subjected to low-level disinfection (LLD) by chemical disinfectants like phenols.

Factors affecting the process of sterilisation and disinfection

1. Presence of any organic matter—blood or tissue on the object affects it adversely
2. Concentration of the chemical agent or degree of heat needs to be optimal
3. Time period of contact with the chemical agent or degree of heat needs to be optimal
4. Nature of the microorganism matters as microbial resistance to chemical disinfectants varies

Microbes in the order of their degree of microbial resistance to disinfectants (from the most resistant to the least) are as follows:

1. Bacterial spores (*Clostridia* spp.)
2. Mycobacteria (Tuberculosis)

3. Non-enveloped viruses (HAV, poliovirus)
4. Fungi (yeast and mycelial)
5. Bacteria without spores (*Pseudomonas* spp. being the most resistant amongst bacteria)
6. Enveloped viruses (HIV, HBV, HCV)

It is evident that enveloped viruses and non-sporing bacteria are more easily killed by disinfectants than mycobacteria or bacterial spores. Only HLD or sterilisation can be used for mycobacteria or bacterial spores. The contact time of HLD for mycobacteria will also be longer than that for other bacteria.

Methods of achieving sterility and disinfection

1. Sterilisation (destroys bacterial spores) — autoclave, peracetic acid, H_2O_2 or plasma, chlorine
2. High-level disinfection (destroys TB but not spores) — aldehydes
3. Intermediate-level disinfection (destroys TB but not spores) — phenols, iodine, chlorine, alcohol
4. Low-level disinfection (bacteria other than TB killed) — quaternary ammonium compounds, cetrimide, bromide

Irrespective of the method of sterilisation or disinfection indicated, cleaning is the single most important step towards making a medical device ready for reuse.

Large instruments can be cleaned by using running water and detergents prior to sterilisation or disinfection, as indicated. Finer instruments with small bores require ultrasonic cleaning washers that are available in the **central sterile supply department** (CSSD).

In a healthcare setting, sterilisation and disinfection processes are employed for:
1. Skin antisepsis
2. Surgical instruments
3. Reusable medical devices
4. Single-use devices (SUDs) being reused for economical reasons
5. Hospital environment disinfection
6. Laboratory disinfection requirements

Surgical instrument sterilisation, reusable medical device HLD and the reprocessing of certain SUDs are undertaken in a centralised facility in the hospital, equipped with ultrasonic washers, autoclaves, ethylene oxide (ETO) and plasma sterilisers, called the 'central sterile supply department'(CSSD). The CSSD is designed to function with quality control measures, to ensure unidirectional workflow from the contaminated items received to the sterile items stored and distributed.

1. Skin antisepsis

The human skin is colonised by normal commensal bacteria that form the resident microflora, and therefore skin cannot be sterilised but disinfected. The commonest skin residents are *Staphylococcus* spp. that are capable of giving rise to infection whenever the patient's skin is breached, as in surgery or intravenous catheterisation. However, the skin also acquires transient bacterial flora by virtue of contact with the environment.

Handwashing with soap and water is most important in making the hands clean, tackling the transient bacteria and preventing cross-infection. The use of an antiseptic agent on the hands reduces the burden of bacteria on the hands, thus reducing the risk of iatrogenic infection due to the resident bacteria.

A skin antiseptic should have 'immediate action' on application, 'residual action' to provide antisepsis during the entire course of the procedure but be devoid of 'cumulative action' of being toxic on repeated applications. The antiseptics in use are 70% ethyl or isopropyl alcohol, and alcohol-based chlorhexidine gluconate and povidone iodine. The advantage of alcohol and alcohol-based preparations over aqueous solutions is that alcohol evaporates, leaving behind the required antiseptic effect and obviating the need for drying. If aqueous solutions are used, drying with a sterile towel is important.

The patient's skin being in closer contact with the surgical incision than a surgeon's gloved hands, the recommended concentration of chlorhexidine gluconate is 0.5% when used on a surgeon's hand and 4% when used for a patient's pre-surgical skin preparation. The recommended concentration of povidone iodine is 7.5% when used on a surgeon's hand and 10% when used for a patient's pre-surgical skin preparation. The recommended contact time for antiseptics to achieve antisepsis is 2 minutes. When compared to povidone iodine, chlorhexidine gluconate has a longer duration of residual action and is not inactivated by organic matter.

Handwashing can be social, hygienic or surgical depending upon situational needs. A social handwash is one practised outside the healthcare facility, with soap and water, in order to clean hands and does not need the use of an antiseptic. A hygienic handwash is practised in a healthcare setting where the HCW comes in contact with the patient's intact skin and needs the use of an antiseptic to prevent cross-contamination. A surgical handwash includes the steps to be taken when coming in contact with the patient's mucosal or sterile tissues. It includes the use of an antiseptic followed by wearing gloves on dry hands. The steps of handwashing based on WHO recommendations and gloving are included in Chapter 44.

Skin antisepsis using 70% isopropyl alcohol for 2 min is also mandated while collecting blood and giving injections or starting an intravenous peripheral line. Additionally, povidone iodine needs to be used while introducing central lines or for wound dressing.

2. Surgical Instruments

Surgical instruments are critical items which, with other steel, plastic, glass articles and liquids that can withstand moist heat, are sterilised by autoclaving. Moist heat kills microorganisms by denaturation and coagulation of proteins. To autoclave is to sterilise with steam at high temperatures by applying pressure on the water. Water boils at 100°C and by increasing the pressure, the boiling point of water is increased in order to obtain high-temperature steam that can also destroy bacterial spores. The most commonly used setting for autoclaving is steam at 121°C at 15 pounds per square inch pressure for 15 min. The quality control on the process is ensured by using chemical or biological indicators during the process. Chemical indicators change colour when the cycle run is optimal and biological indicators are heat-resistant bacteria that are killed by autoclaving (Fig. 45.1).

Surgical instruments can also be sterilised using a plasma steriliser. Plasma, considered to be the fourth state of matter, consists of ionised gas. Hydrogen peroxide (H_2O_2) is an oxidising agent with high-level disinfection action against bacterial spores. H_2O_2 is ionised to HO, O_3, OH and O_2H, with only H_2O water as the residual product leaving no toxic residues on the articles thus sterilised or in the environment. The cycle run time is generally 45 min.

3. Reusable medical device (RMD)

The semi-critical equipment that comes in contact with a patient's mucosa and is reused on a number of patients are the endoscopes. These are heat-labile, thereby requiring high-level disinfection by chemicals. After use, an endoscope first needs to be dismantled and cleaned to remove all organic matter prior to chemical disinfection. Following this, the endoscope is immersed in the disinfecting agent. This comprises: 2% glutaraldehyde for 20 min (60 min if TB is suspected) or 0.5% orthopthaldehyde for 5 min. Oxidising agents with 7.5% hydrogen peroxide or in combination with 0.2% peracetic acid, and plasma or ETO sterilisation are also effective. The disadvantage of aldehydes is their toxicity to the environment.

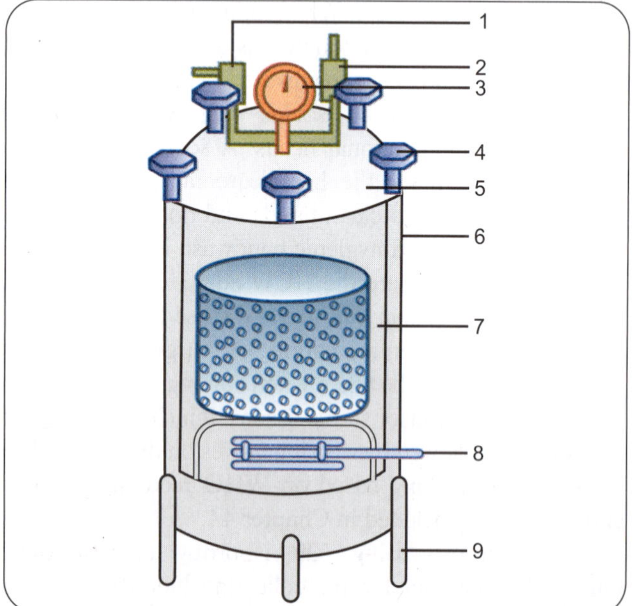

Fig. 45.1 A simple autoclave. 1. release valve; 2. safety valve; 3. pressure gauge; 4. tightening screws; 5. cover; 6. body; 7. bin containing material; 8. heating element; 9. stand

4. Single use device (SUD)

SUDs like syringes and needles, urinary catheters and IV devices are **not to be reused** but to be disposed effectively. However, in low economic settings, SUDs like interventional cardiac catheters may be reused with the informed consent of the patient. The potential risk to the patient is not just infection but pyrogens, toxic residues and the functional integrity of the interventional cardiac catheter. In developed countries, the reuse of intracardiac catheters is either not permitted or is regulated with stringent control by designated authorities and third-party re-processors. In India, it is the responsibility of the hospital's CSSD to ensure integrity and sterility for a pre-decided number of reuses. There are no national regulations or authorised third-party re-processors. Intracardiac catheters and other heat-labile tubings, if reused, are sterilised either by ETO or a plasma steriliser.

Ethylene oxide is a toxic, explosive gas that is used to sterilise heat-labile items. It leaves toxic residues on the sterilised item and therefore needs neutralising and aeration with sterile air before the item is reused on a patient. A safer alternative to ETO is plasma sterilisation.

5. Hospital environment disinfection

Hospital floors, surfaces, bio-spillages, OT air and patient utility items need to be disinfected. For floors and items that come in contact with a patient's intact skin, low-level disinfection with phenolic compounds (phenol, cresol, lysol, chloroxylenol, hexachlorophene) is adequate. These compounds are inexpensive and effective against bacteria, including TB, fungi and viruses. Hospital and laboratory surfaces can also be disinfected with 70% alcohol or 1% sodium hypochlorite (freshly prepared), or quaternary ammonium compounds. Any blood spillage needs to be treated with 10% phenol or 1% sodium hypochlorite for 30 min **before** being cleaned off the surface with gloved hands.

Chlorinated compounds are cheap and fast-acting with high-level disinfection, including that of spores. They are corrosive on prolonged contact and need to be freshly prepared, as they exist in liquid and solid form. The liquid form is 5% sodium hypochlorite equivalent to 50,000 parts per million (ppm) of chlorine. It is used as 1% sodium hypochlorite (10,000 ppm) for blood spillage and at 0.1% (1,000 ppm) for general environmental disinfection. The solid form of chlorine is calcium hypochlorite as bleaching powder or sodium dichloroisocyanurate (NaDCC) as a tablet for disinfecting household water.

Water used in a healthcare facility can be for drinking (potable) or for other disinfection purposes like the immersion of endoscopes in aldehyde solutions or dialysis water. Potable water is surveyed for enteric pathogens, expressed as the total coliform count or most probable number (MPN), to assign the degree of contamination. Drinking water should have a count of 0 MPN/100 mL of water to be satisfactory. Water used for disinfection purposes may be contaminated with hospital environmental bacteria such as non-tuberculous bacteria and gram-negative bacilli including *Pseudomonas*. In-use testing of disinfectants prepared and used in the hospital is a part of the infection control plan. The water used for dialysis needs to be free from microbial contamination and endotoxin.

An **operation theatre** can never be sterilised as bacteria are ubiquitous in the environment and also present as commensals on the skin of HCWs. However, the bacterial load in the operation theatre air can be reduced in order to prevent surgical site infection by the use of aerosolised chemically bound aldehydes or hydrogen peroxide with silver nitrate. The OT is cleaned, sprayed and closed overnight to maintain air quality. Quality control of this process is ensured by testing the air for bacteria by an air sampler or an air settle plate. Non-ionising radiation of UV rays or ozone can additionally be used to disinfect air in OTs or biological safety cabinets in laboratories.

Modern OTs are equipped with a heating ventilation air conditioning (HVAC) system and air handling units (AHUs) that control the optimal microbial load along with the optimal temperature (22°C) and optimal humidity (55%). Clean air with a bacterial load of < 35 CFU/m^3 is adequate for most OTs and is achieved by 25 air changes/hour. Ultra-clean air with < 0.5 CFU/m^3, necessary for certain cardiac, orthopedic and burns surgical units, is achieved by high-efficiency particle arrester (HEPA) filters with 300 air changes/hour.

Routine environmental sampling of hospital surfaces, personnel, or devices is not indicated unless in the case of an outbreak investigation or for designed research studies. Post-disinfection sampling of air is recommended either by using an air sampler or the settle plate method. In the former, a fixed volume of air is inoculated onto a culture medium while in the latter a culture medium plate is exposed to the OT air.

6. Laboratory disinfection requirements

Reusable glassware is sterilised in a hot air oven using dry heat. Laboratory ware made of material other than glass but that which can withstand heat is sterilised in an autoclave by moist heat, including culture media and liquids. Moist heat less than 100°C is employed in sterilising media containing ingredients that could be destroyed by autoclaving. Laboratory environment surfaces are disinfected with phenolic compounds, or sodium hypochlorite, including for bio-spillage and infectious sample discard jars.

Laboratory acquired infections (LAIs) are an occupational hazard following exposure to infectious agents in the laboratory. The route of exposure can be inhalation, percutaneous inoculation, mucous membrane exposure or ingestion. The common microbes that can cause LAIs by inhalation are *M. tuberculosis*, *B. anthracis*, *C. diphtheriae*, influenza and coronaviruses, prevented by the use of PPE and biosafety precautions. Those acquired by percutaneous or mucosal contact are HBV, HCV and HIV, prevented by PPE and standard precautions. Bacterial and parasitic enteric pathogens can be acquired by ingestion and prevented by standard precautions and strict avoidance of drinking or eating in the laboratory.

Advances in disinfection

These include 'contact biocides' utilised in devices and implants wherein an antimicrobial response is triggered only upon microbial contact.

Freshly prepared oxidising agents 0.1% peracetic acid for 5 min or 3% hydrogen peroxide for 20 min or a combination of both have been used for the HLD of immersible, reusable medical devices. Ozone (O_3) is also safe and eco-friendly besides being compatible with a wide range of medical material like glass, stainless steel and silicone.

Super-oxidised water and performic acid are also effective, eco-friendly approaches. Electrolysis of saline results in hypochlorous acid and free chlorine (super-oxidised water) that help achieve HLD. Hydrogen peroxide and formic acid combine to form performic acid with the ability for HLD.

In spite of newer disinfectants and technologies emerging, autoclaving, glutaraldehyde and ETO continue to be widely used for the sterilisation of surgical instruments, endoscopes and intracardiac catheters for cost considerations. The disinfectants and antiseptics with their specific use, optimal concentration and minimum contact time are listed in Table 45.1.

The disinfectants and antiseptics routinely used in a hospital setting need to be checked in the laboratory for efficacy in achieving disinfection and antisepsis by 'in-house disinfectant testing'.

Table 45.1 Practically used disinfectants and antiseptics, their specific use, optimal concentration and minimum contact time

S. No	Disinfectant	Contact time	Use
1.	7% lysol	10 min	Inanimate surfaces, spills, solution in discard jars
2.	Freshly prepared 1% sodium hypochlorite	20 min	Spillages
3.	2% gluteraldehyde	20 min (2 hr if TB is suspected)	Endoscopes
4.	0.5% orthophthaldehyde	5 min	Endoscopes
5.	10% chemically bound formaldehyde	8 hr	OT disinfection
6.	Stabilised 11% hydrogen peroxide with 0.01% (w/w) silver nitrate	60 min	OT disinfection
7.	70% isopropyl or ethyl alcohol	2 min	Skin at blood collection or IV canulation
8.	Alcohol-based chlorhexidine	2 min	Preoperative 0.5% — handwash 4% — patient skin
9.	Alcohol-based povidone iodine	2 min	Preoperative 7.5% — handwash 10% — patient skin
10.	Chloroxylenol	2 min	Handwash

Questions

1. Define sterilisation. Enumerate the methods of sterilisation citing specific uses.
2. Define sterilisation. Discuss the sterilisation of surgical instruments.
3. Elaborate Spaulding's classification of medical devices with methods to achieve sterilisation, high-level disinfection and low-level disinfection.
4. Enumerate the factors affecting the process of sterilisation and disinfection.
5. Define disinfection. Specify the degree of microbial resistance to disinfectants.
6. Define antisepsis. Write a note on pre-operative skin antisepsis.
7. Cite two examples each of reusable and single-use medical devices. How can reusable medical devices be sterilised or disinfected?
8. Name single-use devices that can be reused. How can they be made ready for reuse?
9. Enumerate low-level disinfectants and their use in the hospital general environment.
10. What are the sources of infection of a surgical wound? How is a low microbial load achievable in OT air?

Outbreak Investigation and Environmental Surveillance

Outbreaks of infectious diseases can be local, national (epidemic) or global (pandemic). An outbreak is an increase in the number of cases reported in a specific geographic population. There is no defined number of cases that constitutes an outbreak, as the number depends on the disease under consideration, e.g., with India being declared polio-free since March 2014, even a small number of cases in a given locality would be of concern.

In India, enteric fever, cholera, dengue and malaria, being endemic, have the potential for a community outbreak. An outbreak can also be a hospital-acquired infection (HAI), e.g., post-operative SSI in a cluster of cases.

Surveillance is an essential tool for detecting an outbreak. Surveillance can be **national active, national passive** or **sentinel**. National active surveillance is targeted for the eradication or elimination of the disease, with designated surveillance staff. Presently, poliomyelitis, neonatal tetanus, rubella and measles are selected for active surveillance by the WHO. Regular reporting of cases by all laboratories and healthcare centres constitutes national passive surveillance, selected diseases being diphtheria, pertussis, tetanus, hepatitis B and mumps. Sentinel surveillance is high-quality data gathered through selected reporting sites like medical college laboratories, e.g., HIV testing. The most important surveillance is keen observation by a clinician or laboratory of any increase in the number of specific infectious disease reports.

> Microbes with community outbreak potential in India are:
> 1. Food-borne — *Salmonella, Shigella, Vibrio*
> 2. Respiratory — Influenza virus, SARS-CoV-2
> 3. STD — HIV
> 4. Zoonotic — *Leptospira*
> 5. Vector-borne — *Plasmodium*, dengue virus
> 6. Nosocomial — *Staphylococcus aureus, Klebsiella, Pseudomonas*

An outbreak investigation consists of a systematic approach — define the outbreak microbe, location and number of cases, confirm with biologic samples from the suspected population, identify not just the microbe but the strain, track cases and notify health authorities, analyse data, and determine the source of infection in order to control the outbreak.

In bacterial outbreaks, besides the biologic samples obtained from the patients, environmental sampling—water, food or air—may be indicated in order to identify the source of infection. In order to link any person or inanimate source to the outbreak, it is essential to employ a typing (serotyping or biotyping or phage typing) method to ascertain the identity of both the strains—source and patient. In an HAI, an antibiogram profile serves as a useful tool to establish the source of infection.

The Integrated Disease Surveillance Program (IDSP) covers malaria, typhoid, cholera, tuberculosis, measles, poliomyelitis, HIV, HBV and HCV by regular and sentinel surveillance, through state reference laboratories reporting to the National Centre for Disease Control (NCDC), Delhi.

Water surveillance

Drinking water needs to be potable and free from water-borne diseases like typhoid, cholera, polio, hepatitis A and E. In general, drinking water quality is tested for fecal contamination by detecting a presumptive coliform count in water with human intestinal commensals as indicator bacteria. The WHO prescribes 0 CFU/100 mL as safe potable drinking water. In the event of a local outbreak, specific pathogens—*Salmonella* or *Vibrio*—can be isolated from water.

Food surveillance

Food poisoning outbreak investigation requires bacteriologic culture for specific pathogens from homogenised food samples. Examples of bacteria causing food poisoning are *Brucella* in milk, *S. aureus* in salads, *Bacillus cereus* in fried rice, *Vibrio* in shellfish, *Salmonella* in eggs and enterohemorrhagic *E. coli* (EHEC) in meat.

Air surveillance

The Centers for Disease Control and Prevention, USA recommends targeted air surveillance in the event of an increased HAI rate and not routinely. Microbiologic sampling of air can be done by the air settle plate method (culture plate exposed for 30 min) or by using an air sampler (fixed air volume). This is generally done in an operation theatre to monitor the quality of air. Surface surveillance in hospitals using moistened swabs is not routinely indicated and may be done only in the event of an HAI outbreak being suspected from an inanimate surface.

Bioterrorism

It is the intentional release of hazardous microbes to cause an outbreak in a susceptible population.

The microbe used as the bioweapon should have a high transmissibility, morbidity or mortality, and lack vaccine availability. The bacteria *B. anthracis, Y. pestis, C. botulinum*, *Burkholderia* and *Brucella* species, and the smallpox, influenza, coronaviruses have the potential to be used as bioweapons.

QUESTIONS

1. Define an outbreak and list the types of surveillance.
2. Write briefly on environmental surveillance in an outbreak.

Further Reading

Apurba S. Sastry, Deepashree R. 2019. *Essentials of Hospital Infection Control*. 1st edition. Jaypee Brothers Medical Publishers Pvt. Limited.

Archana Singal, Chander Grover. 2016. *Comprehensive Approach to Infections in Dermatology*. 1st edition. Jaypee Brothers Medical Publishers Pvt. Limited.

Bennett J.E., Raphael Dolin, and Martin J Blaser. 2019. *Mandell, Douglas, and Bennett's Principles and Practice of Infectious Diseases*. 9th edition. Elsevier Health Sciences.

Cary N. Engleberg, Terence Dermody, and Victor DiRita. 2013. *Schaechter's Mechanisms of Microbial Disease*. 5th edition. Wolters Kluwer Health/Lippincott Williams & Wilkins.

Clinical and Laboratory Standards Institute. 2020. *Performance Standards for Antimicrobial Susceptibility Testing, M100*. 30th Informational Supplement.

Chatterjee K.D. 2016. *Parasitology (Protozoology and Helminthology)*. 13th edition. CBS Publishers & Distributors.

Jenni Punt, Sharon Stranford, Patricia Jones and Judith A. Owen. 2019. *Kuby Immunology*. 18th edition. W.H. Freeman & Co Ltd.

Kasper D.L., Fauci A.S., Hauser S.L., Longo D.L., Jameson J.L., and Loscalzo J. 2015. *Harrison's Principles of Internal Medicine*. 20th edition. McGraw Hill Education.

National HIV Counselling and Testing Services (HCTS) Guidelines, National AIDS Control Organisation, Govt of India, 2016.

National Laboratory Guidelines for Testing of Viral Hepatitis, Ministry of Health and Family Welfare, Govt of India, 2018.

National Viral Hepatitis Control Program Operational Guidelines, Ministry of Health and Family Welfare, Govt of India, 2018.

National Guidelines for Diagnosis & Management of Viral Hepatitis, Ministry of Health and Family Welfare, Govt of India, 2018.

National Guidelines for Clinical Management of Dengue Fever, National Vector Borne Disease Control Programme, World Health Organisation (Country office for INDIA), National Health Mission, India, 2015.

Reba Kanungo (ed). 2020. *Ananthanarayan and Paniker's Textbook of Microbiology*. 11th edition. Universities Press.

RNTCP National Strategic Plan for TB Elimination in India (2017–2025) Central TB Division, Directorate General of Health Services, Ministry of Health & Family Welfare, Govt of India.

Seasonal Influenza, Ministry of Health and Family Welfare, Govt of India, 2016.

Suryakantha A.H. 2019. *Community Medicine with Recent Advances*. 5th edition. Jaypee Brothers Medical Publishers Pvt. Limited.

Technical and Operational Guidelines for TB Control in India. Central TB Division, Directorate General of Health Services, Ministry of Health & Family Welfare, Govt of India, 2016.

Treatment Guidelines for Antimicrobial Use in Common Syndromes, Indian Council of Medical Research, New Delhi, India, 2019.

Vashishtha Vipin M. 2019. *IAP Textbook of Vaccines*. 2nd edition. Jaypee Brothers Medical Publishers Pvt. Limited.

Index